Gmelin Handbook of Inorganic Chemistry

8th Edition

Volumes published on "Manganese" (Syst. No. 56)

Manganese A 1 (in German)
History – 1980

Manganese B (in German)
The Element – 1973

Manganese C 1 (in German)
Compounds (Hydrides. Oxides. Oxide Hydrates. Hydroxides) – 1973

Manganese C 2 (in German)
Compounds (Oxomanganese Ions. Permanganic Acid. Compounds and Phases with Metals of the Main and Subgroups I and II) – 1975

Manganese C 3 (in German)
Compounds of Manganese with Oxygen and Metals of the Main and Subgroups III to VI. Compounds of Manganese with Nitrogen – 1975

Manganese C 4 (in German)
Compounds of Manganese with Fluorine – 1977

Manganese C 5 (in German)
Compounds of Manganese with Chlorine, Bromine, and Iodine – 1978

Manganese C 6 (in German)
Compounds of Manganese with Sulfur, Selenium, Tellurium, Polonium – 1976

Manganese C 7
Compounds of Manganese with Boron and Carbon – 1981

Manganese C 8
Compounds of Manganese with Silicon – 1982

Manganese C 9
Compounds with Phosphorus, Arsenic, Antimony – 1983

Manganese D 1 (in German)
Coordination Compounds 1 – 1979

Manganese D 2 (in German)
Coordination Compounds 2 – 1980

Manganese D 3
Coordination Compounds 3 – 1982 (present volume)

Gmelin Handbuch der Anorganischen Chemie

Achte völlig neu bearbeitete Auflage

BEGRÜNDET VON Leopold Gmelin

ACHTE AUFLAGE BEGONNEN im Auftrage der Deutschen Chemischen Gesellschaft von R. J. Meyer

FORTGEFÜHRT VON E. H. E. Pietsch und A. Kotowski
Margot Becke-Goehring

HERAUSGEGEBEN VOM Gmelin-Institut für Anorganische Chemie der Max-Planck-Gesellschaft zur Förderung der Wissenschaften

Direktor: Ekkehard Fluck

Springer-Verlag
Berlin · Heidelberg · New York 1982

Gmelin-Institut für Anorganische Chemie
der Max-Planck-Gesellschaft zur Förderung der Wissenschaften

Gmelin Handbook of Inorganic Chemistry

8th Edition

Mn
Manganese

D 3

Coordination Compounds 3

With 28 illustrations

AUTHORS Karl Koeber, Dietrich Schneider

FORMULA INDEX Helga Köttelwesch

EDITORS Helga Demmer, Ingeborg Hinz, Hildegard List, Edith Schleitzer-Rust

CHIEF EDITOR Edith Schleitzer-Rust

System Number 56

Springer-Verlag
Berlin · Heidelberg · New York 1982

LITERATURE CLOSING DATE: 1980
IN MANY CASES MORE RECENT DATA ARE CONSIDERED

Library of Congress Catalog Card Number: Agr 25-1383

ISBN 3-540-93467-7 Springer-Verlag, Berlin · Heidelberg · New York
ISBN 0-387-93467-7 Springer-Verlag, New York · Heidelberg · Berlin

LN-Druck Lübeck

Preface

The present volume, "Manganese" D 3, continues the description of the manganese complexes. The arrangement of the complexes in these D volumes is based on ligand type. The introduction, on. p. 1, shows the classes of complexes which have already been described in Chapters 1 through 11 in volumes D 1 (1979) and D 2 (1980).

In Chapters 12 to 15 of this volume, the complexes with ammonia, amines, hydrazine, derivatives of hydrazine, and hydroxylamine are described. The manganese(II) complexes with ethylenediamine and with polyamines like triethylenetetraamine are catalysts for the decomposition of hydrogen peroxide. They also accelerate the photopolymerization of vinyl monomers in the presence of carbon tetrachloride.

The complexes with N-heterocycles are placed in Chapter 16. There the arrangement of the complexes is based on the number of nitrogen atoms in the ring and the size of the ring. The part of Chapter 16 in this volume includes all the ligands with one nitrogen in a ring and some of the ligands with two nitrogens per ring (pyrazoles and imidazoles). Numerous papers have appeared on the complexes with pyridine, pyridine derivatives, quinoline, hydroxyquinolines and related compounds, bipyridyl, and phenanthroline. Many of these complexes have been used in chemical analysis. A second important line of research, the use of manganese complexes as models for analogous iron complexes of biological interest, was suggested by the similarity of the coordination chemistries of iron and manganese.

While the majority of the complexes described in this volume are high-spin manganese(II) compounds, there are compounds of Mn^{III} and Mn^{IV} and even binuclear complexes with both oxidation states: Complexes of the type $[L_2Mn^{III}O_2Mn^{IV}L_2]^{3+}$, where L = 2,2'-bipyridyl or 1,10-phenanthroline, have been thoroughly studied in recent years. X-ray crystallography, extensive magnetic measurements, infrared spectra, and electronic spectra all contributed in demonstrating that these are indeed oxygen-bridged structures with both Mn^{III} and Mn^{IV}.

The next volume, "Manganese" D 4, will continue the description of the manganese complexes, including complexes with N-heterocycles containing two or more nitrogen atoms in a ring, e.g., the purine derivatives and the phthalocyanines.

Frankfurt/Main
October 1982

Edith Schleitzer-Rust

Table of Contents

Coordination Compounds of Manganese

(Continued)

Introduction

Arrangement. In Series D coordination compounds of manganese, with the exception of the organometallic compounds, are described. The volumes "Mangan" D 1 (1979) and "Mangan" D 2 (1980) contain the following chapters:

"Mangan" D 1	1)	Review
	2)	Complexes with H_2O
	3)	Complexes with Alcohols
	4)	Complexes and Salts with Phenols and Other Aromatic Hydroxy Compounds
	5)	Complexes with Aldehydes
	6)	Complexes with Ketones
	7)	Complexes with Quinones
	8)	Complexes with Ethers and O-Heterocycles
"Mangan" D 2	9)	Complexes and Salts of Carboxylic Acids and Their Derivatives
	10)	Cyanomanganate Complexes
	11)	Cyanato, Thiocyanato, and Selenocyanato Complexes

This volume, D 3, describes the complexes with ammonia; amines; hydrazine; derivatives of hydrazine; hydroxylamine; and N-heterocycles. The arrangement of the complexes with the N-heterocycles is detailed in the introduction to Chapter 16, p. 77.

Mixed-ligand complexes are generally arranged according to the principle of last position. For example, the complexes containing both diamines or triamines and bipyridyl are placed in the chapter "Complexes with Bipyridyl". The index at the end of this volume, which lists the empirical formulas of the ligands, is intended to expedite locating specific compounds.

Rules and Definitions. Generally the names of the ligands correspond to IUPAC nomenclature; however, trivial names are also used.

The stepwise stability (formation) constants K_n for the formation of complexes in solution from a central atom M and ligands L and the over-all constants β_n are defined as follows:

$$K_n = [ML_n]/[ML_{n-1}] \cdot [L] \text{ in L/mol for equilibria } ML_{n-1} + L \rightleftharpoons ML_n \; (n = 1, 2, 3, \ldots)$$

$$\beta_n = [ML_n]/[M] \cdot [L^n] \text{ in L}^n/\text{mol}^n \text{ for equilibria } M + n\,L \rightleftharpoons ML_n \; (n = 1, 2, 3, \ldots)$$

The formation of complexes with protonated ligands is defined as follows:

$$K^M_{MH_pL} = [MH_pL]/[M] \cdot [H_pL] \text{ for equilibria } M + H_pL \rightleftharpoons MH_pL \; (p = 1, 2, 3, \ldots)$$

All constants given without dimensions are in concentrations of mol/L.

Abbreviations and Dimensions. Temperatures are normally given in °C; K stands for Kelvin. Abbreviations used with temperatures are m.p. for melting point and dec. for decomposition. With thermodynamic data, (s) is used to label solids, (g) is used to designate the gaseous state, and (l) is used for liquids. Thermoanalyses are named DTA (differential thermoanalysis), TG (thermogravimetry), DTG (differential thermogravimetry), or DSC (differential scanning calorimetry).

The vibrational spectra are labeled as IR (infrared) or R (Raman). The assigned bands (frequencies in cm^{-1}) are labeled with the symbols ν, for stretching vibrations, and δ, for deformation vibrations. The intensities are placed in parentheses (w, m, s, vs, etc.); sh means shoulder; br means broad. The UV absorption maxima of the electronic spectra are given in nm (λ_{max}) or cm^{-1} (ν_{max}); the extinction coefficient ε ($L \cdot mol^{-1} \cdot cm^{-1}$) or log ε is in parentheses.

Ligand and Solvent Abbreviations:

Hacac	acetylacetone = 2,4-pentanedione
H_4edta	ethylenediaminetetraacetic acid
bpy	2,2′-bipyridyl
dien	diethylenetriamine
en	ethylenediamine
phen	1,10-phenanthroline
pn	propylenediamine
py	pyridine
tren	2,2′,2″-triaminotriethylamine
trien	triethylenetetraamine
DMF	dimethylformamide
DMSO	dimethyl sulfoxide
THF	tetrahydrofuran

12 Complexes with Ammonia

12.1 $Mn(NH_3)_n^{2+}$ Ions (n = 1 to 6). Formation and Properties in Solution

The complex ions $(Mn(NH_3)_n^{2+}$ with n = 1, 2, 3, and 4 are formed in aqueous solutions containing $MnCl_2$ (0.005 M), NH_3 (0.8 M), and NH_4Cl (0.4 M). The same species are present if a mixed solvent with 43.5 wt% t-butyl alcohol (= $C_4H_{10}O$) is used instead of pure water. The concentrations of these species (together with the simple solvated species $Mn(solv)_6^{2+}$ (solv = H_2O and $(CH_3)_3COH$)) as a function of the ammonia concentration, determined potentiometrically, are shown in **Fig. 1**, p. 4 [1]. In aqueous solutions containing $Mn(NO_3)_2$, NH_3, and NH_4NO_3 the same complex ions are formed, as was determined by potentiometric pH measurement. The concentration curves of the species are similar to those in Fig. 1 [2]. Although Trofimenko et al. [3] found these same complex ions in aqueous solutions containing $MnCl_2$, NH_3, and NH_4Cl by polarography, their concentration curves deviate from those in [1] and [2]. In concentrated aqueous ammonia, $MnCl_2$ is complexed to form $Mn(NH_3)_6^{2+}$ ions [4]. Both $Mn(NH_3)_5^{2+}$ and $Mn(NH_3)_6^{2+}$ have been assumed to be present in 2 to 6 M NH_3 solutions [1]. The formation of the complex ion $Mn(NH_3)_2^{2+}$ during the dissolution of manganese metal in neutral and weakly acidic solutions containing NH_4^+ ions is evident from electrochemical measurement [5]. The formation of the complexes $Mn(NH_3)_n^{2+}$ with n = 2, 3, 4 during the polarographic reduction of Mn^{2+} ions in NH_4Cl solutions containing ammonia is described in "Mangan" B, 1973, p. 303.

The consecutive stability constants, K_n, for the complexes $Mn(NH_3)_n^{2+}$, n = 1, 2, 3, and 4, have been determined on aqueous solution and on t-butyl alcohol-water mixtures from the pH of the solutions at temperatures t and ionic strengths I (in mol/L):

medium	I	t in °C	log K_1	log K_2	log K_3	log K_4	Ref.
H_2O	0.4(NH_4Cl)	25±0.2	0.90	0.67	−0.40	0.30	[1]
H_2O	2(NH_4NO_3)	20	1.0	0.54	0.16	−0.40	[2]
H_2O	0.5 to 5	20 to 30	0.8*)	0.5*)	—	—	[6, 7]
8.1% $(CH_3)_3COH$	0.4(NH_4Cl)	25±0.2	1.08	0.60	−0.30	0.4	[1]
43.5% $(CH_3)_3COH$	0.4(NH_4Cl)	25±0.2	1.30	0.60	0.48	−0.30	[1]

*) Calculated values.

The values given in [2] were adopted by [8]. The consecutive stability constants for a 24.8 wt% t-butyl alcohol-water mixture are also given in [1]. Consecutive instability constants for aqueous 2 M NH_4Cl determined polarographically by [3] deviate sharply from the (converted) values of [1, 2].

The enthalpy of formation of the hypothetical gaseous ion $Mn(NH_3)_6^{2+}$ from gaseous Mn^{2+} and gaseous NH_3 has been calculated to be $\Delta H = -341$ kcal/mol (14.8 eV/molecule) [1] and −316 kcal/mol [3].

The ligand-field splitting energy of the ion $Mn(NH_3)_6^{2+}$ was deduced from calorimetric measurements to be $\Delta = 10\ D_q = 26.0$ kcal/mol or $9.10 \times 10^3\ cm^{-1}$ [11].

Fig. 1

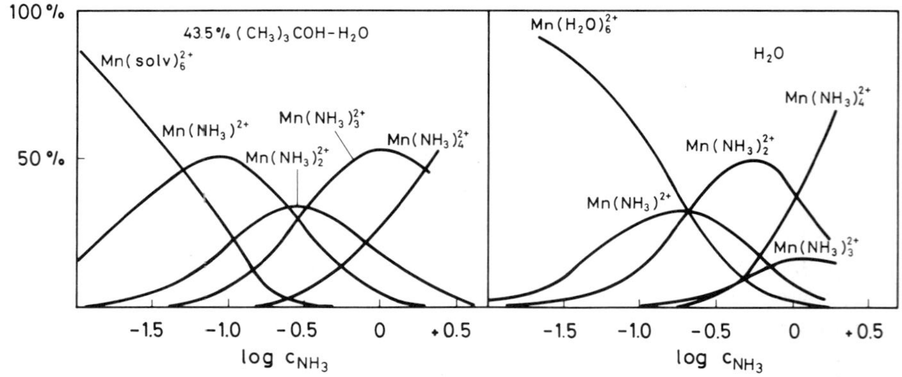

The distribution of the complex ions $Mn(NH_3)_n^{2+}$ in percent of the total manganese(II) concentration as a function of the ammonia concentration c_{NH_3} in aqueous solution and in 43.5 wt% t-butyl alcohol-water mixture.

References:

[1] J. Lipkowski, Z. Galus (J. Electroanal. Chem. Interfacial Electrochem. **48** [1973] 337/52, 345). – [2] V. S. Kublanovskii, V. N. Belinskii (Zh. Neorgan. Khim. **17** [1972] 129/31; Russ. J. Inorg. Chem. **17** [1972] 68/9). – [3] V. V. Trofimenko, I. V. Gamali, A. V. Vasilevskaya (Elektrokhimiya **7** [1971] 1230/5; Soviet Electrochem. **7** [1971] 1185/90, 1188), V. V. Trofimenko, I. V. Gamali (Khim. Tekhnol. [Kharkov] **1971** No. 18, pp. 73/4; C.A. **76** [1972] No. 67360). – [4] K. H. Schmidt, A. Müller (Inorg. Chem. **14** [1975] 2183/7). – [5] Ya. M. Kolotyrkin, T. R. Agladze (Zashch. Metal. **4** [1968] 721/4; Protect. Metals [USSR] **4** [1968] 634/5).

[6] L. G. Sillén, A. E. Martell (Stability Constants of Metal-Ion Complexes, London 1964, p. 150). – [7] K. B. Yatsimirskii (Zh. Obshch. Khim. **24** [1954] 1498/507; J. Gen. Chem. [USSR] **24** [1954] 1485/91, 1491). – [8] R. M. Smith, A. E. Martell (Critical Stability Constants, Vol. 4: Inorganic Complexes, New York 1976, p. 40). – [9] K. B. Yatsimirskii (Zh. Neorgan. Khim. **11** [1966] 2429/36; Russ. J. Inorg. Chem. **11** [1966] 1304/8, 1306). – [10] R. de Hartoulari, P. Barret (Bull. Soc. Chim. France **1960** 1849/67, 1852).

[11] M. Badri, J. W. S. Jamieson (Can. J. Chem. **55** [1977] 3530/5).

12.2 Isolated Manganese(II) Compounds

12.2.1 Fluorides

Moissan, Venturi [1] reported that MnF_2 reacted with liquid ammonia to form the compound $3MnF_2 \cdot 2NH_3$. The crystalline powder lost ammonia at room temperature to form MnF_2 again. Later, Biltz and Rahlfs [2] could not detect any reaction beween anhydrous MnF_2 and liquid ammonia at −78.5°C or ammonia gas between −30 and 100°C. Gaseous NH_3 reacted with solid MnF_2 at 1200°C to yield a brown substance that contained ammonia [1].

If $MnF_2 \cdot 4H_2O$ is treated with liquid ammonia at −78.5°C and then with ammonia gas at 0°C in the absence of oxygen, the aquopentaammine complex, $MnF_2 \cdot 5NH_3 \cdot H_2O$, forms.The compound is almost colorless (with an only slightly yellowish tinge). Its density is1.479 g/cm^3 at −21°C. The decomposition pressures at −21, −11, and 0°C are 39.2, 80.0, and 173 Torr. The ammonia is lost in steps, first $MnF_2 \cdot NH_3 \cdot H_2O$ forms and then $MnF_2 \cdot {}^1/_2 NH_3 \cdot H_2O$. The monoammine complex has a density of 2.291 g/cm³ at 25°C. Its decomposition pressures at 25, 34.4, 46, and 61°C are 10.5, 20.0, 43.0, and 108.0 Torr [2].

References:

[1] H. Moissan, A. Venturi (Compt. Rend. **130** [1900] 1158/62). – [2] W. Biltz, E. Rahlfs (Z. Anorg. Allgem. Chem. **166** [1927] 351/76, 353, 358, 360, 371/2, 376).

12.2.2 Chlorides

The $MnCl_2$-NH_3 System. Pressure isotherms for the system $MnCl_2$-NH_3 show the three ammines $Mn(NH_3)_6Cl_2$, $Mn(NH_3)_2Cl_2$, and $Mn(NH_3)Cl_2$ [1]. Much later, the existence of these ammines was confirmed by a differential thermal analysis of the hexaammine [2]. Under supercritical ammonia, the ammines $MnCl_2 \cdot nNH_3$ with n = 7, 8, 10, 11, and 12, form [5, 6]. The region in which each ammine is stable was determined tensiometrically and is shown in **Fig. 2**. Another ammine, $Mn(NH_3)_4Cl_2$, is reported to have only a small region of stability [6]; see p. 8.

Fig. 2

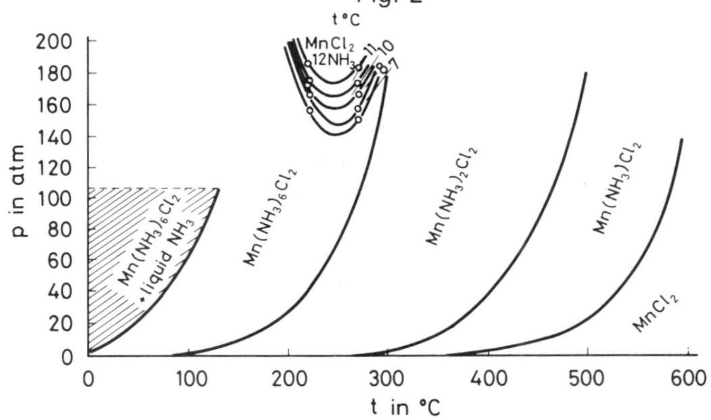

Phase diagram for the system $MnCl_2$-NH_3 [5].

$MnCl_2 \cdot nNH_3$ (n = 7, 8, 10, 11, 12). Hüttig, Kollmann [6] prepared the cited ammine complexes from $MnCl_2$ and supercritical NH_3. They also prepared the lower four ammines by decomposition of the next higher ones. Biltz, Rahlfs [31] synthesized the compounds $MnCl_2 \cdot 12NH_3$ and $MnCl_2 \cdot 10NH_3$ in 1925 by the action of excess liquid ammonia on $MnCl_2 \cdot 6NH_3$ (or $MnCl_2$) at −78.5°C overnight and subsequent cautious warming of the reaction mixture in a closed vessel. The phase diagram is given in Fig. 2. The change in the decomposition pressure with temperature (−78.5 to −50°C) yields the enthalpies of formation (from $MnCl_2$ and NH_3) −125.8 kcal/mol for $MnCl_2 \cdot 12NH_3$ and −111.5 kcal/mol for $MnCl_2 \cdot 10NH_3$. These two compounds are pure white [31].

Along the 220 and 250°C isotherms, the pressures (atm) for the transitions $MnCl_2 \cdot nNH_3 \rightarrow MnCl_2 \cdot n'NH_3 + (n-n')NH_3$ are [6]:

n→n′	12→11	11→10	10→8	8→7	7→6
220°C	187	175	172.5	167	157
250°C	183	172.5	166	157	150

The decomposition pressures of $MnCl_2 \cdot 10NH_3$ and $MnCl_2 \cdot 12NH_3$ at liquid ammonia temperatures are given in [31]:

t in °C	−78.5	−65	−60	−55	−50
$MnCl_2 \cdot 10NH_3$	20.5	72.0	107	160	220
$MnCl_2 \cdot 12NH_3$	29.5	95.7	—	—	—

$Mn(NH_3)_6Cl_2$. Schmidt, Müller [14] prepared $Mn(NH_3)_6Cl_2$ by bubbling gaseous ammonia through an aqueous solution of $MnCl_2 \cdot 4H_2O$ and NH_4Cl. Others, for instance [1, 7, 15, 32], slowly passed dry ammonia gas over anhydrous $MnCl_2$ at room temperature in the absence of air and moisture for several hours [2]. The reaction is exothermic, and the salt swells greatly [1, 17]. The ammine is a white powder [7, 10, 15]. The ammonia content was determined to be 99.77 ± 0.20% of its theoretical value [16]. The reaction took several days at room temperature until the composition $Mn(NH_3)_6Cl_2$ was attained [1, 3]. It could be accelerated by the use of excess liquid ammonia at −30°C [10] or −78°C [1 to 3].

Rossini et al. [8] gave $\Delta H_f^\circ(298.16\ K) = -267.8$ kcal/mol for the enthalpy of formation from the elements under standard conditions, from the results of [9, 10]. The enthalpy of formation from $MnCl_2(s)$ and $NH_3(g)$ calculated from the change in the decomposition pressure with temperature has been given as −82.4 kcal/mol [11]. Enthalpy and kinetic data for the formation of $Mn(NH_3)_6Cl_2$ from $Mn(NH_3)_2Cl_2$ are given on pp. 7 and 9, together with the treatment of the equilibrium $Mn(NH_3)_6Cl_2 \rightleftharpoons Mn(NH_3)_2Cl_2 + 4NH_3$.

$Mn(NH_3)_6Cl_2$ forms cubic crystals of the CaF_2 type with a lattice parameter of $a = 10.19_8$ Å. The space group is Fm3m-O_h^5 (No. 225) [20 to 22]; Z = 4 [3]. A table of d values is given by [3]. The $Mn(NH_3)_6^{2+}$ cation has a regular octahedral structure [23]. Each complexed cation is surrounded by eight chloride ions. The Mn-Cl distance is 4.41 Å, while Mn-Mn nearest neighbors are 7.20 Å apart and Cl-Cl nearest neighbors are 5.099 Å apart [7]. $Mn(NH_3)_6Cl_2$ is isostructural with $Mn(NH_3)_6X_2$ with X = ClO_4, Br, I, the hexammine chlorides, bromides, and perchlorates of Mg, Zn, Cd, Ni, Co, and Fe [20 to 22], and with $K_2[PtCl_6]$ [33]. Its lattice energy (here the energy difference between one mole of the crystalline ammine and its free gaseous anions and complex ammine cations) is U = 323 kcal/mol. Then the coordination energy is $E_{coord} = U - (U_{MnCl_2} + Q) = -374$ kcal/mol, where Q is the heat of formation of the hexaammine from $MnCl_2(s)$ and $NH_3(g)$ [19].

The density measured pycnometrically at 25°C is 1.372 g/cm³ [24]. The compound is paramagnetic at room temperature [7] with a magnetic moment of 5.92 ± 0.04 μ_B [7, 12]. Down to 1.5 K the magnetic susceptibility follows the Curie-Weiss law [12] with the Curie constant $C = 4.35\ cm^3 \cdot K \cdot mol^{-1}$ [7] and the paramagnetic Curie temperature $\Theta_p = 0.2 \pm 0.2$ K. (The Curie-Weiss law was written as $\chi = C/T + \Theta_p$) [12]. The heat capacity vs. temperature curve near absolute zero, given in a figure in [16], has a maximum at 0.60 K which was attributed to an antiferromagnetic phase change.

The ^1H NMR spectrum of $Mn(NH_3)_6Cl_2$ at room temperature has a signal width of 7.0 Gauss at a resonance frequency of 15.5 MHz [23].

The ESR spectrum (X band) of the powdered compound was investigated in the range between 77 and 300 K [25], ~1.4 and 300 K [26]. At room temperature a single line with g = 2.015 was observed. A sudden change of the line width in the neighborhood of 110 K points to a phase transition at this temperature [25, 26]. The ESR spectra of manganese(II) (1 mol%) doped into $Cd(NH_3)_6Cl_2$ recorded at various temperatures in the range between 77 and 300 K show that the g-factor of 2.007 and the hyperfine coupling constant of A = 9.3 mT remain unchanged within this temperature range. A sudden change in the relative intensity $I_{5/2}/I_{1/2}$ (I = integral line intensity) at about 150 K points to a phase transition at this temperature. The zero-field splitting parameters D = 0 for T > 175 K and D = 0.0051 cm^{-1} for T < 140 K have been derived, suggesting a lowering of the crystal-field symmetry below 150 K [25].

In the powder reflectance spectrum at 77 K, there are two weak bands in the visible region, at 24130 and 24240 cm^{-1}, which are assigned to transitions from the $^6A_1$ ground state to the $^4A_1$ and ^4E levels. Each band has a half-height width of 25 cm^{-1} [15]. In the IR spectrum of a suspension in Nujol, the antisymmetric Mn-N stretching frequency ν_3 (F_{1u}) is at 302 cm^{-1} [14] or at 307 cm^{-1} [27]. The antisymmetric N-Mn-N bending frequency ν_4 (F_{1u}) is at 165 cm^{-1} [14], and the deformation frequencies of the NH_3 ligands are at 617, 1134, and 1592 cm^{-1} [27]. Remy [4] gives further IR bands for the ammonia ligands in $MnCl_2 \cdot 5.74 NH_3$. The Raman frequency ν_1 (A_{1g}) of the totally symmetric Mn-N stretching vibration is at 330 cm^{-1}. Thus the force constant of the Mn-N bond can be calculated to be 0.6_7 mdyn/Å. This force constant is the weakest in the series of $M(NH_3)_6Cl_2$ with M = Mn, Fe, Cu, Ni [14].

Heated in a vacuum, $Mn(NH_3)_6Cl_2$ loses four molecules of NH_3 at 290 K (17°C) to form $Mn(NH_3)_2Cl_2$ [7]. Wendlandt, Smith [2] heated the compound in a dynamic helium atmosphere at a rate of 5°C/min and observed the decomposition by differential thermal analysis. The decomposition to the diammine began at 40°C, and the peak temperature was 75°C [2]. The equilibrium decomposition pressure of $Mn(NH_3)_6Cl_2$ was measured several times:

t in °C	−96	−50	−25	0	20	40	45	48.5	59
p in Torr	0.18	1.17	3.30	9.35	23.1	93.7	100	107.4	218
Ref.	[3]	[3]	[3]	[3]	[3]	[3]	[11]	[10]	[1]

t in °C	60	63	66	72.5	76.5	78.5	82.0	86.1
p in Torr	237.2	258	277.5	378.5	495	570.8	616	734
Ref.	[3]	[1]	[10]	[10]	[1]	[3]	[10]	[10]

Results of a recent differential thermoanalytic investigation of the equilibrium $Mn(NH_3)_6Cl_2 \rightleftharpoons Mn(NH_3)_2Cl_2 + 4 NH_3$ in the pressure region from 0.2 to 26 bar [34] are shown in **Fig. 3**, p. 8. In this figure the line with the full circles corresponds to the decomposition reaction of $Mn(NH_3)_6Cl_2$, and that with the open circles corresponds to the reverse reaction. The apparatus used did not allow equilibrium values to be attained because of too low a reaction rate in the region between the two curves. The equilibrium values, according to the data of [1], are given by the broken line, which can be extrapolated to higher pressure regions [34]. Thus, the data of [1] were confirmed by the measurement of [34]. Fig. 3 shows that the Clausius-Clapeyron equation is valid for this system [34]. Equilibrium pressures at higher temperatures are shown in Fig. 2 on p. 5. An older set of presssures is given in [17].

Hart, Partington [10] studied the equilibrium $Mn(ND_3)_6Cl_2 \rightleftharpoons Mn(ND_3)_2Cl_2 + 4 ND_3$:

t in °C	48.5	60.5	66.0	82.0	86.1
p in Torr	91.0	178.7	244.0	556.5	679.5

For the reaction $Mn(NH_3)_6Cl_2 \rightleftharpoons Mn(NH_3)_2Cl_2 + 4 NH_3$ the following enthalpies were determined from the dependence of the decomposition pressure on temperature: $\Delta H = 11.33$ [11] or 11.700 [10] kcal/mol NH_3. The enthalpy was also determined from the peak areas of the DTA curves: $\Delta H = 57$ kJ ± 5% at 70°C (13.6 kcal) and 52 kJ ± 5% at 170°C (12.4 kcal), each per mole NH_3 [34]. Previous values are given in [1, 17]. For the reaction $Mn(ND_3)_6Cl_2 \rightleftharpoons Mn(ND_3)_2Cl_2 + 4 ND_3$ $\Delta H = 12.230$ kcal/mol NH_3 was found [10]. Kinetic studies of the equilibrium $Mn(NH_3)_2Cl_2 + 4 NH_3 \rightleftharpoons Mn(NH_3)_6Cl_2$ are described on p. 9.

Fig. 3

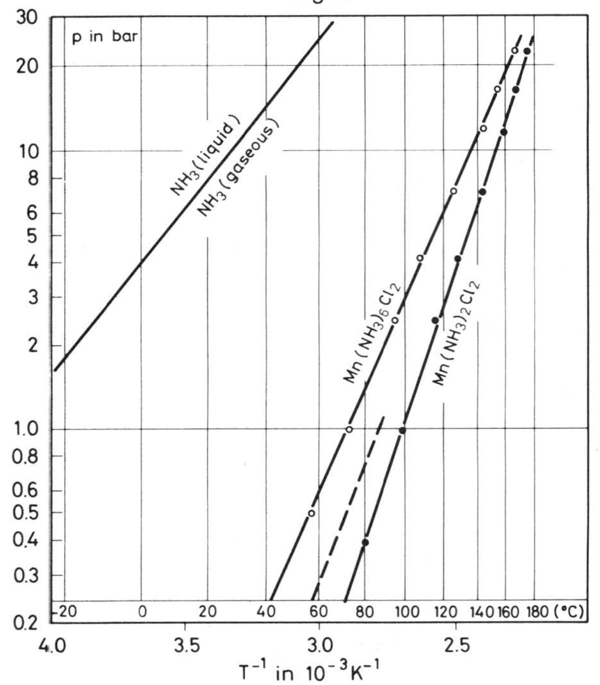

The equilibrium $Mn(NH_3)_6Cl_2 \rightleftharpoons Mn(NH_3)_2Cl_2 + 4\,NH_3$
• reaction in the direction of the upper arrow
○ reverse reaction --- values of [1].

$Mn(NH_3)_6Cl_2$ turns brown in dry air after a few hours as a result of oxidation [15]. In moist air it is partly oxidized and hydrolyzed [3]. The hexaammine is insoluble in liquid ammonia. However, it reacts with a solution of KPH_2 in liquid ammonia as follows: $Mn(NH_3)_6Cl_2 + 4\,KPH_2 \rightarrow K_2Mn(PH_2)_4 \cdot xNH_3 + 2\,KCl + (6-x)NH_3$ [28].

$Mn(NH_3)_6Cl_2 \cdot KCl$ and $Mn(NH_3)_6Cl_2 \cdot 2\,NH_4Cl$. The two double salt ammine complexes can be prepared by passing a rapid stream of ammonia gas over $MnCl_2 \cdot KCl \cdot 2\,H_2O$ or $MnCl_2 \cdot 2\,NH_4Cl \cdot 2\,H_2O$ at room temperature. Both reactions are exothermic. The two products are white crystalline powders, which slowly lose ammonia in air and turn brown from oxidation. The compounds dissolve with decomposition in water, mineral acids, and organic acids. They are insoluble in the usual organic solvents [29, 30].

$Mn(NH_3)_4Cl_2$. The formation of $Mn(NH_3)_4Cl_2$ was observed in the isothermal decomposition of $Mn(NH_3)_6Cl_2$ at 250°C and 114 atm or at 220°C and 60 atm. This compound only exists under supercritical conditions. Its pressure of decomposition to form the diammine is 90 atm at 250°C [6].

$Mn(NH_3)_2Cl_2$. The diammine was prepared by thermally decomposing the hexaammine [1, 7, 10]. Attempts to synthesize the compound by elongated treatment of $MnCl_2$ with ammonia at 132°C led to a product of the composition $MnCl_2 \cdot \sim 1.80\,NH_3$ [3]. Rossini et al. [8] gave the enthalpy of formation from the elements in their standard states as $\Delta H_f^\circ = -175.9$ kcal/mol on the basis of the data of [9, 10]. The enthalpy of formation from

$MnCl_2$(s) and NH_3(g) has been calculated to be −37.08 kcal/mol (130 to 220°C) [11] and −33.50 kcal/mol (−34.02 for the ND_3 complex) (at 180 to 243°C) [10], all from the change in the decomposition pressure as a function of temperature.

The colorless [7] crystals appear on the basis of the diffraction pattern to belong to the orthorhombic system with a = 8.40, b = 8.29, and c = 3.64 Å; Z = 2. The density determined in a pycnometer at 0°C is 2.40 g/cm³ [3]. (The results [3] were obtained by interpolation between those for several heterogeneous mixtures with nonstoichiometric ammonia content.) The complex is paramagnetic, and at room temperature the magnetic moment is 5.84 μ_B [7, 12]. Above 1.8 K the susceptibility obeys the Curie-Weiss law, $\chi = C/T + \Theta_p$, with $C = 4.21\ cm^3 \cdot K \cdot mol^{-1}$ [7] and $\Theta_p = 1 \pm 0.2$ K [12]. Below 1.8 K the susceptibility is somewhat lower than that predicted by the Curie-Weiss law. A susceptibility versus temperature plot in the range 1.5 to 20.4 K is given in [12]. The IR spectrum shows a band at 240 cm^{-1}, which is attributed to the Mn-Cl stretching frequency. The structure of the complex is considered to be polymeric octahedral as is the structure of the corresponding pyridine compound [13]. Ammonia bands for $MnCl_2 \cdot 1.83NH_3$ are given in [4].

When heated, the compound loses ammonia. The measured ammonia pressures for the equilibrium $Mn(NH_3)_2Cl_2 \rightleftharpoons Mn(NH_3)Cl_2 + NH_3$ are:

t in °C	−20	+20	70	100	120	150	180	190	192
p in Torr	0.085	0.280	1.20	3.30	5.80	19.4	72.36	100	129.2
Ref.	[3]	[3]	[3]	[3]	[3]	[3]	[3]	[11]	[10]

t in °C	206	210	211.5	222.5	230	231.5	240	243.5
p in Torr	196	232	273.3	402	448	537	760	777.4
Ref.	[1]	[3]	[10]	[10]	[3]	[10]	[37]	[10]

The pressures in the supercritical region are shown in Fig. 2 on p. 5. For the equilibrium $Mn(ND_3)_2Cl_2 \rightleftharpoons Mn(ND_3)Cl_2 + ND_3$, Hart, Partington [10] report:

t in °C	180.2	192.0	206.5	211.5	222.5	231.5	243.5
p in Torr	75	118.9	205.0	252.5	375.0	511.0	745.0

In moist air, $Mn(NH_3)_2Cl_2$ is partially oxidized and hydrolyzed [3]. In an ammonia atmosphere, ammonia is absorbed and the hexammine is formed [1, 34]. The kinetics of this reaction, $Mn(NH_3)_2Cl_2 + 4NH_3 \rightleftharpoons Mn(NH_3)_6Cl_2$, and the reverse one were studied. **Fig. 4**, p. 10, shows the rate constant as a function of temperature. The upper parts of the curves correspond to absorption of NH_3, while the lower parts correspond to loss of NH_3. The rates of decomposition are plotted as negative reaction rates. Each curve is characterized by a region about the equilibrium temperature where the rates of reaction are extremely slow. Such studies were undertaken in part to see if the reaction would be useful for heat-storage units [34]. The reaction rates were also observed to decrease with temperature in the range from 0 to 50°C [35].

$Mn(NH_3)Cl_2$. The monoammine was prepared by the thermal decomposition of $Mn(NH_3)_2Cl_2$ or $Mn(NH_3)_6Cl_2$ [1, 10] in a vacuum [7]. Rossini et al. [8] gave ΔH_f°(298.16 K) = −148.1 kcal/mol for the enthalpy of formation of the crystalline compound from the elements, all in their standard states, on the basis of the data [9, 10]. Biltz [11] found the enthalpy of formation from $MnCl_2$(s) and NH_3(g) to be −20.12 kcal/mol from the change in the decomposition pressure with temperature in the range between 130 and 230°C.

The X-ray diffraction pattern of the crystals is most consistent with the tetragonal system, and then the lattice parameters are a = 11.25 and c = 4.95 Å; Z = 6. (If rhombohedral symmetry is assumed, the parameters would be a = 5.99 Å and $\alpha = 48°30'$; Z = 4.) The density measured in a pycnometer is around 2.2 g/cm³ at 27°C [3]. (The results [3] were obtained by interpolation between those for several heterogeneous mixtures with non-stoichiometric ammonia content.) The compound is paramagnetic with an effective magnetic moment of 5.74 ± 0.03 μ_B at room temperature. Above 90 K the magnetic susceptibility obeys the Curie-Weiss law (written as $\chi = C/T + \Theta_p$) with C = 4.08 cm³ · K · mol⁻¹ and the paramagnetic Curie temperature $\Theta_p = 3$ K [7]. The magnetic susceptibility is shown as a function of temperature in a figure. There is a maximum at $T_N = 2.6$ K, which indicates a phase transition to the antiferromagnetic state [12].

Fig. 4

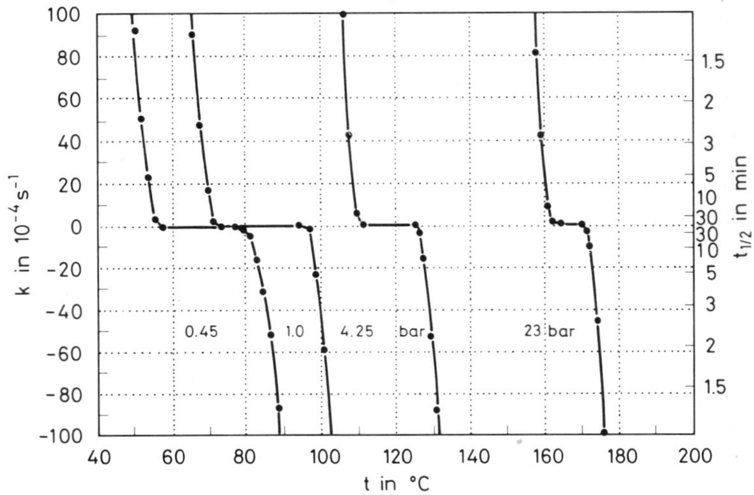

Rate constant k (or half-conversion time $t_{1/2}$) for the reaction $Mn(NH_3)_2Cl_2 + 4NH_3 \rightleftharpoons Mn(NH_3)_6Cl_2$ as a function of temperature for various pressures.

$Mn(NH_3)Cl_2$ is white to pale rose [7]. Ammonia bands of a product with the composition $MnCl_2 \cdot 0.83NH_3$ are given in [4]. If heated, the complex decomposes into $MnCl_2$ and NH_3, producing the following decomposition pressures p at the temperatures t:

t in °C	20	65	103	140	170	200	
p in Torr	0.0095	0.027	0.075	0.203	0.80	5.04	
Ref.	[3]	[3]	[3]	[3]	[3]	[3]	
t in °C	215	228	230	257	273	296	>380
p in Torr	9.4	21.96	28	114.20	100	384.15	760
Ref.	[1]	[3]	[1]	[3]	[11]	[3]	[36]

The pressures in the ammonia supercritical area are shown in Fig. 2, p. 5.

In moist air the complex is partially oxidized and hydrolyzed to form hydroxides [3].

References:

[1] W. Biltz, G. F. Hüttig (Z. Anorg. Allgem. Chem. **109** [1920] 89/110, 93/6). – [2] W. W. Wendlandt, J. P. Smith (Nature **201** [1964] 73/4). – [3] F. Remy (Rev. Chim. Minerale **2** [1965] 693/725, 697/8, 718, 724). – [4] F. Remy (Rev. Chim. Minerale **5** [1968] 935/99, 983, 986). – [5] G. F. Hüttig (Monatsh. Chem. **85** [1954] 98/119, 103/4).

[6] G. F. Hüttig, O. Kollmann (Monatsh. Chem. **82** [1951] 735/41, 741). – [7] K. Kido, T. Watanabe (J. Phys. Soc. Japan **14** [1959] 1217/24, 1218/20). – [8] F. D. Rossini, D. D. Wagman, W. H. Evans, S. Levine, I. Jaffe (NBS-Circ.-500 [1952] 280). – [9] W. Biltz, G. F. Hüttig (Z. Anorg. Allgem. Chem. **109** [1920] 111/25, 118, 124). – [10] A. B. Hart, J. R. Partington (J. Chem. Soc. **1943** 104/18, 110, 115).

[11] W. Biltz (Z. Anorg. Allgem. Chem. **130** [1923] 93/139, 100). – [12] T. Watanabe (J. Phys. Soc. Japan **16** [1961] 1131/41, 1132/4). – [13] R. H. Nuttall (Talanta **15** [1968] 157/69, 162). – [14] K. H. Schmidt, A. Müller (Inorg. Chem. **14** [1975] 2183/7). – [15] D. H. Goode (J. Chem. Phys. **43** [1965] 2830/9, 2831, 2833).

[16] H. Suga, E. Lagendijk, W. J. Huiskamp (Phys. Letters A **32** [1970] 297/8). – [17] F. Ephraim (Z. Physik. Chem. **81** [1913] 513/38, 518, 525, 531, 536). – [18] F. Ephraim (Ber. Deut. Chem. Ges. **45** [1912] 1322/31, 1330). – [19] F. J. Garrick (Phil. Mag. [7] **11** [1931] 741/4). – [20] G. Bödtker-Naess, O. Hassel (Z. Physik. Chem. B **22** [1933] 471/2).

[21] G. Bödtker-Naess, O. Hassel (Avhandl. Norske Videnskaps Akad. Oslo I Mat. Nat. Kl. **1933** No. 4, pp. 1/19, 5, 9; C. **1933** II 3391). – [22] Strukturbericht Vol. 3, 1933/35 [1937], p. 475. – [23] P. H. Kim (J. Phys. Soc. Japan **15** [1960] 445/55, 445, 448). – [24] W. Biltz, E. Birk (Z. Anorg. Allgem. Chem. **134** [1924] 125/36, 132). – [25] P. B. Sczaniecki, D. Kaczyńska, J. Stankowski (Bull. Acad. Polon. Sci. Ser. Sci. Chim. **25** [1977] 319/23; C.A. **87** [1977] No. 109031).

[26] T. Garofano, M. B. Palma-Vittorelli, M. U. Palma, F. Persico (Paramagnetic Resonance Proc. 1st Intern. Conf., Jerusalem 1962 [1963], Vol. 2, pp. 582/6, 583; C.A. **60** [1964] 7569). – [27] L. Sacconi, A. Sabatini, P. Gans (Inorg. Chem. **3** [1964] 1772/4). – [28] G. Uecker, O. Schmitz-DuMont (Z. Anorg. Allgem. Chem. **371** [1969] 318/24, 322). – [29] G. Spacu, L. T. Caton (Bul. Soc. Stiinte Cluj **2** [1924/25] 29/51, 38, 40; C. **1924** I 1650). – [30] L. T. Caton (Ann. Sci. Univ. Jassy **17** [1931/32] 199/204; C. **1933** I 3039).

[31] W. Biltz, E. Rahlfs (Z. Anorg. Allgem. Chem. **148** [1925] 145/51, 149/50). – [32] W. Peters (Ber. Deut. Chem. Ges. **42** [1909] 4826/36, 4835; Z. Anorg. Allgem. Chem. **77** [1912] 137/90, 160). – [33] R. W. G. Wickoff (Crystal Structures, Vol. 3, New York 1965, pp. 783/4). – [34] M. Furrer (EIR-Ber. No. 392 [1980] 1/44, 9, 22/4; C.A. **93** [1980] No. 102014). – [35] M. M. Pavljuchenko, G. A. Laserko (Zh. Fiz. Khim. **29** [1955] 1064/72, 1066; C. **1956** 9653).

[36] F. Müller (Jahrb. Phil. Fak. II Univ. Bern **2** [1922] 80/2).

12.2.3 Perchlorates

$Mn(NH_3)_n(ClO_4)_2$ (n = 6, 2). The complex with n = 6 was prepared by refluxing ~2 g of freshly prepared $Mn(ClO_4)_2 \cdot 6H_2O$ with ~25 mL of anhydrous liquid ammonia [1]. Reported much earlier was the precipitation of an undefined ammine complex by mixing a solution of a manganese(II) salt in aqueous ammonia with a 20% solution of NH_4ClO_4 in aqueous NH_3 of a density of 0.90 g/cm³ [2].

The crystals of $Mn(NH_3)_6(ClO_4)_2$ are cubic, belonging to the CaF_2 type. The lattice parameter is a = 11.57_8 Å [3 to 5]. The IR spectrum is very similar to the spectra of the compounds $M(NH_3)_6(ClO_4)_2$, M = Zn, Cd, Ni, Co, and Cu. The characteristic absorption frequencies (in cm^{-1}) of the coordinated ammonia for $M(NH_3)_6(ClO_4)_2$ are assigned as follows:

frequency	~3400, ~3300	~1620	1270 to 1240	~750	450 to 420
assignment	$\nu(NH_3)$	$\delta(NH_3)$	$\delta(NH_3)$	$\varrho(NH_3)$	ν(M-N)

Absorption bands at ~1100 (ν_3), 940 (ν_1), and 630 (ν_4) cm^{-1} indicate the presence of a perchlorate ion having T_d symmetry [1].

Differential thermal analysis with a heating rate of 12°C/min shows that the compound decomposes in two steps. At 93°C (first peak) it loses four molecules of ammonia in an endothermic reaction to form $Mn(NH_3)_2(ClO_4)_2$. At 234°C (second peak) the diammine explodes, forming manganese oxide [1].

The IR spectrum of the diammine shows a marked difference to that of the hexaammine in the following frequencies: The N-H stretching frequency appears as a single broad band at ~3300 cm^{-1}. The NH_3 deformation band is shifted from ~1250 to ~1400 cm^{-1}. The splittings of the ClO_4 vibrations (ν_3 at 1210, 1150, and 1100 cm^{-1}, ν_4 at 635 and 630 cm^{-1}) indicate the presence of mono- or bidentate perchlorate groups coordinated to the metal ion, changing the T_d symmetry of the perchlorate group to C_{3v} or C_{2v}. The changes of the NH_3 vibrations suggest hydrogen bonding between the hydrogen atoms of the ammonia and the oxygen atoms of the perchlorate. The described structure would explain why it is not possible to completely deammoniate the diammine to yield the pure perchlorate prior to the explosive decomposition [1].

References:

[1] K. C. Patil, V. R. P. Verneker, S. R. Jain (Thermochim. Acta **15** [1976] 257/60). – [2] R. Salvadori (Gazz. Chim. Ital. **40** II [1910] 19/21). – [3] G. Bödtker-Naess, O. Hassel (Z. Physik. Chem. B **22** [1933] 471/2). – [4] G. Bödtker-Naess, O. Hassel (Avhandl. Norske Videnskaps Akad. Oslo I Mat. Nat. Kl. **1933** No. 4, pp. 1/19, 5, 11; C. **1933** II 3391). – [5] Strukturbericht, Vol. 3, 1933/35 [1937], p. 475).

12.2.4 Bromides and Iodides

$MnBr_2 \cdot 10NH_3$. The pure white compound was prepared as described on p. 5 for the chlorides $MnCl_2 \cdot 10NH_3$ and $MnCl_2 \cdot 12NH_3$. The enthalpy of formation from $MnBr_2$(s) and NH_3(g) calculated from the change in the decomposition pressure with temperature (see below) is $\Delta H = -117.8$ kcal/mol. The enthalpy of formation from $Mn(NH_3)_6Br_2$(s) and NH_3(g) is $\Delta H = -28.64$ kcal/mol. The decomposition pressures are:

t in °C	−78.5	−70	−65	−64	−60	−55
p in Torr	27.5	61.5	95.0	100	146.5	212.5

The pressure isotherm at −78.5°C, showing the decomposition of the decaammine to the hexaammine, is given in [10].

$Mn(NH_3)_nBr_2$ (n = 6, 2, 1). $Mn(NH_3)_6Br_2$ is formed when dry gaseous NH_3 is passed over anhydrous $MnBr_2$ in the absence of air [1 to 3]. The reaction is exothermic, and during the reaction the salt swells greatly [3]. The ammonia content reaches 99.98 ± 0.20% of its theoretical value [1]. To assure a complete reaction in a reasonable time, liquid ammonia was reacted with $MnBr_2$ at −78.5°C for several hours [4]. A yield of 90% was achieved by stirring a mixture of $MnBr_2$ and liquid NH_3 for three hours and removing the excess NH_3

under reduced pressure [5]. The hexaammine is also obtained when manganese metal is dissolved in a solution of NH_4Br in liquid ammonia [6]. It is formed by ligand exchange when the ethylenediamine complex, $Mn\,en_3Br_2$, p. 40, or the diethylenetriamine complex, $Mn\,dien_2Br_2$, p. 55, reacts with excess liquid ammonia at −35°C for five hours or when the triethylenetetraamine complex, $Mn\,trien_2Br_2$, p. 58, reacts with liquid ammonia at −70°C for one day [5].

$Mn(NH_3)_2Br_2$ and $Mn(NH_3)Br_2$ are obtained by the stepwise thermal decomposition of the hexaammine [2, 4].

The enthalpies of formation of crystals of the three complexes from the elements in their standard states have been calculated from data in the literature:

complex	ΔH_f°(298.15 K) in kcal/mol Wagman et al. [7]	Rossini et al. [8]
$Mn(NH_3)_6Br_2$	−250	−256.5
$Mn(NH_3)_2Br_2$	−154.7	−155.2
$Mn(NH_3)Br_2$	−124.1	−123.7

The enthalpies of formation for the reactions $MnBr_2(s) + nNH_3(g) \rightleftharpoons Mn(NH_3)_nBr_2(s)$, n = 1, 2, 6, were calculated from the decomposition pressure as a function of temperature for the mono-, di-, and hexaammine to be $\Delta H = -20.03$, −38.44 [9], and −89.16 kcal/mol, respectively [9, 10].

The almost white [3, 5] crystals of $Mn(NH_3)_6Br_2$ are cubic and belong to the CaF_2 type as does the chloride, see p. 6. The lattice parameter is $a = 10.51_9$ Å [11 to 13]. The measured density is 1.784 g/cm³ at 25°C [14]. The heat capacity (measured from 0.1 to 3 K) with a maximum at 0.35 K indicates an antiferromagnetic phase change. This measured heat capacity is shown in a figure in [1]. The IR spectrum shows the strong Mn-N stretching band at 299 cm^{-1}, the NH_3 rocking frequency at 606 cm^{-1}, and the NH_3 deformation frequencies at 1145 and 1589 cm^{-1} [15].

Heated in a dynamic helium atmosphere at a rate of 5°C/min, $Mn(NH_3)_6Br_2$ shows endothermic peaks at 123, 215, and 258°C recorded by differential thermal analysis. The bromides are thus more stable than the corresponding chlorides but less stable than the iodides [2]. The decomposition pressures p as a function of the temperature t of the three complexes have been measured several times.

	For $Mn(NH_3)_6Br_2$:						
t in °C	82	83	99.3	107	120	125.8	131.8
p in Torr	100	86	200	305	468	600	760
Ref.	[9,10]	[3]	[3]	[4]	[3]	[3]	[3]

Selected values were taken from a table of [3].

	For $Mn(NH_3)_2Br_2$:				For $Mn(NH_3)Br_2$:			
t in °C	182	215	230	231	215	230	271	278
p in Torr	13.7	56.4	98.0	100	12.3	22.9	100	114
Ref.	[4]	[4]	[4]	[9]	[4]	[4]	[9]	[4]

Plots of measured pressures versus temperature, in the range between 40 and 120°C for the hexaammine, between 170 and 280°C for the diammine, and between 190 and 320°C for the monoammine are given in [9]. An isothermal decomposition study revealed the following transitions $MnBr_2 \cdot n NH_3 \rightarrow MnBr_2 \cdot n' NH_3 + (n - n')NH_3$ at temperatures t [17]:

transition $n \rightarrow n'$ ...	$6 \rightarrow 5$	$5 \rightarrow 2$	$2 \rightarrow 1$	$1 \rightarrow 0$
t in °C	20 to 128	128	>270	>360

An isotherm at 107.3°C shows the decomposition of the hexaammine to the diammine, and an isotherm at 230.3°C shows the decomposition of the diammine to the monoammine [4].

In air, $Mn(NH_3)_6Br_2$ becomes only slightly brown. Thus it is less sensitive to air than the chloride but more sensitive than the iodide [3]. With moist air, $Mn(NH_3)_6Br_2$ reacts instantaneously [5].

$Mn(NH_3)_nI_2$ (n = 6, 2). $Mn(NH_3)_6I_2$ can be prepared by dissolving MnI_2 in concentrated hot aqueous ammonia. The solution is filtered hot into ice-cold alcohol that has been freed of dissolved air and saturated with ammonia. The crystallization is finished with ice cooling of the reaction mixture. The product is filtered off in the absence of air and dried in a vacuum over NaOH and some NH_4Cl with cooling [4]. The hexaammine is also formed when anhydrous MnI_2 stands in ammonia gas for some hours [2] or in a stream of ammonia gas [3].

Isothermal decomposition studies at 153.6 and 214.8°C show that the hexaammine decomposes into $Mn(NH_3)_2I_2$, which in turn decomposes into MnI_2. There is no monoammine. The isotherms are shown in [4]. The absence of a monoammine has been confirmed by a differential thermal analysis of $Mn(NH_3)_6I_2$ [2].

Rossini et al. [8] gave the enthalpies of formation of the crystals from the elements in their standard states at 298.16 K as $\Delta H_f^\circ = -233.4$ kcal/mol for $Mn(NH_3)_6I_2$ and $\Delta H_f^\circ = -124.7$ kcal/mol for $Mn(NH_3)_2I_2$, calculated from data in the literature. The enthalpy of formation from MnI_2(s) and NH_3(g) from the change in the decomposition pressure with temperature is −96.06 kcal/mol for $Mn(NH_3)_6I_2$(s) and −39.42 kcal/mol for $Mn(NH_3)_2I_2$(s) [9].

Both complexes are white crystalline powders. Under a microscope, $Mn(NH_3)_6I_2$ consists of well formed octahedrons [4]. It has a cubic crystal structure of the CaF_2 type (as has $Mn(NH_3)_6Cl_2$, see p. 6) with a = 11.03_7 Å [11 to 13]. Mössbauer studies using ^{127}I and ^{129}I indicate that it undergoes a phase transition at 25 K. Below 25 K the rotation of the NH_3 molecules is hindered [16]. The density measured pycnometrically is 2.087 g/cm^3 at 25°C [14]. In the IR spectrum, there are bands at 295 cm^{-1} assigned to the Mn-N stretching frequency, at 592 cm^{-1} assigned to the NH_3 rocking frequency, and at 1155 and 1582 cm^{-1} assigned to the NH_3 deformation frequencies [15].

The following ammonia pressures over $Mn(NH_3)_6I_2$ are given by [3] (selected values):

t in °C	125.5	135	142.3	143.5	149.5	156	173.5
p in Torr	105	150	200	225	280	364	720

The following pressures were reported later: 100 Torr at 126°C [9] and 316 Torr at 154°C [4].

The ammonia pressure of $Mn(NH_3)_2I_2$ is 11.1 Torr at 208°C and 14.6 Torr at 215°C [4]. Plots of measured pressure versus temperature are given for both compounds in [9].

The decomposition has also been studied by differential thermal analysis. $Mn(NH_3)_6I_2$ was heated at a rate of 5°C/min in a stream of helium. Observed were an endothermic peak at 139°C, corresponding to the decomposition of the hexaammine to the diammine, and a second endothermic peak at 235°C, corresponding to the conversion of $Mn(NH_3)_2I_2$ to MnI_2 [2]. The decomposition temperature of $Mn(NH_3)_2I_2$ was found to be >200°C in an older isothermal decomposition study [17]. Thus the hexaammine and diammine are thermally more stable than the corresponding chlorides or bromides [2].

$Mn(NH_3)_6I_2$ turns brown after a few minutes in air [4], but it is less sensitive to air than the chloride or bromide [3].

References:

[1] H. Suga, E. Lagendijk, W. J. Huiskamp (Phys. Letters A **32** [1970] 297/8). – [2] W. W. Wendlandt, J. P. Smith (Nature **201** [1964] 73/4). – [3] F. Ephraim (Z. Physik. Chem. **81** [1913] 513/38, 518, 525, 531, 536). – [4] W. Biltz, G. F. Hüttig (Z. Anorg. Allgem. Chem. **109** [1920] 89/110, 96/102). – [5] G. W. Watt, B. S. Manhas (J. Inorg. Nucl. Chem. **28** [1966] 1945/7).

[6] F. W. Bergstrom (J. Phys. Chem. **29** [1925] 160/5, 161). – [7] D. D. Wagman, W. H. Evans, V. B. Parker, I. Halow, S. M. Bailey, R. H. Schumm (NBS-TN-270-4 [1969] 1/141, 112). – [8] F. D. Rossini, D. D. Wagman, W. H. Evans, S. Levine, I. Jaffe (NBS-Circ.-500 [1952] 280). – [9] W. Biltz (Z. Anorg. Allgem. Chem. **130** [1923] 93/139, 100). – [10] W. Biltz, E. Rahlfs (Z. Anorg. Allgem. Chem. **148** [1925] 145/51, 149/50).

[11] G. Bödtker-Naess, O. Hassel (Z. Physik. Chem. B **22** [1933] 471/2). – [12] G. Bödtker-Naess, O. Hassel (Avhandl. Norske Videnskaps Akad. Oslo I Mat. Nat. Kl. **1933** No. 4, pp. 1/19, 5, 9, 10; C. **1933** II 3391). – [13] Strukturbericht. Vol. 3, 1933/35 [1937], pp. 475/6. – [14] W. Biltz, E. Birk (Z. Anorg. Allgem. Chem. **134** [1924] 125/36, 132). – [15] L. Sacconi, A. Sabatini, P. Gans (Inorg. Chem. **3** [1964] 1772/4).

[16] L. Asch, J. P. Sanchez, J. M. Friedt, G. K. Shenoy (Contrib. Papers Intern. Conf. Hyperfine Interactions Stud. Nucl. React. Decay, Uppsala 1974, p. 236; C.A. **83** [1975] No. 199888). – [17] F. Müller (Jahrb. Phil. Fak. II Univ. Bern **2** [1922] 80/2).

12.2.5 Sulfates

$Mn(NH_3)_6SO_4$ was prepared by reacting anhydrous manganese(II) sulfate with anhydrous ammonia gas [1 to 3, 7, 9] at −18°C [2] or 20°C [9], at atmospheric pressure for two days [7]. The manganese(II) sulfate used had been dried at temperatures above 250°C [3, 7, 9]. The hexaammine is unstable at room temperature. It was stored in an atmosphere of ammonia in a sealed tube [7].

Kinetic studies of the formation reaction of $Mn(NH_3)_6SO_4$ at various temperatures between −14 and +90°C showed that the rate of reaction is not constant. At 0°C, the reaction rate goes through a minimum around 2.5 min after the start of the reaction and then goes through a maximum ~5 min after the start. With increasing temperature the start rate and the height of the maximum decrease. The strong decrease of the rate of formation with temperature as well as the occurrence of maxima and minima of the formation rate with time are assumed to be due to the formation and decomposition of mixed crystals between lower ammines and anhydrous manganese(II) sulfate and between lower ammines and the hexaammine. At temperatures below 0°C the rate of formation is not dependent on temperature. Adsorption measurements suggest that adsorbed ammonia rather than gaseous ammonia reacts to form the ammines [3].

Isothermal decomposition studies of the hexaammine gave the following pressures [1]:

t in °C	50.5	56.5	65	67	71.5	73
p in Torr	261	342	476	500	695	760

The dissociation heat is 12.29 kcal/mol NH_3 at 346.5 K. In another experiment appreciably lower pressures and an extremely slow attainment of equilibrium were observed. The disturbances were accounted for by the formation of solid solutions of the ammines and of ammonia in the decomposition products [1].

In [1] and [9] (isothermal decomposition in vacuo) the decomposition product of the hexaammine was found to be the diammine, **$Mn(NH_3)_2SO_4$** [1, 9]. Its decomposition pressures are [1]:

t in °C	102	114	128.5
p in Torr	73	99	130

On steady heating, a curve with the following transitions $Mn(NH_3)_nSO_4 \rightarrow Mn(NH_3)_{n'} + (n-n')NH_3$ was obtained [2]:

$n \rightarrow n'$	$6 \rightarrow 5$	$5 \rightarrow 2$	$2 \rightarrow 0.5$	$0.5 \rightarrow 0$
t in °C	53	70.5	201	276

In an isothermal decomposition study the same intermediate ammines, but the monoammine, instead of the hemiammine, were found [10].

Various mixtures of lower ammines (and of the hexaammine) were prepared by heating the hexaammine at atmospheric pressure or in vacuo at various temperatures. Heats of solution in aqueous 1M HCl were measured and reported together with the ammonia contents and the methods of preparation of the samples. In aqueous ammonia, all the manganese sulfate ammines are hydrolyzed to form manganese oxide hydrates [7].

Another group could not obtain the hexaammine on exposing manganese(II) sulfate to an ammonia atmosphere at room temperature for several hours and following the ammonia uptake by periodic weighings. (No drying temperature for the manganese(II) sulfate was given.) By reacting manganese(II) sulfate with liquid ammonia at −78°C a compound of the composition **$Mn(NH_3)_4SO_4$** was obtained, but no elemental analysis was given [5]. Much earlier, a white compound with a similar ammonia content (theoretical 31.09%) was prepared by reacting manganese(II) sulfate with ammonia [8]. The ESR spectrum of a solution of $Mn(NH_3)_4SO_4$ was reported, but no method of preparation or analysis was given [6]. A differential thermal analysis at a heating rate of 5°C per min in a helium atmosphere gave a curve which showed two endothermic peaks, at 88 and 313°C. The peaks were assigned to the transitions tetraammine → monoammine and monoammine → anhydrous salt [5].

$Mn(NH_3)_5SO_4 \cdot H_2O$ was prepared by the reaction of $MnSO_4 \cdot H_2O$, which had been dried at 160°C, with ammonia at 17°C. The compound decomposed in vacuo to form the diammine, $Mn(NH_3)_2SO_4 \cdot H_2O$ [9].

References:

[1] F. Ephraim (Z. Physik. Chem. **83** [1913] 196/220, 197, 201, 204, 211). – [2] F. Ephraim (Ber. Deut. Chem. Ges. **59** [1926] 1219/31, 1220, 1222, 1228). – [3] M. M. Pavljuchenko, G. A. Laserko (Zh. Fiz. Khim. **29** [1955] 1064/72, 1065; C. **1956** 9653). – [4] F. D. Rossini,

D. D. Wagman, W. H. Evans, S. Levine, I. Jaffe (NBS-Circ.-500 [1952] 281). – [5] W. W. Wendlandt, J. P. Smith (Nature **201** [1964] 73/4).

[6] D. Bösnecker, E. Lutze (Z. Angew. Physik **12** [1960] 354/60, 356/7). – [7] M. Badri, J. W. S. Jamieson (Can. J. Chem. **55** [1977] 3530/5). – [8] H. Rose (Ann. Physik Chem. [2] **20** [1830] 147/64, 148/9). – [9] W. Peters (Z. Anorg. Allgem. Chem. **77** [1912] 137/90, 147, 160/1; Ber. Deut. Chem. Ges. **45** [1909] 4826/36, 4829, 4833, 4835). – [10] F. Müller (Jahrb. Phil. Fak. Univ. Bern **2** [1922] 80/2).

12.2.6 Fluorosulfate and Tetrafluoroborates

$Mn(NH_3)_6(SO_3F)_2$ can be prepared by the precipitation of Mn^{2+} ions with NH_4SO_3F in aqueous ammonia with the exclusion of air. The dry compound is less sensitive to air oxidation than the wet compound. The crystals are cubic, belonging to the CaF_2 type with $a = 11.59_3$ Å. The distance from cation to anion is $a\sqrt{3}/4 = 5.02_0$ Å [1].

$Mn(NH_3)_6(BF_4)_2$ was prepared by a modified method of [2] by mixing a hot aqueous 2M NH_4BF_4 solution which had been saturated with gaseous ammonia and an aqueous 2M $Mn(BF_4)_2 \cdot 6H_2O$ solution. The precipitate was sucked dry on the filter and washed with alcohol saturated with ammonia and then with a small quantity of dry ether. All operations were carried out with exclusion of air. Finally, the compound was dried a short time in a vacuum [3]. It can be recrystallized from an aqueous NH_3 solution which contains some NH_2OH [1].

The nearly colorless and lustrous crystals of the hexaammine show octahedral habits. They are cubic and belong to the CaF_2 type with a lattice parameter of $a = 11.37_4$ Å. The cation-anion distance is 4.92_5 Å [1]. Thermal-gravimetric and differential-thermal plots show that when the compound is heated at a heating rate of 5°C/min it loses four molecules of NH_3 from 105 to ~200°C to form $Mn(NH_3)_2(BF_4)_2$ [3]. Under the action of air, the dry complex oxidizes superficially. In a wet condition, oxidation occurs rapidly. The complex is only sparingly soluble in aqueous ammonia. It is stable in aqueous ammonia which contains some hydroxylamine. In such solutions containing additionally some ammonium tetrafluoroborate the compound remained unchanged [2].

$Mn(NH_3)_2(BF_4)_2$ forms on heating $Mn(NH_3)_6(BF_4)_2$. A DTA peak is observed at 200°C. The diammine is stable up to 290°C and decomposes between 290 and 410°C into MnF_2, BF_3, and NH_3. DTA peaks appear at 360 and 380°C [3].

References:

[1] G. Bödtker-Naess, O. Hassel (Avhandl. Norske Videnskaps Akad. Oslo I Mat. Nat. Kl. **1933** No. 7, pp. 1/12, 4, 7; C. **1934** I 1274). – [2] E. Wilke-Dörfurt, G. Balz (Z. Anorg. Allgem. Chem. **159** [1927] 197/225, 206/7). – [3] F. Kutek (Collection Czech. Chem. Commun. **31** [1966] 4238/44, 4239, 4242).

12.2.7 Acetylide

$Mn(NH_3)_4(C_2H)_2$ was prepared by mixing solutions of 2.9 mmol $Mn(NCS)_2$ and 6.2 mmol KC_2H in liquid NH_3. The white precipitate, which formed immediately, became crystalline and pink on boiling the suspension. It was dried at −40°C for 20 min in a vacuum. The yield was almost quantitative. The compound was also formed by reaction of $Mnpy_4(NCS)_2$ with KC_2H in liquid NH_3. It loses ammonia very easily and should be handled in a counter-stream of gaseous ammonia. It is hydrolyzed and oxidized by moist air instantaneously to a brown product. Acid aqueous solutions decompose it with formation of acetylene. It is stable against impact. In a solution of KC_2H in liquid NH_3 it is soluble, forming $K_2Mn(C_2H)_4$, R. Nast, H. Griesshammer (Z. Anorg. Allgem. Chem. **293** [1957] 322/6, 324).

12.2.8 Palmitate and Stearate

Gaseous ammonia converts pulverized manganese(II) palmitate and manganese(II) stearate into yellowish-white amorphous powders of composition $Mn(C_{16}H_{31}O_2)_2 \cdot NH_3$ and $Mn(C_{18}H_{35}O_2)_2 \cdot NH_3$, S. Fisel (Acad. Rep. Populare Romine Filiala Iasi Studii Cercetari Stiint. Chim. **6** [1955] 295/327, 297, 302/4; C.A. **1959** 14928).

12.2.9 Cyanides

$Mn(NH_3)_6(CN)_2$ was prepared from $Mn(NCS)_2$ and KCN as described for the acetylide. The crystalline and bright pink solid was dried five to ten minutes under a high-vacuum at −40°C. The yield was almost quantitative. Even in the presence of a large excess of KCN, only $Mn(NH_3)_6(CN)_2$ formed.

At room temperature the dissociation pressure of $Mn(NH_3)_6(CN)_2$ is around 100 Torr. In a vacuum the compound loses ammonia to form $Mn(NH_3)_2(CN)_2$. Oxidation by air turns the hexaammine brown after 15 to 30 min, the compound losing ammonia at the same time. In the absence of air, water hydrolyzes the compound with formation of manganese oxide hydrates. The hydrolysis proceeds even faster in dilute acids.

$Mn(NH_3)_2(CN)_2$ was prepared by placing the hexaammine in a good vacuum for one or two hours at room temperature. It is pale yellow green and paramagnetic. The magnetic susceptibility χ has been measured in a high-vacuum at the temperatures T, and the magnetic moments μ_B calculated:

T in K	90	195	293
χ in 10^{-6} cm^3/g	292	154	110.5
μ_B	6.32 ± 0.13	6.26 ± 0.08	6.38 ± 0.14

The susceptibility obeys the Curie-Weiss law with a paramagnetic Curie temperature of $\Theta_p = -30$ K. The magnetic moments are greater than the 5.91 B.M. expected for five unpaired electrons. However, for twice the formula weight the measured susceptibilities yield magnetic moments of 8.90 to 9.03 μ_B, in good agreement with the expected value for 8 unpaired electrons, 8.94 μ_B. Thus, the dinuclear formulation $[Mn(NH_3)_4][Mn(CN)_4]$ may be discussed. It would have five unpaired electrons in the high-spin cation and three in the low-spin anion.

When heated to 250°C, the diammine loses part of its ammonia and turns yellow brown. It turns brown in air, losing HCN. Water (in the absence of air) hydrolyzes the complex completely to a dark green product of composition $Mn(CN)_2$. The complex is almost insoluble in common organic solvents.

Reference:

R. Nast, H. Griesshammer (Z. Anorg. Allgem. Chem. **293** [1958] 322/6).

12.2.10 Fulminates

A compound of the composition $Mn(NH_3)_n(ONC)_2$ with $n \approx 4$ was prepared by shaking an anhydrous-methanol $Hg(ONC)_2$ solution that had been saturated with ammonia with manganese amalgam in the absence of oxygen. After removing the brown precipitate initially formed the complex separated on further shaking with manganese amalgam. The colorless compound, which was washed with ether and dried in a vacuum, lost NH_3 and became yellow. After three days of standing in a vacuum the substance became dark brown, and its composition was $Mn(ONC)_2 \cdot 1.64NH_3$. It was therefore formulated $Mn(ONC)_2 \cdot 2NH_3$. Upon heating, the fulminate deflagrates with a dull pop, L. Wöhler, A. Berthmann (Ber. Deut. Chem. Ges. **62** [1929] 2748/58, 2750/1), see also L. Wöhler, A. Weber, A. Berthmann (AD-633399 [1966] 1/28, 13/4; C.A. **66** [1967] No. 81945).

12.2.11 Isothiocyanates

Colorless crystals of $\mathbf{Mn(NH_3)_4(NCS)_2}$ were obtained by dissolving $Mn(NCS)_2$ in liquid ammonia and evaporating the solvent. It reacts with KPH_2 in liquid ammonia at −78°C as follows: $Mn(NH_3)_4(NCS)_2 + 2KPH_2 \rightarrow Mn(NH_3)_3(PH_2)_2 + 2KSCN + NH_3$ [1].

Manganese metal dissolves in a solution prepared by saturating NH_4SCN with dry ammonia gas. From this solution the complex $\mathbf{Mn(NH_3)_4(NCS)_2 \cdot 2NH_4SCN}$ was isolated [2].

References:

[1] G. Uecker, O. Schmitz-DuMont (Z. Anorg. Allgem. Chem. **371** [1969] 318/24, 318, 324). – [2] A. Stasiewicz (Ann. Univ. Mariae Curie-Sklodowska Lublin-Polonia AA **14** [1959/61] 49/57, 56; C.A. **57** [1962] 6681).

12.2.12 Dihydrogen Phosphides

Colorless $\mathbf{Mn(NH_3)_3(PH_2)_2}$ is obtained when two moles of KPH_2 are reacted with one mole of $Mn(NH_3)_4(NCS)_2$ in liquid ammonia at −75°C and the solvent is evaporated. The complex is very soluble in liquid ammonia, as is KSCN. Therefore, it cannot be separated from the KSCN formed in the reaction. Above −79°C, it decomposes as follows: $3Mn(NH_3)_3(PH_2)_2 \rightarrow Mn_3P_2 + 4PH_3 + 9NH_3$ (the black Mn_3P_2 may be a mixture of MnP and Mn_2P).

$\mathbf{K_2[Mn(PH_2)_4] \cdot 2NH_3}$ forms when four moles of KPH_2 react with one mole of $Mn(NH_3)_4(NCS)_2$ or $Mn(NH_3)_6Cl_2$ in liquid ammonia at −75°C and the solvent is evaporated. It is also very soluble and cannot be separated from the other reaction products. At −60°C it begins to lose ammonia and at higher temperatures loses PH_3 and H_2.

Reference:

G. Uecker, O. Schmitz-DuMont (Z. Anorg. Allgem. Chem. **371** [1969] 318/24, 318, 319, 322).

12.2.13 Chromate

$\mathbf{Mn(NH_3)_6CrO_4}$ is prepared from stoichiometric amounts of $MnCrO_4$ and NH_3 in 95% ethanol by shaking the reaction mixture for three days. The yellow-brown complex is filtered off, washed with alcohol and ether, and dried over P_2O_5. It is paramagnetic; the magnetic moment is $\mu_{eff} = 5.90\ \mu_B$. The IR spectrum (in KBr pellets) shows the NH stretching frequency of the compound at 3260 cm^{-1}. This frequency is observed at 3414 cm^{-1} in free ammonia, the large shift of 154 cm^{-1} indicating a distinct weakening of the nitrogen-hydrogen bonds caused by the Mn-N coordination bond. The shift is larger than that for the corresponding alkylamine complexes (p. 22), which suggests that the ammonia complex is more stable than the alkylamine complexes.

The complex dissolves in formamide and nitrobenzene. A 0.01M solution in formamide shows absorption bands in the visible and near-UV region at 17860, 23520, 25010, 25150, 27780, 28570, and 33340 cm^{-1}. The electrical conductivity of a 0.001M solution in nitrobenzene is $\Lambda = 20.5\ \Omega^{-1} \cdot cm^2 \cdot mol^{-1}$, corresponding to that of a 1:1 electrolyte. The chromate anion can be removed by anion exchange.

Reference:

G. Narain, P. Shukla (J. Indian Chem. Soc. **43** [1966] 694/7).

12.3 Ammine Complexes of Manganese in Other Oxidation States

$Mn(NH_3)(NO)_3$

The compound s prepared by reaction of ammonia with $Mn(NO)_3CO$ in cyclohexane. The IR spectrum of the cyclohexane solution shows NO stretching frequencies at 1778.5 cm^{-1} (A_1) and at 1678.5 cm^{-1} (E). From these frequencies the force constant $f_{NO} = 12.90$ mdyn/Å was calculated [1].

$Mn(NH_3)_3Cl_3$

Conducting dry NH_3 gas through an ether solution of $MnCl_3$ at −20°C immediately produces a voluminous brown precipitate of $Mn(NH_3)_3Cl_3$. Several hours of standing at 20°C gives a fine crystalline product. It is dried in a stream of nitrogen gas. The compound is stable at room temperature in the absence of moisture. It decomposes upon heating above 50°C. Water decomposes the compound immediately. It is insoluble in organic solvents [2].

References:

[1] W. Beck, K. Lottes (Chem. Ber. **98** [1965] 2657/73, 2667, 2673). – [2] H. Funk, H. Kreis (Z. Anorg. Allgem. Chem. **349** [1967] 45/9, 46).

13 Complexes with Amines

Mixed-ligand complexes with amines and 2,4-pentanedione (acetylacetone) are described in "Mangan" D 1, 1979, pp. 76/7.

13.1 With Primary and Secondary Aliphatic Monoamines

ligand L	name	formula	ligand L	name	formula
CH_3NH_2	methylamine	CH_5N	$(CH_3)_2NH$	dimethylamine	C_2H_7N
$C_2H_5NH_2$	ethylamine	C_2H_7N	$(C_2H_5)_2NH$	diethylamine	$C_4H_{11}N$
$C_3H_7NH_2$	propylamine	C_3H_9N	$[CH_3(CH_2)_3]_2NH$	dibutylamine	$C_8H_{19}N$
$(CH_3)_2CHNH_2$	iso-propylamine	i-C_3H_9N	$[(CH_3)_2CHCH_2]_2NH$	diiso-butylamine	$C_8H_{19}N$
$CH_3(CH_2)_3NH_2$	butylamine	$C_4H_{11}N$			

13.1.1 Complexes in Solution

When methylamine, ethylamine, propylamine, butylamine, dimethylamine, or diethylamine is added in excess to aqueous $MnSO_4$ solutions buffered with $(NH_4)_2SO_4$, color changes and shifts of the maxima in the visible absorption spectra indicate complex formation. Each complex exhibits six sharp absorption bands, with shifts between 2 and 15 cm^{-1} from those of the $MnSO_4$ solution. In the case of the solution with methylamine the bands are at 352, 375, 395, 402, 425, and 560 nm. The spectra of the complexes with the other amines are also given. All of the observed bands can be fit by taking the ligand field parameter $\Delta = 10$ Dq to be 8600 cm^{-1}. The spectra suggest that in each case only the hexaamine is formed, P. Shukla (J. Prakt. Chem. [4] **38** [1968] 253/6).

13.1.2 Isolated Manganese(II) Compounds

Halides MnL_nX_2 (n = 6, 4, 2). Complexes of the type $Mn(CH_5N)_6X_2$ and $Mn(C_2H_7N)_4X_2$ are formed by passing gaseous methylamine or ethylamine over well dehydrated MnX_2 salts at room temperature. At low temperatures MnI_2 takes up six moles of ethylamine to give $Mn(C_2H_7N)_6I_2$. The compounds are white, or brownish white due to partial oxidation. The following table lists the temperature T_{diss} at which the dissociation pressure reaches 1 atm and the dissociation enthalpy ΔH_{diss} calculated by the Nernst formula [1]:

complex	T_{diss} in K	ΔH_{diss} in kcal/mol	complex	T_{diss} in K	ΔH_{diss} in kcal/mol
$Mn(CH_5N)_6Cl_2$	346	12.25	$Mn(C_2H_7N)_4Br_2$	305	10.66
$Mn(CH_5N)_6Br_2$	387	13.85	$Mn(C_2H_7N)_4I_2$	320	11.23
$Mn(CH_5N)_6I_2$	395	14.16	$Mn(C_2H_7N)_6I_2$	247	9.36
$Mn(C_2H_7N)_4Cl_2$	289	10.05			

The temperature dependence of the dissociation pressure of the compounds cited is given in **Fig. 5**, p. 22.

Fig. 5

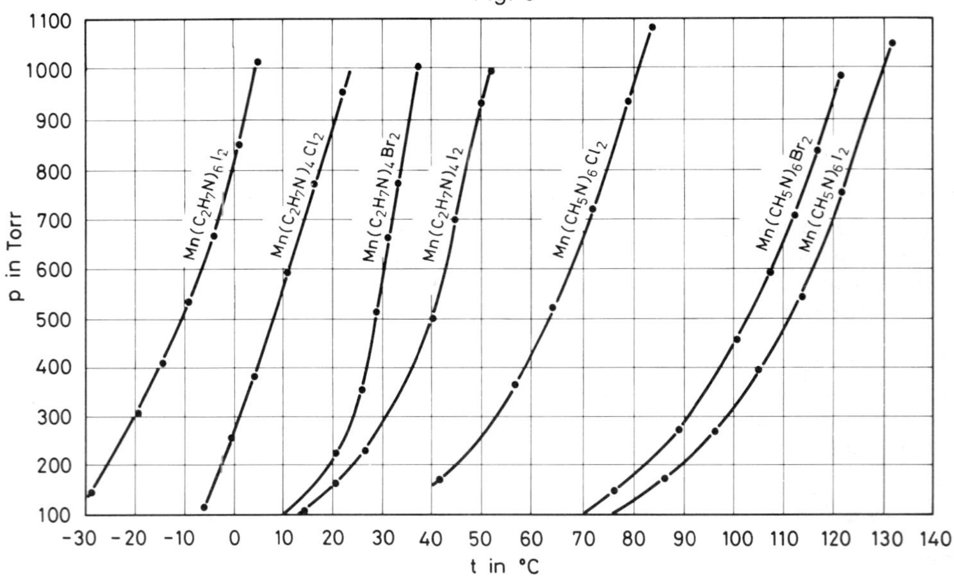

Temperature dependence of the dissociation pressure p of manganese(II) complexes with methylamine and ethylamine.

The following bis- and mono(amine) complexes were prepared by mixing approximately stoichiometric amounts of anhydrous manganese(II) chloride and amine in alcohol and, if necessary, adding benzene to induce precipitation. The compounds were washed with benzene or alcohol and dried in a vacuum over fused calcium chloride.

ligand	complex	color	ligand	complex	color
ethylamine	$Mn(C_2H_7N)_2Cl_2$	black	dibutylamine	$Mn(C_8H_{19}N)Cl_2$	reddish brown
dimethylamine	$Mn(C_2H_7N)Cl_2$	chocolate brown	diisobutylamine	$Mn(C_8H_{19}N)Cl_2$	earth colored

The compounds are fairly stable in a dry atmosphere. When heated, they do not melt, decomposing instead. They are insoluble in water but soluble in mineral acids [4].

Palmitates and Stearates. The yellowish-white $\mathbf{Mn(RNH_2)_n(C_{15}H_{31}COO)_2}$ ($R=CH_3$, C_2H_5, for $n=2$; $R=C_3H_7$, for $n=0.5$) or coffee-brown $\mathbf{Mn(RNH_2)_{0.5}(C_{17}H_{35}COO)_2}$ ($R=CH_3$, C_2H_5, C_3H_7) are formed by reaction of powdered manganese(II) palmitate or stearate with gaseous methylamine, ethylamine, or propylamine [2].

Chromates. $\mathbf{MnL_6CrO_4}$. The yellowish-brown complexes listed in the table below are prepared by shaking $MnCrO_4$ and the amine in a 1:6 ratio in 95% ethanol for 3 days. The magnetic moments μ_{eff} (in μ_B) and the NH stretching frequencies (IR, KBr discs, in cm^{-1}) are:

ligand L	complex	μ_{eff}	ν(NH)
methylamine	$Mn(CH_5N)_6CrO_4$	5.89	3285
ethylamine	$Mn(C_2H_7N)_6CrO_4$	5.88	3288
propylamine	$Mn(C_3H_9N)_6CrO_4$	5.92	3330
isopropylamine	$Mn(i\text{-}C_3H_9N)_6CrO_4$	5.90	3320
butylamine	$Mn(C_4H_{11}N)_6CrO_4$	5.91	3340

The strong negative shift of the ν(NH) band (free amines, ~3400 cm^{-1}) shows the weakening of the N-H bonds due to the formation of a highly covalent Mn-N bond. The shift decreases with the molecular weight of the amine, an indication that the covalency of the Mn-N bond and the stability of the complex decrease in the same order [3].

The chromates dissolve in formamide and nitrobenzene. The 0.01M solutions in formamide exhibit six to seven absorption bands at positions similar to those given on p. 20 for the ammine complexes. The electrical conductivities of 0.001M solutions in nitrobenzene ($\Lambda = 20$ to 28 $\Omega^{-1} \cdot cm^2 \cdot mol^{-1}$) indicate that the complexes are 1:1 electrolytes. Solutions of the complexes passed through an anion-exchange resin contain only $Mn(RNH_2)_6^{2+}$, and no CrO_4^{2-} ions [3].

References:

[1] F. Ephraim, R. Linn (Ber. Deut. Chem. Ges. **46** [1913] 3742/62, 3742, 3745, 3751, 3755). – [2] S. Fisel (Acad. Rep. Populare Romine Filiala Iasi Studii Cercetari Stiint. **6** [1955] 295/327, 297, 305, 307/9; C.A. **1959** 14928). – [3] G. Narain, P. Shukla (J. Indian Chem. Soc. **43** [1966] 694/7). – [4] S. Prasad, P. J. Sarma (J. Proc. Inst. Chemists [India] **37** [1965] 246/8).

13.1.3 Isolated Manganese(III) Compounds

$Mn(RNH_2)_3Cl_3$ ($R = CH_3$, C_2H_5, C_3H_7, C_4H_9) and **$Mn(R'_2NH)_3Cl_3$** ($R' = CH_3$, C_2H_5) are formed by reacting $MnCl_3$ with the corresponding primary or secondary amines in ether at −20 to −10°C. The complexes are brown crystalline powders (those with RNH_2) or brown leaflets (those with R'_2NH). All of them are stable at room temperature. The complexes with methylamine and ethylamine decompose above 50°C. All the compounds are sensitive to moisture and hydrolyze in water, H. Funk, H. Kreis (Z. Anorg. Allgem. Chem. **349** [1966] 45/9).

13.2 With Tertiary Aliphatic Monoamines

13.2.1 With Trimethylamine $N(CH_3)_3$ (= C_3H_9N) and Its N-Oxide $(CH_3)_3NO$ (= C_3H_9NO)

The action of gaseous trimethylamine on thin coatings of solid $MnCl_2$, $MnBr_2$, and MnI_2 has been studied by measurement of the IR absorption in a special reaction cell for trimethylamine pressures from 400 down to 10^{-6} Torr. While MnI_2 does not react, $MnCl_2$ and $MnBr_2$ form undefined reaction products with IR bands (in cm^{-1}) which are shifted in a characteristic manner. The following assignments are given [1]:

assignment	ν(CH)	δ(CH_3)
chloride	3080, 2940	1460, 1435, 1395
bromide	3130, 2960, 2900	1445
$(CH_3)_3N$ gas*)	2967, 2822, 2777	1466, 1402

*) See reference [6].

assignment	$\varrho(CH_3)$ antisym.	sym.	$\nu(CN)$ antisym.	sym.	$\delta(C_3N)$
chloride	1045, 975	1180	1255	—	—
bromide	1052, 1035	1180	—	—	520
$(CH_3)_3N$ gas*)	1104, 1043	1183	1272	826	425, 365

*) See reference [6]. – ϱ = rocking vibration, ν = stretching vibration, δ = deformation vibration.

The complexes are stable at a vacuum of 10^{-5} Torr. They are very hygroscopic [1].

Addition of an excess of trimethylamine to the alcohol complexes $Mn(ROH)_xX_2$ with $R = CH_3$, C_2H_5, or C_5H_{11}, x = 6 to 9, and $X = ClO_4$ or BF_4 gave products of the general formula $[(CH_3)_3NH]_rMn[N(CH_3)_3]_m(RO)_nX_2$, n = 1 to 8, and m = 0 to 3, rather than the simple amine complexes. The compounds were washed with dry diethyl ether, dried at room temperature, and identified by chemical analysis, X-ray powder patterns, IR spectra, etc. (no data given). Separation of the products by solvent extraction was not possible. This suggests that they are not mixtures of salts [2].

$Mn(C_3H_9NO)_2Cl_2$ and **$Mn(C_3H_9NO)_2Br_2$** were prepared by mixing solutions of trimethylamine N-oxide and $MnCl_2$ or $MnBr_2$ in absolute 2-propanol or acetone. After recrystallization, the complexes are dried at 70°C in vacuo [3].

The colorless crystals of $Mn(C_3H_9NO)_2Br_2$ melt at 96 to 98°C. The magnetic moment, $\mu_{eff} = 5.64\ \mu_B$ at 291.5 K, suggests a tetrahedral structure. The solid complex is unstable in air; it has the characteristic odor of trimethylamine oxide. It is decomposed by water or ethanol, is soluble in acetonitrile and phenol, and is insoluble in ether, benzene, or tetrahydrofuran. The electrical conductivity of the solution in CH_3CN and the molecular weight in phenol (180, calculated 365) indicate partial dissociation in these solvents [3].

$Mn(C_3H_9NO)_4(ClO_4)_2$ was prepared by reaction of trimethylamine N-oxide with $Mn(ClO_4)_2$ under completely anhydrous conditions [5] in 2-propanol, methanol [4], or dichloromethane [5] after dehydration of each component by reaction with 2,2-dimethoxypropane in 2-propanol or methanol [4]. Precipitation was completed by adding a large excess of anhydrous ether. Occasionally an oil formed. Crystallization was then induced by dissolution in nitromethane and reprecipitation from a more dilute solution [4].

The X-ray powder pattern of $Mn(C_3H_9NO)_4(ClO_4)_2$ is nearly identical with that of the analogous zinc or cobalt compounds. A table of d values is given. The structure of these complexes is assumed to be tetrahedral since the IR spectrum (Nujol) shows no coordinated perchlorate in the solid form. A single broad band around 1100 cm^{-1} is assigned to the ν_3 vibration of the tetrahedral uncoordinated perchlorate group. The magnetic moment of the solid manganese complex is $\mu_{eff} = 5.29 \pm 0.04\ \mu_B$ at 25°C [4]. The IR spectrum (Nujol) shows a strong band of the ν(Mn-O) vibration at 547 cm^{-1}, strong ν(NO) or ν(CN) bands at 971, 961, and 949 cm^{-1}, and other ligand bands at 490, 477, 402, and 386 cm^{-1} [5]. The complex is slightly hygroscopic [4].

References:

[1] G. G. Guilbault, S. M. Billedeau (J. Inorg. Nucl. Chem. **33** [1971] 1411/5). – [2] M. G. W. de Bolster, W. L. Groeneveld (J. Inorg. Nucl. Chem. **35** [1973] 2106/7). – [3] K. Issleib, A. Kreibich (Z. Anorg. Allgem. Chem. **313** [1961] 338/48, 341, 346). – [4] R. S. Drago, J. T. Donoghue, D. W. Herlocker (Inorg. Chem. **4** [1965] 836/9). – [5] S. H. Hunter, V. M. Langford, G. A. Rodley, C. J. Wilkins (J. Chem. Soc. A **1968** 305/8).

[6] J. R. Barcelo, J. Bellanato (Spectrochim. Acta **8** [1956/57] 27/40, 37).

13.2.2 With Triethylamine $(C_2H_5)_3N$ (= $C_6H_{15}N$)

A white, unstable complex of composition $Mn(C_6H_{15}N)Cl_2$ is formed by mixing alcohol solutions of the ligand and anhydrous $MnCl_2$. It does not melt sharply. It is decomposed by boiling water and by solutions or mineral acids or NaOH [1]. According to [2] the addition of an excess of triethylamine to the alcohol complexes $Mn(ROH)_xX_2$ with $R=CH_3$, C_2H_5, or n-C_5H_{11}, $x=6$ to 9, and $X=ClO_4$ or BF_4 gave products of the general formula $[(C_2H_5)_3NH]_nMn[N(C_2H_5)_3]_m(RO)_nX_2$ with n = 1 to 8 and m = 0 to 3 (compare to the reaction with trimethylamine on p. 24).

The manganese(III) complex $Mn(C_6H_{15}N)_3Cl_3$ was prepared by adding an ether solution of the ligand cooled to −5°C dropwise to an ether solution of manganese(III) chloride cooled to about −10°C until the violet color of the salt solution vanished. The complex is dark brown and stable even in moist air [3].

A mixed-ligand complex, $Mn(C_6H_{15}N)(C_{22}H_{22}N_4)$, containing triethylamine and 7,16-dihydro-6,8,15,17-tetramethyldibenzo[b,i][1,4,8,11]tetraazocyclotetradecine (= $C_{22}H_{24}N_4$) is described in "Manganese" D 4.

References:

[1] S. Prasad, K. P. Kacker (J. Indian Chem. Soc. **35** [1958] 719/21). – [2] M. W. G. de Bolster, W. L. Groeneveld (J. Inorg. Nucl. Chem. **35** [1973] 2106/7). – [3] H. Funk, H. Kreis (Z. Anorg. Allgem. Chem. **349** [1966] 45/9).

13.2.3 With Other Tertiary Aliphatic Amines

Manganese(II) can be extracted with trioctylamine from aqueous HCl into organic solvents. The maximum extractability, 35%, for extraction into xylene occurs with 8 M HCl [1]. The maximum distribution ratio for extraction into 1:1 trioctylamine-benzene, D ≈ 10, occurs with 7 M HCl [2]. However, the maximum distribution ratio into 0.3 M tridodecylamine in xylene is only D ≈ 1, occurring with ~10 M HCl. The distribution ratio is increased by addition of ethanol and isopropyl alcohol [3].

Trace amounts of manganese(II) were effectively extracted from 2 M $Co(NO_3)_2$ by mixtures of trioctylamine (= $C_{24}H_{51}N$) and cobalt naphthenate (CoA_2). The composition of the complex which passed into the organic phase was determined to be $4[Mn(NO_3)_2\text{-}MnA_2]\cdot C_{24}H_{51}N$ (A = anion of naphthenic acid) [4].

References:

[1] G. Nakagawa (Nippon Kagaku Zasshi **82** [1961] 1042/5 from C.A. **57** [1962] 15782). – [2] B. Neef, H. Grosse-Ruyken (J. Chromatog. **79** [1973] 275/85, 278, 280). – [3] A. Alian (Microchem. J. **18** [1973] 377/92, 388). – [4] I. N. Kremenskaya, E. V. Goncharov, T. S. Balanevskaya (Zh. Neorgan. Khim. **17** [1972] 201/3; Russ. J. Inorg. Chem. **17** [1972] 105/7).

13.3 With Benzylamine $C_6H_5CH_2NH_2$ (= C_7H_9N)

$Mn(C_7H_9N)_2Cl_2$. The brown complex was prepared in a manner similar to that described for the complexes with primary and secondary aliphatic amines on p. 22. Its properties are also similar to the properties of those compounds [1].

$(NH_4)_2Mn(C_7H_9N)_2(SO_4)_2\cdot 2H_2O$ was obtained by mixing powdered $(NH_4)_2Mn(SO_4)_2\cdot 6H_2O$ ("Mangan" C 6, 1976, p. 203) with benzylamine and alcohol, and refluxing the mixture for 20 min. The white crystalline powder was washed with alcohol and ether, and dried over P_2O_5. It is stable in air, but is decomposed by water [2].

References:

[1] S. Prasad, P. J. Sarma (J. Proc. Inst. Chemists [India] **37** [1965] 246/8). – [2] G. Spacu, R. Ripan (Bul. Soc. Stiinte Cluj **1** [1922] 473/96, 494; C.A. **1924** 949).

13.4 With Dibenzylamine $(C_6H_5CH_2)_2NH$ $(=C_{14}H_{15}N)$ and Tribenzylamine $(C_6H_5CH_2)_3N$ $(=C_{21}H_{21}N)$

By mixing solutions of anhydrous manganese(II) chloride and the corresponding amine in absolute alcohol, brown and white products of composition $Mn(C_{14}H_{15}N)Cl_2$ and $Mn(C_{21}H_{21}N)Cl_2$ were obtained. The complexes are unstable. They do not melt sharply. They are decomposed by boiling water or by solutions of mineral acids or NaOH [1].

The manganese(III) complex, $Mn(C_{21}H_{21}N)_3Cl_3$, with tribenzylamine was prepared like the complex with triethylamine, $Mn(C_6H_{15}N)_3Cl_3$ (p. 25). It is dark green and can be stored only in a nitrogen atmosphere [2].

References:

[1] S. Prasad, K. P. Kacker (J. Indian Chem. Soc. **35** [1958] 719/21). – [2] H. Funk, H. Kreis (Z. Anorg. Allgem. Chem. **349** [1966] 45/9).

13.5 With Aniline $C_6H_5NH_2$ $(=C_6H_7N)$

$Mn(C_6H_7N)_nX_2$ Compounds

$Mn(C_6H_7N)_2(NO_3)_2$ was prepared by mixing manganese(II) nitrate and aniline (without any solvent). Spectroscopic studies suggest that the structure is octahedral. An IR band (mulls, no detailed data given) near 1100 cm^{-1}, which originates from the nitrate groups, is broad but does not appear to be markedly split. This would support a regular coordination of all the oxygen atoms of each nitrate group, to form coordination polymers. The far-IR spectrum shows a band at 258 cm^{-1} assigned to the Mn-O vibration. The low position of the Mn-O stretching vibration is consistent with the presence of oxo-anions as bridges. The diffuse visible reflectance spectrum shows a band at 18200 cm^{-1}, which is assigned to a transition from the ground state $^6A_{1g}(S)$ to $^4T_{1g}(G)$. The complex decomposes explosively when heated [1].

$Mn(C_6H_7N)_2X_2$ (X = Cl, Br, I). The halogeno complexes can be prepared by mixing alcohol solutions of equivalent amounts of anhydrous manganese(II) chloride and aniline, washing the colorless crystals with diethyl ether, and drying them 2 to 4 h at 110°C [3, 4]. Preparation by mixing hot solutions of the initial compounds in absolute ethanol and drying the complexes at 70°C is reported by [11, 12]. The complexes also can be synthesized by mixing the initial compounds without any solvent [3] or by heating or refluxing a mixture of excess aniline with anhydrous manganese(II) chloride under an N_2 atmosphere [4, 6]. The compounds are also formed by thermal decomposition of $Mn(C_6H_7N)_2X_2 \cdot 2\,C_2H_5OH$ (p. 29) or $Mn(C_6H_7N)_4I_2$ (p. 28). N-sulfinylaniline, C_6H_5NSO, and $MnCl_2$ in ethanol solution form $Mn(C_6H_7N)_2Cl_2$ and diethyl sulfite, $(C_2H_5O)_2SO$ [7]. The complex of $MnCl_2$ with deuterated aniline, $C_6H_5ND_2$ $(=C_6H_5D_2N)$, $Mn(C_6H_5D_2N)_2Cl_2$ was prepared in ether solution [4].

The following enthalpies of formation of solid $Mn(C_6H_7N)_2X_2$ from two moles of aniline and one mole of manganese(II) halogenide, ΔH_l (liquid aniline) and ΔH_g (gaseous aniline), at 25°C were calculated from the calorimetric heats of dissolution [8]:

enthalpy of formation	$Mn(C_6H_7N)_2Cl_2$	$Mn(C_6H_7N)_2Br_2$	$Mn(C_6H_7N)_2I_2$
$-\Delta H_l$ in kJ/mol	70.96	78.4	72.0
$-\Delta H_g$ in kJ/mol	173.6	181.1	174.7

$\Delta H_g = -30.1 \pm 0.7$ kcal/mol at 520 ± 20 K was determined for the chloro complex in a previous work by differential scanning calorimetry [5].

The IR spectrum was determined with the KBr disc technique at room temperature; the bands were assigned as follows [4]:

compound	$\nu_{as}(NH_2)$	$\nu_s(NH_2)$	$\delta(NH_2)$	$\omega(NH_2)$	ring + ν(CN)
$Mn(C_6H_7N)_2Cl_2$	3320(m)	3268(m)	1582(w)	1010(s)	1238(w)
$Mn(C_6H_7N)_2Br_2$	3300(m)	3247(m)	1580(w)	1050(s)	1238(w)
aniline	3440(m)	3360(m)	1619(m)	—	1278(w)

The ND_2 antisymmetric stretching vibration band of $Mn(C_6H_5D_2N)_2Cl_2$ is observed at 2434 cm^{-1}. The considerable shifts of the NH_2 bands from their positions for the free ligand indicate coordination via the NH_2 group [4]. The far-IR spectra were recorded using pressed polythene discs at 30 K and room temperature [10] and Nujol mulls at room temperature [4, 6]. Only the data at 30 K are given. Assignments of the ν(Mn-N) and ν(Mn-X) modes have been made with the aid of the perdeuterated complexes, $Mn(C_6D_7N)_2X_2$ [10]. The bands were assigned as follows:

complex/ assignment	ν(Mn-N)	ν(Mn-X)	δ(N-Mn-N)	Ref.
$Mn(C_6H_7N)_2Cl_2$	388sh, 384ms, 325m, 311mw	244sh, 233s, 216m, 164m, 155sh	204s, 190s	[10]
	386s, 374s	318 [4], 227	—	[6]
$Mn(C_6H_7N)_2Br_2$	384sh(?), 375ms, 332ms, 325ms	162s, 152s, 122ms	206s, 199s	[10]
	367s, 350s	—	—	[6]
$Mn(C_6H_7N)_2I_2$	361s, 355s	—	—	[6]

The following bands, found in the farthest IR region measured at 30 K, could not be assigned: 144m, 116ms, and 96s for X = Cl and 100s, 89sh, and 79sh for X = Br [10].

Bands in the visible reflectance spectra of the complexes (wavelengths in nm) were assigned to transitions from the ground state $^6A_{1g}$ to the following excited states [6]:

complex	$^4T_{1g}(G)$	$^4T_{2g}(G)$	$^4E_g(G)$, $^4A_{1g}(G)$
$Mn(C_6H_7N)_2Cl_2$	550	460	400
$Mn(C_6H_7N)_2Br_2$	530	450	—
$Mn(C_6H_7N)_2I_2$	560	440	—

The spectroscopic properties indicate an octahedral arrangement of neutral ligands and coordinated anions around the manganese atom [4, 6, 10]. To attain the coordination number six, halogen bridges between two manganese atoms must be present [6, 10]. Thus linear polymeric chains are formed. The compounds are isostructural with the analogous complexes of Cd^{2+} and Ni^{2+} [10].

Differential thermal curves (DTA, TG, DTG) at a heating rate of 3.3°C per min for the chloride and 1.2°C per min for the bromide in air show endothermal effects at 195, 214, and 247°C for the chloride, corresponding to the decomposition $Mn(C_6H_7N)_2Cl_2 \rightarrow Mn(C_6H_7N)Cl_2 \rightarrow Mn(C_6H_7N)_{1/2}Cl_2 \rightarrow MnCl_2$ [11], and endothermal effects at 212 and 220°C for the bromide, corresponding to the loss of 1.5 and 0.5 mol aniline, respectively [12]. The three chloride decomposition steps had been observed earlier by differential scanning calorimetry [5]. The enthalpies of dissolution of the complexes with X = Cl, Br, and I in aqueous 2N HCl were determined calorimetrically to be $\Delta H = -64.19$, -64.12 kJ/mol [8], or -61.87 kJ/mol, respectively, at 25°C [9].

$Mn(C_6H_7N)_4I_2$. The complex can be prepared by heating anhydrous manganese(II) iodide directly with aniline [6] or by desolvation of the ethanol adduct, $Mn(C_6H_7N)_4I_2 \cdot 2C_2H_5OH$, p. 29 [8, 9]. The enthalpy of formation for the reaction of four moles of liquid or gaseous aniline with one mole MnI_2 to form solid $Mn(C_6H_7N)_4I_2$ at 25°C has been calculated to be $\Delta H_l = -121.4$ or $\Delta H_g = -326.8$ kJ/mol from the calorimetric heats of dissolution. These enthalpies, large compared to those of $Mn(C_6H_7N)_2X_2$ (p. 27), indicate easy formation of the complex compound [8]. In the far-IR spectrum (Nujol mull), a band at 357 cm^{-1} is assigned to the Mn-N vibration. The visible reflectance spectrum shows two bands at 560 and 490 nm, corresponding to the transitions from the ground state $^6A_{1g}$, to the excited states $^4T_{1g}(G)$ and $^4T_{2g}(G)$, respectively. The spectroscopic properties show that the complex is octahedral with the iodine atoms attached to the manganese atom [6].

The complex decomposes at 130°C to form $Mn(C_6H_7N)_2I_2$, p. 26 [6]. Differential thermal analysis with a heating rate of 3.3°C per min in air showed two endothermic effects, at 105 and 135°C, corresponding to the loss of 0.5 and 1.5 mol aniline, respectively [9]. The enthalpy of dissolution in aqueous 2N HCl was determined calorimetrically at room temperature to be $\Delta H = -79.02$ kJ/mol [9].

$Mn(C_6H_7N)_4Cr_2O_7$ was prepared by successively adding an ethanol-aniline solution and an 1% $K_2Cr_2O_7$ solution to aqueous manganese(II) nitrate. After 10 min the complex began to crystallize. A second fraction could be obtained by adding an additional amount of dichromate solution to the filtrate. The chocolate-brown crystals were washed with water, ethanol, and acetone [14].

The magnetic moment is 6.07 μ_B at room temperature [13]. An octahedral structure is assumed on account of the similarity with the structures of nickel and cobalt complexes, four corners being occupied by the N atoms of the amino groups and the other two by O from the bidentate $Cr_2O_7^{2-}$ [13]. The IR spectrum of the complex shows bands at 879 and 922 cm^{-1} from the $Cr_2O_7^{2-}$ group, at 1030 cm^{-1} from the CN, and at 3205 and 3288 cm^{-1} from the symmetric and antisymmetric NH vibrations. The CN and NH modes are distinctly shifted to lower frequencies (free aniline: $\nu(CN)$ at 1282 cm^{-1}, $\nu_s(NH)$ at 3362 cm^{-1}, $\nu_{as}(NH)$ at 3435 cm^{-1}) [15]. The force constant of the N-H bond, $f_{NH} = 5.83$ mdyn/Å, and an H-N-H valence angle of 112° have been calculated from the IR frequencies. The force constant is lower than that of the free aniline owing to the decrease of the electron density at the N atom [16].

The complex is sparingly soluble in water, ethanol, and acetone. The molar conductance in aqueous solution, $\Lambda = 111\ \Omega^{-1} \cdot cm^2 \cdot mol^{-1}$, suggests that the compound dissociates into two ions [14].

$Cd_2MnCl_6 \cdot 6C_6H_7N$ (?). The white complex is reported to have been obtained by adding a solution of aniline in an alcohol-benzene mixture to the powdered double salt $Cd_2MnCl_6 \cdot 12H_2O$ (?), see "Manganese" C 5, 1978, p. 221. The formula $[MnCd(C_6H_7N)_6][CdCl_6]$ is discussed as the most probable. The compound is relatively stable in air but is light sensitive. It is only slightly soluble in water, acetone, and benzene. In hot water it decomposes. It dissolves well in acids [17].

Ethanol Adducts

The complexes were prepared by mixing ethanol solutions of the manganese(II) salts and aniline [2, 6, 9], except the sulfate, for which an aqueous solution of $MnSO_4 \cdot 5H_2O$ was reacted with an ethanol solution of aniline [2]. Hot solutions of the anhydrous manganese(II) halide and aniline in absolute ethanol were used by [11, 12]. While the chloride and the bromide crystallized immediately on cooling in an inert atmosphere [11, 12], the bright crystals of the sulfate and thiocyanate formed only after a few days [2].

$Mn(C_6H_7N)(NO_3)_2 \cdot 2C_2H_5OH$. The IR spectrum of the solid complex shows the bands of the NH_2 stretching vibrations at 3245, 3277, 3308, 3330 cm^{-1}, and those of the NO_3^- stretching vibrations at 1490, ~1310, and ~1040(?) cm^{-1}. The splitting of the IR bands of the NO_3^- stretching vibrations indicates that the NO_3^- groups are in the inner coordination sphere. The comparatively small degree of splitting suggests that the Mn-ONO_2 bond is only weak.

When heated, the complex decomposes at 100°C. It is readily soluble in water, methanol, ethanol or acetone, and sparingly soluble in hexane, CCl_4, or benzene. When the complex dissolves in water the NO_3^- groups are detached. The electrical conductivity of the aqueous solution, $\Lambda = 220\ \Omega^{-1} \cdot cm^2 \cdot mol^{-1}$, corresponds to that of three-ion electrolytes [2].

$Mn(C_6H_7N)_4I_2 \cdot 2C_2H_5OH$. The adduct is very unstable and loses ethanol on storage over calcium chloride. It is oxidized and turns black in air [9].

$Mn(C_6H_7N)_2X_2 \cdot 2C_2H_5OH$ (X = Cl, Br, I). The enthalpy of formation of the adducts from two moles of liquid or gaseous ethanol and one mole $Mn(C_6H_7N)_2Cl_2$ (p. 26) at 25°C has been calculated to be $\Delta H_l = -45.9$ or $\Delta H_g = -113$ kJ/mol from the calorimetric heats of dissolution [8].

The far-IR spectra (Nujol) show Mn-O bands at 461 cm^{-1} for the chloride and 462 cm^{-1} for the bromide. The bands assigned to the Mn-N modes are at 386 and 370 for the chloride or 383, 370, and 350 cm^{-1} for the bromide [6]. Bands in the visible reflectance spectra (in nm) are assigned to transitions from the ground state $^6A_{1g}$ to the following excited states [6]:

complex	$^4T_{1g}(G)$	$^4T_{2g}(G)$	$^4E_g(G)$, $^4A_{1g}(G)$
$Mn(C_6H_7N)_2Cl_2 \cdot 2C_2H_5OH$	550	470	410
$Mn(C_6H_7N)_2Br_2 \cdot 2C_2H_5OH$	—	—	420

The spectroscopic studies suggest an octahedral coordination of the various ligands around the manganese atom [6].

When the complexes are heated to 70°C [8] or 80°C [6], the ethanol is removed completely. DTA peaks (heating rate 3.3°C per min) are observed at 90°C for the chloride [11] and 97°C for the bromide [12]. The enthalpy of dissolution of $Mn(C_6H_7N)_2Cl_2 \cdot 2C_2H_5OH$ in 2N HCl was calorimetrically determined to be $\Delta H = -34.43$ kJ/mol at 25°C [8].

$Mn(C_6H_7N)SO_4 \cdot C_2H_5OH \cdot 1.5H_2O$ and $Mn(C_6H_7N)_2(NCS)_2 \cdot 0.5C_2H_5OH$. In the IR spectrum of the sulfate, the characteristic bands (in cm^{-1}) were assigned as follows: 3328 ($\nu_{as}NH_2$), 3265 (ν_sNH_2), 1180, ~1145, ~1130, 1106 ($\nu_{as}SO_4$), and 1005 (ν_sSO_4). For the isothiocyanate, the bands were assigned as follows: 3330 ($\nu_{as}NH_2$), 3257 (ν_sNH_2), 2108 (νCN), 783 (νCS). The IR spectra show that the SO_4 and the NCS groups enter the inner spheres of the complexes. The sulfate group is appreciably distorted; however, it is uncertain whether this ligand is unidentate, bidentate, or bridging. The NCS^- ligand is unidentate and coordinated to the manganese atom through its nitrogen atom [2].

When heated, the sulfate decomposes at 208°C, and the isothiocyanate decomposes at 180°C. The complexes are readily soluble in water, methanol, ethanol, or acetone, and sparingly soluble in hexane, carbon tetrachloride, or benzene. On dissolving the compounds in water, the SO_4^{2-} or NCS^- ions are detached, as is shown by the molar conductances of the aqueous solutions, $\Lambda = 109$ or 200 $\Omega^{-1} \cdot cm^2 \cdot mol^{-1}$, respectively [2].

References:

[1] I. S. Ahuja, D. H. Brown, R. H. Nuttall, D. W. A. Sharp (J. Chem. Soc. A **1966** 938/41). – [2] N. I. Pirtskhalava, Yu. Ya. Kharitonov, M. V. Bregvadze (Zh. Neorgan. Khim. **20** [1975] 1852/6; Russ. J. Inorg. Chem. **20** [1975] 1034/6). – [3] A. R. Leeds (J. Am. Chem. Soc. **3** [1981] 134/51, 141, 148). – [4] M. A. J. Jungbauer, C. Curran (Spectrochim. Acta **21** [1965] 641/8, 642, 644). – [5] G. Beech, C. T. Mortimer, E. G. Tyler (J. Chem. Soc. A **1967** 1111/5).

[6] I. S. Ahuja, D. H. Brown, R. H. Nuttall, D. W. A. Sharp (J. Inorg. Nucl. Chem. **27** [1965] 1105/10). – [7] N. C. Collins, W. K. Glass (Chem. Commun. **1967** 211). – [8] A. V. Ablov, Z. P. Burnasheva, T. B. Konunova (Zh. Neorgan. Khim. **18** [1973] 2802/6; Russ. J. Inorg. Chem. **18** [1973] 1489/91). – [9] Z. P. Burnasheva (Poverkhn. Yavleniya Adsorbtsiya Koord. Vzaimodeistvie **1972** 9/12; C.A. **79** [1973] No. 61057). – [10] M. Goldstein, R. J. Hughes (Inorg. Chim. Acta **40** [1980] 229/37, 232).

[11] A. V. Ablov, Z. P. Burnasheva, V. Ya. Ivanova (Issled. Khim. Koord. Soedin. Fiz. Khim. Metod. Anal. **1969** 5/12, 9; C.A. **74** [1971] No. 106657). – [12] A. V. Ablov, Z. P. Burnasheva, V. Ya. Ivanova, N. H. Chi (Tr. Tekhnol. Fak. Kishinev. Politekhn. Inst. **1971** No. 24, pp. 79/86, 83, 85; C.A. **78** [1973] No. 42924). – [13] M. Genchev (Natura [Plovdiv, Bulg.] **6** [1973] 81/4; C.A. **83** [1975] No. 201177). – [14] M. Genchev, I. Pojarliev [Pozharliev] (Compt. Rend. Acad. Bulgare Sci. **12** [1959] 219/21; C.A. **1960** 20610). – [15] M. Genchev (Nauchn. Tr. Vissh. Pedagog. Inst. Plovdiv Mat. Fiz. Khim. Biol. **4** [1966] 83/96, 86, 88/9; C.A. **67** [1966] 77588).

[16] M. Genchev, K. Davarski (Nauchn. Tr. Vissh. Pedagog. Inst. Plovidv Mat. Fiz. Khim. Biol. **6** [1968] 49/54, 50; C.A. **71** [1969] No. 95050). – [17] G. Spacu, L. Caton (Bul. Soc. Stiinte Cluj **2** [1924/25] 332/53, 334, 343; C.A. **1925** 1995). – [18] E. Bruninx (J. Inorg. Nucl. Chem. **23** [1961] 317/22).

13.6 With Derivatives of Aniline

13.6.1 With Halogenoanilines, Toluidines, and Xylidines

R	R′	name	formula
2-CH_3	H	o-toluidine	C_7H_9N
3-CH_3	H	m-toluidine	C_7H_9N
4-CH_3	H	p-toluidine	C_7H_9N
3-CH_3	4-CH_3	3,4-xylidine	$C_8H_{11}N$

X	name	formula
3-Cl	3-chloroaniline	C_6H_6ClN
4-Cl	4-chloroaniline	C_6H_6ClN
3-Br	3-bromoaniline	C_6H_6BrN
4-Br	4-bromoaniline	C_6H_6BrN
2-I	2-iodoaniline	C_6H_6IN
4-I	4-iodoaniline	C_6H_6IN

Nitrates

The following complexes are formed by reaction of $Mn(NO_3)_2$ with o-, m-, and p-toluidine or 3,4-xylidine in the absence of a solvent: $2Mn(NO_3)_2 \cdot 3C_7H_9N$ (with o-toluidine), $Mn(NO_3)_2 \cdot 2C_7H_9N$ (with m-toluidine), $4Mn(NO_3)_2 \cdot 9C_7H_9N$ (with p-toluidine), and $Mn(NO_3)_2 \cdot 2C_8H_{11}N$ (with 3,4-xylidine). The complexes with m-toluidine and 3,4-xylidine show in their IR spectra Mn-O stretching vibrations at 253 and 256 cm^{-1}, respectively. A broad, nearly unsplit band near 1100 cm^{-1} originates from the NO_3 group. This band together with the low-energy position of the Mn-O stretching vibration band suggests a polymeric structure with oxoanion bridges, as was proposed for the aniline complex $Mn(C_6H_7N)_2(NO_3)_2$ (p. 26) [1].

Chlorides and Bromides

The complexes listed in the table on p. 32 have been prepared by methods I to VI described after the table. The enthalpies of formation from one mole of MnX_2 and two moles of liquid or gaseous ligand at 25°C, ΔH_l and ΔH_g in kJ/mol, were calculated from the enthalpies of dissolution [2]. The band maxima in the UV and visible reflectance spectra (in nm) are given together with the corresponding excited states (ground state $^6A_{1g}$) [3]. The enthalpies of dissolution in aqueous 2N HCl at 25°C, ΔH_{diss} in kJ/mol, were determined calorimetrically [2].

Most of the complexes are pale pink [3, 5]. The o- and m-toluidine complexes, $Mn(C_7H_9N)_2Cl_2$, are gray [17]. The p-toluidine complex, $Mn(C_7H_9N)_2Cl_2$, is white [7], and has the magnetic moment 5.87 μ_B at 303 K [8]. The far-IR spectra of the complexes $Mn(C_7H_9N)_2Cl_2$ with m- and p-toluidine have a band at 227 cm^{-1}. Those with 3,4-xylidine, $Mn(C_8H_{11}N)_2Cl_2$ and the mono complex with o-toluidine, $Mn(C_7H_9N)Cl_2$, show a band at 233 cm^{-1}. These bands were assigned to the Mn-Cl valence vibration. Further maxima in the region 250 to 430 cm^{-1} for all the complexes prepared by [3] are reported, but were not assigned. The reflectance spectra as given in the table and the IR spectra are consistent in each complex with an octahedral environment about the manganese atom. Thus the MnL_2X_2 complexes are suggested to have polymeric structures with bridging halogen atoms [3].

The o- and m-toluidine complexes, $Mn(C_7H_9N)_2Cl_2$, are fairly stable in a dry atmosphere. They do not melt, but decompose when heated. They are insoluble in cold water but soluble in mineral acids [17].

ligand	complex	method	$-\Delta H_l$	$-\Delta H_g$	$^4T_{1g}$	$^4T_{2g}$	^4E_g, $^4A_{1g}$	ΔH_{diss}
o-toluidine	$Mn(C_7H_9N)_2Cl_2$	I, II	88.9	195.1	—	—	—	48.86
o-toluidine	$Mn(C_7H_9N)Cl_2$	II	—	—	565	480	420	—
m-toluidine	$Mn(C_7H_9N)_2Cl_2$	I, II	—	—	560	470	425	—
p-toluidine	$Mn(C_7H_9N)_2Cl_2$	II to V	39.6	180.8	550	—	425	66.43
3,4-xylidine	$Mn(C_8H_{11}N)_2Cl_2$	II	—	—	565	—	430	—
3-chloroaniline	$Mn(C_6H_6ClN)_2Cl_2$	III	62.1	179.2	—	—	—	53.38
4-chloroaniline	$Mn(C_6H_6ClN)_2Cl_2$	III	34.6	186.8	—	—	—	51.18
3-bromoaniline	$Mn(C_6H_6BrN)_2Cl_2$	III	62.8	187.3	—	—	—	53.95
4-bromoaniline	$Mn(C_6H_6BrN)_2Cl_2$	III	41.8	233.8	—	—	—	47.10
o-toluidine	$Mn(C_7H_9N)_2Br_2$	II, III	95.1	201.3	—	—	—	50.78
m-toluidine	$Mn(C_7H_9N)_4Br_2$	II	—	—	550	470	425	—
m-toluidine	$Mn(C_7H_9N)_2Br_2$	VI	—	—	530	—	430	—
p-toluidine	$Mn(C_7H_9N)_2Br_2$	II, III	46.6	187.8	530	490	430	66.69
3,4-xylidine	$Mn(C_8H_{11}N)_2Br_2$	II	—	—	550	—	440	—
3-chloroaniline	$Mn(C_6H_6ClN)_2Br_2$	III	70.9	188.0	—	—	—	52.46
4-chloroaniline	$Mn(C_6H_6ClN)_2Br_2$	III	31.7	183.9	—	—	—	61.82
3-bromoaniline	$Mn(C_6H_6BrN)_2Br_2$	III	72.2	196.7	—	—	—	52.44
4-bromoaniline	$Mn(C_6H_6BrN)_2Br_2$	III	38.74	192.4	—	—	—	53.59

Methods of Preparation:

I) Solutions of anhydrous manganese(II) chloride and a slight excess of the amine are mixed, and if necessary benzene is added to induce precipitation [17]. – II) $MnCl_2$ or $MnBr_2$ is heated with the ligand. The resultant solid is washed with ether [3]. – III) The ethanol adducts $Mn(L)_2X_2 \cdot 2C_2H_5OH$, X = Cl, Br (p. 34) are heated at 60 to 70°C to constant weight [4, 5]. – IV) $MnCl_2$ is added to fused p-toluidine [6]. – V) A solution of excess p-toluidine in benzene is mixed with an alcohol solution of anhydrous manganese(II) chloride [7]. – VI) The complex with m-toluidine and ethanol, $Mn(C_7H_9N)_nBr_2 \cdot xC_2H_5OH$ (p. 34), is heated at 160°C to constant weight [3].

Thermoanalyses (DTA, TG, DTG) at a heating rate of 3.3°C per min for the chlorides [4] and bromides [5] show the following endothermal effects:

ligand L	DTA peaks,	t in °C	thermal-decomposition steps
p-toluidine	213	268	$MnL_2Cl_2 \rightarrow Mn(L)_{1/2}Cl_2 \rightarrow MnCl_2$
3-chloroaniline	150 to 166	204, 237*)	$MnL_2Cl_2 \rightarrow Mn(L)_{4/3}Cl_2 \rightarrow Mn(L)_{2/3}Cl_2 \rightarrow MnCl_2$
4-bromoaniline	215	252	$MnL_2Cl_2 \rightarrow Mn(L)_{3/4}Cl_2 \rightarrow MnCl_2$
p-toluidine	220	240	$MnL_2Br_2 \rightarrow Mn(L)_{1/2}Br_2 \rightarrow MnBr_2$
3-chloroaniline	205	228	$MnL_2Br_2 \rightarrow Mn(L)_{3/4}Br_2 \rightarrow MnBr_2$
4-chloroaniline	209	234	$MnL_2Br_2 \rightarrow Mn(L)_{3/4}Br_2 \rightarrow MnBr_2$
4-bromoaniline	230	245	$MnL_2Br_2 \rightarrow Mn(L)_{3/4}Br_2 \rightarrow MnBr_2$

*) Heating rate of 2.2°C per min.

$Mn(C_7H_9N)_2Br_2 \cdot 8H_2O$ was obtained by mixing wet $MnBr_2 \cdot 4H_2O$ with an alcohol solution of toluidine (probably a mixture of o-, m-, and p-toluidine). The gray crystalline mass, which precipitated after 28 h, was washed with alcohol and ether. The complex is formulated as $Mn(C_7H_9N)_2(H_2O)_4Br_2 \cdot 4H_2O$. It turns green after a time. It is insoluble in methanol, ethanol, ether, and acetone. Diluted mineral acids dissolve and decompose it [9].

Isothiocyanates and Isoselenocyanates

The tetrakis complex with p-toluidine, $Mn(C_7H_9N)_4(NCS)_2$, and the bis complex with m-toluidine, $Mn(C_7H_9N)_2(NCS)_2$, were prepared by mixing ethanol solutions of freshly prepared manganese(II) isothiocyanate and the ligands. These compounds formed even when the reactant ratios were varied from one to four moles of amine per mole of salt [11].

The bis (toluidine) complex, $Mn(C_7H_9N)_2(NCS)_2$, where toluidine probably is a mixture of o-, m-, and p-toluidine, was obtained in an older work by dehydrating $Mn(C_7H_9N)_2(NCS)_2 \cdot 5H_2O$, p. 34 [10].

The IR spectra of the white or yellow-green m- and p-toluidine complexes were obtained from mulls in liquid paraffin. The assignments of the most pertinent bands are given together with the temperatures at which thermal decomposition begins, t_{dec} [11].

ligand L	complex	IR frequencies in cm^{-1}	t_{dec} in °C
p-toluidine	$Mn(C_7H_9N)_4(NCS)_2$	$\nu_{as}(NH_2) = 3325$; $\nu_s(NH_2) = 3260$; $\nu(CN)_{NCS} = 2082, 2113$; $\nu(CS) = 785$	150
m-toluidine	$Mn(C_7H_9N)_2(NCS)_2$	$\nu_{as}(NH_2) = 3327$; $\nu_s(NH_2) = 3260$; $\nu(CN)_{NCS} = 2110$; $\delta(NCS) = 458, 471$	195

The IR spectra indicate that the NCS groups are in the inner coordination sphere. The NCS^- ligand is unidentate and coordinated by its N atom. The p-toluidine ligands probably are arranged in cis positions in the octahedral complex, as shown by the splitting of the ν(CN) band [11].

The complexes are readily soluble in water, methanol, ethanol, or acetone and sparingly soluble in hexane, CCl_4, or benzene. When they are dissolved in water, the NCS groups must ionize since the electrical conductivities of the aqueous solutions correspond to those of two- or three-ion electrolytes [11].

$Mn(C_7H_9N)_2(NCS)_2 \cdot 5H_2O$ was obtained by mixing solutions of $Mn(NCS)_2$ and toluidine in 95% ethanol (toluidine is probably a mixture of o-, m-, and p-toluidine). After 5 to 6 days gray-violet crystals form. These were washed with alcohol and ether. The complex was formulated as $Mn(C_7H_9N)_2(H_2O)_4(NCS)_2 \cdot H_2O$. It was completely dehydrated at 75°C in vacuo over P_2O_5. It is soluble in methanol, ethanol, acetone, and insoluble in water, ether, benzene, and chloroform [10].

$Mn(C_7H_9N)_2(H_2O)_4(NCSe)_2$, obtained by mixing solutions of toluidine and freshly prepared $Mn(NCSe)_2$ in 96% ethanol, is a beige crystalline powder. The four water molecules are not lost in vacuo over P_2O_5 after two days. It is therefore assumed that they are coordinated to Mn. The compound is soluble in methanol, acetone, and pyridine, only sparingly soluble in ethanol, and insoluble in benzene, chloroform, and petroleum ether [12].

Dichromate

$Mn(C_7H_9N)_4Cr_2O_7$. The tetrakis complex with p-toluidine is prepared like the complex with aniline, $Mn(C_6H_7N)_4Cr_2O_7$ (p. 28), and has similar properties as that compound [13 to 16].

Ethanol Adducts

The complexes were prepared like those with aniline and ethanol (p. 29) [3 to 5, 11].

Halogenides of composition $MnL_2X_2 \cdot 2C_2H_5OH$ with X = Cl for L = p-toluidine, 3- and 4-chloroaniline, 3-bromoaniline and X = Br for L = o-, p-toluidine, 3-, 4-chloroaniline, 3-, 4-bromoaniline, 4-iodoaniline are generally not stable and lose some ethanol even at room temperature. Only the complexes of $MnBr_2$ with 4-chloroaniline or 4-bromoaniline are somewhat more stable and do not lose ethanol even on standing over $CaCl_2$. But all the complexes lose their ethanol quantitatively on heating at 70°C, forming MnL_2X_2 [4, 5].

A complex with m-toluidine, $Mn(C_7H_9N)_nBr_2 \cdot xC_2H_5OH$, decomposes at 160°C yielding $Mn(C_7H_9N)_2Br_2$ (p. 31) [3].

For the listed nitrates, sulfates, and isothiocyanates, the most pertinent maxima in the IR spectra (mulls) are given together with the molar conductivity of the aqueous solutions, Λ in $\Omega^{-1} \cdot cm^2 \cdot mol^{-1}$, and the temperatures at which decomposition begins, t_{dec} in °C [11]:

ligand	complex	IR bands in cm^{-1}	Λ	t_{dec}
o-toluidine	$Mn(C_7H_9N)(NO_3)_2 \cdot C_2H_5OH$	$\nu(NH_2) = 3318, 3308, 3243$; $\nu(NO_3) = 1450, 1305$	270	92
m-toluidine	$Mn(C_7H_9N)(NO_3)_2 \cdot 2C_2H_5OH$	$\nu(NH_2) = 3330, 3305, 3262, 3242$	225	108
p-toluidine	$Mn(C_7H_9N)_2(NO_3)_2 \cdot C_2H_5OH \cdot 1.5H_2O$	$\nu(NH_2) = 3335, 3280$; $\nu(NO_3) = 1480, 1315, \sim 1040$?	231	150

ligand	complex	IR bands in cm^{-1}	Λ	t_{dec}
m-toluidine	$Mn(C_7H_9N)SO_4 \cdot C_2H_5OH$	$\nu(NH_2) \sim 3300$; $\nu_s(SO_4) = 1010$; $\nu_{as}(SO_4) = 1103, 1130, \sim 1145, 1182$	143.5	226
p-toluidine	$Mn(C_7H_9N)SO_4 \cdot C_2H_5OH$	$\nu_{as}(NH_2) = 3335$; $\nu_s(NH_2) = 3272$; $\nu_{as}(SO_4) = 1180, \sim 1145, \sim 1130, 1106$	148.5	215
o-toluidine	$Mn(C_7H_9N)_2(NCS)_2 \cdot C_2H_5OH$	$\nu(NH_2) = 3370, 3335, 3300, \sim 3275$; $\nu(CN)_{NCS} = 2109, 2102$; $\nu(CS) = 787$	260	140
2-iodoaniline	$Mn(C_6H_6IN)(NCS)_2 \cdot C_2H_5OH$	$\nu(NH_2) = 3395, 3295, 3190$ $\nu(CN)_{NCS} = 2098$	258	125

The splitting of the anion bands suggests that the anions are coordinated to the manganese atom, just as for the aniline and ethanol complexes on pp. 29 and 30. All the complexes are readily soluble in water, methanol, ethanol, or acetone and sparingly soluble in hexane, benzene, or carbon tetrachloride. The electrical conductivities of the aqueous solutions corresponding to those of two- or three-ion electrolytes, indicate that the anions are ionized in water [11].

References:

[1] I. S. Ahuja, D. H. Brown, R. H. Nuttall, D. W. A. Sharp (J. Chem. Soc. A **1966** 938/41). – [2] A. V. Ablov, Z. P. Burnasheva, T. B. Konunova (Zh. Neorgan. Khim. **18** [1973] 2802/6; Russ. J. Inorg. Chem. **18** [1973] 1489/91). – [3] I. S. Ahuja, D. H. Brown, R. H. Nuttall, D. W. A. Sharp (J. Inorg. Nucl. Chem. **27** [1965] 1625/34, 1626, 1628). – [4] A. V. Ablov, Z. P. Burnasheva, V. Ya. Ivanova (Issled. Khim. Koord. Soedin. Fiz. Khim. Metod. Anal. **1969** 5/12, 9; C.A. **74** [1971] No. 106657). – [5] A. V. Ablov, Z. P. Burnasheva, V. Ya. Ivanova, N. H. Chi (Tr. Tekhnol. Fak. Kishinev. Politekhn. Inst. **1971** No. 24, pp. 79/86, 83, 85; C.A. **78** [1973] No. 42924).

[6] A. R. Leeds (J. Am. Chem. Soc. **3** [1881] 134/51, 141, 148). – [7] A. Syamal (J. Inst. Chem. [India] **40** [1968] 102/4; C.A. **69** [1968] No. 83043). – [8] S. Prasad, V. R. Reddy (J. Indian Chem. Soc. **35** [1958] 722/4). – [9] G. Spacu (Bul. Soc. Stiinte Cluj **4** [1928/29] 210/24, 220; C.A. **1929** 1362). – [10] G. Spacu, C. G. Macarovici (Bul. Soc. Stiinte Cluj **7** [1933] 227/47, 231, 245; C.A. **1933** 5672).

[11] N. I. Pirtskhalava, Yu. Ya. Kharitonov, M. V. Bregvadze (Zh. Neorgan. Khim. **20** [1975] 1852/6; Russ. J. Inorg. Chem. **20** [1975] 1034/6). – [12] G. Spacu, C. G. Macarovici (Bul. Soc. Stiinte Cluj **6** [1931] 95/138, 108, 135; C.A. **1932** 934). – [13] M. Genchev (Natura [Plovdiv, Bulg.] **6** [1973] 81/4; C.A. **83** [1975] No. 201177). – [14] M. Genchev, I. Pojarliev [Pozharliev] (Compt. Rend. Acad. Bulgare Sci. **12** [1959] 219/21; C.A. **1960** 20610). – [15] M. Genchev (Nauchn. Tr. Vissh. Pedagog. Inst. Plovdiv Mat. Fiz. Khim. Biol. **4** [1966] 83/96, 86, 88/9; C.A. **67** [1966] No. 77588).

[16] M. Genchev, K. Davarski (Nauchn. Tr. Vissh. Pedagog. Inst. Plovdiv Mat. Fiz. Khim. Biol. **6** [1968] 49/54, 50; C.A. **71** [1969] No. 95050). – [17] S. Prasad, P. J. Sarma (J. Proc. Inst. Chem. [India] **37** [1965] 246/8).

13.6.2 With 4-Aminobenzenesulfonic Acid (= Sulfanilic Acid = $C_6H_7NO_3S$)

Spectrophotometric studies of the ion $Mn(H_2P_2O_7)_3^{3-}$ in an aqueous solution of the ligand indicate the formation of complexes with an ion : ligand mole ratio of 1:1 and 1:2.

The complexes disproportionate in a first-order reaction into manganese(II) and manganese(IV) complexes. The rate constant of the disproportionation reaction was determined spectrophotometrically to be $k = 0.0213\ s^{-1}$ at 21°C. The activation energy is 19.84 kcal/mol. Complex formation and disproportionation occurs for a solution of Mn_2O_3 in the aqueous ligand as well.

Spectrophotometric, polarographic, and paper-chromatographic studies showed the manganese(IV) complexes formed by disproportionation to be 1:2 and 1:4 complexes. Complexes with the same manganese-to-ligand ratios were obtained by dissolving MnO_2 in aqueous ligand. The overall stability constant of the 1:4 complex at room temperature, $\beta_4 = 2.80 \times 10^{16}\ (L/mol)^4$, was spectrophotometrically determined. The electronic absorption spectrum of the 1:4 complex in solution shows maxima at 203 ($\pi \rightarrow \pi^*$), 249 ($\pi \rightarrow \pi^*$), 283 ($n \rightarrow \pi^*$), 391 ($^4A_{2g} \rightarrow {}^4T_{1g}(P)$), 543 ($^4A_{2g} \rightarrow {}^4T_{1g}$), and 885 nm ($^4A_{2g} \rightarrow {}^4T_{2g}$). The assignments were made for a d^3 ion in O_h symmetry. The ligand-field parameter $\Delta = 10\ Dq = 11300\ cm^{-1}$ and the nephelauxetic parameter $\beta_{35} = 0.63$ were evaluated from the spectrum.

Reference:

N. Calu, A. Popa-Rang, M. Calenciuc (Rev. Roumaine Chim. **19** [1974] 1889/97, 1891, 1895; C.A. **83** [1975] No. 52598).

13.6.3 With N-Alkyl Derivatives of Aniline and Related Compounds

The following complexes were obtained by reacting anhydrous manganese(II) chloride with a slight excess of the ligand in absolute ethanol [1, 2] and adding benzene to induce precipitation if necessary [2].

No.	ligand	formula	complex	color	Ref.
I	N-methylaniline	C_7H_9N	$Mn(C_7H_9N)Cl_2$	brown	[2]
II	N-ethylaniline	$C_8H_{11}N$	$Mn(C_8H_{11}N)Cl_2$	violet	[2]
III	N-benzylaniline	$C_{13}H_{13}N$	$Mn(C_{13}H_{13}N)Cl_2$	light green	[2]
IV	N,N-dimethylaniline	$C_8H_{11}N$	$Mn(C_8H_{11}N)Cl_2$	dark blue	[1]
V	N,N-diethylaniline	$C_{10}H_{15}N$	$Mn(C_{10}H_{15}N)Cl_2$	gray	[1]
VI	N,N-dimethyl-o-toluidine	o-$C_9H_{13}N$	Mn(o-$C_9H_{13}N)Cl_2$	gray	[1]
VII	N,N-dimethyl-p-toluidine	p-$C_9H_{13}N$	Mn(p-$C_9H_{13}N)Cl_2$	brick red	[1]

The complexes with the ligands I to III are fairly stable in a dry atmosphere. When heated, they do not melt but rather decompose. They are insoluble in cold water but soluble in mineral acids [2]. The other complexes are unstable. They do not give sharp melting points and are decomposed by boiling water, mineral acids, or NaOH solutions [1].

References:

[1] S. Prasad, K. P. Kacker (J. Indian Chem. Soc. **35** [1958] 719/21). – [2] S. Prasad, P. J. Sarma (J. Proc. Inst. Chem. [India] **37** [1965] 246/8).

13.7 With 1-Naphthylamine and 2-Naphthylamine (= α- and β-Naphthylamine = $C_{10}H_9N$)

Mn(α-$C_{10}H_9N)_2Cl_2$ and **Mn(β-$C_{10}H_9N)_2Cl_2$** were prepared by adding excess naphthylamine in benzene solution to a solution of anhydrous manganese(II) chloride in absolute alcohol. The complex with α-naphthylamine is white; the complex with β-naphthylamine violet [1]. The magnetic moments of the complexes at 303 K are 5.92 or 5.95 μ_B, respectively. Both complexes are stable in a dry atmosphere at room temperature. They are decomposed by boiling water or NaOH solution [1].

$Mn(\beta\text{-}C_{13}H_9N)_2SO_4$ was obtained by reacting manganese(II) sulfate and the ligand in water/ethanol mixtures (1:4 or 3:1) in the presence of hydrochloric acid. The pinkish-beige complex is unstable. It melts at 350°C.

Its IR spectrum shows the bands of the symmetric and antisymmetric stretching vibrations of the NH_2 group at 3440 cm^{-1} and 3245 cm^{-1}. (The free-ligand bands are at 3472 and 3390 cm^{-1}, respectively.) The strong shifts indicate a donor-acceptor bond between the amino group and the manganese atom. Strong bands at 1260, 1160, 1065, 765, 625, and 615 cm^{-1} were assigned to vibrations of the SO_4 group. The positions of these bands suggest that the SO_4^{2-} ions are bidentate bridges. Further IR bands were not assigned. The complex is only slightly soluble in the usual solvents. This is consistent with a polymeric structure with SO_4 bridges. The molar conductivities are 88.00 $\Omega^{-1}\cdot cm^2\cdot mol^{-1}$ for aqueous solutions and 4.10 $\Omega^{-1}\cdot cm^2\cdot mol^{-1}$ for ethanol solutions [3].

References:

[1] S. Prasad, V. R. Reddy (J. Indian Chem. Soc. **35** [1958] 722/4). – [2] A. Syamal (J. Inst. Chem. [India] **40** [1968] 102/4). – [3] E. Butuceanu (Bul. Inst. Politehn. George Georghiu-Dej Bucuresti Ser. Chim. Metal. **39** [1977] 71/6, 72/3; C.A. **88** [1978] No. 44308).

13.8 With Ethylenediamine $H_2NCH_2CH_2NH_2$ ($=C_2H_8N_2=$en)

A mixed-ligand complex with ethylenediamine and 2,4-pentanedione (acetylacetone) is described in „Mangan" D 1, 1979, p. 78. For a mixed-ligand complex with ethylenediamine and bipyridyl see p. 216.

13.8.1 $Mnen_n^{2+}$ Ions (n = 1, 2, 3). Formation and Properties in Solution

The following stepwise stability constants were determined in aqueous solutions at temperature t and ionic strength I:

t in °C	I in mol/L	method*)	log K_1	log K_2	log K_3	Ref.
20	1 KCl	calc	2.8	2.1	0.9	[2]
25	1.4	gl	2.77	2.10	0.92	[3, 4]
25	0.1 KNO_3	gl	2.85	~1.9	—	[5]
25	0.1 $NaClO_4$	gl	2.74 ± 0.06	—	—	[6]
30	1 KCl	H	2.73	2.06	0.88	[4, 7]

*) Methods: calc = calculated from the data of [7], gl = glass electrode, H = hydrogen electrode.

In mixed aqueous organic solutions values of log K_1 were determined for the various contents of the organic solvent in the mixture at 25°C and I = 0.1 mol/L KNO_3 with the glass electrode [5]:

mixture	vol% solvent				
	20	40	60	70	80
aqueous methanol	—	—	3.07	3.15	3.26
aqueous dioxane	2.93	3.16	3.57	3.90	4.27
aqueous acetamide	2.98	3.29	3.65	3.90	~4.4
aqueous DMF	2.98	3.14	3.40	3.59	3.95

In solutions of 60, 70, and 80 vol% methanol, log K_2 is 2.10, 2.15, 2.18, or 2.31, respectively, at 25°C and $I = 0.1$ mol/L KNO_3. In solutions with 20, 60, 70, and 80 vol% dioxane log K_2 is 2.01, 2.61, 2.95, or 3.29, respectively. Values of log K_2 and log K_3 in aqueous DMF at 25°C and $I = 0.1$ mol/L KNO_3 are reported in [5]:

vol% DMF	40	50	60	70	80
log K_2	2.26	2.34	2.47	2.66	2.89
log K_3	—	~1.6	1.85	1.96	2.17

The stability constants increase with decreasing dielectric constants of the solvents [5].

For dimethyl sulfoxide solutions, the overall stability constants were determined potentiometrically at 25°C and $I = 0.1$ mol/L $KClO_4$: log $\beta_1 = 3.7 \pm 0.2$, log $\beta_2 = 6.9 \pm 0.2$, and log $\beta_3 = 10.1 \pm 0.2$ [8]. For N-methylpyrrolidone log $\beta_2 = 14.8$ was determined polarographically at room temperature and $I = 0.1$ mol/L $NaClO_4$ or $(C_2H_5)_4NClO_4$ [9].

The enthalpies of formation of the ions $Mnen_n^{2+}$ from the elements in aqueous solution at 298.15 K under standard conditions were calculated from literature data [1]:

n	1	2	3
$-\Delta H_f^\circ$ in kcal/mol	68.9	85.4	103.8

The listed values for the enthalpy, ΔH in kcal/mol, free enthalpy, ΔG in kcal/mol, and entropy, ΔS in cal·mol^{-1}·K^{-1}, for the complex formation from Mn^{2+} and 1 to 3 molecules of ethylenediamine in aqueous solution at 25°C and $I = 1$ mol/L KCl are reported; the enthalpy was measured calorimetrically. Oxidation was prevented by bubbling nitrogen through the solution [10, 11]:

reaction	$-\Delta H$	$-\Delta G$	ΔS
$Mn^{2+} + en \rightarrow Mnen^{2+}$	2.80 (2.80 ± 0.12)*)	3.75	3.2
$Mn^{2+} + 2en \rightarrow Mnen_2^{2+}$	6.00 (6.07 ± 0.11)*)	6.65	2.2
$Mn^{2+} + 3en \rightarrow Mnen_3^{2+}$	11.05 (10.98 ± 0.13)*)	7.90	−10.5

*) Values in parentheses were calculated by [12] from the experimental values of [10, 11].

From the ΔH value for $Mnen_3^{2+}$, the coordinative bond energy for the bond between Mn^{2+} and one NH_2 group of the ethylenediamine was calculated to be 186.1 kJ/mol (44.46 kcal/mol) [13].

The visible electronic absorption spectrum of 1M aqueous $Mnen_3^{2+}$ solutions (containing excess ethylenediamine (1M) and a few drops of hydrazine to prevent oxidation) shows maxima at the following wavelengths λ, which were assigned to transitions from the ground state $^6A_{1g}$ to the excited states given:

excited state	a^4T_{1g}	a^4T_{2g}	$^4A_{1g}, a^4E_g$				b^4T_{2g}		b^4E_g
λ in nm	638	498	423.2	421.5	419.0	417.3	375.5	371.5	354.5

The ligand-field parameter is $\Delta = 10\,Dq = 9500$ cm^{-1} [14]. In an earlier investigation less resolved maxima were observed [15].

The complex solutions are sensitive to oxidation [10, 11, 14]. For the replacement reactions $Mnen_2^{2+} + L \rightarrow MnL^{2+} + 2en$, with a polyamine L, the following thermodynamic values, ΔH and ΔG in kcal/mol, ΔS in cal·mol^{-1}·K^{-1}, at 25°C are reported:

L	ΔH	ΔG	ΔS	Ref.
triethylenetetramine = trien	3.70	−0.05	12.5	[16]
2,2′,2″-triaminotriethylamine = tren	3.00	−1.20	14	[17]

(The complexes $Mntrien^{2+}$ and $Mntren^{2+}$ are described on pp. 56 and 59.)

The complex $Mnen^{2+}$ reacts with malonic acid, $HOOCCH_2COOH$ (= $C_3H_4O_4$), in aqueous solution to give the mixed-ligand complex $Mnen(C_3H_2O_4)$. The following equilibrium constants log K for the various formation reactions of this complex were determined for aqueous solutions at 25°C and I = 0.1 mol/L $NaClO_4$ with a glass electrode [6]:

stability constant K	log K
$[Mnen(C_3H_2O_4)] / [Mnen^{2+}][C_3H_2O_4^{2-}]$	2.56
$[Mnen(C_3H_2O_4)] / [Mn(C_3H_2O_4)][en]$	2.58
$[Mnen(C_3H_2O_4)] / [Mn^{2+}][en][C_3H_2O_4^{2-}]$	5.3 ± 0.2

$Mnen_3^{2+}$ reacts with ethylenediaminetetraacetic acid (= H_4edta), in aqueous solution, $2Mnen_3^{2+} + edta^{4-} \rightarrow Mn_2en_4edta + 2en$. Mn_2en_4edta is also formed by the reaction of $Mnen_3^{2+}$ with $Mnenedta^{2-}$ [18]. These mixed-ligand complexes are described in "Manganese" D 4.

Manganese(II) complexes with ethylenediamine show a catalytic activity on the decomposition of hydrogen peroxide in aqueous solutions [19, 20]. They also accelerate the photopolymerization of vinyl monomers in the presence of carbon tetrachloride [21].

References:

[1] D. D. Wagman, W. H. Evans, V. B. Parker, I. Halow, S. M. Bailey, R. H. Schumm (NBS-TN-270-4 [1969] 114). – [2] G. Schwarzenbach (Helv. Chim. Acta **35** [1952] 2344/59, 2357). – [3] R. L. Pecsok, J. Bjerrum (Acta Chem. Scand. **11** [1957] 1419/21). – [4] L. G. Sillén, A. E. Martell (Stability Constants of Metal-Ion Complexes, 2nd Ed., Chem. Soc., London 1964, p. 372). – [5] K.-K. Mui, W. A. E. McBryde, E. Nieboer (Can. J. Chem. **52** [1974] 1821/33, 1825/6).

[6] H. Sigel, B. E. Fischer, B. Prijs (J. Am. Chem. Soc. **99** [1977] 4489/96, 4490). – [7] J. Bjerrum (Metal Ammine Formation in Aqueous Solutions, Haase and Son, Copenhagen 1941, p. 293). – [8] K. H. Pool, D. E. Sandberg (Talanta **16** [1969] 1319/23, 1320). – [9] P. Mayne, C. Sinicki (Compt. Rend. C **281** [1975] 233/4). – [10] M. Ciampolini, P. Paoletti, L. Sacconi (J. Chem. Soc. **1960** 4553/61, 4557).

[11] M. Ciampolini, P. Paoletti, L. Sacconi (Nature **186** [1960] 880/1). – [12] G. Berthon, G. Valensi (Bull. Soc. Chim. France **1972** 479/85, 483). – [13] C. Glidewell (J. Coord. Chem. **6** [1977] 189/92). – [14] C. K. Jørgensen (Inorg. Chim. Acta **3** [1969] 313/4). – [15] C. K. Jørgensen (Acta Chem. Scand. **11** [1957] 53/72, 63).

[16] L. Sacconi, P. Paoletti, M. Ciampolini (J. Chem. Soc. **1961** 5115/20, 5120). – [17] P. Paoletti, M. Ciampolini, L. Sacconi (Ric. Sci. **30** [1960] 1791/4). – [18] N. N. Barkhanova, A. Ya. Fridman, N. M. Dyatlova (Zh. Neorgan. Khim. **17** [1972] 2982/8; Russ. J. Inorg. Chem. **17** [1972] 1569/73, 1571). – [19] R. P. Pantaler, L. D. Alfimova, A. M. Bulgakova (Zh. Analit. Khim. **30** [1975] 1584/9; J. Anal. Chem. [USSR] **30** [1975] 1332/5). – [20] M. Augustin, W. Langenbeck (J. Prakt. Chem. **19** [1963] 186/91).

[21] T. Okimoto, M. Takahashi, Y. Inaki, K. Takemoto (Angew. Makromol. Chem. **38** [1974] 81/90, 84).

13.8.2 Isolated Manganese(II) Compounds

Zn(Mn)en$_3$(NO$_3$)$_2$. Single crystals of Znen$_3$(NO$_3$)$_2$ with up to 10 mol% manganese were grown by slow evaporation from an N_2-purged saturated aqueous solution of Znen$_3$(NO$_3$)$_2$ containing small amounts of Mn(NO$_3$)$_2$ and a small excess of ethylenediamine [1, 2]. The ESR spectra yield $g_{\parallel} = 2.004 \pm 0.005$, $g_{\perp} = 2.021 \pm 0.005$ (magnetic field parallel or perpendicular to the c axis), the hyperfine coupling constant $A = (-77 \pm 1) \times 10^{-4}$ cm^{-1}, and the zero-field splitting parameter $D = (197 \pm 10) \times 10^{-4}$ cm^{-1} [1].

The crystals exhibit macroscopic optical activity. The ligand-field bands (in cm^{-1}) of the axial absorption spectrum (abs) and axial circular dichroism (CD) of the Mnen$_3^{2+}$ ions, measured at room temperature, were assigned on the basis of O_h symmetry as follows:

transition	$^6A_{1g} \rightarrow {}^4T_{1g}$(G)	$^4T_{2g}$(G)	(^4E_g, $^4A_{1g}$)(G)	$^4T_{2g}$(D)
ν_{max}(abs)	15000	19900	23570, 23700, 23850, 23950	26500
ν_{max}(CD)	—	19800	23600	26500

Dipole strengths D, rotational strengths R, and anisotropy factors are given in a table. The ligand-field parameters deduced are Dq = 1080, B = 860, and C = 3010 cm^{-1} [1]. Similar axial absorption and axial circular dichroism data are reported in [2]. A positive R for all d-d transitions is found to be associated with the Λ configuration for the complex ions by correlation with the crystal and solution circular dichroism of Ruen$_3^{2+}$ [2]. The trigonal-field splitting in the electronic spectrum correlates with the magnitude and sign of the zero-field splitting parameter from the ESR spectrum. On cooling, the crystals become opaque, but become transparent again upon warming [1].

MnenCl$_2$ was prepared by adding excess ethylenediamine in benzene to an absolute-alcohol solution of anhydrous manganese(II) chloride [3]. The magnetic moment of the gray [3, 4] complex has been determined to be 5.93 μ_B at 303 K [4]. The compound is stable at room temperature in a dry atmosphere. It is decomposed by boiling water or by heating with NaOH solution [3].

Mnen$_3$Br$_2$ was prepared by adding finely divided MnBr$_2 \cdot$H$_2$O to ethylenediamine in an He atmosphere and stirring the mixture at 60°C for 3 h. The gray-white powder is washed with ethylenediamine and dried in vacuo over Mg(ClO$_4$)$_2$. The yield is 95%.

The complex melts at 238 to 255°C with decomposition. It has the magnetic moment 5.7 μ_B at room temperature. X-ray diffraction d-values are given in [5]. A diagram of the IR spectrum between 200 and 4000 cm^{-1}, but no assignments, is given in [6]. Mnen$_3$Br$_2$ is unstable in moist air and is decomposed by water. It reacts with liquid ammonia at −35°C quantitatively to form Mn(NH$_3$)$_6$Br$_2$, p. 12. Mnen$_3$Br$_2$ is insoluble in dimethylformamide, bis(2-methoxyethyl)ether, tetrahydrofuran, and γ-butyrolactone [5, 6].

Mnen$_3$S$_2$O$_3$ and **Mnen$_3$SeSO$_3$** were prepared by mixing 0.5 M aqueous solutions of MnSO$_4$ and K$_2$S$_2$O$_3$ or K$_2$SeSO$_3$ with the calculated amount of 20% ethylenediamine solutions. The crystalline, pale brownish precipitates were washed with a 1:5 ethylenediamine-methanol mixture, then with methanol, and dried in a vacuum [10].

The compounds have a long prismatic, needle-like shape. Mnen$_3$S$_2$O$_3$ is trigonal and isostructural with Men$_3$S$_2$O$_3$, M = Zn, Cd, Ni, Co, and Fe, and Znen$_3$SeSO$_3$. X-ray analysis gave the lattice parameters $a = 9.38 \pm 0.02$, $c = 9.71 \pm 0.02$ Å; Z = 2. Mnen$_3$SeSO$_3$ is pseudotrigonal with $a = 16.56 \pm 0.03$ and $c = 29.42 \pm 0.05$ Å; Z = 18. Its structure is probably a derivative of the structure of the thiosulfate ($a' \approx \sqrt{3}a$, $c' \approx 3c$). The following experimental and calculated X-ray densities at room temperature, d_{exp} and d_{calc} in g/cm^3, and indexes of refraction, n_o and n_e at $\lambda = 589$ nm, were obtained:

complex	d_{exp}	d_{calc}	n_o	n_e
$Mnen_3S_2O_3$	1.58	1.54	1.609 ± 0.002	1.591 ± 0.002
$Mnen_3SeSO_3$	1.72	1.69	1.626 ± 0.002	1.602 ± 0.002

The crystals of both complexes are optically uniaxial and negative ($n_o > n_e$) [7].

The thiosulfate is more stable than the selenosulfate. The latter turns dark brown rather quickly at room temperature; the thiosulfate becomes brown more slowly. On heating, the thiosulfate decomposes at 237 to 239°C; the selenosulfate decomposes at 220 to 225°C [7].

$Mnen_3(BF_4)_2$ was prepared by mixing solutions of 1.5 g $Mn(BF_4)_2 \cdot 6H_2O$ and 1 mL of ethylenediamine hydrate in absolute ethanol. The white crystalline precipitate turns light brown during filtration and washing with absolute ethanol. The complex is soluble in water, but insoluble in alcohol or ether [8].

$Mnen_3(C_{14}H_8O_4)$ was prepared by heating a mixture of ethylenediamine and the manganese salt of 2,2′-biphenyldicarboxylic acid (diphenic acid = $C_{14}H_{10}O_4$) gently to 80°C while it is stirred and maintained at that temperature for several days. The white microcrystalline precipitate was washed with ether and dried over sulfuric acid. The complex is very hygroscopic.

The IR spectrum shows strong absorption bands. The maxima (in cm^{-1}) are given together with the corresponding bands of ethylenediamine and diphenic acid. They were assigned as follows:

assignment	$\nu(NH)$	$\nu(CH_2)$	$\nu_{as}(COO)$[1)]
complex	3300	2950, 2890	1610, 1580, 1560
ethylenediamine	3500, 3200	2926, 2853	—
diphenic acid	—	—	1650, 1580

assignment	$\nu_s(COO)$	$\delta(CO)$[2)]	$\nu(CN)$
complex	1390, 1280, 1260	1280, 1260	1110, 1010
ethylenediamine	—	—	1120, 1020
diphenic acid	1410	1290	—

1) Presumably combined with $\delta(NH_2)$ of ethylenediamine. – 2) In the COH plane.

The shifts of the pertinent bands in the complex indicate complexation with ethylenediamine. Other transition metal complexes investigated show similar IR spectra, suggesting similar structures for all these compounds [9].

Other Complexes. A compound of composition $Mn(NO_3)_2 \cdot 8en \cdot 4HNO_3 \cdot 4H_2O$ was prepared by treating a concentrated solution of manganese(II) nitrate with strong nitric acid, adding ammonium nitrate and ethylenediamine until the solution was slightly acidic, and crystallizing the product. Similar methods were used to prepare $MnF_2 \cdot 3en \cdot 3HF$, $Mn(CH_3COO)_2 \cdot en \cdot CH_3COOH$, and $MnC_2O_4 \cdot 10en \cdot 4H_2C_2O_4$. From the thermal behavior of the nitrate and oxalate it is concluded that six molecules of ethylenediamine are bound stronger than the others. The compounds are therefore formulated $[Mnen_6](NO_3)_2 \cdot 2en \cdot H_4(NO_3)_4 \cdot 4H_2O$ and $[Mnen_6](C_2O_4) \cdot 4en \cdot H_4(C_2O_4)_2$.

The colorless or pale rose microcrystalline compounds are soluble in water and insoluble in ethanol, ether, and benzene. The electrical conductivities of the aqueous solutions indicate that the compounds completely ionize in water [10, 11].

References:

[1] R. A. Palmer, M. Chin-Lan Yang, J. C. Hempel (Inorg. Chem. **17** [1978] 1200/3). – [2] R. A. Palmer, M. Chin-Lan Yang (Chem. Phys. Letters **31** [1975] 492/7, 493). – [3] S. Prasad, V. R. Reddy (J. Indian Chem. Soc. **35** [1958] 722/4). – [4] A. Syamal (J. Inst. Chem. [India] **40** [1968] 102/4). – [5] G. W. Watt, B. S. Manhas (J. Inorg. Nucl. Chem. **28** [1966] 1945/7).

[6] B. S. Manhas (Diss. Univ. Texas 1965, pp. 1/96, 6, 10, 65; Diss. Abstr. **26** [1965] 685). – [7] V. L. Varand, N. V. Podberezskaya, V. M. Shul'man, V. V. Bakakin, E. D. Ruchkin (Izv. Sibirsk. Otd. Akad. Nauk SSSR Ser. Khim. Nauk **12** No. 5 [1967] 44/7; Sib. Chem. J. **12** [1967] 595/8). – [8] G. Spacu, L. Dima (Z. Anorg. Allgem. Chem. **223** [1935] 185/91, 189). – [9] A. Ara Blesa (Ion [Madrid] **37** [1977] 355/7). – [10] O. E. Zvyagintsev, M. E. Chkoniya (Izv. Gruz. Ind. Inst. **1939** No. 11, pp. 79/85; C.A. **1943** 1344).

[11] O. E. Zvyagintsev, M. Chkoniya (Zh. Obshch. Khim. **10** [1940] 1647/52; C.A. **35** [1941] 3187).

13.8.3 Isolated Manganese(III) Compound

Mnen$_3$(ClO$_4$)$_3$ was prepared by addition of 0.002 mol of the solid urea complex Mnur$_6$(ClO$_4$)$_3$ to a solution of ethylenediamine (0.04 mol) in acetonitrile. After the tan precipitate immediately formed had been removed, addition of absolute ethanol to the green solution gave the complex. It was washed with absolute ethanol, filtered under a stream of dry N_2, and dried in vacuo over P_2O_5. The green complex has the magnetic moment $\mu_{eff} = 4.85\ \mu_B$ at room temperature ($\mu_{calc} = 5.10\ \mu_B$). The g factor was determined to be 2.08. The diffuse reflectance spectrum shows maxima at 17400 and 27600 cm^{-1}. The ligand-field parameter was evaluated to be $\Delta = 10\ Dq = 20100\ cm^{-1}$. A 10^{-3} M solution of the complex in dimethylformamide has a specific electrical conductivity of $186 \times 10^{-6}\ \Omega^{-1} \cdot cm^{-1}$ at room temperature, J. C. Summers (Diss. Univ. Florida 1968, pp. 1/132, 17, 27, 30, 66; Diss. Abstr. Intern. B **30** [1969] 109).

13.9 With Other Aliphatic and Alicyclic Diamines

Complexes in Solution. The stability constants of manganese(II) complexes were determined potentiometrically (glass electrode) at 25°C and the ionic strengths I (in mol/L):

ligand	formula	I	log K_1	log K_2	Ref.
N,N'-dimethyl-1,2-ethanediamine					
$CH_3NHCH_2CH_2NHCH_3$	$C_4H_{12}N_2$	not given	2.48	—	[7]
N,N'-diethyl-1,2-ethanediamine					
$C_2H_5NHCH_2CH_2NHC_2H_5$	$C_6H_{16}N_2$	*)	2.3	—	[1]
N,N,N',N'-tetramethyl-1,2-ethanediamine					
$(CH_3)_2NCH_2CH_2N(CH_3)_2$	$C_6H_{16}N_2$	*)	<0.9	—	[1]
meso-2,3-butanediamine					
$H_2NC(CH_3)HCH(CH_3)NH_2$	$C_4H_{12}N_2$	0.46	2.72	1.60	[1, 8]
		0.1 (KNO_3)	2.64(5)	—	[2]

ligand	formula	I	log K_1	log K_2	Ref.
racemic 2,3-butanediamine $NH_2CH(CH_3)CH(CH_3)NH_2$	$C_4H_{12}N_2$	0.1 (KNO_3)	2.94(2)	—	[2]
2,3-dimethyl-2,3-butanediamine $H_2NC(CH_3)_2C(CH_3)_2NH_2$	$C_6H_{16}N_2$	*)	2.09	1.69	[1]
2-ethyl-1,2-butanediamine $H_2NCH_2C(C_2H_5)(NH_2)CH_2CH_3$	$C_6H_{16}N_2$	0.5 (KNO_3)	2.34	1.52	[6]
trans-1,2-cyclohexanediamine $C_6H_{10}(NH_2)_2$	$C_6H_{14}N_2$	0.46	2.94	2.39	[1, 8]

*) The initial manganese dichloride concentration was 0.030 M, and 0.368 mol of amine hydrochloride was added per liter to suppress precipitation of $Mn(OH)_2$.

Although a high concentration of amine hydrochloride was used, it was found that $Mn(OH)_2$ precipitation interfered in the experiments with N-substituted ethylenediamines [1]. Stepwise stability constants of the complexes with N-methyl-1,2-ethanediamine ($=C_3H_{10}N_2$), N-ethyl-1,2-ethanediamine ($=C_4H_{12}N_2$), and 1,3-propanediamine ($=C_3H_{10}N_2$) estimated by comparative calculation are given in [5]. The kinetics of the catalytic activity of manganese(II) complexes with N,N′-dimethyl-1,2-ethanediamine or N,N,N′,N′-tetramethyl-1,2-ethanediamine for the decomposition of hydrogen peroxide [3] and the decarboxylation of β-keto acids [4] have been studied.

References:

[1] D. L. Leussing (Inorg. Chem. **2** [1963] 77/81, 79). – [2] L. D. Pettit, J. L. M. Swash (J. Chem. Soc. Dalton Trans. **1977** 697/701, 698). – [3] M. Augustin, W. Langenbeck (J. Prakt. Chem. **19** [1963] 186/91). – [4] J. V. Rund, K. G. Claus (Inorg. Chem. **7** [1968] 860/5). – [5] M. Kh. Karapet'yants, E. V. Sklenskaya (Zh. Fiz. Khim. **38** [1964] 1312/6; Russ. J. Phys. Chem. **38** [1964] 715/7).

[6] A. Kircheiss, W. Jabs, J. Bremer (Z. Chem. [Leipzig] **9** [1969] 116/7). – [7] J. Bjerrum, P. Andersen (from D. L. Leussing, Inorg. Chem. **2** [1963] 77/81, 79). – [8] R. M. Smith, A. E. Martell (Critical Stability Constants, Vol. 2, Amines, Plenum, New York 1975, pp. 42, 46).

13.10 With N,N-Dimethyl-1,2-ethanediamine N-Oxide

$(CH_3)_2N(O)CH_2CH_2NH_2$ ($=C_4H_{12}N_2O$)

$Mn(C_4H_{12}N_2O)Cl_2$. The light brown complex was prepared by reacting $MnCl_2 \cdot 4H_2O$ with the ligand in ethanol. The product formed was dissolved in methanol and reprecipitated with ethyl ether. The complex melts at about 123°C with decomposition. The magnetic moment is 5.86 μ_B at room temperature. The IR spectrum (KBr discs) shows a strong band at 915 cm^{-1}, assigned to the ν(NO) vibration (free ligand, 948 cm^{-1}). The solution in dimethylformamide (4.507 mmol/L) has a molar conductivity of 10.6 $\Omega^{-1} \cdot cm^2 \cdot mol^{-1}$ at 25°C indicating the complex to be a nonelectrolyte. The results of the physical measurements suggest that the bidentate organic ligand is coordinated by its nitrogen and oxygen atoms, forming a six-membered chelate ring.

Reaction of $Mn(ClO_4)_2 \cdot 6H_2O$ with the ligand in ethanol gave a product which could not be obtained analytically pure.

Reference:

J. T. Summers, J. V. Quagliano (Inorg. Chem. **3** [1964] 1767/72, 1769).

13.11 With Aromatic Diamines

13.11.1 With o-, m-, and p-Phenylenediamine $C_6H_4(NH_2)_2$ (=o-, m-, p-$C_6H_8N_2$=L)

$Mn(p\text{-}C_6H_8N_2)_n^{2+}$ Ions (n=1, 2). **Formation in Solution**

The stability constants of the manganese(II) complexes with p-phenylenediamine in aqueous solution have been determined potentiometrically (glass electrode) at the ionic strength I=0.1 mol/L KNO_3 and the given temperatures [1]:

t in °C	20	23	26	29	32
log K_1	3.12	3.12	3.06	3.04	2.96
log K_2	3.12	3.08	2.99	2.90	2.88

From these results, the enthalpy and entropy values for the stepwise formation of the complex ions have been calculated, ΔH in kJ/mol, ΔS in $J \cdot mol^{-1} \cdot K^{-1}$ [1]:

$-\Delta H_1$	$-\Delta H_2$	$-\Delta S_1$	$-\Delta S_2$
22.78	37.74	17.58	68.89

Isolated Compounds MnL_nX_2

Apart from the complexes described in this section, compounds of the composition MnL_nX_2 with n=1 or 2 and X=NO_3, Cl, ClO_4, $^1/_2SO_4$, and NCS form as the first products in the thermal decomposition of the ethanol adducts $MnL_nX_2 \cdot C_2H_5OH$. These products have not been isolated; see p. 45.

$Mn(o\text{-}C_6H_8N_2)(NO_3)_2$. The complex with o-phenylenediamine was obtained by mixing ethanolic solutions of the stoichiometric amounts of manganese(II) nitrate and the ligand.

In the IR spectrum (paraffin or fluorinated-oil mulls) of the pale brown complex, the most important maxima are assigned as follows: $\nu_s(NH_2)$=3260 and 3235, $\nu_{as}(NH_2)$=3346, 3320, and 3296, $\delta(NH_2)$=1620 and 1580, $\nu(NO_3)$=1445, 1300, and 1020, and $\nu(NO_3^-)$=1372 cm^{-1}. The anion bands show that there are both inner-sphere nitrato groups and nitrate ions. The $\nu(NH_2)$ bands are those of coordinated NH_2 groups.

When heated in air, the compound decomposes between 95 and 128°C, as was shown thermogravimetrically. The compound is soluble in water, methanol, and ethanol, but only sparingly soluble in benzene, hexane, and CCl_4. The molar conductance of a freshly prepared 10^{-3}M solution in water was 221 $\Omega^{-1} \cdot cm^2 \cdot mol^{-1}$ at 20°C. It increased to 239 $\Omega^{-1} \cdot cm^2 \cdot mol^{-1}$ in one hour. The conductivity of a 10^{-3}M solution in methanol increased from 225 to 229 $\Omega^{-1} \cdot cm^2 \cdot mol^{-1}$ in one hour [6].

$Mn(o\text{-}C_6H_8N_2)Cl_2$ and **$Mn(p\text{-}C_6H_8N_2)Cl_2$** were obtained by adding a solution of excess diamine in anhydrous benzene to an absolute alcohol solution of anhydrous manganese(II) chloride [2]. The complex **$Mn(m\text{-}C_6H_8N_2)Cl_2$** was prepared by mixing ethanol solutions of manganese(II) chloride and the ligand and adding benzene to induce precipitation if necessary [27]. $Mn(o\text{-}C_6H_8N_2)Cl_2$ is greenish white [2, 3], $Mn(m\text{-}C_6H_8N_2)Cl_2$ is black [27], and $Mn(p\text{-}C_6H_8N_2)Cl_2$ is violet [2, 3]. The effective magnetic moments, μ_{eff}=5.99 μ_B for $Mn(o\text{-}C_6H_8N_2)Cl_2$ and μ_{eff}=5.85 μ_B for $Mn(p\text{-}C_6H_8N_2)Cl_2$ were determined at 303 K [3]. The complexes are stable at room temperature in a dry atmosphere [2, 27]. $Mn(m\text{-}C_6H_8N_2)Cl_2$ decomposes when heated. It is insoluble in cold water, but soluble in mineral acids [27]. All the compounds decompose readily in boiling water or on heating with NaOH solution [2].

Mn(o-$C_6H_8N_2$)(HSO_3)$_2$ and **Mn(p-$C_6H_8N_2$)(HSO_3)$_2$.** The complexes with o- and p-phenylenediamine were prepared by mixing aqueous solutions of $MnCl_2 \cdot 4H_2O$, $KHSO_3$, and the phenylenediamine. Mn(o-$C_6H_8N_2$)(HSO_3)$_2$ is stable at ordinary temperature in air. It is sparingly soluble in cold water, soluble in warm water, and insoluble in alcohol, ether, and CS_2. The electrical conductivity of the aqueous solution indicates that the complex is dissociated into three ions. Mn(p-$C_6H_8N_2$)(HSO_3)$_2$ is a white crystal powder with a yellowish tinge; it is insoluble in water and other common solvents [4].

Mn(m-$C_6H_8N_2$)$_2SO_4$ was prepared by mixing an aqueous solution of manganese(II) sulfate with an ethanol solution of m-phenylenediamine in stoichiometric amounts.

The IR spectrum of the gray complex has maxima (in cm^{-1}) assigned as follows: $\nu_s(NH_2)$ 3275, $\nu_{as}(NH_2)$ 3320, $\delta(NH_2)$ 1616 and 1583, and $\nu_{as}(SO_4)$ 1140, 1108, 990, and 960. The appearance of additional bands of the noncoordinated NH_2 group in the IR spectrum (at 3581 and 3415 cm^{-1}) suggests that the ligand in this complex is unidentate. The splitting of the $\nu_{as}(SO_4)$ band into four frequencies suggests inner-sphere coordination of the sulfate group.

When heated in air, the complex decomposes between 188 and 310°C. It is soluble in water, methanol, and ethanol but only sparingly soluble in benzene, hexane, and CCl_4. The molar conductivity of a freshly prepared 10^{-3}M solution in water is $\Lambda = 132\ \Omega^{-1} \cdot cm^2 \cdot mol^{-1}$ at 20°C. It increases to 145 $\Omega^{-1} \cdot cm^2 \cdot mol^{-1}$ in one hour [6].

Mn(p-$C_6H_8N_2$)SO_4. The compound with p-phenylenediamine was prepared like the preceding compound, Mn(m-$C_6H_8N_2$)$_2SO_4$.

The observed bands (in cm^{-1}) in the IR spectrum were assigned as follows: $\nu_s(NH_2)$ 3232, $\nu_{as}(NH_2)$ 3338, $\delta(NH_2)$ 1585, and $\nu_{as}(SO_4)$ 1112. The appearance of the $\nu_{as}(SO_4)$ band as a singlet and the fact that its frequency is characteristic for the free SO_4^{2-} ion suggest a chiefly ionic character of the sulfate group in this compound. The p-phenylenediamine is apparently bridging.

The thermal decomposition and the solubility of the compound are like those of Mn(m-$C_6H_8N_2$)$_2SO_4$. The molar conductivity of a freshly prepared 10^{-3}M solution in water was 209 $\Omega^{-1} \cdot cm^2 \cdot mol^{-1}$ at 20°C. After one hour it has increased to 216 $\Omega^{-1} \cdot cm^2 \cdot mol^{-1}$ [6].

Mn(o-$C_6H_8N_2$)$_2$(NCS)$_2$. The complex with o-phenylenediamine, slurried in acetonitrile, reacts on addition of the equimolar amount of 7,16-dihydro-6,8,15,17-tetramethyldibenzo[b,i][1,4,8,11]tetraazocyclotetradecine (=$C_{22}H_{24}N_4$) and four moles of anhydrous triethylamine to form the mixed-ligand complex Mn($C_6H_{15}N$)($C_{22}H_{22}N_4$) [5].

Ethanol Adducts $MnL_nX_2 \cdot C_2H_5OH$

The following complexes were prepared by reactions of the stoichiometric amounts of the appropriate phenylenediamine and manganese(II) salt in ethanol [6, 7]. The finely crystalline precipitates were washed with ethanol and dried over $CaCl_2$ in vacuo. The following compositions, colors, and pertinent infrared frequencies (in cm^{-1}) are reported [6]:

complex*)	color	$\nu_s(NH_2)$	$\nu_{as}(NH_2)$	$\delta(NH_2)$	anion bands
$Mn(L_2)_2(NO_3)_2 \cdot C_2H_5OH$	gray	3277, 3245	~3315	1615, 1590	$\nu(NO_3)$ = 1353
$Mn(L_3)(NO_3)_2 \cdot C_2H_5OH$	pale brown	~3133	~3322	1665, 1603	$\nu(NO_3)$ = 1365
$Mn(L_2)_2Cl_2 \cdot C_2H_5OH$ [7]	—	—	—	—	—
$Mn(L_3)Cl_2 \cdot C_2H_5OH$ [7]	—	—	—	—	—
$Mn(L_1)(ClO_4)_2 \cdot C_2H_5OH$	dark red	3285, 3120	3775, 3340	1616	$\nu(ClO_4)$ = 1150, 1116, 1092
$Mn(L_2)_2(ClO_4)_2 \cdot C_2H_5OH$	gray	~3280, 3265	3350, 3315	1610	$\nu(ClO_4)$ = 1150, 1118, 1093
$Mn(L_3)(ClO_4)_2 \cdot C_2H_5OH$	pale brown	3250, 3232	3230, (3425 ?)	1616	$\nu(ClO_4)$ = 1152, 1117, 1092
$Mn(L_1)SO_4 \cdot 0.5C_2H_5OH$	orange	broad band (3400 to 2800)	3360	1620, 1597	$\nu(SO_4)$ = 1156, ~985, 965
$Mn(L_3)SO_4 \cdot C_2H_5OH$	lilac	—	—	—	—
$Mn(L_1)(NCS)_2 \cdot C_2H_5OH$	beige	3390, 3370, 3295, 3260	3342, 3230	1638, 1615	$\nu(CN)_{NCS}$ = 2065
$Mn(L_2)_2(NCS)_2 \cdot C_2H_5OH$	gray	3335, 3328, 3268	—	1618, 1610, 1585	$\nu(CN)_{NCS}$ = 2120, 2086
$Mn(L_3)(NCS)_2 \cdot C_2H_5OH$	dark violet	3336, 3323, 3276, 3255, 3245	—	1600	$\nu(CN)_{NCS}$ = 2121, 2090

*) L_1 = o-$C_5H_6N_2$, L_2 = m-$C_5H_6N_2$, L_3 = p-$C_5H_6N_2$.

As shown by the IR spectra, the anions in the perchlorates, sulfates, and thiocyanates are coordinated, whereas the nitrate groups are in the outer sphere [6, 7].

For the compounds with m-phenylenediamine, $Mn(L_2)_2(NCS)_2 \cdot C_2H_5OH$ and $Mn(L_2)_2(ClO_4)_2 \cdot C_2H_5OH$, the appearance of additional $\nu(NH_2)$ bands in the range 3400 and 3600 cm^{-1}, that is, in the range characteristic of noncoordinated amino groups, suggests the ligands to be unidentate. The complexes with p-phenylenediamine apparently have a chain structure since only $\nu(NH_2)$ bands were observed for coordinated NH_2 groups. Therefore the ligand is assumed to be bridging.

Thermoanalysis shows that the compounds are rather stable with respect to the loss of amine molecules. When heated in air, they decompose in the ranges (in °C): 95 to 128 (nitrates), 105 to 142 (perchlorates), 188 to 310 (sulfates), and 108 to 235 (isothiocyanates). The ethanol molecules are lost first, then the amine ligands are lost stepwise. All the complexes are soluble in water, methanol, and ethanol but are only poorly soluble in hexane, benzene, and CCl_4. The electrical conductivities of the aqueous and methanol solutions, which increase with time, indicate that all the complexes dissociate in these solutions [6, 7].

13.11.2 With N,N-Dimethyl-1,4-benzenediamine
$(CH_3)_2NC_6H_4NH_2$ (= p-Aminodimethylaniline = $C_8H_{12}N_2$)

$Mn(C_8H_{12}N_2)_2Cl_2$ was prepared by mixing solutions of anhydrous manganese(II) chloride and the ligand (in excess) in alcohol and adding benzene to induce precipitation if the compound does not form immediately. The black compound is fairly stable in a dry atmosphere. When heated, it decomposes. It is insoluble in cold water and hydrolyzes on boiling. It is soluble in mineral acids [27].

13.11.3 With N-Phenyl-1,4-benzenediamine $C_6H_5NHC_6H_4NH_2$ (= $C_{12}H_{12}N_2$)

$Mn(C_{12}H_{12}N_2)_2(HSO_3)_2$ was obtained by mixing solutions of $MnCl_2\cdot4H_2O$, $KHSO_3$, and the ligand. The complex is a pink powder, which is stable on the air at ordinary temperature. It is insoluble in water and the common organic solvents. It dissolves in acids, decomposing and forming SO_2 [4].

13.11.4 With 2,2'- and 3,3'-Biphenyldiamine $H_2NC_6H_4C_6H_4NH_2$ (= $C_{12}H_{12}N_2$)

$Mn(L_1)_nX_2$ and **$Mn(L_2)_nX_2$**. The complexes with L_1 = 2,2'-biphenyldiamine (n = 3 for X = Cl, Br, I; n = 2 for X = Cl, Br) and the complexes with L_2 = 3,3'-biphenyldiamine (n = 1 for X = Cl, Br; n = 3 for X = I) were prepared by mixing alcohol solutions of the stoichiometric amounts of MnX_2 and the ligands [11, 22] and boiling the solutions if necessary [11]. The magnetic moments μ_{eff} (in μ_B) at room temperature, the g factors from ESR, the characteristic IR frequencies (in cm^{-1}), and the decomposition temperatures t_{dec} are reported in the table on p. 48 [11].

Most of the complexes have an octahedral structure, as shown by the ESR spectra. The IR spectra suggest that the bis and tris complexes with 2,2'-biphenyldiamine form seven-membered chelate rings. In the compounds with 3,3'-biphenyldiamine six coordination is achieved by the formation of ligand bridges. The low magnetic moments of $Mn(L_2)Cl_2$ and $Mn(L_2)Br_2$ indicate antiferromagnetic interaction between neighboring manganese atoms, probably through anion bridges. That iodo complexes with less than three biphenyldiamine molecules do not exist is explained by the inability of the I^- ion to form such bridges [11].

$Mn(L_1)_2Cl_2$ is sparingly soluble in water, and is decomposed by mineral acids [22].

$Mn(L_1)_2Br_2\cdot0.75CH_3COCH_3$ was prepared by mixing solutions of $MnBr_2$ and 2,2'-biphenyldiamine in dry acetone and washing the white microcrystalline precipitate with acetone and ether. The magnetic moment is 5.90 μ_B at 293.5 K. The IR spectrum shows a band at 1715 cm^{-1}, which is assigned to the CO stretching vibration of the acetone. On heating, the acetone is lost almost quantitatively at 100°C. Water causes the complex to lose the organic ligands, even at room temperature. It is suggested that the ligands are bidentate, forming two seven-membered chelate rings [23].

$Mn(L_1)_2X_2$ and **$Mn(L_2)X_2$** (X = NCS and NCSe). The colorless compounds were obtained by reacting concentrated alcohol solutions of the metal salts and the ligands. The complexes form air-stable prismatic microcrystals. The effective magnetic moments μ_{eff} (in μ_B) at 296 K, decomposition temperatures t_{dec} (in °C), and important maxima (in cm^{-1}) in the IR spectra (KBr discs) are reported in the table on p. 49 [13].

complex	color	μ_{eff}	g	$\nu_{as}(NH_2)$	$\nu_s(NH_2)$	$\nu(CN)$	$\delta(NH_2)$	t_{dec} in °C
$Mn(L_1)_2Cl_2$	pale yellow [22]	5.81	2.05	3345, 3315	3252, 3232	1240, 1260	1006	—
$Mn(L_1)_3Cl_2$	yellowish white	5.87	2.05	3332	3240	1238	1067	~283
$Mn(L_2)Cl_2$	beige [22]	5.49 [13]	2.05	—	—	—	—	—
$Mn(L_1)_2Br_2$	white	—	—	3330, 3305	3250, 3220	1255, 1239	1011	—
$Mn(L_1)_3Br_2$	yellow rose	—	—	3318	3235	1238	1058	~304
$Mn(L_2)Br_2$	beige maroon	5.53	0.035	3325	3254	1240	1080	~296
$Mn(L_1)_3I_2$	beige maroon	5.82	2.046	3312	3240	1275	1011	~265
$Mn(L_2)_3I_2$	maroon	6.20	2.05	3348	3280	1245	992	~310

L_1 = 2,2′-biphenyldiamine; L_2 = 3,3′-biphenyldiamine.

compound	μ_{eff}	t_{dec}	$\nu_{as}(NH_2)$	$\nu_s(NH_2)$	$\nu(CN)_X$	$\delta(NH_2)$	$\nu(CN)_L$
$Mn(L_1)_2(NCS)_2$	5.92	250	3330, 3301	3258, 3215	2080, 2035	1632, 1605, 1582	1293 1278, 1222
$Mn(L_1)_2(NCSe)_2$	5.91	<200	3323, 3300	3248, 3212	2085, 2025	1622, 1608, 1580	1293, 1278, 1222
2,2′-biphenyl-diamine = L_1	—	—	3395, 3378	3280	—	1625	1300, 1290
$Mn(L_2)(NCS)_2$	5.62	210	3318	3248	2111, 2065	1648, 1600, 1585	1233
$Mn(L_2)(NCSe)_2$	5.59	198	3318	3243	2115, 2071	1608, 1600, 1585	1232
3,3′-biphenyl-diamine = L_2	—	—	3432, 3400	3325	—	1620, 1600	1262

The magnetic susceptibilities of the complexes with 3,3′-biphenyldiamine were measured in the range 157 to 296 K. The inverse molar susceptibility $1/\chi$ is 44.77 mol/cm^3 at 157.5 K (μ_{eff} = 5.32 μ_B) and 75.76 mol/cm^3 at 296 K (μ_{eff} = 5.62 μ_B) for the isothiocyanate and 43.04 mol/cm^3 at 157.5 K (μ_{eff} = 5.41 μ_B) and 66.04 mol/cm^3 at 296 K (μ_{eff} = 5.60 μ_B) for the isoselenocyanate. The Curie-Weiss law is obeyed; the paramagnetic Curie temperature is $\Theta_p = -36$ K for the isothiocyanate and $\Theta_p = -20$ K for the isoselenocyanate. The negative values of Θ_p indicate an antiferromagnetic interaction through NCS or NCSe bridges.

The IR spectra suggest that the complexes with 2,2′-biphenyldiamine are monomeric, with the NCS and NCSe groups coordinated to Mn through the N atom. The appearance of two $\nu_s(NH_2)$ and two $\nu_{as}(NH_2)$ frequencies indicates that the coordinated amino groups do not occupy equivalent positions in these complexes. The complexes with 3,3′-biphenyldiamine show ν(CN) frequencies, which suggest that the NCS and NCSe groups are bridge-bonded to form polymeric chains. The coordinated amino groups of the ligand must occupy equivalent positions around Mn since there are only one $\nu_s(NH_2)$ and only one $\nu_{as}(NH_2)$ band.

In the reflection spectra, maxima at about 16500 and 24000 cm^{-1} for the compounds with L_1 and at ~16600 and 26300 cm^{-1} for the compounds with L_2 were assigned to manganese d-d transitions. A maximum around 31000 cm^{-1} for all the complexes may be due to an L→Mn charge transfer or a transition in the ligand. The positions of the maxima suggest that the manganese atom is six-coordinated [13].

13.11.5 With 4,4′-Biphenyldiamine $H_2NC_6H_4C_6H_4NH_2$ (= Benzidine = $C_{12}H_{12}N_2$)

A mixed-ligand complex with benzidine and pyridine is described on p. 104.

$Mn(C_{12}H_{12}N_2)Cl_2$ was prepared by mixing solutions of stoichiometric amounts of manganese(II) chloride and benzidine in absolute ethanol and concentrating the mixture [9]. The complex forms greenish-white [2, 3], white [9], or cream colored [10] crystals and has the magnetic moment 5.87 μ_B at 303 K [3] and 5.76 μ_B [10], or 6.02 μ_B [9] at room temperature.

The IR frequencies of the symmetric and antisymmetric NH_2 stretching vibration bands are 3240 or 3298 cm^{-1} (free benzidine: 3320 or 3390 cm^{-1}) [9]. Summarized IR data for this complex and a series of related metal complexes with benzidine are given in [10]. The compound is assumed to have a dimeric pseudotetrahedral structure with bridging benzidine molecules on the basis of its IR spectrum [9, 10] and the electronic spectra of the corresponding cobalt and nickel complexes [9].

The complex is stable in a dry atmosphere at room temperature [2]. On heating, it decomposes at 260°C [10] or 280°C [9]. It is decomposed by boiling water or on heating with NaOH solutions [2]. In dimethyl sulfoxide solutions, the solvent molecules probably enter the coordination sphere, displacing the coordinated Cl^- ions, as is indicated by the molar conductivity of the 10^{-3} M solution, $\Lambda = 48.80\ \Omega^{-1} \cdot cm^2 \cdot mol^{-1}$ [10].

$Mn(C_{12}H_{12}N_2)Cl_2 \cdot 2H_2O$ was synthesized like the anhydrous compound, but using 76% ethanol. The beige complex has the magnetic moment 5.89 μ_B at room temperature. In the IR spectrum, the symmetric and antisymmetric NH_2 stretching vibration bands of benzidine are shifted to lower frequencies, 3298 and 3350 cm^{-1}, on complexation. These shifts are smaller than the shifts for the anhydrous complex, probably due to the weakening of the Mn-N bonds by the coordination of the H_2O molecules to the manganese atom. The complex is assumed to have a dimeric octahedral structure with bridging benzidine molecules. On heating, it decomposes at 192°C to form a gray mass [9].

$Mn(C_{12}H_{12}N_2)Br_2$ and **$Mn(C_{12}H_{12}N_2)Br_2 \cdot 2H_2O$** were prepared like the corresponding chloro complexes. The following colors, magnetic moments μ_{eff} (in μ_B) at room temperature, IR frequencies (in cm^{-1}), and decomposition temperatures t_{dec} (in °C) are reported:

complex	color	μ_{eff}	$\nu_s(NH_2)$	$\nu_{as}(NH_2)$	$\nu(OH)$	t_{dec}
$Mn(C_{12}H_{12}N_2)Br_2$	yellow	6.01	3240	3298	—	237
$Mn(C_{12}H_{12}N_2)Br_2 \cdot 2H_2O$	pale pink	5.93	3300	3360	3615	105

The properties suggest that the bromides have the same structures as the corresponding chlorides [9].

$Mn(C_{12}H_{12}N_2)_3I_2$, forming gray crystals, has the magnetic moment 5.91 μ_B at room temperature. The ESR spectrum is consistent with an octahedral structure: the g factor of $Mn(C_{12}H_{12}N_2)_3I_2$ is 2.00 at room temperature. The IR spectrum (KBr discs) exhibits bands at $\nu_{as}(NH_2) = 3425$, 3290, $\nu_s(NH_2) = 3385$, 3245, $\nu(CN) = 1271$, 1250, and $\delta(NH_2) = 1082$, 1050 cm^{-1} (free benzidine: 3390, 3320, and 1280 cm^{-1}). The compound decomposes at ~286°C [11].

$Mn(C_{12}H_{12}N_2)SO_4$ was prepared by adding a warm aqueous solution of benzidine to a boiling solution of manganese(II) sulfate heptahydrate and boiling the mixture for some time with stirring. The yellow microcrystalline complex precipitated after a brief period of standing. It is not hygroscopic. It is only sparingly soluble in water, but somewhat more soluble in alcohol [12].

$Mn(C_{12}H_{12}N_2)_3(NCS)_2$ and **$Mn(C_{12}H_{12}N_2)_3(NCSe)_2$** were prepared according to the methods of [14, 15] by mixing alcoholic solutions of the manganese(II) salts and benzidine. The isothiocyanate is yellow green, the isoselenocyanate is yellow pink. Each complex forms prismatic microcrystals and is air stable. The magnetic moments μ_{eff} (in μ_B) at 296 K, decomposition temperatures t_{dec} (in °C), and pertinent IR frequencies in cm^{-1} (KBr discs) are listed in the table on p. 51:

complex	μ_{eff}	t_{dec}	$\nu_{as}(NH_2)$	$\nu_s(NH_2)$	$\nu(CN)_X$	$\delta(NH_2)$	$\nu(CN)_L$
$Mn(L)_3(NCS)_2$	5.87	<100	3375, 3325	3310, 3265	2100, 2072, 2045	1615, 1608	1270, 1248
$Mn(L)_3(NCSe)_2$	5.89	<110	3378, 3362	3342, 3260	2093 2045	1615, 1578	1264, 1245
benzidine = L	—	—	3390	3320, 3305	—	1615	1268

The stretching vibration bands of the amino groups in benzidine are split and shifted to lower wave numbers on coordination. The splitting might suggest the fact that not all the amino groups participate in coordination.

In the reflection spectra, maxima appear at about 16500, 19500, 26000, and 31000 cm^{-1}. The first three were assigned to manganese d-d transitions, while the last was assigned to an L→Mn charge transfer or a ligand transition. The experimental results suggest an octahedral configuration [13].

In an older work a tan compound of the composition $Mn(C_{12}H_{12}N_2)_3(NCSe)_2$ was prepared by reacting manganese(II) isoselenocyanate and benzidine in absolute alcohol. This compound is reported to be very unstable. It decomposed during filtration to give elemental selenium [15].

$Mn(C_{12}H_{12}N_2)_2Cr_2O_7$ was prepared by adding aqueous $K_2Cr_2O_7$ to a mixture of aqueous $Mn(NO_3)_2$ and a solution of benzidine in ethanol/acetone (50 vol%). After an oxidation product of benzidine had been separated, the chocolate-brown complex slowly crystallized [20]. The magnetic moment is 6.18 μ_B at ordinary temperature [19]. The IR spectrum shows two bands of the $Cr_2O_7^{2-}$ group, 884 and 922 cm^{-1}, one band of the CN vibration, 1022 cm^{-1}, and the symmetric and antisymmetric NH_2 vibration bands, 3196 or 3290 cm^{-1}, respectively. The CN and NH_2 modes are shifted to significantly lower frequencies from their free-benzidine values (1267, 3389, and 3430 cm^{-1}) [21]. The complex is assumed to have an octahedral structure, with four N atoms of bidentate benzidine and two O atoms of the dichromate group surrounding the manganese atom [19]. It is sparingly soluble in water, ethanol, and acetone [20].

Other Compounds. The following compounds were reported in the older literature, but have not been confirmed. They were prepared by mixing solutions of the manganese(II) salt and benzidine in 96% alcohol, unless stated otherwise.

$Mn(C_{12}H_{12}N_2)_3(NO_3)_2 \cdot 4H_2O$ is insoluble in water, only slightly soluble in alcohol, and is decomposed by acids [8]. $Mn(C_{12}H_{12}N_2)_2Cl_2 \cdot 2H_2O$ forms white crystals, which are stable in air, light sensitive, insoluble in water and common organic solvents, and are decomposed by acids [8]. $Mn(C_{12}H_{12}N_2)_2Br_2$ forms colorless, stable crystals, which lose benzidine when dissolved in water or acids [8]. The greenish-beige isothiocyanate $Mn(C_{12}H_{12}N_2)_3(NCS)_2 \cdot H_2O$ is soluble in methanol and acetone, slightly soluble in ethanol, and insoluble in ether, benzene, and chloroform [14]. The isoselenocyanate $Mn(C_{12}H_{12}N_2)_3(NCSe)_2 \cdot H_2O$ is altered in air and is light sensitive. In hot water it loses benzidine. It is soluble in methanol, ethanol, acetone, aniline, and pyridine but insoluble in ether, benzene, and chloroform [15].

$Mn(C_{12}H_{12}N_2)[SiF_6] \cdot H_2O$ was obtained by adding aqueous $Mn[SiF_6] \cdot 6H_2O$ to a boiling alcoholic benzidine solution. The white precipitate, dried in the dark, is then very

References on p. 53

stable and insoluble in water, ethanol, and ether [16]. $MnCd(C_{12}H_{12}N_2)_3[CdCl_6]$ is a white powder, which is insoluble in common solvents [17]. The white, finely crystalline compound $Mn(C_{12}H_{12}N_2)_6[SnBr_6]\cdot 2H_2O$ has only a low stability in air and is light sensitive. It is decomposed by hot water and HNO_3. The complex is insoluble in CH_3OH, C_2H_5OH, C_4H_9OH, $CHCl_3$, ether, acetone, benzene, pyridine, and petroleum ether [18]. $Mn(C_{12}H_{12}N_2)_5(C_3H_6O)[SnBr_6]\cdot 2H_2O$ was obtained by reacting $Mn[SnBr_6]\cdot 6H_2O$ and benzidine in acetone ($=C_3H_6O$). It forms a yellow powder and is much more stable than the preceding complex. It is oxidized by air only slowly, and decomposes on heating, in hot water, and in acids. It is insoluble in common solvents [18].

3.11.6 With 3,3′-Dimethyl-[1,1′-biphenyl]-4,4′-diamine $NH_2(CH_3)C_6H_3C_6H_3(CH_3)NH_2$ (= o-Tolidine = $C_{14}H_{16}N_2$)

$Mn(C_{14}H_{16}N_2)Cl_2$ was prepared by mixing solutions of excess diamine in dry benzene and anhydrous mangan chloride in absolute alcohol [2]. The complex is gray [2, 3] and has the magnetic moment 5.93 μ_B at 303 K [3]. It is stable in a dry atmosphere at room temperature [2].

$Mn(C_{14}H_{16}N_2)_2(NCS)_2$ and **$Mn(C_{14}H_{16}N_2)_2(NCS)_2\cdot 5H_2O$ (?).** The gray-violet hydrate was prepared by mixing solutions of mangan isothiocyanate and o-tolidine in 95% ethanol and concentrating the mixture. The hydrate loses all its water of hydration on standing 24 h over P_2O_5 at 75°C to give the gray anhydrous complex. The two compounds are soluble in ethanol, methanol, and acetone but insoluble in water, ether, benzene, and chloroform [14].

13.11.7 With 4,4′-Methylenebis(benzenamine) $H_2NC_6H_4CH_2C_6H_4NH_2$ ($=C_{13}H_{14}N_2=L$)

$Mn(C_{13}H_{14}N_2)_2X_2$ and **$Mn(C_{13}H_{14}N_2)_2X_2\cdot 2H_2O$** ($X=NO_3$, Cl). The hydrates were prepared by mixing ethanolic solutions of the ligand and the hydrated metal salts. The pale rose, prismatic microcrystals were dried in a vacuum at room temperature [24, 25]. The refractive indices of $Mn(C_{13}H_{14}N_2)_2Cl_2\cdot 2H_2O$ are $n_\alpha=1.606$, $n_\beta=1.694$, and $n_\gamma=1.730$ at room temperature. The X-ray spectrum differs clearly from that of the reactants [25]. $Mn(C_{13}H_{14}N_2)_2(NO_3)_2\cdot 2H_2O$ is sparingly soluble in cold water, and readily soluble in diluted acids. The hydrates were dehydrated in a vacuum over P_2O_5 at 100°C [24].

The IR spectrum of $Mn(C_{13}H_{14}N_2)_2Cl_2$ was obtained for KBr discs. The characteristic maxima (in cm^{-1}) were assigned as follows: $\nu_{as}(NH_2)$ 3450, 3300, $\nu_s(NH_2)$ 3321, 3256, $\delta(NH_2)$ 3100, 1620, δ(CH) 1580, ν(CN) 1290, 1277, 1245. Of these, the maxima at 3450, 3321, 1620, 1290, and 1277 cm^{-1} were ascribed to vibrations of the uncoordinated ligand, and the others, which had been considerably shifted to frequencies lower than those of the free ligand, to the coordinated ligand. The IR frequencies suggest a monomeric structure with two unidentate ligands [26]. On heating, $Mn(C_{13}H_{14}N_2)_2(NO_3)_2$ begins to decompose about 200°C; $Mn(C_{13}H_{14}N_2)Cl_2$ decomposes at 240°C [24].

$Mn(C_{13}H_{14}N_2)_3(NCS)_2$ and **$Mn(C_{13}H_{14}N_2)_3(NCSe)_2$** were prepared by reacting concentrated alcohol solutions of $Mn(NCS)_2$ or $Mn(NCSe)_2$ and the ligand. The beige or yellow complexes form prismatic microcrystals. The magnetic moments μ_{eff} (in μ_B) at 296 K and characteristic IR frequencies in cm^{-1} (KBr discs) are listed on p. 53.

compound	μ_{eff}	$\nu_{as}(NH_2)$	$\nu_s(NH_2)$	$\nu(CN)$*)	$\delta(NH_2)$	$\nu(CN)_L$
$MnL_3(NCS)_2$	6.20	3365	3300	2070	1630 to 1620	1287
$MnL_3(NCSe)_2$	6.06	3375, 3322	3312, 3253	2102, 2082	1640, 1610,1585	1260, 1246
ligand L	—	3446, 3415	3338	—	1630	1290, 1279

*) ν(CN) of NCS^- or $NCSe^-$.

These frequencies are shifted considerably from those of the free ligand, indicating complexation via the NH_2 groups of the ligand. The values of the anion ν(CN) frequencies suggest unidentate NCS^- or $NCSe^-$ ligands.

The complexes are stable in air but not very stable if heated up to 100°C [13].

13.11.8 With 1-Imino-1H-phenalen-9-amine

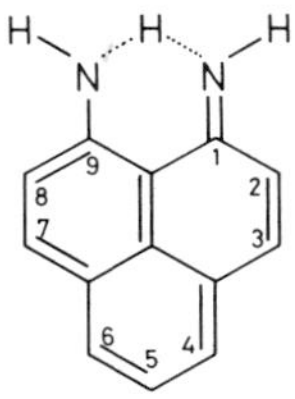

($=C_{13}H_{10}N_2$)

$Mn(C_{13}H_9N_2)_2 \cdot 2H_2O$ was prepared by adding a methanol solution of 0.5 mmol manganese(II) acetate dropwise to a hot solution of 1 mmol $C_{13}H_{10}N_2$ in methanol. The solution immediately turned dark, and a dark violet substance precipitated. This was washed with hot water and hot ethanol. The yield was 98%. The melting point is above 350°C [28].

References for 13.11 (With Aromatic Diamines):

[1] F. J. Andres-Ordax, J. M. Merino, A. de Diego (Anales Quim. A **76** [1980] 235/9). – [2] S. Prasad, V. Rama Reddy (J. Indian Chem. Soc. **35** [1958] 722/4). – [3] A. Syamal (J. Inst. Chem. [India] **40** [1968] 102/4). – [4] I. C. Popescu (Acad. Rep. Populare Romine Filiala Cluj Studii Cercetari Stiint. **1** [1950] 269/314, 273, 283, 291). – [5] M. C. Weiss, B. Bursten, S.-M. Peng, V. L. Goedken (J. Am. Chem. Soc. **98** [1976] 8021/31, 8022).

[6] M. V. Bregvadze, N. I. Pirtskhalava, Yu. Ya. Kharitonov (Zh. Neorgan. Khim. **23** [1978] 1539/45; Russ. J. Inorg. Chem. **23** [1978] 849/52). – [7] N. I. Pirtskhalava, Yu. Ya. Kharitonov, M. V. Bregvadze (Koord. Khim. 1 [1975] 1574/5; Soviet J. Coord. Chem. **1** [1975] 1300). – [8] G. Spacu (Bul. Soc. Stiinte Cluj **3** [1926/27] 285/307, 286, 293/5; C.A. **1928** 1349). – [9] R. Constantinescu, R. Micu-Semeniuc (Syn. Reactiv. Inorg. Metal-Org. Chem. **7** [1977] 399/412, 402, 406, 409; C.A. **87** [1977] No. 145082). – [10] S. M. F. Rahman, N. Ahmad, V. Kumar (Indian J. Chem. **13** [1975] 86/8).

[11] R. Micu-Semeniuc, R. Constantinescu (Studia Univ. Babes Bolyai Ser. Chem. **23** [1978] 17/22, 18/20; C.A. **89** [1978] No. 172781). – [12] A. Tettamanzi (Atti Accad. Sci. Torino Classe Sci. Fis. Mat. Nat. **69** [1933/34] 225/30; C.A. **1935** 5765). – [13] C. G. Macarovici, R. Micu-Semeniuc (Rev. Roumaine Chim. **16** [1971] 1367/76, 1369, 1374; C.A. **76** [1972] No. 20788). – [14] G. Spacu, C. G. Macarovici (Bul. Soc. Stiinte Cluj **7** [1933] 227/47, 239; C.A. **1933** 5672). – [15] G. Spacu, C. G. Macarovici (Bul. Soc. Stiinte Cluj **6** [1931] 95/138, 102, 121; C.A. **1932** 934).

[16] G. Spacu, L. Dima (Bul. Soc.Stiinte Cluj **8** [1937] 549/55, 549, 552; C.A. **1938** 71). – [17] G. Spacu, C. G. Macarovici (Bul. Soc. Stiinte Cluj **8** [1935] 286/95, 288, 290; C.A. **1936** 2873). – [18] G. Spacu, J. Dick (Bul. Soc. Stiinte Cluj **4** [1928] 187/210, 192, 197, 205). – [19] M. Genchev (Natura [Plovdiv, Bulg.] **6** [1973] 81/4; C.A. **83** [1975] No. 201177). – [20] M. Genchev, I. Pozharliev (Compt. Rend. Acad. Bulgare Sci. **12** [1959] 219/21; C.A. **1960** 20610).

[21] M. Genchev (Nauchn. Tr. Vissh. Pedagog. Inst. Plovdiv Mat. Fiz. Khim. Biol. **4** No. 1 [1966] 83/96, 86, 88/9; C.A. **67** [1966] No. 77588). – [22] C. G. Macarovici, G. Schmidt (Rev. Roumaine Chim. **9** [1964] 397/404, 402/3; C.A. **62** [1965] 7355). – [23] F. Hein, W. Jehn (Z. Anorg. Allgem. Chem. **341** [1965] 244/51, 248). – [24] C. G. Macarovici, A. Dorutiu (Studia Univ. Babes Bolyai Cluj Ser. Chem. **8** [1963] 75/83, 80/1; C.A. **62** [1965] 217). – [25] T. K. Dzhashiashvili, Yu. P. Smol'nikov, G. V. Tsintsadze, E. I. Abashidze (Izv. Akad. Nauk Gruz.SSR Ser. Khim. **4** [1978] 89/91; C.A. **89** [1978] No. 122160).

[26] C. G. Macarovici, A. Dorutiu (Studia Univ. Babes Bolyai Cluj Ser. Chem. **11** [1966] 135/40, 137; C.A. **68** [1968] No. 25271). – [27] S. Prasad, P. J. Sarma (J. Proc. Inst. Chem. **37** [1965] 246/8). – [28] R. Neidlein, Z. Behzadi (Chemiker-Ztg. **102** [1978] 150/2).

13.12 With Triamines

13.12.1 With Bis(2-aminoethyl)amine $HN(CH_2CH_2NH_2)_2$ (= Diethylenetriamine = $C_4H_{13}N_3$ = dien)

Mndien$_n^{2+}$ Ions. Formation and Properties in Solution

Formation constants for the ions $Mndien^{2+}$ and $Mndien_2^{2+}$ in aqueous solution with the ionic strength $I = 1 (KNO_3 + KCl)$ were determined potentiometrically (glass electrode): $\log K_1 = 3.99$ and $\log K_2 = 2.83$ at 30°C; $\log K_1 = 3.89$ and $\log K_2 = 2.72$ at 40°C [1, 2]. From these constants the thermodynamic quantities have been calculated: $\Delta H_1 = -4$ and $\Delta H_2 = -5$ kcal/mol at 35°C [1, 3]; $\Delta G_1 = -5.5$ and $\Delta G_2 = -3.9$ kcal/mol at 30°C [1]; $\Delta S_1 = 5.2$ and $\Delta S_2 = -3.7$ cal·mol^{-1}·K^{-1} at 35°C [4]. A calorimetric measurement gave the enthalpy of formation of $Mndien_2^{2+}$ from Mn^{2+} and the ligand of $\Delta H = -6.95$ kcal/mol at 25°C in an aqueous solution with $I = 0.1$ mol KCl/L [2, 5].

In the ion $Mndien_2^{2+}$ the Mn is octahedrally surrounded by the six N atoms of the tridentate ligands [1, 6]. The partial entropies of $Mndien^{2+}$ and $Mndien_2^{2+}$ in a hypothetical ideal aqueous solution of unit molality are $\overline{S}° = 55$ or 120 cal·mol^{-1}·K^{-1} at 25°C, as calculated by Cobble [4] from the data of Jonassen et al. [1]. The absorption spectrum of the $Mndien_2^{2+}$ ion in the visible region was obtained from a pale violet aqueous solution containing $MnCl_2 \cdot 4H_2O$, excess diethylenetriamine, and a few drops of hydrazine to prevent oxidation. The maxima observed, ν_{max} (in cm^{-1}), were assigned as follows (ground state $^6A_{1g}$):

excited state	$^4T_{1g}$	$^4T_{2g}$	$^4A_{1g}$, ^4E_g	$^4T_{2g}$	^4E_g
ν_{max}	15700	20100	23630, 23720, 23870, 23960	26630, 26920	28200

The ligand-field parameter $\Delta = 10\,Dq = 8800$ cm^{-1} was evaluated [6].

The complex solutions are air sensitive [6]. $Mndien^{2+}$ reacts with 2,2'-bipyridyl in aqueous solution to form $Mndienbpy^{2+}$, pp. 215/7. The manganese chelates with diethylenetriamine are catalysts for the decomposition of hydrogen peroxide, but they are less efficient than the chelate with triethylenetetramine [7]; see p. 57. They also accelerate the photopolymerisation of vinyl monomers in the presence of carbon tetrachloride [10].

Isolated Compounds

Mndien$_2$(ClO$_4$)$_2$. White crystals of the complex were obtained by adding aqueous 5 M $NaClO_4$ solution to 50 mL of aqueous solution containing $MnCl_2 \cdot 4H_2O$ (10 g), diethylene triamine (16 mL, 6.5 M), and a few drops of hydrazine. Electronic absorption maxima (ν_{max} in cm^{-1}) observed in the reflection spectrum were assigned as follows (ground state $^6A_{1g}$) [6]:

excited state	$^4T_{1g}$	$^4T_{2g}$	$^4A_{1g}$, ^4E_g	$^4T_{2g}$	^4E_g
ν_{max}	16700	20200	23500	26000	27800

Mndien$_2$Br$_2$ was prepared by adding finely divided $MnBr_2 \cdot H_2O$ (8 g) to diethylenetriamine (50 g) cooled in an ice bath in a helium atmosphere. After the initial vigorous exothermal reaction had finished, the mixture was stirred at 60°C for five hours. The white solid was washed with ethanol and dried in vacuo over $Mg(ClO_4)_2$. The yield was 90%. X-ray diffraction d values are given. The complex melts at 210 to 265°C with decomposition. The magnetic moment is 6.1 μ_B at room temperature [8, 9]. A diagram of the IR spectrum in the range of 4000 to 200 cm^{-1} is given in [9]. The compound is unstable in air. It reacts with anhydrous liquid ammonia at −35°C to form $Mn(NH_3)_6Br_2$, p. 12. It is insoluble in methanol, ethanol, acetonitrile, dioxane, and γ-butyrolactone but has a solubility of ~1% by weight in dimethyl sulfoxide [8, 9].

References:

[1] H. B. Jonassen, G. G. Hurst, R. B. LeBlanc, A. W. Meibohm (J. Phys. Chem. **56** [1952] 16/9). – [2] L. G. Sillén, A. E. Martell (Stability Constants of Metal-Ion Complexes, The Chemical Society, London 1964, p. 422). – [3] L. G. Sillén, A. E. Martell (Stability Constants of Metal-Ion Complexes, Suppl. No. 1, The Chemical Society, London 1971, p. 334). – [4] J. W. Cobble (J. Chem. Phys. **21** [1953] 1451/6, 1454). – [5] M. Ciampolini, P. Paoletti, L. Sacconi (J. Chem. Soc. **1961** 2994/9, 2996).

[6] C. K. Jørgensen (Inorg. Chim. Acta **3** [1969] 313/4). – [7] W. Langenbeck, M. Augustin, G. Thrumann, D. Eschrich (J. Prakt. Chem. [4] **27** [1965] 64/8). – [8] G. W. Watt, B. S. Manhas (J. Inorg. Nucl. Chem. **28** [1966] 1945/7). – [9] B. S. Manhas (Diss. Univ. Texas 1965, pp. 1/96, 9, 65; Diss. Abstr. **26** [1965] 685). – [10] T. Okimoto, M. Takahashi, Y. Inaki, K. Takemoto (Angew. Makromol. Chem. **38** [1974] 81/90, 84).

13.12.2 With N-Methyl-N,N-bis[2-(methylamino)ethyl]amine $CH_3N(CH_2CH_2NHCH_3)_2$ (= N,N′,N″-Trimethyldiethylenetriamine = $C_7H_{19}N_3$)

The catalytic activity of manganese(II) ions on the decomposition of hydrogen peroxide is enhanced by the formation of the chelate with the ligand. But the complex is less active than that with diethylenetriamine, see p. 54, W. Langenbeck, M. Augustin, F. Richter (Chem. Ber. **94** [1961] 831/3).

13.12.3 With N,N′-Bis[2-(dimethylamino)ethyl]-N-methylamine $CH_3N[CH_2CH_2N(CH_3)_2]_2$ (= $C_9H_{23}N_3$)

Mn(C$_9$H$_{23}$N$_3$)X$_2$ with X = Cl, Br were prepared by mixing hot solutions of $MnCl_2$ or $MnBr_2$ and a slight excess of the triamine in 1-butanol and recrystallizing the white or buff precipitates from chloroform-petroleum ether. All the manipulations were carried out in a pure N_2 atmosphere [1]. The enthalpy changes, in kcal/mol, for the formation of solid $Mn(C_9H_{23}N_3)Br_2$ from $MnBr_2$ and the ligand in condensed phase or in aqueous solution (no temperatures given) were determined calorimetrically [2]:

	reaction	ΔH
condensed phase	$MnBr_2$(solid) + $C_9H_{23}N_3$(liquid) → $Mn(C_9H_{23}N_3)Br_2$(solid)	−28.74 ± 0.11
aqueous solution	$MnBr_2$ + $C_9H_{23}N_3$ → $Mn(C_9H_{23}N_3)Br_2$(solid)	+6.44 ± 0.20

X-ray powder photographs show that the complex with X = Cl is isomorphous with the analogous complexes of iron, cobalt, and zinc. The complex with X = Br is isomorphous with those of cobalt, nickel, copper, and zinc [1]. The chemical analysis, the conductivity, and the spectra [1], including the far-IR spectrum [3], suggest that the metal atom is five-coordinate in these complexes, three coordination sites being occupied by the tridentate amine and two by the halogen atoms. A structure intermediate between trigonal bipyramidal and square pyramidal was proposed [1].

The magnetic moment of the complex with X = Cl is $\mu_{eff} = 5.85\ \mu_B$ at 22°C. The complexes are high spin [1, 4]. In the far-IR spectra (dispersions of the complexes in polythene at liquid-nitrogen temperature), three strong bands were assigned to the manganese-halogen bond. The frequencies for $Mn(C_9H_{23}N_3)Cl_2$ are 304, 291, and 259 cm^{-1}; those for $Mn(C_9H_{23}N_3)Br_2$ are 242, 231, and 213 cm^{-1} [3]. The reflectance spectrum of $Mn(C_9H_{23}N_3)Cl_2$ shows maxima at 23000 and 27000 cm^{-1} [1].

The complexes are fairly stable in dry air [4] and can be stored in a desiccator for at least several months [1]. They are hygroscopic [1, 4]. The complexes are very soluble in alcohols, nitroethane, nitrobenzene, and chloroform [4]. They are nonconductors in chloroform and nitrobenzene. The electrical conductivities of the 10^{-3} M solutions of the complexes in nitroethane are 2.0 $\Omega^{-1}\cdot cm^2\cdot mol^{-1}$ for X = Cl and 5.8 $\Omega^{-1}\cdot cm^2\cdot mol^{-1}$ for X = Br, both at 25°C, i.e., the solutions are practically nonconducting on account of the coordination of the halogen ions to the manganese atom. As shown by the conductivity and molecular-weight measurements the complexes are predominantly nonionic and monomeric in all the solutions. Only in chloroform solutions is there a small fraction of associated species [1].

References:

[1] M. Ciampolini, G. P. Speroni (Inorg. Chem. **5** [1966] 45/9). – [2] P. Paoletti, M. Ciampolini (Inorg. Chem. **6** [1967] 64/8). – [3] D. Bryson, R. H. Nuttall (Spectrochim. Acta A **26** [1970] 2275/80, 2279). – [4] M. Ciampolini, N. Nardi, G. P. Speroni (Coord. Chem. Rev. **1** [1966] 222/33, 224, 226, 228).

13.13 With Tetraamines

13.13.1 With Triethylenetetraamine $[H_2NCH_2CH_2NHCH_2]_2$ (= $C_6H_{18}N_4$ = trien)

Complex Ions. Formation and Properties in Solution

The stability constants K_1 and thermodynamic quantities for the reaction Mn^{2+} + trien ⇌ Mntrien^{2+} have been reported; the constants were measured for aqueous solutions:

t in °C	20	25	25	30	35	40
ionic strength I in mol/L	0.1 KCl	1 $NaClO_4$	0.1 KCl		1 (KCl + KNO_3)	
log K_1	4.9*)	5.61*)	—	5.43*)	—	5.31*)
$-\Delta H_1$ in kcal/mol	—	—	2.30 ± 0.05**)	—	4*)	—
$-\Delta G_1$ in kcal/mol	—	—	6.70 ± 0.07**)	—	7.5	—
ΔS_1 in cal·$mol^{-1}\cdot K^{-1}$		—	15.0 ± 0.5**)	—	12***)	—
Ref.	[1, 2]	[3]	[2, 4]	[2, 5]	[5 to 7]	[2, 5]

*) glass electrode. – **) calorimetry. – ***) calculation.

References on p. 62

In 50% (v/v) aqueous dioxane, potentiometric titration with a glass electrode yielded log $K_1=5.61\pm0.10$ at 25°C and $I=0.10$ mol/L KNO_3 [8]. Thermodynamic values for the formation of $Mn\,trien^{2+}$ from the complex with ethylenediamine, $Mn\,en_2^{2+}$, and triethylenetetraamine are given on p. 39. The partial entropy for the ion $Mn\,trien^{2+}$ in a hypothetical ideal aqueous solution of unit molality, $\overline{S}°=77$ cal·mol^{-1}·K^{-1} at 25°C, was calculated from results reported in [7].

The aqueous solutions of $Mn\,trien^{2+}$ are colorless in an acid medium and yellow green in strongly alkaline medium. They are very oxygen sensitive [3]. $Mn\,trien^{2+}$ reacts with 2,2′-bipyridyl in aqueous solution to form the mixed-ligand complex $Mn\,trien\,bpy^{2+}$, pp. 215/7.

For the reaction $2Mn^{2+}+3trien \rightleftharpoons Mn_2trien_3^{4+}$ in aqueous solution at $I=1$ $(KCl+KNO_3)$ mol/L the stability constant of the dinuclear complex ion was determined to be log K = 2.84 at 30°C and log K = 2.72 at 40°C, measured with the glass electrode. The corresponding thermodynamic values are $\Delta H=-4$ and $\Delta G=-3.9$ kcal/mol at 35°C [5]. The same values were adopted by [6], but for a trinuclear complex ion, $Mn_3trien_2^{6+}$. However, no dinuclear or trinuclear complex was ever found by others [3].

Conductometric and potentiometric titrations of aqueous solutions containing equimolar amounts of a manganese(II) salt and trien·4HCl with sodium hydroxide indicate the presence of the bis(hydroxo) complex $Mn\,trien(OH)_2$. The titrations and some additional spectrophotometric and magnetochemic studies (for which no results are given) suggest that the manganese is octahedrally surrounded by four nitrogen and two oxygen atoms, the latter in cis position to one another [14]. A mono(hydroxo) complex $Mn\,trien(OH)^+$ was reported earlier by [12].

The chelate complexes of manganese(II) with triethylenetetramine show remarkable catalytic activity for the decomposition of hydrogen peroxide [12]. In these chelates four coordination sites of manganese are occupied by the amine, while the remaining two sites are only loosely occupied by water molecules or hydroxy groups, and thus are available for the addition of the substrate. The coordinative unsaturation is assumed to produce the catalytic activity of the complex [11, 13]. The kinetics of the catalysis are described in [9 to 12]. The complexes were studied as model catalysts to elucidate the complicated mechanism of the very high catalytic efficiency [12]. Manganese(II) complexes with the ligand also accelerate the photopolymerization of vinyl monomers in the presence of carbon tetrachloride. The complexes were found to be more effective than those with other polyamines and more effective than the corresponding complexes of other metals, except those of iron [27].

Isolated Compounds

$Mn\,trien(ClO_4)_2$ and **$Mn\,trien(ClO_4)_2\cdot nH_2O$.** The hydrate was prepared by mixing 1:1 mole ratios of manganese(II) perchlorate and the ligand in ethanol. The complex crystallized within several weeks. The crystals were dried over P_2O_5 under vacuum at 120°C to give the anhydrous complex.

The yellow $Mn\,trien(ClO_4)_2$ has the magnetic moment $\mu_{eff}=6.00\ \mu_B$ at 295 K. The IR spectrum (Nujol mulls) exhibits bands at ~1100 and ~950 cm^{-1}, which are typical of ionic perchlorate. The molar electrical conductivity of the ~10^{-3}M nitromethane solution, $\Lambda=89.3\ \Omega^{-1}\cdot cm^2\cdot mol^{-1}$ at room temperature, is that of a 1:2 electrolyte [15].

Mntrien$_2$Br$_2$ was prepared by adding finely divided $MnBr_2 \cdot H_2O$ (43 mmol) to triethylenetetraamine (400 mmol) cooled in an ice-bath in an He atmosphere. After the initial vigorous exothermal reaction had finished, the mixture was stirred at 60°C for 60 h. The product was washed with ethanol and dried in vacuo over $Mg(ClO_4)_2$ (yield 95%). X-ray diffraction d values are given. The complex melts at 210 to 222°C with decomposition. The magnetic moment is 5.5 μ_B at room temperature [16]. A graph of the IR spectrum in the region 4000 to 600 cm^{-1} is given in [17]. When the complex is stirred with liquid ammonia for 2 h at −35°C, both $Mn(NH_3)_6Br_2$ and the dinuclear complex $Mn_2(trien)_3Br_4$ form [16].

MntrienBr$_2$ was obtained as described for the perchlorate hydrate Mntrien-$ClO_4 \cdot nH_2O$. The magnetic moment of the yellow crystals is $\mu_{eff} = 5.87\ \mu_B$ at 295 K. The $\sim 10^{-3}$ M solution in nitromethane is nearly nonconducting ($\Lambda = 2.1\ \Omega^{-1} \cdot cm^2 \cdot mol^{-1}$ at room temperature). The compound is assumed to be a normal octahedral complex with the anions coordinated in addition to the tetradentate amine [15]. Although expected, this complex was not obtained by treatment of manganese(II) bromide with triethylenetetraamine, rather the bis complex, Mntrien$_2$Br$_2$, see above, formed [16].

Mn$_2$trien$_3$Br$_4$ was obtained by adding a solution of triethylenetetraamine (36 mmol) in deoxygenated ethanol in small portions to an ethanolic solution of $MnBr_2 \cdot H_2O$ in a dry nitrogen atmosphere within 0.5 h and stirring the mixture for 2.5 h at 25°C. It was also formed by reaction of Mntrien$_2$Br$_2$ with liquid ammonia as described above. Mn$_2$trien$_3$Br$_4$ melts at 135 to 170°C with decomposition. It has the magnetic moment $\mu_{eff} = 5.7\ \mu_B$. X-ray diffraction d values are given in [16], and a graph of the IR spectrum is given for the range 4000 to 200 cm^{-1} in [17]. It is assumed that the complex is a ligand-bridged dinuclear species of the type suggested earlier for the nickel(II) complexes [16].

Mn$_2$trien$_3$[B(C$_6$H$_5$)$_4$]$_4$ and **Mn$_2$trien$_3$[B(C$_6$H$_5$)$_4$]$_4 \cdot$nH$_2$O.** The hydrate was prepared by careful addition of an ethanol solution of trieethylenetetraamine to an aqueous solution containing the stoichiometric amount of $Mn[B(C_6H_5)_4]_2$. The fine crystalline deposit was dried by heating it at 120°C over P_2O_5 to give the yellow anhydrous compound. The magnetic moment of Mn$_2$trien$_3$[B(C$_6$H$_5$)$_4$]$_4$ is 6.00 μ_B at 295 K. The $\sim 10^{-3}$ M solution in nitromethane exhibits a molar conductivity of $\Lambda = 89.1\ \Omega^{-1} \cdot cm^2 \cdot mol^{-1}$ at room temperature, which is typical of 1:2 electrolytes. The complex is assumed to be a ligand-bridged dimer [15].

13.13.2 With N,N′,N″,N‴-Tetramethyltriethylenetetraamine
$[CH_3NHCH_2CH_2N(CH_3)CH_2]_2$ ($= C_{10}H_{26}N_4$)

The manganese(II) chelate is an effective catalyst for the decomposition of hydrogen peroxide. At pH 7 and an amine:metal ratio of 1:1, this complex is more active than the triethylenetetraamine complex, p. 56, while with an excess of amine the complex with triethylenetetraamine is more active than the methylated compound [13].

13.13.3 With N,N′-Bis(2-aminoethyl)-1,3-propanediamine
$(H_2NCH_2CH_2NHCH_2)_2CH_2$ ($= C_7H_{20}N_4$)

The complex $Mn(C_7H_{20}N_4)^{2+}$ reacts with 2,2′-bipyridyl in aqueous solution to give the mixed-ligand complex $Mn(C_7H_{20}N_4)bpy^{2+}$, pp. 215/7.

References on p. 62

13.13.4 With Tris(2-aminoethyl)amine

$N(CH_2CH_2NH_2)_3$ (= 2,2′,2″-Triaminotriethylamine = $C_6H_{18}N_4$ = tren)

13.13.4.1 The Mntren^{2+} Ion in Solution

The stability constant in aqueous solution, log $K_1 = 5.8 \pm 0.05$, was determined at 20°C for the ionic strength I = 0.1 mol KCl/L with a glass electrode [2, 18, 19]. For the reaction $Mn^{2+} + H_3tren^{3+} \rightleftharpoons Mntren^{2+} + 3H^+$ the equilibrium constant is log K = −22.6 [18]. The thermodynamic quantities for the reaction $Mn^{2+} + tren \rightarrow Mntren^{2+}$ are $\Delta H_1 = -3.00$, $\Delta G_1 = -7.85$ kcal/mol, and $\Delta S_1 = 16.5$ cal·mol^{-1}·K^{-1}; ΔH_1 was measured calorimetrically at 25°C and I = 0.1 mol KCl/L [2, 20, 21]. A comparison of these values with those for Mntrien^{2+} and Mnen$_2^{2+}$ suggests a cis-diaqua octahedral structure for the complex Mntren^{2+} [21]. Thermodynamic quantities for the formation of Mntren^{2+} from the complex with ethylenediamine Mnen$_2^{2+}$, and 2,2′,2″-triaminotriethylamine are given on p. 39. Mntren^{2+} reacts with 2,2′-bipyridyl in aqueous solution to form Mntrenbpy^{2+}; see pp. 215/7.

13.13.4.2 Isolated Compounds

$[Mn_2tren_2X_2][B(C_6H_5)_4]_2$ (X = N_3, Cl, Br, CN, NCO, NCS). The complexes were prepared by successively adding aqueous solutions of the ligand (1.5 mL in 25 mL H_2O), NaX (10 mmol in 25 mL H_2O), and $Na[B(C_6H_5)_4]$ (0.25 g in 25 mL H_2O) to aqueous manganese(II) sulfate hydrate (10 mmol in 25 mL H_2O) in a nitrogen atmosphere by using a Schlenk tube technique. Crystals could be grown by slow evaporation of a degassed acetonitrile solution in a dry box. The crystals are clear rectangular prisms, which may have slight coloration, a faint greenish tint in the case of the bromo complex or a slight brownish tint in the case of the isothiocyanato compound.

X-ray investigations for the complexes with X = NCO and NCS indicated the space group to be $P2_1/c-C_{2h}^5$ (No. 14), belonging to the monoclinic system. The following lattice parameters were obtained at 20 to 21°C [7]:

complex	a in Å	b in Å	c in Å	β	Z
$Mn_2tren_2(NCO)_2[B(C_6H_5)_4]_2$	14.167(7)	10.753(8)	19.654(11)	96.62(3)°	2
$Mn_2tren_2(NCS)_2[B(C_6H_5)_4]_2$	14.882(6)	9.843(4)	21.109(6)	98.04(3)°	2

Positional parameters and thermal parameters are given in [7]. Both compounds are composed of $B(C_6H_5)_4^-$ anions and two discrete outer-sphere bridged [MntrenX]$^+$ cations. As can be seen from **Fig. 6,** p. 60, in each cation the Mn atom is surrounded by the 4N atoms of the tetradentate organic ligand and the N atom of the NCO or NCS group, forming a distorted trigonal bipyramid. The cations are associated by hydrogen bonding between the O atom of the NCO^- ion (or the S atom of the NCS^- ion), coordinated to one Mn atom, and two N atoms of the tren molecule, which is coordinated to the second Mn atom. In the complex with X = NCO the Mn-N_{NCO} distance is 2.051(5) Å, the Mn-N_{tren} distances are 2.186 to 2.412 Å, and the Mn-Mn distance is 5.893(1) Å. In the complex with X = NCS the corresponding distances are: 2.100(4), 2.194 to 2.343, and 5.8729(8) Å; other atomic distances and angles are given in [22].

The densities are 1.25(2) g/cm^3 for the complex with X = NCO and 1.26(2) g/cm^3 for the complex with X = NCS at 20 or 21°C, as observed by flotation in toluene/bromotoluene. The same values were calculated from the lattice parameters. The magnetic moments of the complexes are: 5.64 μ_B at 296 K for X = Cl; an essentially constant 6.00 $\mu_B \pm 1\%$ down to 7.9 K for X = CN; 5.91 μ_B at 296 K, and 4.85 μ_B at 4.2 K for X = NCO; and 5.85 μ_B at 267 K and 5.05 μ_B at 4.2 K for X = NCS. A graph of the temperature dependence of the molar susceptibility and the magnetic moment of the complex with X = NCO was published. The susceptibility data were analyzed with an isotropic-exchange-interaction model ($S_1 = S_2 = {}^5/_2$) to give the exchange parameter $J = -15\ cm^{-1}$ and the zero-field splitting parameter $D = -29\ cm^{-1}$ on the assumption that g is isotropic and equal to 2.0. A similar-magnitude antiferromagnetic exchange interaction is found for the complex with X = NCS, whereas the interaction in the complex with X = CN is negligible. ESR spectra were taken for powdered samples [22, 23], frozen glasses, and a single crystal of the complex with X = NCO. The most intense feature is at g = 2.0. The complex spectra were compared with the simpler spectra of the monomeric complexes of the permethylated ligand, $[MnX(C_{12}H_{30}N_4)]X$ (see below), but were not interpreted.

Fig. 6

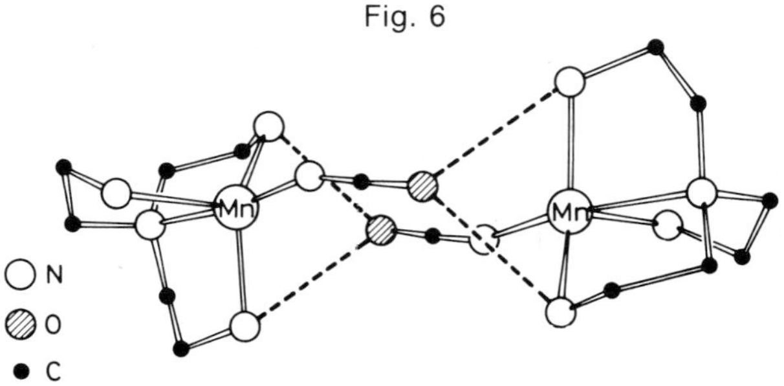

Structure of the $Mn_2(tren)_2(NCO)_2^{2+}$ ion. Hydrogen atoms are omitted. Possible hydrogen-bonding contacts for the oxygen atom of each NCO^- group are indicated by dashed lines.

The IR spectra of the six complexes (no frequencies given) indicate that in addition to the structurally characterized complexes with X = NCO and NCS, the complexes with X = Cl, Br, and CN also have five-coordinate manganese ions and that in the complex with $X = N_3$ the manganese ion is most likely six-coordinate as a result of inner-sphere bridging by the N_3^- ion.

All the complexes are indefinitely stable when dry, but sensitive towards air in the presence of moisture. Then they change from white to tannish brown within seconds [22].

13.13.5 With Tris[2-(dimethylamino)ethyl]amine $N[CH_2CH_2N(CH_3)_2]_3$ ($= C_{12}H_{30}N_4$)

$[MnBr(C_{12}H_{30}N_4)]Br$ was prepared by adding a butanol solution of the amine in slight excess to a hot solution of anhydrous manganese(II) bromide in butanol. On cooling and slowly adding petroleum ether, the greenish white tetrahedral crystals precipitated. These were washed with a mixture of butanol/petroleum ether and then with petroleum ether. The complexes were recrystallized by dissolving them in chloroform containing a few drops of the ligand and reprecipitating with petroleum ether. They were dried at 80°C. All operations were carried out in an atmosphere of pure N_2 [24].

The enthalpy changes, in kcal/mol, for the formation of the solid complex from $MnBr_2$ and the ligand in condensed phases or in aqueous solution (no temperature given) were determined calorimetrically [25]:

References on p. 62

state	reaction	ΔH
in condensed phases	$MnBr_2$(solid) + $C_{12}H_{30}N_4$(liquid) → [$MnBr(C_{12}H_{30}N_4)$]Br(solid)	-33.51 ± 0.11
in aqueous solution	$MnBr_2 + C_{12}H_{30}N_4$ → [$MnBr(C_{12}H_{30}N_4)$]Br(solid)	$+6.81 \pm 0.22$

[$MnBr(C_{12}H_{30}N_4)$]Br crystals are cubic. They belong to the space group $P2_13\text{-}T^4$ (No. 198) and have the lattice parameter $a = 12.216 \pm 0.007$ Å; Z = 4 [26]. The complex is isomorphous with [$MnI(C_{12}H_{30}N_4)$]I and [$MBr(C_{12}H_{30}N_4)$]Br, M = Fe, Co, Ni, Cu, and Zn, and with the analogous iodides [24, 26]. The structure consists of $[MnBr(C_{12}H_{30}N_4)]^+$ and Br^- ions, arranged in a distorted NaCl type lattice. As shown in **Fig. 7,** the manganese atom is surrounded by the four nitrogen atoms of the tetradentate ligand and the bromide ion, the five ligands located at the apexes of a trigonal bipyramid. The deviation of the manganese atom from the equatorial plane of the bipyramid towards the Br atom is 0.36 Å. The Mn-N(1) distance is 2.19(3) Å, the Mn-N(2) distance is 2.27(2) Å, and the Mn-Br distance is 2.491(6) Å. The angle N(1)-Mn-N(2) is 80.7(1.0)°, N(2)-Mn-Br is 99.3(6)°, and N(2)-Mn-N(2′) is 117.4(9)°. Further distances, angles, and the positional parameters, with temperature factors, are given in [26].

Fig. 7

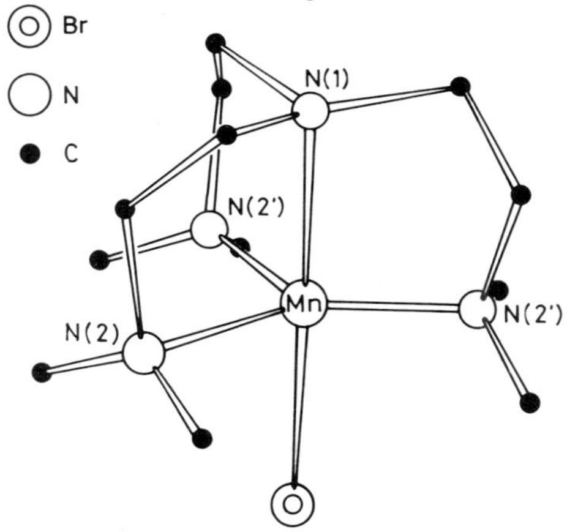

Structure of the $[MnBr(C_{12}H_{30}N_4)]^+$ ion.

The magnetic moment is 6.01 μ_B at 20°C. The measured magnetic susceptibility is plotted as a function of the temperature down to near zero K. The effective magnetic moment is relatively constant except for a drop at low temperatures. Least-squares fitting of the data for a high-spin d^5 monomer with axial symmetry gave the values $g_{\parallel} = 1.97$, $g_{\perp} = 2.07$, and the zero-field splitting parameter $D = 0.24$ cm^{-1}. ESR spectra were taken on powdered samples but were not analyzed. The spectra are similar to those of known distorted manganese(II) compounds [22]. The diffuse reflectance spectrum of the complex shows bands at 15400, 21150, 22700, 24900, and 26700 cm^{-1}, and it resembles closely the absorption spectrum of the solution in CH_2Cl_2. This indicates that there is no gross change in the stereochemistry of the complex on dissolution.

The complex does not oxidize appreciably in dry air over several weeks, but it does oxidize quickly in damp air or in solution. It is very hygroscopic. It is soluble in alcohols, methylene chloride, chloroform, nitroethane, nitrobenzene and insoluble in nonpolar

References on p. 62

solvents. The electrical conductivity of the ~10^{-3}M solution in nitroethane is 93 $\Omega^{-1}\cdot cm^2\cdot mol^{-1}$ at 25°C [24].

[MnI($C_{12}H_{30}N_4$)]I was prepared like the bromide. The iodide is isomorphous with the bromide and the bromides and iodides of Fe, Co, Ni, Cu, and Zn. The magnetic moment is 5.95 μ_B at 18°C [24]. ESR spectra were taken on powdered samples of [MnI($C_{12}H_{30}N_4$)]I and Mn^{2+} doped into [ZnI($C_{12}H_{30}N_4$)]I. The features of the spectra for the doped compound were assigned. The features can be accounted for by taking the axial zero-field splitting parameter D = 0.20 cm^{-1} and the rhombic zero-field splitting parameter E = 0 [22]. The diffuse reflectance spectrum of the green complex shows bands at 15200, 21200, 22750, 24800, and 26600 cm^{-1}. It resembles closely the absorption spectrum of the solution in CH_2Cl_2. The chemical and solubility properties are like those of [MnBr($C_{12}H_{30}N_4$)]Br. The electrical conductivity of the ~10^{-3}M solution in nitroethane is 79 $\Omega^{-1}\cdot cm^2\cdot mol^{-1}$ at 25°C [24].

References for 13.13 (With Tetraamines):

[1] G. Schwarzenbach (Helv. Chim. Acta **33** [1950] 974/85, 981). – [2] L. G. Sillén, A. E. Martell (Stability Constants of Metal-Ion Complexes, The Chemical Society, London 1964, pp. 492/3). – [3] Y. Bizri, M. Cromer-Morin, J.-P. Scharff (Compt. Rend. [2] **292** [1981] 447/50). – [4] L. Sacconi, P. Paoletti, M. Ciampolini (J. Chem. Soc. **1961** 5115/20, 5116, 5120). – [5] H. B. Jonassen, G. G. Hurst, R. B. LeBlanc, A. W. Meibohm (J. Phys. Chem. **56** [1952] 16/9).

[6] L. G. Sillén, A. E. Martell (Stability Constants of Metal-Ion Complexes, Suppl. No. 1, The Chemical Society, London 1971, p. 430). – [7] J. W. Cobble (J. Chem. Phys. **21** [1953] 1451/6, 1454). – [8] W. A. E. McBryde, H. K. J. Powell (Can. J. Chem. **57** [1979] 1785/91, 1787). – [9] A. Ya. Sychev, V. G. Isak (Zh. Fiz. Khim. **47** [1973] 244/6; Russ. J. Phys. Chem. **47** [1973] 138/40). – [10] R. P. Pantaler, L. D. Alfimova, A. M. Bulgakova (Zh. Analit. Khim. **30** [1975] 1584/9; J. Anal. Chem. [USSR] **30** [1975] 1332/5).

[11] W. Langenbeck, M. Augustin, G. Thrumann, D. Eschrich (J. Prakt. Chem. [4] **27** [1965] 64/8). – [12] J. H. Wang (J. Am. Chem. Soc. **77** [1955] 4715/9). – [13] W. Langenbeck, M. Augustin, F. H. Richter (Chem. Ber. **94** [1961] 831/3). – [14] H. B. Jonassen, H. Thielemann (Z. Anorg. Allgem. Chem. **320** [1963] 274/82, 280). – [15] B. Chiswell (Inorg. Chim. Acta **12** [1975] 195/8).

[16] G. W. Watt, B. S. Manhas (J. Inorg. Nucl. Chem. **28** [1966] 1945/7). – [17] B. S. Manhas (Diss. Univ. Texas 1965, pp. 1/96, 9, 13; Diss. Abstr. **26** [1965] 685). – [18] J. E. Prue, G. Schwarzenbach (Helv. Chim. Acta **33** [1950] 963/74, 968, 970). – [19] H. Ackermann, J. E. Prue, G. Schwarzenbach (Nature **163** [1949] 723/4). – [20] P. Paoletti, M. Ciampolini, L. Sacconi (Ric. Sci. **30** [1960] 1791/4).

[21] P. Paoletti, M. Ciampolini, L. Sacconi (J. Chem. Soc. **1963** 3589/93). – [22] E. J. Laskowski, D. N. Hendrickson (Inorg. Chem. **17** [1978] 457/70, 457/8, 460, 463/4, 467). – [23] D. N. Hendrickson, D. M. Duggan (ACS Symp. Ser. No. 5 [1974] 76/93, 90, 92/3; C.A. **82** [1975] No. 132402). – [24] M. Ciampolini, N. Nardi (Inorg. Chem. **5** [1966] 1150/4). – [25] P. Paoletti, M. Ciampolini (Inorg. Chem. **6** [1967] 64/8).

[26] M. Di Vaira, P. L. Orioli (Acta Cryst. B **24** [1968] 1269/72). – [27] T. Okimoto, M. Takahashi, Y. Inaki, K. Takemoto (Angew. Makromol. Chem. **38** [1974] 81/90, 83, 84).

13.14 With Pentaamines and Hexaamines

13.14.1 With 3,6,9-Triazaundecane-1,11-diamine
$(H_2NCH_2CH_2NHCH_2CH_2)_2NH$ (= Tetraethylenepentaamine = $C_8H_{23}N_5$)

The stability constant $K_1 = [Mn(C_8H_{23}N_5)^{2+}]/[Mn^{2+}] \cdot [C_8H_{23}N_5]$ was determined with a glass electrode at temperature t and ionic strength I on aqueous solutions; the values for log K_1 are as follows:

t in °C	25	25	25	35	45
log K_1	6.55	7.0 ± 0.1	7.62	7.55	7.46
I in mol/L	0.1 (KCl)	0.1 (KNO_3)	0.15 ($KClO_4$)	0.15 ($KClO_4$)	0.15 ($KClO_4$)
Ref.	[1,2,6]	[3,4]	[4,5]	[4,5]	[4,5]

The following thermodynamic values for the formation reaction $Mn^{2+} + C_8H_{23}N_5 \rightleftharpoons Mn(C_8H_{23}N_5)^{2+}$ in aqueous solution were reported:

t in °C	25	25	26.5	25 to 35	35 to 45
$-\Delta H$ in kcal/mol	3.70	—	5.16	3.67	4.08
$-\Delta G$ in kcal/mol	8.95	10.41	—	10.62 (35°C)	10.85 (45°C)
ΔS in cal·mol⁻¹·K⁻¹	17.5	22.6	17.5	21.8 (35°C)	21.3 (45°C)
method*)	I, II	II	I, II	II	II
Ref.	[2,6]	[4,5]	[5]	[4,5]	[5]

*) I = calorimetry; II = calculation.

$Mn(C_8H_{23}N_5)^{2+}$ is irreversibly oxidized by oxygen; therefore, the aqueous solutions have to be saturated with nitrogen and kept under an inert gas. Attempts to prepare salts of the manganese(II) complex ion or to obtain a manganese(III) complex ion have been unsuccessful since a partially oxidized product or hydroxide of the oxidized manganese ion forms [5].

Conductometric and potentiometric titrations of an equimolar mixture of $Mn(ClO_4)_2$ and $C_8H_{23}N_5 \cdot 5HCl$ with NaOH in aqueous solution indicate the formation of the ions $Mn(C_8H_{26}N_5)^{5+}$ and $Mn_2(C_8H_{24}N_5)_2(OH)^{5+}$. The latter is assumed to be a dinuclear complex with octahedral arrangements around the manganese atoms, four positions being occupied by the tetradentate ligand, the fifth by the bridging hydroxo group, and the sixth by a water molecule. A magnetic titration showed the paramagnetic susceptibility to be a minimum at the equivalence point of the dinuclear species. The magnetic moment calculated for this point corresponds to about two unpaired electrons per manganese(II) atom [7]. Manganese(II) complexes with the ligand were found to accelerate the photopolymerization of vinyl monomers in the presence of carbon tetrachloride [14].

13.14.2 With N,N,N′,N′-Tetrakis(2-aminoethyl)-1,2-ethanediamine
$[(H_2NCH_2CH_2)_2NCH_2]_2$ (= $C_{10}H_{28}N_6$)

Potentiometric and calorimetric measurements yielded the following values of the stability constant K_1 and the thermodynamic functions of formation of the 1:1 complex ion at temperature t and the ionic strength I:

t in °C	20	25	25
I in mol/L	0.1 (KCl)	0.1 (KNO_3)	0.1 (KCl)
log K_1	9.37	9.24	9.30
$-\Delta H_1$ in kcal/mol	—	8.85	8.85
$-\Delta G_1$ in kcal/mol	—	12.57	12.65 [11]
ΔS_1 in cal·mol⁻¹·K⁻¹	—	12.5	12.5 ± 0.2
Ref.	[4,10]	[8,13]	[2,9,11]

The measurements suggest that in aqueous solution containing manganese(II) ions, this hexadentate ligand forms the species $Mn(C_{10}H_{28}N_6)(H_2O)^{2+}$, with one water molecule left in the coordination sphere, i.e., a complex with coordination number 7 [8, 9].

13.14.3 With N,N,N′,N′-Tetrakis(2-aminoethyl)-1,3-propanediamine $[(H_2NCH_2CH_2)_2NCH_2]_2CH_2$ (= $C_{11}H_{30}N_6$)

The hexadentate ligand forms the complex ion $Mn(C_{11}H_{30}N_6)^{2+}$ and the dinuclear species $Mn_2(C_{11}H_{30}N_6)^{4+}$ in aqueous solution containing manganese(II) ions at pH ≧ 8. The six-membered chelate ring comprising the two tertiary nitrogen atoms of the ligand in the mononuclear complex $Mn(C_{11}H_{30}N_6)^{2+}$ is more strained than the five-membered chelate rings in the dinuclear complex $Mn_2(C_{11}H_{30}N_6)^{4+}$. Therefore, the formation of the dinuclear complex ion, is energetically possible. The stability constant of the 1:1 complex in aqueous solution was determined potentiometrically to be $\log K_1 = 5.3 \pm 0.05$ at 20°C and ionic strength I = 0.1 mol/L (KNO_3); the equilibrium constant for the formation reaction of the dinuclear complex ion, $Mn(C_{11}H_{30}N_6)^{2+} + Mn^{2+} \rightleftharpoons Mn_2(C_{11}H_{30}N_6)^{4+}$, is $\log K = 2.2 \pm 0.15$ under the same conditions. The thermodynamic data of the reaction $Mn^{2+} + C_{11}H_{30}N_6 \rightleftharpoons Mn(C_{11}H_{30}N_6)^{2+}$ are $\Delta H_1 = -2.57 \pm 0.1$, $\Delta G_1 = -7.2 \pm 0.07$ kcal/mol, and $\Delta S_1 = 15.6 \pm 0.5$ cal·mol^{-1}·K^{-1} at 25°C and I = 0.1 mol KNO_3/L, as measured calorimetrically [8, 13].

13.14.4 With 1,1,1-Propanetris[2-(methylamino)ethylamine] $CH_3CH_2C(CH_2NHCH_2CH_2NH_2)_3$ (= $C_{12}H_{32}N_6$)

In aqueous solution manganese(II) ions form with the hexadentate ligand the complex ion $Mn(C_{12}H_{32}N_6)^{2+}$ and with the monoprotonated amine, which is also hexadentate, the species $Mn(C_{12}H_{33}N_6)^{3+}$. The stability constant of $Mn(C_{12}H_{32}N_6)^{2+}$, $\log K_1 = 8.2$, was determined at 25°C and ionic strength I = 0.1 mol/L ($NaClO_4$) in aqueous solution with a glass electrode; the equilibrium constant for $Mn^{2+} + C_{12}H_{33}N_6^+ \rightleftharpoons Mn(C_{12}H_{33}N_6)^{3+}$ (protonated-ligand complex) is $\log K = 5.1$. The equilibrium constant for the reaction of the unprotonated complex with hydrogen ion to give the protonated complex, $Mn(C_{12}H_{32}N_6)^{2+} + H^+ \rightleftharpoons Mn(C_{12}H_{33}N_6)^{3+}$, is $\log K = 7.3$ [4, 12].

References:

[1] P. Paoletti, A. Vacca, I. Giusti (Ric. Sci. Rend. A [2] **3** [1963] 523/5). – [2] L. G. Sillén, A. E. Martell (Stability Constants of Metal-Ion Complexes, Suppl. No. 1, The Chemical Society, London 1971, pp. 546, 607). – [3] C. N. Reilley, J. H. Holloway (J. Am. Chem. Soc. **80** [1958] 2917/9). – [4] L. G. Sillén, A. E. Martell (Stability Constants of Metal-Ion Complexes, The Chemical Society, London 1964, pp. 576, 623, 668). – [5] H. B. Jonassen, A. Schaafsma, L. Westerman (J. Phys. Chem. **62** [1958] 1022/4).

[6] P. Paoletti, A. Vacca (J. Chem. Soc. **1964** 5051/7, 5056). – [7] H. B. Jonassen, V. V. Ramanujam (J. Phys. Chem. **65** [1961] 176/7). – [8] P. Paoletti, R. Walser, A. Vacca, G. Schwarzenbach (Helv. Chim. Acta **54** [1971] 243/65, 246, 255/6, 262). – [9] L. Sacconi, P. Paoletti, M. Ciampolini (J. Chem. Soc. **1964** 5046/51, 5047/8). – [10] G. Schwarzenbach, P. Moser (Helv. Chim. Acta **36** [1953] 581/97, 589, 595).

[11] M. Ciampolini, P. Paoletti, L. Sacconi (Ric. Sci. **30** [1960] 2148/51). – [12] R. W. Green, K. W. Catchpole, A. T. Phillip, F. Lions (Inorg. Chem. **2** [1963] 597/600). – [13] D. D. Perrin (Stability Constants of Metal-Ion Complexes, Pt. B, The Chemical Society, London 1979, p. 730). – [14] T. Okimoto, M. Takahashi, Y. Inaki, K. Takemoto (Angew. Makromol. Chem. **38** [1974] 81/90, 84).

14 Complexes with Hydrazine and Its Derivatives

The hydrazine carboxylic acid complexes are described in "Manganese" D 4, together with complexes of amine-N carboxylic acids.

14.1 With Hydrazine

14.1.1 The $Mn(N_2H_4)^{2+}$ Ion. Formation in Solution

Mn^{2+} ions at a concentration of about 0.001M in aqueous hydrazine solutions form only the complex $Mn(N_2H_4)^{2+}$ [1]. The formation constant is log $K_1 = 4.76$ at 30°C and an ionic strength of 1 mol/L ($NaClO_4$), determined potentiometrically with a glass electrode [1], also see [2]. A polarographic determination of the constant at 25°C for an undefined ionic strength gave log $K_1 = 1.93$ [3, 4].

References:

[1] D. Banerjea, I. P. Singh (Z. Anorg. Allgem. Chem. **349** [1967] 213/9). – [2] L. G. Sillén, A. E. Martell (Stability Constants of Metal-Ion Complexes, Suppl. No. 1, London 1971, p. 86). – [3] R. L. Causey, R. M. Mazo (Anal. Chem. **34** [1962] 1630/3). – [4] L. G. Sillén, A. E. Martell (Stability Constants of Metal-Ion Complexes, London 1964, p. 159).

14.1.2 Anhydrous Complexes $Mn(N_2H_4)_nX_2$ (n = 3, 2, 1)

$Mn(N_2H_4)_3(NO_3)_2$ and **$Mn(N_2H_4)_2(NO_3)_2$ (?).** The tris(hydrazine) complex was prepared by a method of [2] by adding $N_2H_4 \cdot H_2O$ in excess to a hot solution of $Mn(NO_3)_2$ in concentrated aqueous ammonia. It forms colorless hexagonal prisms, which are isostructural with $M(N_2H_4)_3(NO_3)_2$, M = Zn, Cd, Ni, Co (space group $P\overline{6}2c$-D_{3h}^4 (No. 190); Z = 6). In these complexes, the metal atom is surrounded pseudo-octahedrically by six nitrogen atoms. A structure in which metal atoms are joined by three bridging hydrazine ligands is most probable. The density has been measured by flotation to be 2.026 g/cm³ at 18°C. The compound is easily decomposed [1].

A bis(hydrazine) complex, $Mn(N_2H_4)_2(NO_3)_2$, was reported in an older paper. However, the analysis is that for the tris complex, $Mn(N_2H_4)_3(NO_3)_2(MnH_{12}O_6N_8)$. The compound was prepared by adding 10 mL of hydrazine hydrate to 10 g of manganese(II) nitrate in 50 mL of 95% ethanol. It precipitated as a fine white powder, which appeared to be unaffected by water. On heating, it deflagrated [2]. A bis(hydrazine) complex, $Mn(N_2H_4)_2(NO_3)_2$, was also reported in another earlier work. It was obtained by slowly adding 20 to 30% alcoholic manganese(II) nitrate to 40% alcoholic hydrazine maintained at 10 to 20°C until 4 to 5% excess hydrazine remained. The precipitate was washed with alcohol and dried at 100°C. The complex was obtained as hexagonal ocher colored crystals, which were stable when dry at room temperature but easily ignited at about 150°C. The complex did not appear to be sensitive to shock. It is only slightly soluble in water [3].

$Mn(N_2H_4)_2F_2$ and **$Mn(N_2H_4)F_2$.** $Mn(N_2H_4)_2F_2$ is prepared by reaction of MnF_2 and anhydrous hydrazine in excess. The excess hydrazine has to be removed at low temperatures (−30 to 0°C) to prevent loss of coordinated N_2H_4. The formation of $Mn(N_2H_4)_2F_2$ is much slower, requiring 2 to 5 months, than that of $Mn(N_2H_4)F_2$. The bis(hydrazine) complex $Mn(N_2H_4)_2F_2$ is stable in a vacuum and in a dry atmosphere when closed in small vessels. But it quickly decomposes in air, giving off white fumes and leaving an $Mn(N_2H_4)F_2$ residue [4].

$Mn(N_2H_4)F_2$ was obtained by heating $Mn(N_2H_4)_2F_2$ to 90°C [5]. It was prepared in a few hours by reaction of MnF_2 with 80% of the stoichiometric amount of $N_2H_4 \cdot H_2O$. The compound is not sensitive to moisture. Evidently no hydrolysis occurs although some water is bound or adsorbed in the product [4].

The important IR frequencies of the white compounds are given as follows:

complex	$\nu(NH_2)$	$\delta(NH_2)$	$\tau(NH_2)$*)	ν(N-N)	$\varrho(NH_2)$**)	ν(Mn-N)
$Mn(N_2H_4)_2F_2$	3292, 3248, 3189	1646, 1611	1228, 1171	963	618, 570	—
$Mn(N_2H_4)F_2$	3308, 3264, 3192	1667, 1637	1222	958	592, 546	416

*) τ = twisting. – **) ϱ = rocking.

The position of the N-N stretching vibration bands indicates the presence of hydrazine bridges in the complexes. The thermal decomposition of $Mn(N_2H_4)F_2$ proceeds to produce pure MnF_2. Three DTA peaks are observed, two exothermic peaks at 170 and 230°C, and one endothermic peak at 270°C. The decomposition in argon is complete at about 300°C [4].

$Mn(N_2H_4)_2Cl_2$. The compound was prepared according to a method of [2] by mixing ethanol solutions of manganese(II) chloride hydrate and hydrazine hydrate in an approximately 1:2 mole ratio, filtering off the precipitate, washing it with ethanol, then ether, and drying in vacuo [9].

The crystals are monoclinic [6, 7], belonging to the space group C2/m-C^3_{2h} (No. 12) [6]. The lattice constants have been determined by X-ray diffraction to be $a=9.08 \pm 0.01$, $b=8.01 \pm 0.01$, $c=4.29 \pm 0.01$ Å, and $\beta=105°30' \pm 12'$; Z=2. $Mn(N_2H_4)_2Cl_2$ is isomorphous with the analogous zinc and cadmium complexes [6, 7]. As shown in **Fig. 8,** each Mn atom is surrounded by four nitrogen atoms and two chlorine atoms, the coordination polyhedron forming a tetragonally distorted octahedron. Each of the four nitrogen atoms belongs to a different N_2H_4 molecule. Thus each N_2H_4 ligand is bidentate and forms bridges between two manganese atoms [6]. This bridging is presumably the cause for twinning [7]. The infinite $>Mn(N_2H_4)_2Mn<$ chains stretch in the [001] direction, as shown in **Fig. 9.** The hydrazine ligands have a staggered configuration, resulting in tilting of each octahedron with respect to the chain axis [9]. Fig. 8 shows the distances around the Mn atom. The N-N distance in the bridges is 1.459 Å; the angle Mn-N-N is $117° \pm 3°$ and the angle Cl-Mn-N is $90° \pm 22'$ [6]. The atomic coordinates are given in [6, 8]. Twinning of the crystals is observed; twin laws are defined by (100), (001), and (110) as twin planes [7, 8]. These laws reflect either different packing of the chains [(100) and (110) laws] or change of stacking within each chain [(001) law] [8]. The experimental density, 2.096 g/cm^3 at room temperature, which was measured by flotation, is the same as that calculated from the X-ray data [6, 7].

Important bands (in cm^{-1}) in the IR spectrum of Nujol or hexachlorobutadiene mulls were assigned as follows: 3298, 3237, and 3151, NH_2 stretching; 1606 and 1575, NH_2 bending; 1156, NH_2 twisting; 960, N-N stretching; and 590 and 518, NH_2 rocking. Assignments were made with the aid of the isotopic shifts of the deuterated compounds [10]. Combination bands of the NH_2 vibrations in the 6500 to 3700 range were recorded and assigned by [11]. The pertinent maxima in the far-IR spectrum of the compound on polyethylene discs were assigned as follows (in cm^{-1}): 342 [9] or 343 [10], Mn-N stretching; 210, Mn-Cl stretching; and 225 and 220(sh), N-Mn-N deformation. Assignments were made with the aid of low-temperature data (~ -150°C) on the basis of a C_{2h} site symmetry [9].

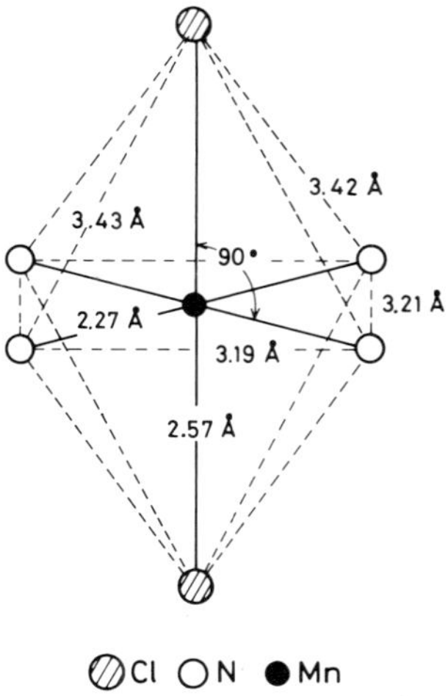

Fig. 8

Molecular structure of $Mn(N_2H_4)_2Cl_2$.

Fig. 9

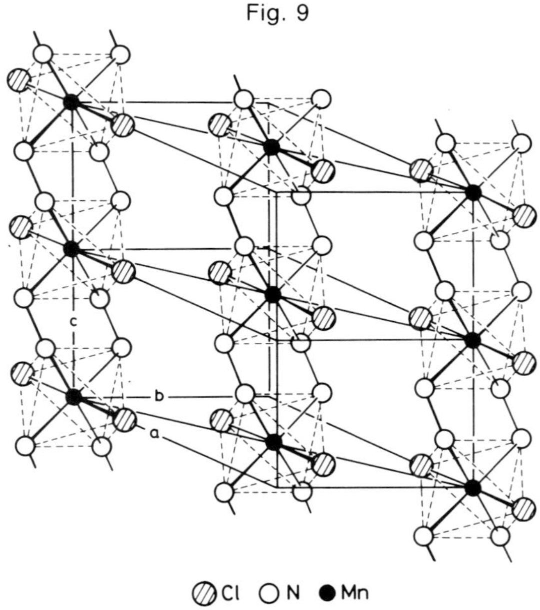

Crystal structure of $Mn(N_2H_4)_2Cl_2$ projected on (110). Only chains with axes passing through points, 0 0 z, $^1/_2$ $^1/_2$ z, and 1 1 z have been drawn.

$Mn(N_2H_4)_2Br_2$. The bromide was obtained by mixing ethanol solutions of hydrated $MnBr_2$ and $N_2H_4 \cdot H_2O$ in stoichiometric amounts, washing the precipitate with ethanol, then ether, and drying it in vacuo. The far-IR spectrum of the complex is consistent with a structure similar to that of $Mn(N_2H_4)_2Cl_2$. The bands were assigned as follows (in cm^{-1}): 341 ν(Mn-N), 175 ν(Mn-Br), and 247 and 209(sh) δ(N-Mn-N) [9].

$Mn(N_2H_4)_2(HCOO)_2$. The formate was prepared by dissolving $Mn(HCOO)_2 \cdot 2H_2O$ in hot $N_2H_4 \cdot H_2O$ solution. On addition of ethanol, white microcrystals formed. These were washed with ethanol and dried in vacuo. All operations had to be carried out in an inert atmosphere. The complex melts at 163 to 165°C with decomposition. The IR spectrum of the solid compound (KBr disc) shows strong or medium-strong maxima at the following frequencies: 3330, 3290, 3250, 3150, 1635, 1620, 1585, 1330, 1295, 1152, 954, 620, 530, and 335 cm^{-1}. Of these, the bands in the 3300 to 3100 cm^{-1} range were assigned to the $\nu(NH_2)$ vibrations (hydrazine gas 3350, 3325, 3280 cm^{-1}). The bands around 1600 cm^{-1} correspond to $\nu_{as}(OCO)$ combined with $\delta(NH_2)$. The band at 1330 cm^{-1} is assigned to $\nu_s(OCO)$, and the band at 954 cm^{-1} is assigned to ν(N-N) (free hydrazine 885 cm^{-1}). The shift of the ν(N-N) band relative to that of the free ligand indicates a bidentate coordination of the hydrazine to manganese. The compound is air sensitive. It forms brown oxidation products on standing in air [12].

$Mn(N_2H_4)_2(CH_3COO)_2$ was obtained by adding hydrazine to a warm ammoniacal solution of manganese(II) acetate and allowing the mixture to cool. It forms triclinic crystals with the lattice constants $a = 6.63 \pm 0.01$, $b = 8.60 \pm 0.01$, $c = 4.28 \pm 0.02$ Å, $\alpha = 90°32' \pm 3'$, $\beta = 90°39' \pm 7'$, and $\gamma = 96°42' \pm 12'$; Z = 1. The structure is assumed to consist of $>Mn(N_2H_4)_2Mn<$ chains. $Mn(N_2H_4)_2(CH_3COO)_2$ is isomorphous with the analogous zinc and cadmium compounds. The crystals are inclined to twin, but this tendency is not so marked as for the chloride and isothiocyanate. The observed twin law can be defined by (010) as the twin plane. At room temperature, the experimental density is 1.681 g/cm^3; the calculated density is 1.624 g/cm^3 [13]. When an aqueous solution of the compound is exposed to air, carbon dioxide is adsorbed and $Mn(N_2H_3COO)_2 \cdot 2H_2O$ forms [14].

$Mn(N_2H_4)_nC_2O_4$ (n = 1, 2). The mono(hydrazine) oxalate was prepared by reacting 1 g $MnC_2O_4 \cdot 2H_2O$ with 0.9 mL of 30% N_2H_4 in the absence of air. The mixture was stirred for 24 h at room temperature, and the precipitate washed with alcohol and ether. The bis(hydrazine) complex was prepared from 1 g $MnC_2O_4 \cdot 2H_2O$ and 3 mL 60% N_2H_4 in the same way [15]. Both compounds are white [17]. At 20°C the pycnometric densities are 2.109 g/cm^3 for $Mn(N_2H_4)C_2O_4$ and 2.302 g/cm^3 for $Mn(N_2H_4)_2C_2O_4$. Their X-ray diffraction patterns were determined, and the d values and their intensities are given in [15].

The magnetic susceptibility, χ_{mol} in 10^{-6} cm^3/mol, of the $Mn(N_2H_4)_nC_2O_4$ compounds was studied by the Faraday method [16]:

T in K	80	120	160	200	240	280	300
n = 1	41110	29390	23270	19420	16440	14170	11720
n = 2	49810	32460	24850	20270	16960	14040	12670

The experimental g factor is 1.981 ± 0.007 for $Mn(N_2H_4)C_2O_4$ and 1.985 ± 0.007 for $Mn(N_2H_4)_2C_2O_4$. Use of the Heisenberg linear chain model gives g = 2.00, the exchange parameter J/k = −2 K, and the paramagnetic Curie temperature $\Theta_p = -23$ K for $Mn(N_2H_4)C_2O_4$ and g = 1.98, J/k = −86 K, and $\Theta_p = -10$ K for $Mn(N_2H_4)_2C_2O_4$. Graphically: $\Theta_p = -26$ K for the complex with n = 1 and $\Theta_p = -8$ K for the complex with n = 2. The data suggest that in the mono(hydrazine) complex the metal ions are joined in linear chains of indefinite length through quadridentate bridging oxalato groups in the equatorial (xy) plane and through bidentate N_2H_4 molecules in the axial direction so that an infinite planar network is produced. The antiferromagnetic exchange occurs to an equal extent in the (xy) plane and along the z axis. The bis(hydrazine) complex consists of -Mn-N_2H_4-

Mn-chains along the z axis, but only dimers with hydrazine bridges or $-Mn-C_2O_4-Mn-$ chains of finite length in the (xy) plane. The low value of J/k for this compound is assumed to be due to structural disorder in the equatorial plane. The structure of $Mn(N_2H_4)_2C_2O_4$ differs from that of the analogous Fe, Co, and Ni complexes [16].

The IR spectra were obtained for suspensions in liquid paraffin or KBr tablets. The important maxima were assigned as follows:

complex	ν(NH)	ν_{as}(CO)	ν_{as}(OCO)	δ(NH_2)	ν_s(OCO)	ν(N-N)
$Mn(N_2H_4)C_2O_4$	3340, 3330, 3265, 3195	1695	1630	1610, 1590	1375, 1320, 1300	975
$Mn(N_2H_4)_2C_2O_4$	3345, 3330, 3290, 3260, 3190, 3170	1665	1632	1614, 1595	1380, 1320, 1300	974, 902

Assignments were made with the aid of both isotopic shifts of the deuterated compounds and published data. For $Mn(N_2H_4)C_2O_4$ the IR spectrum indicates a structure with quadridentate C_2O_4 groups in the equatorial plane and bridging N_2H_4 molecules in the axial direction. The spectrum of $Mn(N_2H_4)_2C_2O_4$ shows that there are both quadridentate and bidentate C_2O_4 groups in the equatorial plane. Thus the oxalato group leaves less than two coordination sites for the second N_2H_4 molecule, and consequently a part of the hydrazine should be unidentate or attached by hydrogen bonds. In fact, the IR maximum at 902 cm^{-1} can be assigned to N_2H_4 attached by hydrogen bonds [15].

$Mn(N_2H_4)C_2O_4$ is stable at room temperature. But $Mn(N_2H_4)_2C_2O_4$ loses hydrazine spontaneously in the air. Within 24 hours it is converted entirely into $Mn(N_2H_4)C_2O_4$. On heating, the mono complex is stable up to 208°C. Then it decomposes in an exothermal reaction to give MnC_2O_4. The loss of the first molecule of N_2H_4 from $Mn(N_2H_4)_2C_2O_4$ takes place in two stages, as is shown by DTA and DTG curves. Up to 140°C, 0.3 mol N_2H_4 is lost. The IR band at 902 cm^{-1} disappears as a result. The remaining 0.7 mol N_2H_4 is lost in the range 140 to 170°C [15].

$Mn(N_2H_4)_2(C_6H_4N_3O_5)_2$. The picramate(2-amino-4,6-dinitrophenolate) was prepared by adding ~10% solution of hydrazine hydrate in ethanol to an ethanol suspension of manganese(II) picramate. After stirring for about 6 h, the orange compound precipitated. It was washed with ethanol and dried at 80°C.

When heated, the complex does not melt until 250°C, but it explodes above 250°C. The magnetic moment is $\mu_{eff} = 6.14\ \mu_B$ at room temperature. In the IR spectrum (Nujol), the important bands (in cm^{-1}) were assigned as follows: 3470, 3385, 3315, 3270, 3190, and 3110 ν(NH), 1625, 1618, and 1615 δ(NH_2), 915 ν(N-N), 430 ν(Mn-O), and 330 ν(Mn-N). The bands at 3315 and 3270 cm^{-1} in the NH stretching frequency region were assigned to the coordinated NH_2 group of the picramate anion; the remaining four were assigned to unidentate hydrazine. The position of the ν(N-N) band at 915 cm^{-1} confirms that the hydrazine molecule acts as an unidentate ligand. The IR and magnetic results indicate that the complex has an octahedral configuration, the manganese atom being surrounded by two nitrogen atoms of the hydrazine molecule and two nitrogen and two oxygen atoms of the picramate anion. The electronic spectrum in the UV and visible region for a

Nujol mull shows bands at 25000, 27400, and 40000 cm^{-1}, which are due to intraligand transitions.

The complex explodes violently, not only on heating but also on addition of concentrated sulfuric acid. $Mn(N_2H_4)_2(C_6H_4N_3O_5)_2$ is insoluble in methanol, ethanol, benzene, chloroform, and carbon tetrachloride [18].

$Mn(N_2H_4)_2(NCS)_2$ was prepared by reacting a solution of $MnCl_2$, excess NH_4SCN, and NH_4Cl in concentrated aqueous ammonia with $N_2H_4 \cdot H_2O$ at elevated temperature. The ammonia was evaporated on a water bath and the solution cooled, the complex precipitating as small prisms [19]. A second method added a solution of hydrazine sulfate in aqueous ammonia to a mixture of concentrated $MnCl_2$ and NH_4SCN solutions. The precipitate was washed with absolute alcohol and dried in a vacuum [20].

The crystals are monoclinic and belong to the space group $P2_1/a\text{-}C^5_{2h}$ (No. 14). The lattice constants are a = 7.21 ± 0.01, b = 14.82 ± 0.01, c = 4.37 ± 0.01 Å, and β = 105°59′ ± 21′; Z = 2. The crystals form twins. The twin plane is (100), and the twin axis is normal to this [19]. The structure probably consists of $>Mn(N_2H_4)_2Mn<$ chains which run in the direction of the c axis [7, 13]. The complex is isomorphous with the compounds $M(N_2H_4)_2(NCS)_2$, M = Zn, Cd. The experimental density at room temperature is 1.707 g/cm^3; the calculated density is 1.735 g/cm^3 [19].

$Mn(N_2H_4)_2(NCS)_2$ is white [19, 20]. In dry air it is reasonably stable. In the presence of water, even at room temperature, it decomposes to form manganese hydroxide. It dissolves in dilute acids, and it is converted into manganese hydroxide in dilute aqueous ammonia. At room temperature, in the absence of air it dissolves unchanged in aqueous solutions containing high concentrations of NH_4SCN and N_2H_4. The resulting solutions are stable for an extended period of time [20].

References:

[1] A. Ferrari, A. Braibanti, A. M. Lanfredi (Ann. Chim. [Rome] **48** [1958] 1238/43, 1240). – [2] H. Franzen, O. v. Mayer (Z. Anorg. Allgem. Chem. **60** [1908] 247/91, 285/6). – [3] L. Médard, J. Barlot (Mem. Poudres **34** [1952] 159/66, 161; C.A. **1954** 6125). – [4] P. Glavič, A. Bole, J. Slivnik (J. Inorg. Nucl. Chem. **35** [1973] 3979/81). – [5] P. Glavič, J. Slivnik (IJS-R-613 [1972] 1/10, 6, 7, 9; C.A. **79** [1973] No. 12982).

[6] A. Ferrari, A. Braibanti, G. Bigliardi, F. Dallavalle (Z. Krist. **119** [1963] 284/96, 285, 288). – [7] A. Ferrari, A. Braibanti, G. Bigliardi (Z. Krist. **117** [1962] 241/8, 247). – [8] A. Braibanti, G. Bigliardi, A. M. Lanfredi, M. Camellini (Z. Krist. **120** [1964] 261/77, 260). – [9] M. Goldstein, W. D. Unsworth (Spectrochim. Acta A **28** [1972] 1297/304, 1303). – [10] L. Sacconi, A. Sabatini (J. Inorg. Nucl. Chem. **25** [1963] 1389/93).

[11] A. F. Reid (Australian J. Chem. **19** [1966] 1785/99, 1792). – [12] F. J. Arnaiz Garcia (Anales Quim. **73** [1977] 1121/3). – [13] A. Ferrari, A. Braibanti, G. Bigliardi, A. M. Lanfredi (Gazz. Chim. Ital. **93** [1963] 937/41). – [14] A. Braibanti, G. Bigliardi, A. M. Manotti-Lanfredi (Ateneo Parmense II **1** [1965] 81/6; C.A. **65** [1966] 16174). – [15] T. M. Zhdanovskikh, E. I. Krylov, V. A. Sharov, N. G. Fedotovskikh (Zh. Neorgan. Khim. **18** [1973] 1234/9; Russ. J. Inorg. Chem. **18** [1973] 651/4).

[16] V. A. Sharov (Zh. Neorgan. Khim. **22** [1977] 2762/7; Russ. J. Inorg. Chem. **22** [1977] 1500/2). – [17] T. B. Shkodina, T. M. Zhdanovskikh, E. A. Nikonenko (Tr. Khim. Met. Inst. Akad. Nauk Kaz.SSR **25** [1974] 164/70, 166; C.A. **83** [1975] No. 201384). – [18] R. C. Aggarwal, N. K. Singh (Def. Sci. J. **27** [1977] 109/14). – [19] A. Ferrari, A. Braibanti, A. M. Lanfredi (Gazz. Chim. Ital. **91** [1961] 69/76, 70/1). – [20] P. Rây, P. V. Sarkar (J. Chem. Soc. **117** [1920] 321/3).

14.1.3 Hydrates $Mn(N_2H_4)_2X_2 \cdot nH_2O$

The complex hydrates were prepared by two methods: I) Hydrazine hydrate was added dropwise to the manganese salt MnX_2 in saturated aqueous solution until a small precipitate of basic salt was formed. The precipitate was removed and the solution left to crystallize. II) Saturated solutions of MnX_2 and the corresponding hydrazine salts in stoichiometric proportions were mixed, and the resulting solutions left to crystallize [1, 3]. The compounds were dried at 40 to 45°C [1] or 50 to 60°C [3].

X	n	method	D_{20} *)	Ref.	X	n	method	D_{20} *)	Ref.
NO_3	2	I, II	2.42	[1]	$^1/_2SO_4$	2	I, II	2.13	[1,3]
Cl	2	I, II	2.12	[1,3]	CH_3COO	2	II	1.9267	[2]
Br	1	II	2.72	[1]	$^1/_2C_2O_4$	1	I, II	2.96	[1]
I	1	II	3.6845	[2]	NCS	1	II	2.76	[1]

*) D_{20} = density in g/cm³ at 20°C determined pycnometrically.

The compounds were obtained as fine white square crystals (for X = NO_3), pale green hexagonal crystals (for X = NCS), or white or pale pink needles or platelets (for the remaining anions) [1 to 3]. The chloro, iodo, and acetato complexes are soluble in water, while the oxalate and isothiocyanate are only slightly soluble. Almost all the complexes are soluble in 2N mineral acid (HNO_3, HCl, or H_2SO_4), but they are insoluble in the usual organic solvents, such as alcohols, acetone, or benzene [1 to 3]. Ammonium sulfide does not precipitate MnS from solutions of the compounds with X = Cl, Br, I. This suggests that the Mn-N bonds are fairly strong. However, silver nitrate slowly precipitates AgCl from solutions of $Mn(N_2H_4)_2Cl_2 \cdot H_2O$ and AgI from $Mn(N_2H_4)_2I_2 \cdot H_2O$. Thus the halogenides appear to be slightly dissociated. A $BaCl_2$ solution does not precipitate barium sulfate from the sulfato complex or barium oxalate from the oxalato complex. The isothiocyanato and acetato complexes do not give the test for the free ions with $FeCl_3$ [1], also see [2]. $FeCl_3$ oxidizes the chloro and sulfato complexes in aqueous solution with evolution of N_2 [3].

The sebacate hydrates, $Mn(CH_2)_8(COO)_2 \cdot nH_2O$ with n = 0.5 or 1, were prepared by reacting an aqueous solution of sebacic acid and hydrazine hydrate with aqueous manganese(II) sulfate. The precipitates were washed with alcohol and dried in air. At a 1:1 mole ratio of manganese(II) sulfate to sebacic acid the monohydrate was obtained; for 1:2 and 1:4 mole ratios of manganese(II) sulfate to sebacic acid, the hemihydrate precipitated. The compounds are crystalline powders, which are soluble in water. The molar conductivity of a 0.001M aqueous solution, $\Lambda \approx 150\ \Omega^{-1} \cdot cm^2 \cdot mol^{-1}$ at 25°C, is that of a 1:1 electrolyte [4].

The picrate, $Mn(N_2H_4)_2(C_6H_2N_3O_7)_2 \cdot H_2O$, was prepared by mixing ~5% solutions of hydrazine hydrate and manganese(II) picrate. The yellow precipitate was washed with ethanol and dried in air. The complex melts at 210°C with decomposition. The magnetic moment is 5.92 μ_B at room temperature, indicating a high-spin complex with an octahedral structure.

In the IR spectrum (Nujol), the important maxima (in cm^{-1}) were assigned as follows: 3450 and 3360, ν(OH); 3285 and 3230, ν(NH); 1628 and 1604, $\delta(NH_2)$ combined with δ(HOH); 962, ν(N-N); 392, ν(Mn-O); and 243, ν(Mn-N).

The strong broad ν(NH) bands are shifted to frequencies lower than those of the free hydrazine. These shifts and the occurrence of the ν(N-N) band in the 980 to 948 cm^{-1} region (free hydrazine ~873 cm^{-1}) show that the hydrazine acts as a bidentate bridging ligand. The UV and visible spectrum (Nujol) exhibit five bands at frequencies between 23000 and 43000 cm^{-1}, which are assigned to intraligand transitions.

On addition of concentrated sulfuric acid, $Mn(N_2H_4)_2(C_6H_2N_3O_7)_2 \cdot H_2O$ explodes violently. The complex is slightly soluble in tetrahydrofuran or dimethylformamide, but it is insoluble in methanol, ethanol, benzene, chloroform, or carbon tetrachloride. The 10^{-3} M solution in tetrahydrofuran shows a molar conductivity of 33 $\Omega^{-1} \cdot cm^2 \cdot mol^{-1}$ [5].

References:

[1] R. Ya. Aliev, A. D. Kuliev, N. G. Klyuchnikov (Zh. Neorgan. Khim. **17** [1972] 3282/6; Russ. J. Inorg. Chem. **17** [1972] 1726/8). – [2] R. Ya. Aliev (Zh. Obshch. Khim. **43** [1973] 2572; J. Gen. Chem. [USSR] **43** [1973] 2554/5). – [3] R. Ya. Aliev, A. D. Kuliev, N. G. Klyuchnikov (Zh. Prikl. Khim. **45** [1972] 2743/4; J. Appl. Chem. [USSR] **45** [1972] 2878/9). – [4] L. D. Tsitsishvili, P. V. Gogorishvili, T. I. Tsutsunava (Issled. Obl. Khim. Kompleksn. Prostykh Soedin. Nekot. Perekhodnykh Redk. Metal. No. 2 [1974] 113/6; C.A. **82** [1975] No. 38080). – [5] R. C. Aggarwal, N. K. Singh (Def. Sci. J. **27** [1977] 109/14).

14.2 With 1,2-Dimethylhydrazine $H_3CNH\text{-}NHCH_3$ (= $C_2H_8N_2$)

$Mn(C_2H_8N_2)Cl_2$. The complex was prepared by adding 0.01 mol $C_2H_8N_2$ to an anhydrous solution of 0.005 mol $MnCl_2$ in 25 mL methanol in an atmosphere of N_2. Then the solution was evaporated at room temperature under reduced pressure to half of its original volume. Adding dried ether, which had been saturated with N_2, precipitated the compound as a colorless crystalline powder. It is supposed to have a distorted octahedral structure (D_{4h}). In this proposed structure, the Mn is surrounded by two N and four Cl atoms. The compound is sensitive to hydrolysis and oxidation, P. Barz, H. P. Fritz (Z. Naturforsch. **27b** [1972] 1131/6, 1135).

14.3 With 1,1′-(1,2-Ethanediyl)bis(1-methylhydrazine) $H_2N(CH_3)NCH_2CH_2N(CH_3)NH_2$ (= $C_4H_{14}N_4$)

$Mn(C_4H_{14}N_4)Cl_2$ and **$2MnCl_2 \cdot C_4H_{14}N_4$** were prepared in absolute butyl alcohol at −10°C by reaction of a 50% $C_4H_{14}N_4$ solution with the stoichiometric amount of 5% $MnCl_2$ solution. $Mn(C_4H_{14}N_4)Cl_2$ formed a white crystalline precipitate, which was washed with hexane (yield 50%). For the compound $2MnCl_2 \cdot C_4H_{14}N_4$ a linear structure was proposed. The two $MnCl_2$ molecules are on opposite sides of the extended organic ligand. Both complexes gradually darken. Upon heating at 100°C, they lose 1 to 4% of their weight. The compounds are soluble in alcohol and dimethylformamide, G. S. Gol'din, S. G. Fedorov, N. I. Kirilina (Zh. Obshch. Khim. **43** [1973] 2789; J. Gen. Chem. [USSR] **43** [1973] 2764).

14.4 With Phenylhydrazine $C_6H_5NH\text{-}NH_2$ (= $C_6H_8N_2$)

$Mn(C_6H_8N_2)_2X_2$ (X = Cl, Br, I). The chloro complex was prepared at room temperature by adding a stoichiometric amount of phenylhydrazine in dry ethanol to an ethanol solution of hydrated manganese chloride [1]. For the preparation of the bromo and iodo complexes anhydrous manganese halides in dry ethanol were used. In the case of $Mn(C_6H_8N_2)_2I_2$ the reaction temperature is 60°C [1] or 20 to 100°C [2]. The precipitated complexes were washed with ethanol and n-pentane, and dried in a vacuum desiccator over silica gel [2]. A greater yield of $Mn(C_6H_8N_2)_2I_2$ was obtained when ether was added to an alcohol solution of MnI_2 and phenylhydrazine in slight excess (to prevent precipitation of MnI_2). Then the iodo complex, which forms small prismatic needles, was washed with a mixture of alcohol and ether and finally with ether [2].

The structure of the complexes is assumed to be octahedral and polymeric. Color, magnetic moment (in μ_B), melting point, and bands of the diffuse reflectance spectra are given below:

complex	color	μ_B at 20°C	melting point in °C	wavelengths in nm
$Mn(C_6H_8N_2)_2Cl_2$	pale pink	6.04	280	350, 420, 520, 615
$Mn(C_6H_8N_2)_2Br_2$	pale pink	5.96	258 to 260	360, 380, 425, 515
$Mn(C_6H_8N_2)_2I_2$	buff	5.91	206 to 208 (dec.)	360

All the bands are weak [1].

The complexes are generally stable in air, but are slowly hydrolyzed [1]. On heating $Mn(C_6H_8N_2)_2I_2$ in air, it becomes black at 220°C and then decomposes, evolving iodine and leaving a residue of manganese oxide. The iodo complex is very soluble in water and alcohol but nearly completely insoluble in ether or chloroform [2].

References:

[1] W. K. Glass, J. O. McBreen (J. Inorg. Nucl. Chem. **36** [1974] 747/50). – [2] J. Moitessier (Compt. Rend. **124** [1897] 1529/31).

14.5 With 2- and 4-Nitrophenylhydrazine $O_2NC_6H_4NH\text{-}NH_2$ (= L_1 and $L_2 = C_6H_7N_3O_2$)

$Mn(L_1)_2Cl_2 \cdot 2H_2O$. The preparation of the complex, which is isostructural with the analogous complexes of Ni, Co, and Fe, is reported in [1]. The IR spectrum of this complex and of complexes with 4-nitrophenylhydrazine (mulls in liquid paraffin and a fluorinated hydrocarbon oil) was studied. Wave numbers of the absorption band maxima with their assignments are listed below [2]:

assignment	$\nu(NH)$	$\nu_{as}(NH_2)$	$\nu_s(NH_2)$	$\delta(NH_2)$	$\nu_{as}(NO_2)$
$Mn(L_1)_2Cl_2 \cdot 2H_2O$	3335	3285	3145	1620	1525
$Mn(L_2)_4Cl_2$	3354, ~3325	3275, 3235	3150	1635	1510
$Mn(L_2)Cl_2$	3322	3260	3140	1635	1504

assignment	$\nu_s(NO_2)$	$\tau(NH_2)$	$\delta(NO_2)$	$\nu(Mn\text{-}N)$
$Mn(L_1)_2Cl_2 \cdot 2H_2O$	1340	1160, 1140	~860, 840	~635(?)
$Mn(L_2)_4Cl_2$	1346, 1315	1175, 1115	865, 840	—
$Mn(L_2)Cl_2$	1350, ~1316	1134, 1120	865, 840	—

The shifts of the IR bands from the values for free 2-nitrophenylhydrazine suggest that L_1 is bidentate and forms a chelate ring. It is coordinated via the nitrogen atom of the amino group and via one oxygen atom of the nitro group. The metal-nitrogen bond is less strong than the corresponding bond in the complexes of Ni and Co as shown by comparison of the IR spectra [2]. In solution, the compound behaves as a 1:2 electrolyte [1].

$Mn(L_2)_4Cl_2$. A solution of 0.5 g $MnCl_2 \cdot 4H_2O$ in 10 mL of alcohol was mixed with a solution of 4-nitrophenylhydrazine (1.54 g) in 70 mL of a hot 1:1 alcohol-benzene mixture. The hot solution was filtered. On cooling, the compound separated from the solution as a yellow precipitate. It was washed with a 1:1 benzene-ether mixture, then with ether, and dried in vacuo. Yield: 72% [3].

The compound belongs to the monoclinic crystal system. Its refractive indexes were determined to be 1.681 and >1.740 [3]. Bands and assignments of the IR spectrum are shown in the table above. The displacements of the $\nu_s(NH_2)$ and $\nu_{as}(NH_2)$ bands, compared with the spectrum of free 4-nitrophenylhydrazine, suggest the ligand to be unidentate and coordinated only via the nitrogen atom of the amino group (and not via the NH group) [2].

The dry complex is stable at room temperature. Two exothermic effects are observed in the DTA diagram, the first between 170 and 218°C, and the second at 420°C, where the decomposition is complete [3]. The solubility of $Mn(L_2)_4Cl_2$ in water is 0.138 g/100 mL H_2O at 25°C. The compound shows good solubility in alcohol, but is insoluble in ether or benzene [3].

$Mn(L_2)Cl_2$ was prepared by mixing hot alcoholic solutions of 4 g $MnCl_2 \cdot 4H_2O$ and 0.77 g of 4-nitrophenylhydrazine. The yellow precipitate was washed with a 1:1 ether-alcohol mixture and dried in vacuo. Yield: 96% [3].

The compound belongs to the orthorhombic crystal system. Its refractive indexes were measured to be 1.629 and ~1.718 [3]. The band frequencies in the IR spectrum as shown in the table above suggest the 4-nitrophenylhydrazine to be unidentate and coordinated via the amino group [2].

$Mn(L_2)Cl_2$ is stable at room temperature but decomposes upon heating. A slight endothermic effect is observed in the DTA diagram at 115°C. Between 115 and 210°C the weight loss is 18.86%, and at ~240°C rapid decomposition takes place, with a strong exothermic effect [3]. The solubility in water is 0.29 g/100 mL H_2O. The complex dissolves only slightly in alcohol or benzene and is insoluble in ether [3].

References:

[1] O. Kh. Shamilishvili, P. V. Gogorishvili, M. V. Karkarashvili (Tr. Gruz. Politekhn. Inst. **1971** No. 1, pp. 39/45 from C.A. **77** [1972] No. 96391). – [2] R. I. Machkhoshvili, Yu. Ya. Kharitonov, P. V. Gogorishvili, O. Kh. Shamilishvili (Zh. Neorgan. Khim. **17** [1972] 2713/8; Russ. J. Inorg. Chem. **17** [1972] 1421/5). – [3] P. V. Gogorishvili, O. Kh. Shamilishvili, M. V. Karkarashvili (Issled. Obl. Khim. Kompleksn. Prostykh Soedin. Nekot. Perekhodnykh Redk. Metal. **1970** 46/57, 48/51; C.A. **74** [1971] No. 60338).

15 Complexes with Hydroxylamine

Complexes with derivatives of hydroxylamine, e.g. derivatives of nitrosohydroxylamine and hydroxamic acids, are described in "Manganese" D 4.

15.1 With Hydroxylamine NH_2OH (= H_3NO)

15.1.1 $Mn(NH_2OH)_n^{2+}$ Ions (n = 1 to 4). Formation in Solution

In aqueous solutions, Mn^{2+} ions form with NH_2OH the four exceptionally stable complex ions $Mn(NH_2OH)_n^{2+}$ with n = 1 to 4. The formation constants have been determined by potentiometric pH investigations to be log K_1 = 4.04, log K_2 = 3.52, log K_3 = 3.28, log K_4 = 3.16, and log β_4 = 14.00 at 25°C for the ionic strength 1 (KNO_3) [1, 2]. In a previous pH investigation (glass electrode) log K_1 = 0.5 at 20.0 ± 0.1°C for I = 0.5 M ($NaNO_3$) was obtained [3, 4]. In this work the concentration of manganese (0.05 M) was tenfold greater than the ligand concentration (~0.005 M) [3].

References:

[1] E. M. Ivashkovich, M. I. Skoblei (Zh. Neorgan. Khim. **19** [1974] 760/5; Russ. J. Inorg. Chem. **19** [1974] 411/5). – [2] E. Högfeldt (Stability Constants of Metal-Ion Complexes, Pt. A: Inorganic Ligands, Pergamon, Oxford 1982, p. 115). – [3] L. Szilárd (Acta Chem. Scand. **17** [1963] 2674/80, 2678). – [4] L. G. Sillén, A. E. Martell (Stability Constants of Metal-Ion Complexes, Suppl. No. 1, Chem. Soc., London 1971, p. 87).

15.1.2 Isolated Complexes

$Mn(NH_2OH)_2Cl_2$ was prepared by slowly adding $MnCO_3$ to a great excess of boiling aqueous $NH_2OH \cdot HCl$ while stirring. The resulting solution was evaporated; the pale pink crystals were washed with a little cold water and dried at room temperature [1, 2]. The complex was also prepared by a modified method of [4]: the reaction of alcoholic 1M $MnCl_2$ with a sixfold excess of alcoholic $NH_2OH \cdot H_2O$ [3]. The product can be freed from adsorbed water by drying at 120°C [4].

X-ray diffraction patterns show no isomorphism with the complexes $M(NH_2OH)_2Cl_2$ for M = Ca, Ba, Cd, and Zn. A table of d values is given in [5]. The magnetic moment, μ_{eff} = 5.7 μ_B, is close to the spin-only value for a compound containing five unpaired electrons [1].

The IR spectra were obtained on specimen in KBr discs [3] or on suspensions in vaseline or fluorinated oils [5]. The principal vibration frequencies (in cm^{-1}) were assigned as follows; frequencies for free $NH_2OH \cdot HCl$ are given in parentheses: $\delta(NH_2)$ 1640 (1570), δ(NOH) 1200 (1190), ν(NO) 1000 (985), NH_2 rocking (ϱ) not observed for the complex (580 to 520) [3]. In a previous work are given: $\nu(NH_2)$ ~2955, ~2885, 2712, $\nu(NH_2)$ and δ(HNO) ~1632, δ(HNO) and $\delta(NH_2)$ 1200, ν(NO), $\delta(NH_2)$, and δ(HNO) 983, and ν(Mn-O) and δ(Mn-ON) 420 [5]. Assignment was based on analysis of the normal vibrations of $ZnONX_3$ (X = H or D) and $MnONX_3$ models. The spectrum of the deuterated manganese complex $MnOND_3$ was not actually recorded but only calculated [5]. The absence of absorption at frequencies corresponding to NH_2 rocking in free $NH_2OH \cdot HCl$ (580 to 520 cm^{-1}) and the extreme closeness of δ(HNO) at 1200 cm^{-1} to that for protonated hydroxylamine at 1190 cm^{-1} indicate that $Mn(NH_2OH)_2Cl_2$ contains an Mn-O bond [3]. Coordination through the oxygen atom in a complex of the type $M(ONH_3)_2Cl_2$ had been proposed earlier in an analysis of the normal vibrations [5]. The visible reflectance spectrum (plot given) contains no d-d transition bands, but a charge-transfer band begins in the blue region of the spectrum. The spectrum is reported to be typical for an octahedral manganese(II) complex. Sharing of ligands between adjacent metal ions for the complex in the solid state is assumed [1].

The complex is stable when heated moderately. But it is reported to decompose suddenly at 150 to 160°C [4] or to explode at 190°C [2]. If it is heated with a rate of 10°C/min in a nitrogen stream, it loses hydroxylamine exothermically around 230°C to form the intermediate $Mn(NH_2OH)Cl_2$. This loss of NH_2OH was indicated by the TG and DTA curves and was confirmed by mass spectroscopy [6]. The fact that the loss of one hydroxylamine molecule is exothermic, and not endothermic as expected for a simple dissociation reaction, indicates concurrent decomposition of hydroxylamine. The manganese complex is thermally more stable than the analogous zinc and cadmium complexes but less stable than the cobalt complex [6].

$Mn(NH_2OH)_2Cl_2$ is soluble in water but only sparingly soluble in alcohols, acetone, and dioxane [1, 3]. The molar conductivity of a 0.01M solution in water is 240 $\Omega^{-1}\cdot cm^2\cdot mol^{-1}$ at room temperature. Such a conductivity corresponds to dissociation into three ions [3]. H_2O_2 oxidizes the hydroxylamine in the dissolved complex, thus destroying it [1].

$Mn(NH_2OH)SO_4\cdot 2H_2O$ was synthesized by adding aqueous hydroxylamine to a hot aqueous solution containing manganese(II) sulfate and hydroxylammonium sulfate in a 1:4 mole ratio. The compound gradually precipitated as a white powder which is insoluble in water [4].

$Mn(NH_2OH)(NCS)_2\cdot C_2H_5OH$. The compound was prepared by treating crystalline $Mn(NCS)_2$ with alcoholic NH_2OH in sixfold excess. The yellow precipitate, which gradually precipitated, was washed with alcohol and ether, and dried over $CaCl_2$.

In the IR spectrum (vaseline oil), the principal maxima (frequencies in cm^{-1}) were assigned as follows: ν(CN) 2075, $\delta(NH_2)$ 1560 to 1610 (split into two components), δ(NOH) 1200, ν(NO) 990, and NH_2 rocking (ϱ) 440 and lower. The position of the ν(CN) band indicates that the NCS^- is not coordinated. An insignificant shift of the δ(NOH) band (+10 cm^{-1}) and the position of the $\varrho(NH_2)$ band (free $NH_2OH\cdot HCl$ 580 to 520 cm^{-1}) suggest that the hydroxylamine is coordinated through the oxygen atom. The electronic spectrum recorded in the 400 to 700 nm region on a solution in DMF shows a maximum near 400 nm, which is shifted to the long-wave region with respect to the corresponding band of manganous isothiocyanate. Plots of both spectra are given. The spectrum differs sharply from that of the octahedral $Mn(H_2O)_6^{2+}$ ion. On account of this and the yellow color of the compound, a tetrahedral configuration is assumed.

When the compound is heated at 100 to 105°C for 4 h, a 2% weight loss is observed. The compound is soluble in water and dimethylformamide, only slightly soluble in ethanol, and insoluble in ether, acetone, dioxane, chloroform, carbon tetrachloride, and benzene. The molar conductivity of a 0.001M aqueous solution, $\Lambda = 263\ \Omega^{-1}\cdot cm^2\cdot mol^{-1}$, corresponds to that of a ternary electrolyte [7].

References:

[1] E. L. Simmons, W. W. Wendlandt (J. Inorg. Nucl. Chem. **33** [1971] 3955/7). – [2] C. Kjellin (Svensk. Kem. Tidskr. **33** [1921] 213/28, 221; C.A. **16** [1922] 2111). – [3] E. M. Ivashkovich, N. I. Lazorshak, O. D. Gumanitskaya, M. Yu. Shpontak (Zh. Neorgan. Khim. **21** [1976] 2125/9; Russ. J. Inorg. Chem. **21** [1976] 1170/2). – [4] W. Feldt (Ber. Deut. Chem. Ges. **27** [1894] 401/6, 405). – [5] Ya. Yu. Kharitonov, M. A. Sarukhanov, I. B. Baranovskii (Zh. Prikl. Spektrosk. **15** [1971] 739/43; J. Appl. Spectrosc. [USSR] **15** [1971] 1396/9).

[6] E. L. Simmons, W. W. Wendlandt (Thermochim. Acta **2** [1971] 471/5). – [7] A. M. Golub, E M. Ivashkovich (Ukr. Khim. Zh. **44** [1978] 14/9; Soviet Progr. Chem. **44** [1978] 13/7).

16 Complexes with N-Heterocycles

The complexes with N-heterocyclic ligands are arranged by the number of N atoms in the heterocyclic ring, from one N up. Ligands with the same number of N atoms are arranged by ring size. This volume treats all complexes with ligands containing one N atom in the ring. In addition, complexes with pyrazole, imidazole, and their derivatives are described. Complexes with further N-heterocycles are treated in "Manganese" D 4.

16.1 Complexes with N-Heterocycles Containing One N Atom in the Ring

16.1.1 With Aziridine 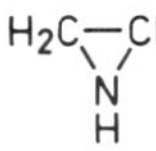 (= Ethylenimine = C_2H_5N)

$Mn(C_2H_5N)_5I_2$ was prepared by adding aziridine (1.5 g) dropwise to an ice-cold aqueous solution (10 mL) of manganese(II) iodide (0.6 g). A crystalline precipitate formed on rubbing the walls of the vessel with a rod. It was collected and redissolved in water at room temperature. The solution was filtered, and the complex was recrystallized by adding 1 g of aziridine and cooling in ice. The crystals were washed first with cold aqueous aziridine, then with a little cold water, and dried in vacuo over KOH. The complex melts at 132 to 133°C with decomposition and turns brownish in the air due to oxidation. It is soluble in water, slightly soluble in methanol, and insoluble in ether or acetone. Aqueous solutions are decomposed by boiling, by prolonged exposition to air, or by mineral acids.

$Mn(C_2H_5N)_4Cl_2$ was obtained by adding aziridine (2 g) to a methanol solution (5 mL) of manganese(II) chloride (1 g) cooled in ice. The white crystalline precipitate which forms on rubbing was collected, washed with cold methanol, and redissolved in room-temperature methanol. The solution was filtered and cooled in ice, aziridine (1 g) was added, and then ether was added to complete the crystallization. The crystals were washed with a 50:50 mixture of methanol and ether containing a few drops of aziridine, then with ether, and dried in vacuo over KOH. The compound decomposes at 135°C. It is very soluble in water and methanol, soluble in ethanol and acetone, and insoluble in ether.

Reference:

T. B. Jackson, J. O. Edwards (J. Am. Chem. Soc. **83** [1961] 355/60, 356).

16.1.2 With 2-Pyrrolidinone 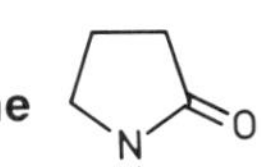 (= 4-Butanelactam = C_4H_7NO)

$Mn(C_4H_7NO)_nX_2$ ($n=6$ for $X=ClO_4$; $n=1$, 2, 3 for $X=Cl$). For the preparation of $Mn(C_4H_7NO)_6(ClO_4)_2$, the hydrate $Mn(ClO_4)_2 \cdot 6H_2O$ was dehydrated by stirring with 2,2-dimethoxypropane (50% excess) for 2 h at room temperature, and then 2-pyrrolidinone was added to the resulting solution. Upon mixing with an equal volume of anhydrous ether, colorless crystals precipitated. $Mn(C_4H_7NO)_3Cl_2$ crystallized as the trihydrate when a hot (100°C) saturated solution of $MnCl_2 \cdot 4H_2O$ in 2-pyrrolidinone was cooled to room temperature. The anhydrous compound formed on vacuum drying of the crystals [2]. To prepare $Mn(C_4H_7NO)_2Cl_2$, a solution of $MnCl_2$ and 2-pyrrolidinone (1:2 mole ratio) in ethanol was evaporated and the resulting pasty mass was vacuum-desiccated over anhydrone until the mass solidified [2]. According to [1], $MnCl_2 \cdot 4H_2O$ was dehydrated by stirring with 2,2-dimethoxypropane (sixfold excess) for 2 h at room temperature. Anhydrous ethanol was added early during the dehydration to dissolve the $MnCl_2 \cdot 4H_2O$ completely. Then 2-pyrrolidinone was added with stirring. Addition of anhydrous ether (excess) gave an oil, which on further stirring for 30 min yielded beige $Mn(C_4H_7NO)_2Cl_2$. The compound is soluble in H_2O, methanol, and ethanol, slightly soluble in acetone, and insoluble in $CHCl_3$ [1]. $Mn(C_4H_7NO)Cl_2$ was prepared much like $Mn(C_4H_7NO)_2Cl_2$, but on addition of anhydrous ether it formed immediately as a light pink precipitate [1].

The enthalpy changes for the formation reactions $MnCl_2(cryst) + nC_4H_7NO(cryst) \rightarrow Mn(C_4H_7NO)_nCl_2(cryst)$ at 25°C, $\Delta H = -20.7 \pm 1$ kcal/mol for $Mn(C_4H_7NO)_3Cl_2$ (n=3) and -16.8 ± 0.1 kcal/mol for $Mn(C_4H_7NO)_2Cl_2$ (n=2), were calculated from the enthalpies of solution (see below) [3].

The X-ray diffraction patterns of the compounds $M(C_4H_7NO)_6(ClO_4)_2$ with M = Mg, Ca, Mn, Fe, Co, Ni, Cu, Zn, and Cd and the 1-methyl-2-pyrrolidinone compounds with M = Co, Ni, and Cd are similar and suggest that these compounds are isomorphous [1].

The following table lists melting points (in °C), magnetic moments (in μ_B) [1], and IR maxima (in cm^{-1}) [2]:

compound	m.p.	μ_{eff}	ν(NH)	ν(CO)	ν(CN)	ν(Mn-O)	ν(Mn-Cl)
$Mn(C_4H_7NO)_6(ClO_4)_2$	156	5.95	—	—	—	—	—
$Mn(C_4H_7NO)_3Cl_2$	—	—	3242	1657, 1646	1289	500	280, 268
$Mn(C_4H_7NO)_2Cl_2$	185	5.94	3275	1667	1287	504	280, 266
$Mn(C_4H_7NO)Cl_2$	280(dec.)	5.77	—	—	—	—	—

The magnetic moments of the compounds are those of 6-coordinate Mn^{II} species. The electronic spectra of the complexes, which were recorded for solid mulls and various solutions (band positions and assignments given in tables) are very similar to each other and are similar to those observed for other octahedrally coordinated Mn^{II} species. The chloro complexes exhibit a bright fluorescence when activated by long- or short-wave UV energy. Their spectra show no triplets in the 400 to 500 nm region, and the band at 500 nm is quite broad and difficult to locate [1].

The IR spectra of $Mn(C_4H_7NO)_3Cl_2$ and $Mn(C_4H_7NO)_2Cl_2$ are very similar to each other. The shifts of the ν(CO) and ν(CN) ligand bands ($\Delta\nu(CO) \approx -100$ cm^{-1}, $\Delta\nu(CN) \approx +30$ cm^{-1}) and the presence of a band at about 500 cm^{-1} indicate the formation of a $CO \rightarrow Mn$ coordination bond [2], also see [1]. The shift of the ν(NH) bands ($\Delta\nu(NH) \approx -200$ cm^{-1}) and δ(NH) bands may be due to bridging hydrogen bonds between the NH groups of the ligand and the Cl^- ions. In $Mn(C_4H_7NO)_3Cl_2$, all 2-pyrrolidinone molecules must be equally coordinated as no ν(CO) or ν(NH) free-ligand bands occur. The nearly identical positions of the ν(Mn-Cl) bands in $Mn(C_4H_7NO)_3Cl_2$ and $Mn(C_4H_7NO)_2Cl_2$ indicate the same coordination number in both complexes [2].

The molar conductivity of 0.002 M $Mn(C_4H_7NO)_6(ClO_4)_2$ in nitromethane is $\Lambda = 143$ $\Omega^{-1} \cdot cm^2 \cdot mol^{-1}$ at 25°C, appropriate for a 1:2 electrolyte. $Mn(C_4H_7NO)_2Cl_2$ and $Mn(C_4H_7NO)Cl_2$, which are insoluble in nitromethane, are probably nonelectrolytes [1].

All the compounds are stable towards air, light, and heat up to 120°C, but hydrolyze readily in water [1]. For $Mn(C_4H_7NO)_3Cl_2$ and $Mn(C_4H_7NO)_2Cl_2$ calorimetric measurements in aqueous 2N HCl at 25°C yield the solution enthalpies $\Delta H_s = -2.13 \pm 0.02$ and -3.71 ± 0.04 kcal/mol (at dilutions of 26500 and 25200 mol H_2O/mol complex) [3].

References:

[1] S. K. Madan, J. A. Sturr (J. Inorg. Nucl. Chem. **29** [1967] 1669/83, 1672, 1675/6, 1679, 1681). – [2] M. S. Barvinok, L. G. Lukina (Zh. Neorgan. Khim. **24** [1979] 2429/34; Russ. J.

Inorg. Chem. **24** [1979] 1347/50). – [3] M. S. Barvinok, L. G. Lukina (Zh. Neorgan. Khim. **26** [1981] 253/5; Russ. J. Inorg. Chem. **26** [1981] 135/7).

16.1.3 With Derivatives of 2-Pyrrolidinone

16.1.3.1 With 1-Methyl-2-pyrrolidinone (= N-Methyl-4-butanelactam = C_5H_9NO)

Mn^{II} solutions in 1-methyl-2-pyrrolidinone have been studied by ESR line widths and proton spin relaxation measurements, both as a function of temperature. Relaxation mechanisms are discussed and rate parameters given. High-temperature ESR data suggest a marked distortion of the octahedral symmetry of the solvated Mn^{II} ion [1].

$Mn(C_5H_9NO)_6(ClO_4)_2$ and **$Mn(C_5H_9NO)Cl_2$.** The two compounds are similar to the 2-pyrrolidinone compounds, both in preparation and properties [1 to 3], except that they are hygroscopic [1, 3]. According to a recent paper, $Mn(ClO_4)_2 \cdot 6H_2O$ is stirred for 20 h with 2,2-dimethoxypropane (100% excess), and the ligand added in a 15:1 mole ratio to the manganese solution to yield $Mn(C_5H_9NO)_6ClO_4$ [1].

For $Mn(C_5H_9NO)_6(ClO_4)_2$, $\mu_{eff} = 5.95\ \mu_B$ was found at 302 K. Its solution in $CHCl_3$ shows the ν(CO) band at 1645 cm^{-1}. The molar electrical conductivity of a 0.00215 M solution in nitromethane is $\Lambda = 167\ \Omega^{-1} \cdot cm^2 \cdot mol^{-1}$ at 25°C [2]. $Mn(C_5H_9NO)Cl_2$ is pink and melts at 240°C with decomposition; $\mu_{eff} = 5.85\ \mu_B$ at 297 K. The electronic spectra of a solid mull and several solutions have been recorded [3].

References:

[1] Tzeng-ming Chen, L. O. Morgan (J. Phys. Chem. **76** [1972] 1973/81, 1974). – [2] R. J. Niedzielski, G. Znider (Can. J. Chem. **43** [1965] 2618/21). – [3] S. K. Madan, J. A. Sturr (J. Inorg. Nucl. Chem. **29** [1967] 1669/83, 1675/6, 1679).

16.1.3.2 With 1-Acetyl-2-pyrrolidinone (= N-Acetyl-4-butanelactam = $C_6H_9NO_2$)

$Mn(C_6H_9NO_2)_3(SbCl_6)_2$ can be prepared by adding the ligand (0.03 mol) to a solution of $Mn(CH_3NO_2)_6(SbCl_6)_2$ (0.01 mol) in nitromethane (20 mL) or adding nitromethane to a mixture of anhydrous $MnCl_2$ (0.01 mol) and $SbCl_5 \cdot CH_3NO_2$ (0.02 mol) and stirring the mixture for 24 h at room temperature. Unreacted solid was filtered off, and the solution was concentrated by evaporation in vacuo and allowed to crystallize. The colorless solid is washed in a dry atmosphere with n-pentane, which has been dried with sodium. The compound melts at 197°C with decomposition. The IR spectrum (recorded for Nujol mulls in the 4000 to 180 cm^{-1} region) shows absorption bands at 1734 cm^{-1} assigned to carbonyl stretching, at 346 cm^{-1} assigned to the anion $SbCl_6^-$ (octahedral), and at 270 cm^{-1} tentatively assigned to Mn-O (ligand) stretching vibrations. Other ligand bands (strongest at 1311 cm^{-1}) are given in the paper. The IR spectrum suggests isomorphism with the related complexes of Mg, Zn, Co, and Ni, which are octahedrally coordinated with 1-acetyl-2-pyrrolidinone, acting as an oxygen-bonded bidentate ligand, W. L. Driessen, P. L. A. Everstijn (J. Coord. Chem. **10** [1980] 155/8).

16.1.3.3 With Polymers and Copolymers of 1-Vinyl-2-pyrrolidinone (= N-Vinylpyrrolidone = C_6H_9NO)

$[MnCl_2 \cdot (C_6H_9NO)_2]_n$. A dilute solution of the polymeric ligand in a mixture of CH_2Cl_2 and acetone is added to a saturated solution of $MnCl_2$ in ethanol [1], also see [2]. The brown compound [1] appears amorphous in the X-ray diffraction pattern. Its IR absorption spectrum shows bands at ν = 1655, 1304, and 284 cm^{-1}, which are assigned to the CO, CN, and Mn-Cl stretching modes, respectively; Δν(CO) is −17 cm^{-1} and Δν(CN) is +19 cm^{-1}. The manganese atom is probably four-coordinate, to two Cl ions and the oxygen atoms of two pyrrolidone units [2]. The compound is not reduced by H_2 at 170°C [1]. It is soluble without decomposition in various alcohols, in 1,2-dichloroethane, and in CH_2Cl_2-acetone mixtures. Both the manganese complex and the analogous cobalt complex show marked catalytic activity [1].

Complexes formed by $MnCl_2$ and various copolymers of N-vinylpyrrolidinone with lupinine acrylate (5 to 30%) have been studied by potentiometry, spectrophotometry, and viscosimetry. The stability constants of the manganese complexes increase with lupinine acrylate content. The compounds exhibit physiological activity [3].

References:

[1] H. G. Biedermann, W. Graf, F. Steininger (Chemiker-Ztg. **100** [1976] 141/2). – [2] S. Sh. Rashidova, M. D. Inoyatov (Uzbeksk. Khim. Zh. **1977** No. 3, pp. 32/40, 35/7; C.A. **87** [1977] No. 102711). – [3] R. V. Reshetnikova, A. G. Muftakhov, N. V. Kapitsa (Nauchn. Tr. Tashkent. Gos. Univ. No. 502 [1976] 157/60; C.A. **87** [1977] No. 168580).

16.1.4 With 1-[4-(1-Methylethoxy)phenyl]-2,5-pyrrolidinedione

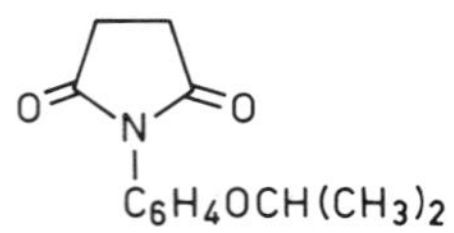

(= N-(4-Isopropoxyphenyl)succinimide = $C_{13}H_{15}NO_3$)

$Mn(C_{13}H_{15}NO_3)Cl_2 \cdot H_2O$ was formed when a saturated solution of $MnCl_2$ in absolute ethanol is added slowly (15 to 20 min) to an equimolar quantity of the ligand dissolved in ethanol. The compound is a colorless to pinkish powder, which is stable under ordinary conditions. On heating, it decomposes, first losing the H_2O and then the organic ligand. The IR spectrum (no frequencies given) suggests a possible coordination to Mn via the lone electron pair of the N atom, L. E. Pogosyan, S. N. Avakyan (Mater. 4th Resp. Soveshch. Neorgan. Khim., Yerevan 1975 [1976], pp. 40/1; C.A. **88** [1978] No. 31358).

16.1.5 With Derivatives of 2,2′-Dipyrromethene

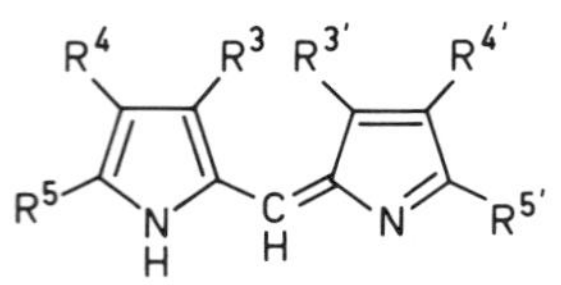

The reaction of manganese(II) chloride with the deprotonated 5,5′-unsubstituted dipyrromethenes L_I^- and L_{II}^- (see p. 81) results in the formation of N-coordinated Mn^{III} tris chelates [1], while the same salt reacts with the deprotonated 5,5′-substituted ligands L_{III}^- and L_{IV}^- to give 4-coordinate Mn^{II} bis chelates [1, 4].

HL_I: $R^3 = R^{3'} = R^4 = CH_3$, $R^{4'} = R^5 = R^{5'} = H$: 3,3′,4-Trimethyl-2,2′-dipyrromethene (= $C_{12}H_{14}N_2$)

HL_{II}: $R^3 = R^{3'} = R^4 = R^{4'} = CH_3$, $R^5 = R^{5'} = H$: 3,3′,4,4′-Tetramethyl-2,2′-dipyrromethene (= $C_{13}H_{16}N_2$)

HL_{III}: $R^3 = R^{3'} = R^5 = R^{5'} = CH_3$, $R^4 = R^{4'} = H$: 3,3′,5,5′-Tetramethyl-2,2′-dipyrromethene (= $C_{13}H_{16}N_2$)

HL_{IV}: $R^3 = R^{3'} = R^4 = R^{4'} = Cl$, $R^5 = R^{5'} = COOC_2H_5$: 3,3′,4,4′-Tetrachloro-5,5′-bis(ethoxycarbonyl)-2,2′-dipyrromethene (= $C_{15}H_{12}Cl_4N_2O_4$)

$Mn(C_{12}H_{13}N_2)_3$ and $Mn(C_{13}H_{15}N_2)_3$. The tris chelates formed as crystalline precipitates upon dropwise addition of dilute aqueous ammonia to suspensions of $HL_I \cdot HCl$ or $HL_{II} \cdot HCl$ and $MnCl_2 \cdot 4H_2O$ in methanol. The products have a green luster. For $Mn(C_{12}H_{13}N_2)_3$, osmometry in $CHCl_3$ gives a molecular weight of 610 (610.7 calculated) [1]. The molar magnetic moment, $\mu_{eff} = 4.82\ \mu_B$ at 293.7 K, corresponds to a high-spin Mn^{III} complex (d^4) [1, 2]. The electronic absorption spectrum shows three ligand-field bands in the 9000 to 15000 cm^{-1} range, with molar extinction coefficients between 80 and 370 $L \cdot mol^{-1} \cdot cm^{-1}$. The spectrum can be interpreted on the basis of effectively D_{4h} symmetry, with an axial elongation as a consequence of a Jahn-Teller effect [1]. In the IR spectrum (Nujol) strong bands appear at 1610 and 1617 cm^{-1}. These bands, due to the skeletal stretching mode of the ligand, are shifted only slightly relative to the free ligand [1].

$Mn(C_{13}H_{15}N_2)_2$. The compound precipitated upon dropwise addition of concentrated ammonia to a mixture of an aqueous solution of $MnCl_2 \cdot 4H_2O$ and an ethanolic solution of HL_{III}. The bis chelate forms a fine crystalline solid with a green luster [1]. The magnetic moment at 297.0 K, $\mu_{eff} = 5.87\ \mu_B$, corresponds to a high-spin Mn^{II} complex (d^5) [1, 2]. The electronic absorption spectrum taken for a solution in $CHCl_3$ and the diffuse reflectance spectrum show three ligand-field bands, at 13800, 14900, and 16100 cm^{-1}, with $\varepsilon = 165$, 168, and ≈ 97(sh) $L \cdot mol^{-1} \cdot cm^{-1}$, respectively. The bands are assigned on the basis of T_d symmetry. The exceedingly high intensity of the observed spin-forbidden bands is explained by the highly covalent nature of the coordination bonds [1]. The K-band ESR spectra for solid samples yielded the anisotropy parameters $D = 0.12$ and $E = 0.032\ cm^{-1}$, values considerably larger than expected for a distorted Mn^{II} complex formed with four equivalent donor atoms. The reasons for the distortion from the regular tetrahedral to planar geometry are discussed [3].

In the IR spectrum (in Nujol) the skeletal stretching band of L_{III} is observed at 1598 cm^{-1}. The shift on coordination amounts to $-15\ cm^{-1}$. The ν(Mn-N) band appears at 379 cm^{-1} [1].

$Mn(C_{15}H_{11}Cl_4N_2O_4)_2$. A solution of $Ca(C_{15}H_{11}Cl_4N_2O_4)_2$ in THF was treated with manganese(II) acetate tetrahydrate to give light green crystals of $Mn(C_{15}H_{11}Cl_4N_2O_4)_2$. The compound is dichroic, appearing red when viewed against bright light. It melts to a red liquid, which cools to a green solid, m.p. 218 to 220°C. At temperatures above the melting point, sublimation is observed. It is very sensitive to moisture. In the ^1H NMR spectrum the large paramagnetic shift of the methine protons ($\Delta\tau = 5.1$) confirms the paramagnetic nature of the complex. In the electronic absorption spectrum maxima are observed at 540 nm (18520 cm^{-1}), 530 nm (18870 cm^{-1}), and 450 nm (22220 cm^{-1}). The 540-nm band corresponds to $\pi \rightarrow \pi^*$, the other two are charge-transfer transitions. The IR absorption spectrum (CCl_4 solution) shows a single ν(CO) band at 1695 cm^{-1} and two ν(C=C) bands at 1605 and 1488 cm^{-1}. The shift of the ν(CO) band to lower frequencies (free-ligand bands at 1730 and 1715 cm^{-1}) indicates an electrostatic interaction of the CO group with the Mn atom. The spectra lead to the conclusion that the complex is nearly a regular tetrahedron, like the Ca and Zn complexes, but unlike the Ni and Cu compounds, which are distorted tetrahedrons [4].

References:

[1] Y. Murakami, K. Sakata, K. Harada, Y. Matsuda (Bull. Chem. Soc. Japan **47** [1974] 3021/4). – [2] Y. Murakami, Y. Matsuda, K. Sakata (Proc. 16th Intern. Conf. Coord. Chem., Dublin 1974, Abstr. R 61, pp. 1/3; C.A. **85** [1976] No. 40267). – [3] Y. Matsuda, Y. Murakami (Bull. Chem. Soc. Japan **50** [1977] 2321/4). – [4] R. J. Motekaitis, A. E. Martell (Inorg. Chem. **9** [1970] 1832/9, 1834).

16.1.6 With Piperidine 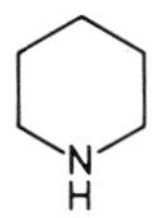 (= $C_5H_{11}N$)

A manganese(II) complex with 2-(3-pyridyl)-piperidine (= anabasine) is described on p. 144. Complexes with other ligands containing piperidine rings (codeine and berberine) are described on pp. 270/7.

$Mn(C_5H_{11}N)Cl_2$. The compound was obtained on mixing solutions of anhydrous $MnCl_2$ and piperidine (slight excess) in absolute ethanol as a dark brown precipitate [1, 2]. Susceptibility measurements at 20 to 21°C yield the effective magnetic moment $\mu_{eff} = 5.92$ μ_B, which indicates a sextet ground state for the Mn atom ($3d^5$) [1]. The compound is stable on exposure to the atmosphere and does not melt sharply as it probably decomposes. It is decomposed (hydrolyzed) by boiling water, mineral acids, or soda lime (with liberation of the piperidine base) [2]. The mixed-ligand complex with acetylacetone [$Mn acac_2(C_5H_{11}N)_2$] is treated in "Mangan" D 1, 1979, p. 78.

References:

[1] R. L. Mital (Indian J. Chem. **3** [1965] 88). – [2] S. Prasad, K. P. Kacker (J. Indian Chem. Soc. **35** [1958] 719/21).

16.1.7 With Pyridine  (= C_5H_5N = py)

Mixed complexes with pyridine and ketones, or diketones are described in "Mangan" D 1, 1979, pp. 57, 61/2, 70, 78/9, 105, 110, 114, 116/7, 120, 121, 123/4, 131, 132. For mixed complexes with pyridine and 8-hydroxyquinoline or methyl-8-hydroxyquinolines, see pp. 171, 175, 181.

16.1.7.1 Complexes in Solution

Potentiometric pH investigations on aqueous solutions of a manganese(II) salt and pyridine indicate the presence of four complexes, $[Mnpy_n]^{2+}$ with n = 1 to 4. The stability constants (K_n in L/mol) in the following table were obtained by measurements with a glass electrode at 25°C [1, 2 to 4] for I = 1.0 ($NaClO_4$) [1] and I=0.5 ([pyH]NO_3) with several concentrations of manganese(II) salt and pyridine [2 to 4]:

n	1	1	2	2	3	4
log K_n	1.86	0.14	1.59	−0.5	0.90	0.60
−ΔG in kcal/mol	2.54	—	2.17	—	1.22	0.82
ΔS in cal·mol^{-1}·K^{-1}	0.5	—	−1.5	—	−3.9	−5.3
Ref.	[1]	[2 to 4]	[1]	[2 to 4]	[1]	[1]

An average enthalpy of formation, ΔH = −2.39 kcal/mol, was determined calorimetrically [1]. Further K_1 and K_2 values from potentiometric measurements are given in [5]. For stability and rate constants estimated from ESR line widths and intensity measurements in aqueous solutions of manganese(II) salt and pyridine, see [1, 11]. The reaction between pyridine and $MnCl_2$ (0.5 M) dissolved in D-sorbitol was investigated by gas-chromatographic measurements. The stability constant K_1 = 2.0 L/mol was estimated from the gas-liquid distribution of pyridine with and without $MnCl_2$ in the liquid phase [6].

An equilibrium [Mnpy$_2$X$_2$] + 2py ⇌ [Mnpy$_4$X$_2$] (X = Cl, Br) in saturated solutions of MnX_2 in mixtures of pyridine and various noncoordinating diluents has been assumed, the four-coordinate complex being pseudotetrahedral and the six-coordinate complex being pseudooctahedral [7 to 10]. The equilibrium constant for X = Cl in pyridine-chlorobenzene mixtures was calculated from the solubility of $MnCl_2$ in these solvent mixtures to be $K = (2.7 \pm 0.6) \times 10^3$ at 20°C and $K = (2.0 \pm 0.3) \times 10^3$ at 30°C [8]. For the equilibrium [Mnpy$_4$Br$_2$] ⇌ [Mnpy$_2$Br$_2$] + 2py in pyridine-chlorobenzene mixtures K is $(5.5 \pm 3.3) \times 10^{-4}$ at 20°C and $(5.7 \pm 4.2) \times 10^{-4}$ at 30°C [10]. The activity coefficients of the complexes [Mnpy$_4$X$_2$] for several mole fractions of pyridine in the solvent were calculated from the solubilities: for the complexes with X = Cl in benzene [7], chlorobenzene [7, 8], o-dichlorobenzene [7], ethyl acetate, and n-heptane diluents [7, 9] and for the complexes with X = Br in chlorobenzene diluent [10]. The activity coefficients of the complexes [Mpy$_4$Cl$_2$] (including M = Mn) are independent on the central metal ion for a particular mixed-solvent composition. Graphs of activity coefficients and solubility isotherms as a function of the mole fraction of pyridine are shown in [7 to 10].

References:

[1] G. Atkinson, J. E. Bauman (Inorg. Chem. **2** [1963] 64/7). – [2] J. Bjerrum (Chem. Rev. **46** [1950] 381/401, 394). – [3] L. G. Sillén, A. E. Martell (Stability Constants of Metal-Ion Complexes, London 1964, p. 441). – [4] J. Bjerrum (Acta Chem. Scand. **27** [1973] 970/6, 972), A. E. Martell, R. M. Smith (Critical Stability Constants, Vol. 2, New York 1973, p. 165). – [5] P. S. Relan, P. K. Bhattacharya (J. Indian Chem. Soc. **46** [1969] 534/6).

[6] M. Leddet, D. Cuzin, F. Coussemant (Bull. Soc. Chim. France **1977** 866/72, 869). – [7] W. Libuś, M. Kluczkowski, A. Kłonowksi, W. Nierzwicki (J. Chem. Soc. Faraday Trans. I **70** [1974] 1057/66, 1061). – [8] W. Libuś, I. Uruska (Inorg. Chem. **5** [1966] 256/64, 257, 260). – [9] W. Libuś, A. Kłonowski, W. Nierzwicki (Bull. Acad. Polon. Sci. Ser. Sci. Chim. **21** [1973] 297/305, 300; C.A. **79** [1973] No. 70658). – [10] W. Libuś, M. Kluczkowski (J. Chem. Soc. Faraday Trans. I **73** [1977] 1870/81, 1871, 1876).

[11] B. R. McGarvey (J. Phys. Chem. **61** [1957] 1232/7, 1235, 1237).

16.1.7.2 Isolated Manganese(II) Complexes

16.1.7.2.1 Hexakis(pyridine) Compounds

Mnpy$_6$Br$_2$ separated when a hot solution of $MnBr_2$ (anhydrous [3] or hydrated [2]) in pyridine was cooled [2, 3] after refluxing [1]. The colorless [1, 2] or slightly green, lustrous [3] crystals were washed with ether [1] or recrystallized from pyridine and stored wet [2] as they easily lose pyridine [2, 3]. The crystals are quadrilateral, rhombohedral tablets [3], which on heating to 140°C [1], or even at room temperature [2], lose four molecules of pyridine to give $Mnpy_2Br_2$ [1, 2]. Susceptibility measurements at 18°C yield the magnetic moment $\mu_{eff} = 5.96\ \mu_B$ [4]. The solubility of $Mnpy_6Br_2$ in pyridine is 0.003 mol/kg at 20°C and 0.007 mol/kg at 30°C (from a figure). The solubility increases to about 0.3 mol/kg on addition of 77 mol% chlorobenzene. The heat of dissolution in pyridine is (65 ± 1) kJ/mol [2].

Mnpy$_6$(NCO)$_2$. Freshly prepared powdered $Mnpy_4(NCO)_2$ (p. 86) was treated with anhydrous pyridine added in small portions with stirring and gentle warming until all solids were dissolved. On cooling, well formed colorless crystals precipitated. They were washed with anhydrous pyridine and dried in vacuo over P_2O_5 and pyridine. $Mnpy_6(NCO)_2$ easily loses pyridine, turning turbid. It is readily soluble in pyridine and aqueous pyridine but less soluble in water. It is hydrolyzed by warm water [5].

Mnpy$_6$[SnBr$_6$] precipitated when 2 g of $Mn[SnBr_6]\cdot 6H_2O$ was stirred with 15 mL of anhydrous pyridine. The precipitate was triturated several times with pyridine, then with ether. The compound was dried over P_2O_5 and pyridine. The colorless complex is very hygroscopic, and it deliquesces in air. On storing, it gradually turns yellow. It is completely hydrolyzed by water on heating and is also decomposed by methanol, 95% ethanol, chloroform, acetone, and mineral acids. $Mnpy_6[SnBr_6]$ is insoluble in benzene and slightly soluble in amyl alcohol [6].

References:

[1] J. R. Allan, D. H. Brown, R. H. Nuttall, D. W. A. Sharp (J. Inorg. Nucl. Chem. **27** [1965] 1865/7). – [2] W. Libuś, M. Kluczkowksi (J. Chem. Soc. Faraday Trans. I **73** [1977] 1870/81, 1870). – [3] H. Grossmann (Ber. Deut. Chem. Ges. **37** [1904] 559/69, 564). – [4] R. W. Asmussen (Proc. Symp. Coord. Chem., Copenhagen 1953 [1954], pp. 27/30; C.A. **1955** 15470). – [5] R. Ripan (Bul. Soc. Stiinte Cluj **3** [1926/27] 176/92, 183; C. **1927** II 2388).

[6] G. Spacu, J. Dick (Bul. Soc. Stiinte Cluj **4** [1928/29] 84/103, 98; C.A. **1928** 3105).

16.1.7.2.2 Tetrakis(pyridine) Compounds

The tetrakis complexes $Mnpy_4X_2$ have a distorted octahedral structure with both anions coordinated to the manganese atom as monodentate ligands. This has been shown by the IR spectra, magnetic moments, and in some cases also by the electronic spectra. The IR data indicate a trans-configuration for the compounds with $X = NCO^-$, NCS^-, $NCSe^-$, and ReO_4^-. On heating, the tetrakis complexes lose pyridine to give the bis complexes, $Mnpy_2X_2$. The thermal decomposition has been investigated thoroughly for $Mnpy_4(NCS)_2$ in connection with its suitability for the gravimetric determination of manganese.

Halides

Mnpy$_4$X$_2$ (X = Cl, Br, I). $Mnpy_4Cl_2$ was prepared by precipitation from a saturated aqueous solution of $MnCl_2$ with pyridine [1], by the action of excess pyridine on solid

$Mnpy_2Cl_2$ [2], by refluxing $MnCl_2$ in pyridine [3], or by triturating $MnCl_2 \cdot 4H_2O$ (1 g) with pyridine (20 mL) until it was discolored [4]. $Mnpy_4Br_2$ was prepared by refluxing $MnBr_2$ in pyridine [3]. $Mnpy_4I_2$ was obtained by reacting MnI_2 and pyridine in ethanol [5].

$Mnpy_4Cl_2$ and $Mnpy_4Br_2$ form colorless crystals [2 to 4]. $Mnpy_4Cl_2$ is unstable in air [1, 2, 4], losing two molecules of pyridine [1, 4]. $Mnpy_4I_2$ forms bright brown crystals, which melt at 152°C. Its magnetic moment was determined from susceptibility measurements to be $\mu_{eff} = 6.60$ μ_B. Its molar electrical conductivity ($\Lambda = 78.61$ $\Omega^{-1} \cdot cm^2 \cdot mol^{-1}$ in acetone solution) indicates a nonelectrolyte [5]. The IR spectrum of $Mnpy_4I_2$ (Nujol or hexachlorobutadiene mulls) shows the low-lying pyridine ring vibration bands (ν_{6a} and ν_{16b}) at 631 and 423 cm^{-1} (pyridine, 601 and 403 cm^{-1}). The following complex bands show shifts up to +25 cm^{-1} relative to those of the free ligand: 1592 (ν_{8a}), 1170 (ν_{15}), 1004 (ν_1), and 764 (ν_4). A new band appeared at 1237 cm^{-1} [11]. It is assumed that the complexes $Mnpy_4X_2$ (X = Cl, Br, I) have octahedral symmetry [2, 3, 5].

On heating, the compounds $Mnpy_4X_2$ lose two molecules of pyridine to give $Mnpy_2X_2$. The temperatures of the maximum rate of decomposition are 120°C for the chloro complex [3] and 115°C for the bromo complex [6] as shown by TGA [3, 6], DTA, and DTG [6].

$Mnpy_4Cl_2$ dissolves readily in water [4]. Its solubility in pyridine is $S = (70.7 \pm 0.5) \times 10^{-3}$ mol/L at 20°C and $(84.5 \pm 0.5) \times 10^{-3}$ mol/L at 30°C [2]. $Mnpy_4Cl_2$ is the stable solid phase in equilibrium with a saturated solution of $MnCl_2$ in pyridine below 42.2°C, while $Mnpy_2Cl_2$ is the stable phase above this temperature. Plots of log S versus 1/T are given for the two compounds. The differential heats of dissolution are 16.2 ± 0.5 kJ/mol for $Mnpy_4Cl_2$ and -13.5 ± 0.4 kJ/mol for $Mnpy_2Cl_2$. With these the thermodynamic quantities for the reaction $Mnpy_4Cl_2(solid) \rightleftharpoons Mnpy_2Cl_2(solid) + 2\,pyridine(solution)$ were calculated: $\Delta H = 29.7 \pm 0.9$ kJ, $\Delta G^\circ_{303} = 1.15 \pm 0.06$ kJ, and $\Delta S = 94 \pm 3$ J/K [8]. $Mnpy_4Cl_2$ is also a stable solid phase in the three-component systems of $MnCl_2$ and pyridine with benzene [8], chlorobenzene [2, 8], o-dichlorobenzene [8], ethyl acetate, n-heptane [9], chloroform, or 1,2-dichloroethane [10] in the temperature range 20 to 30°C. The equilibrium between $Mnpy_4Cl_2$ and $Mnpy_2Cl_2$ depends on the pyridine content and the temperature of the special solvent [2, 8 to 10]. $Mnpy_4I_2$ is soluble in acetone [5].

$Mnpy_4Cl_2 \cdot KCl$. The compound was obtained by triturating finely powdered $KMnCl_3 \cdot 2H_2O$ with anhydrous pyridine until the pinkish tinge disappears. In moist air the colorless crystals easily lose all the pyridine, which is replaced by water, forming pink $KMnCl_3 \cdot 2H_2O$ again. $Mnpy_4Cl_2 \cdot KCl$ dissolves in water and acids with instant hydrolysis. It is insoluble in other usual solvents [2].

References:

[1] D. Fătu, E. Segal (Rev. Roumaine Chim. **14** [1969] 709/15). – [2] W. Libuś, I. Uruska (Inorg. Chem. **5** [1966] 256/64, 257). – [3] J. R. Allan, D. H. Brown, R. H. Nuttall, D. W. A. Sharp (J. Inorg. Nucl. Chem. **27** [1965] 1865/7). – [4] G. Spacu, L. Caton (Bul. Soc. Stiinte Cluj **2** [1924/25] 332/53, 352; C. **1925** I 2435). – [5] K. C. Dash, D. V. Ramana Rao (Indian J. Chem. **5** [1967] 333/4).

[6] G. Liptay, K. Burger, E. Mocsári-Fülöp, I. Porubszky (J. Therm. Anal. **2** [1970] 25/36, 30; Magy. Kem. Folyoirat **77** [1971] 84/90, 85; C.A. **74** [1971] No. 150550). – [7] G. Spacu, L. Caton (Bul. Soc. Stiinte Cluj **2** [1924/25] 29/51, 33, 41; C. **1924** I 1650). – [8] W. Libuś, M. Kluczkowski, A. Kłonowski, W. Nierzwicki (J. Chem. Soc. Faraday Trans. I **70** [1974] 1057/66, 1061). – [9] W. Libuś, A. Kłonowski, W. Nierzwicki (Bull. Acad. Polon. Sci. Ser. Sci. Chim. **21** [1973] 297/305, 300; C.A. **79** [1973] No. 70658). – [10] W. Libuś, B. Janaszewski, R. Pastewski, H. Twardowska (J. Solution Chem. **9** [1980] 737/52, 741, 747).

[11] N. S. Gill, R. H. Nuttall, D. E. Scaife, D. W. A. Sharp (J. Inorg. Nucl. Chem. **18** [1961] 79/87, 82/3).

Perchlorate and Tetrafluoroborate

Mnpy$_4$(ClO$_4$)$_2$ and **Mnpy$_4$(BF$_4$)$_2$** were prepared by evaporating a solution of the hydrated manganese salts and excess pyridine in absolute ethanol, redissolving the residues in absolute ethanol, and precipitating the product with anhydrous diethyl ether [2]. The perchlorate is formed also on storing $Mn(ClO_4)_2 \cdot 5py \cdot 3H_2O$ (see p. 107) for several weeks in a vacuum desiccator over H_2SO_4 [1]. The complexes are colorless and moisture sensitive. The electronic reflectance spectra of both complexes show bands at 22700 and 25000 cm^{-1}, which are assigned to transitions from the ground state to the $^4T_{2g}(G)$ and $^4E_g(G)$ states. The spectra suggest the manganese ion to be in an octahedral environment. This is confirmed by the magnetic moments, $\mu_{eff} = 5.85$ μ_B for $X = ClO_4$ and $\mu_{eff} = 5.68$ μ_B for $X = BF_4$. In the IR spectra (Nujol and hexachlorobutadiene mulls) the bands of the pyridine ring deformation modes appear at 609, 414 cm^{-1} for the perchlorate and 609, 410 cm^{-1} for the tetrafluoroborate (free ligand, at 600, 405 cm^{-1}). The Mn-N bands are observed at 222 and 233 cm^{-1} for the perchlorate and 220 cm^{-1} for the tetrafluoroborate. Shifts and splittings of the free anion bands indicate that the anions have a C_{3v} symmetry and are coordinated as unidentate ligands [2]. On exposure to air $Mnpy_4(ClO_4)_2$ quickly turns dark, losing pyridine. It is readily soluble in water and alcohol but only slightly soluble in pyridine [1].

References:

[1] R. Weinland, K. Effinger, V. Beck (Arch. Pharm. **265** [1927] 352/79, 371). – [2] D. H. Brown, R. H. Nuttall, J. McAvoy, D. W. A. Sharp (J. Chem. Soc. A **1966** 892/6, 894).

Isocyanate, Isothiocyanate, and Isoselenocyanate

Mnpy$_4$(NCO)$_2$ was prepared by reaction of an aqueous solution of hydrated manganese(II) nitrate with an aqueous solution of potassium cyanate in the ratio 1:2 and dropwise addition of the stoichiometric amount of pyridine in aqueous ethanol [2]. Alternatively, a solution of 4 g KOCN and 4 mL of pyridine in 30 mL of water was added to a solution of 4 g $MnCl_2 \cdot 4H_2O$ in 20 mL of water. The crystalline precipitate [2] was washed with water containing pyridine [35] and ether [1] or ethanol [2].

Mnpy$_4$(NCS)$_2$ was precipitated by adding pyridine to a solution containing Mn^{2+} and NCS^- ions [4, 10, 38], by adding a solution of NH_4SCN to a solution of a manganese(II) salt containing pyridine and subsequently cooling the mixture in an ice bath [13], or by reacting stoichiometric quantities of $Mn(NCS)_2$ and pyridine in ethanol [5]. The compound crystallized in colorless, glittering, monoclinic prisms [4, 10].

To prepare **Mnpy$_4$(NCSe)$_2$** a solution of KSeCN was added to a mixture of a solution of pyridine (slight excess) in aqueous ethanol and an aqueous solution of hydrated $Mn(NO_3)_2$ [32, 34]. Alternatively, a manganese(II) salt and KSeCN were reacted in aqueous [31] or ethanolic medium, and pyridine was added after the precipitated potassium salt was filtered off. The complex crystallized on shaking or after refluxing the mixture for 30 min and allowing it to stand for six to seven days. The solids obtained were washed with ethanol and petroleum ether [30].

In the table the most important physical properties are listed: IR frequencies in cm^{-1}, the magnetic moment μ_{eff} in μ_B, and the molar electrical conductivity in $\Omega^{-1} \cdot cm^2 \cdot mol^{-1}$:

compound	ν(CN)	ν(CO), ν(CS), or ν(CSe)	μ_{eff}	Λ	Ref.
$Mn\,py_4(NCO)_2$	2190	1352	5.56	6.0	[2]
$Mn\,py_4(NCS)_2$	2066	798	6.04 [5]	22.92 [5]	[14]
	2061	800	5.98	—	[16]
$Mn\,py_4(NCSe)_2$	2070	—	—	—	[34]
	2061	—	5.60	7.3	[33]

The melting points reported, >250°C for $Mn\,py_4(NCO)_2$ [2], 256°C for $Mn\,py_4(NCS)_2$ [5], and 250°C for $Mn\,py_4(NCSe)_2$ [30], are probably those of the decomposition products, see below.

$Mn\,py_4(NCS)_2$ is isomorphous with the analogous Fe, Co, and Ni compounds as shown by the X-ray powder diagrams [14]. The electronic reflectance spectrum exhibits three absorption bands, at ≈27000, 22900, and ≈16300 cm^{-1} [16]. The solutions in $CHCl_3$ show an absorption maximum at $\lambda=300$ nm (≈33300 cm^{-1}) [17]. In the IR spectrum of $Mn\,py_4(NCS)_2$ (Nujol mulls) the bands of the low-lying pyridine ring vibrations (C-C in-plane bend and C-C out-of-plane bend [12]) are observed at 414, 420 cm^{-1} (doublet), and 625 cm^{-1} (liquid pyridine at 405 and 603 cm^{-1}) [14]. Similar data are given by [12] and [15]. Further prominent bands of free pyridine, at 704, 748, 1029, and 1214 cm^{-1}, occur as doublets in the spectrum of the complex. A new, weak, but well-defined band is present at 1230 cm^{-1} [14]. The ν(Mn-NCS) band (ν_3) appears at about 256 cm^{-1} [12, 14, 15], and the ν($Mn\text{-}N_{py}$) band (ν_5) appears at about 173 cm^{-1} [12, 15]. Another band at 188 [15], or 195 cm^{-1} [12], is assigned to ν($Mn\text{-}N_{py}$) coupled with ν(Mn-NCS) (ν_4) [15]. In the IR spectrum of $Mn\,py_4(NCSe)_2$ the ν(CSe) bands apparently are masked by the absorption bands of the coordinated pyridine. Wave numbers of the maxima appearing in the range 4000 to 400 cm^{-1} (without assignment of the pyridine bands) and discussion, especially of the bands of the NCSe group, are given in [33].

The IR spectra indicate the anions to be trans in the complex of symmetry D_{4h} and to be bonded to the manganese atom via the N atoms. The compounds are monomeric. The low molar conductivities show them to be nonelectrolytes [2, 5, 33].

On heating, the compounds decompose: $Mn\,py_4X_2 \rightarrow Mn\,py_2X_2 + 2\,py$. The temperatures of decomposition determined by thermal analyses are recorded in the following table:

compound	temperature range of decomposition	peak temperature in °C (method)	heating rate	medium	Ref.
$Mn\,py_4(NCO)_2$	50 to 125°C	115 (TG, DTA, DTG)	3°/min	air	[1, 3]
$Mn\,py_4(NCS)_2$	—	125 (DTA, DTG)	3°/min	air	[7, 21] a)
	77 to 162°C	152 (DEA) b)	16°/min	N_2	[6]
	80 to 180°C	160 (DTA)	~5°/min	air	[19]
	—	162 (DTA, DTG)	4°/min	air	[9]
	—	162 (TG)	3°/min	Ar	[20]
$Mn\,py_4(NCSe)_2$	110 to 180°C	170 (TG, DTA, DTG)	3°/min	air	[3, 34]

a) Also see [22]. – b) Differential enthalpimetric analysis.

$Mn py_4(NCS)$ starts losing pyridine above 40°C [13], also see [18, 20]. But complete conversion to $Mn py_2(NCS)_2$ is attained only at much higher temperatures (see the table). For the reaction $Mn py_4(NCS)_2(cryst) \rightarrow Mn py_2(NCS)_2(cryst) + 2py(g)$, the enthalpy $\Delta H = 31.0 \pm 0.6$ kcal/mol was derived from DEA data [6]. The activation energy $E_A = 113.3$ kJ/mol and entropy $\Delta S = -34.42\ J \cdot mol^{-1} \cdot K^{-1}$ were extracted from the DTG and DTA data with the Coats-Redfern relation ("reaction order" n = 1) [9]. The E_A value is in good agreement with the value of 114.9 kJ/mol obtained by the Mampel equation on the basis of a random nucleation as the rate-controlling process [23]. $E_A \approx 16$ kcal/mol ($\triangleq$67 kJ/mol) was calculated from differential enthalpimetric data (heating rate 16 K/min), $n \approx 0.5$ [24]. $E_A = 14.28$ kcal/mol ($\triangleq$59.78 kJ/mol) is the result from DTA and TG data (heating rate 10 K/min) and the Coats-Redfern method, the reaction rate being k = 2920 g/s. The Freeman-Carroll method yields $E_A = 14.18$ kcal/mol, k = 15000 g/s, and n = 0 [25]. The thermal stability of $Mn py_4(NCS)_2$ is lower than that of the compounds of Co, Ni, and Cu [21] or Fe, Ni, Zn, and Co [24], also see [22], but higher than that of $Mn py_4Cl_2$ [6].

$Mn py_4(NCO)_2$ is soluble in pyridine and pyridine-water mixtures, but only slightly soluble in water [35]. $Mn py_4(NCO)_2$ reacts with pyridine on warming to give $Mn py_6(NCO)_2$, p. 84 [35].

$Mn py_4(NCS)_2$ is almost insoluble in water containing NCS^- ions and pyridine. Therefore it is suitable for the gravimetric determination of manganese [10]. (For the application of $Mn py_4(NCS)_2$ in titrimetric determinations, see [8, 36]; for spectrophotometric determinations, see [11, 17].) $Mn py_4(NCS)_2$ is soluble in ethanol, aqueous ethanol [18], acetone [5], chloroform [13, 17, 26], and toluene [37]. The solubility in aqueous ethanol at 25°C is given as S in g/100 g solution; it depends on the ethanol content (in vol%) of the solvent (10-h equilibration time):

% ethanol	10	20	40	60	80	90	99.5
S	1.66	2.02	3.48	5.54	7.15	6.50	4.20

The maximum solubility is near 80 vol% ethanol. The solubility is higher than that of the Co, Ni, and Cd compounds [18]. $CHCl_3$ dissolves 42.20 g of $Mn py_4(NCS)_2$ per liter at room temperature. From almost neutral aqueous 0.7 M NH_4SCN, it may be extracted quantitatively by $CHCl_3$; the optimum extraction occurs for solutions 0.15 M in NH_4SCN and 0.85 M in pyridine (distribution coefficient D = 1.5) [26]. The manganese complex that can be extracted by toluene from aqueous solutions containing Mn^{2+} ions, pyridine, and thiocyanate (composition not given) exhibits a more positive reduction potential than the aquo complex during polarographic experiments [37]. The pyridine ligands in $Mn py_4(NCS)_2$ are phenylated on reaction with N-nitroso-N,N'-diphenylurea (as a phenyl source) in dimethylformamide. About equal amounts of the three isomeric phenylpyridine complexes are formed [29].

$Mn py_4(NCS)_2$ reacts with an ethanolic solution of benzidine ($C_{12}H_{12}N_2$) to give the complex $Mn(C_{12}H_{12}N_2)_3py(NCS)_2$ [28]. In liquid ammonia, $Mn py_4(NCS)_2$ reacts with potassium acetylide to give $Mn(NH_3)_4(C_2H)_2$, see p. 17. Excess K_2CH yields $K_2Mn(C_2H)_4$ [27]. $Mn py_4(NCSe)_2$ is soluble in water, pyridine [31], and in most of the usual organic solvents [30]. It is decomposed by hot water and by acids [31].

References:

[1] I. Porubszky, A. Németh, P. Horvath, G. Liptay (Periodica Polytech. [Budapest] **16** [1972] 381/94, 383, 386, 391; C.A. **78** [1973] No. 131551). – [2] A. Das, D. V. Ramana Rao (Current Sci. India **42** [1973] 56/7). – [3] I. Porubszky, A. Németh, G. Liptay (Therm. Anal. Proc. 3rd Intern. Conf., Davos, Switz., 1971 [1972], Vol. 2, pp. 679/88, 687; C.A. **78** [1973]

No. 167951). – [4] H. Grossmann (Ber. Deut. Chem. Ges. **37** [1904] 559/69, 563/4). – [5] K. C. Das, D. V. Ramana Rao (Indian J. Chem. **5** [1967] 333/4).

[6] G. Beech, G. B. Kauffman (Thermochim. Acta **1** [1970] 93/7, 95). – [7] L. Erdey, G. Liptay (Periodica Politech. [Budapest] **7** [1963] 223/36, 227/9; C. A. **61** [1964] 2694). – [8] P. Indrasenan, C. G. R. Nair (Current Sci. [India] **43** [1974] 683/4). – [9] C. G. R. Nair, P. Indrasenan (Indian J. Chem. A **14** [1976] 402/5). – [10] G. Spacu, J. Dick (Z. Anal. Chem. **74** [1928] 188/91).

[11] K. Eröss-Kiss, E. Pungor (Periodica Polytech. [Budapest] **18** [1974] 25/32, 27/8; C. A. **81** [1974] No. 85553); K. Eröss-Kiss, E. Pungor, B. Schoket (Periodica Polytech. [Budapest] **20** [1976] 139/44; C. A. **86** [1977] No. 37274). – [12] C. W. Frank, L. B. Rogers (Inorg. Chem. **5** [1966] 615/22, 616/9). – [13] G. B. Kauffman, R. A. Albers, F. L. Harlan (Inorg. Syn. **12** [1970] 251/6, 253). – [14] R. J. H. Clark, C. S. Williams (Spectrochim. Acta **22** [1966] 1081/90, 1084, 1089). – [15] C. Engelter, D. A. Thornton (J. Mol. Struct. **42** [1977] 51/8).

[16] G. Liptay, K. Burger, E. Papp-Molnár, Sz. Szebeni (J. Inorg. Nucl. Chem. **31** [1969] 2359/66, 2363/5). – [17] G. H. Ayres, S. S. Baird (Talanta **7** [1960/61] 237/47, 244). – [18] A. Dinescu, I. Strugaru, A. M. Lupu, G. Teodorescu (Bul. Inst. Politeh. Gheorghe Gheorgiu-Dej Bucuresti Ser. Chim. Metal. **39** No. 4 [1977] 55/63, 59; C. A. **89** [1978] No. 118516). – [19] W. W. Wendlandt, S. Iftikhar Ali (Z. Anorg. Allgem. Chem. **337** [1965] 6/13, 10). – [20] T. A. Mal'kova (Issled. Khim. Khelatnykh Soedin. **1971** 37/42; C. A. **77** [1972] No. 83120).

[21] G. Liptay, K. Burger, E. Mocsári-Fülöp, I. Porubszky (J. Therm. Anal. **2** [1970] 25/36, 30; Magy. Kem. Folyoirat **77** [1971] 84/90, 85; C. A. **74** [1971] No. 150550). – [22] G. Liptay, E. Papp-Molnár, K. Burger (J. Inorg. Nucl. Chem. **31** [1969] 247/52, 248, 251). – [23] C. G. R. Nair, P. M. Madhusudanan (Thermochim. Acta **14** [1976] 373/82, 375/7). – [24] G. B. Kauffman, G. Beech (Thermochim. Acta **1** [1970] 99/102, 101). – [25] I. G. Murgulescu, D. Fătu, E. Segal (J. Therm. Anal. **1** [1969] 97/106, 102), E. Segal, S. Vasile (An. Univ. Bucuresti Ser. Stiint. Nat. Chim. **15** [1966] 17/23; C. A. **69** [1968] No. 90180).

[26] A. T. Rane (J. Radioanal. Chem. **8** [1971] 117/22). – [27] R. Nast, H. Griesshammer (Z. Anorg. Allgem. Chem. **293** [1957] 322/6, 324). – [28] G. Spacu, T. I. Pirtea (Rev. Univ. C. I. Parhon Politeh. Bucuresti Ser. Stiint. Nat. **1** [1952] 104/14, 105, 112; C. A. **1956** 16518). – [29] R. J. Gritter, A. W. Godfrey (J. Am. Chem. Soc. **86** [1964] 4724/5). – [30] A. K. Das, D. V. Ramana Rao (Indian J. Chem. **12** [1974] 898/9).

[31] G. Spacu, R. Ripan (Bul. Soc. Stiinte Cluj **4** [1928/29] 3/22, 11/2; C. A. **1928** 3104). – [32] S. M. Nelson, T. M. Shepherd (Inorg. Chem. **4** [1965] 813/7, 814). – [33] Yu. Ya. Kharitonov, V. V. Skopenko (Zh. Neorgan. Khim. **10** [1965] 1803/15; Russ. J. Inorg. Chem. **10** [1965] 984/90, 986, 988). – [34] I. Porubszky, A. Németh, G. Liptay (Periodica Polytech. [Budapest] **17** [1973] 365/71, 367/8; C. A. **80** [1974] No. 140736). – [35] R. Ripan (Bul. Soc. Stiinte Cluj **3** [1926] 176/92; C. **1927** II 2388).

[36] K. S. Bhatki, A. T. Rane (Proc. Indian Acad. Sci. A **58** [1963] 177/81; C. A. **60** [1964] 4784). – [37] B. K. Afghan, R. M. Dagnall (Talanta **13** [1966] 1097/103, 1101). – [38] A. I. Vogel (A Text-Book of Quantitative Inorganic Analysis, London 1961, p. 132).

Trihalogenoacetates

$Mnpy_4(CX_3COO)_2$ (X = F, Cl) were obtained by adding a slight excess of pyridine to a solution of $Mn(CX_3COO)_2$ in dry diethyl ether while warming. They were recrystallized from chloroform. The UV reflectance spectra show a strong charge-transfer band at about 25000 cm^{-1}. The following IR data (Nujol) (ν_{as} = antisymmetric, ν_s = symmetric stretching mode in cm^{-1}) and magnetic moments at 293 K have been recorded:

compound	$\nu_{as}(COO)$	$\nu_s(COO)$	$\Delta(\nu_{as}-\nu_s)$	$\nu(Mn\text{-}N_{py})$	μ_{eff}
$Mnpy_4(CF_3COO)_2$	1722, 1710	1420	290*)	204	5.94
$Mnpy_4(CCl_3COO)_2$	1690	1315	375	216	5.92

*) In original paper: 270 cm^{-1}.

Reference:

S. Amasa, D. H. Brown, A. Sharp (J. Chem. Soc. A **1969** 2892/4).

Other Tetrakis(pyridine) Compounds

$Mnpy_4[C(CN)_2NO]_2$ was prepared by reacting 0.01 mol $Mn(NO_3)_2 \cdot 6H_2O$ and 0.02 mol $K[C(CN)_2NO]$ with 4 mol of pyridine in 50 mL of water. The brick red complex will crystallize after standing several hours. Its magnetic moment, μ_{eff} = 6.00 μ_B, indicates a high-spin complex. The electronic spectrum of a solution in acetone exhibits a maximum at 21050 cm^{-1} (ε = 830 $L \cdot mol^{-1} \cdot cm^{-1}$). This may be an intraligand band of the anion $[C(CN)_2NO]^-$ (corresponding band for the free anion, 24600 cm^{-1}). It is assumed that the anion $[C(CN)_2NO]^-$ is coordinated to Mn through the N atom of its nitrosyl group as suggested by the IR spectra of the Co, Ni, and Cu compounds [1].

$Mnpy_4[SbCl_6]_2$ was obtained by treating a suspension of anhydrous $MnCl_2$ in CCl_4 containing pyridine with a solution of $SbCl_5$ in CCl_4 (dilution ratio 1:10) with exclusion of moisture. X-ray studies show the cream-pink solid to be crystalline and to be isostructural with the Ni, Cd, and Ca compounds. $Mnpy_4[SbCl_6]_2$ is only slightly soluble in H_2O [2].

$Mnpy_4[HgCl_4]$. To a solution of $Mn[HgCl_4] \cdot 4H_2O$ in 5 mL of absolute alcohol, 10 to 30 drops of anhydrous pyridine and 10 mL of absolute alcohol were added. After ten minutes the colorless precipitate was sucked off and dried over P_2O_5. The air-stable compound forms thin leaflets, which are readily soluble in water and pyridine at room temperature or in boiling alcohol and mineral acids and insoluble in acetone and benzene [3].

$Mnpy_4Cr_2O_7$. The compound precipitated from a solution containing $MnSO_4$ and dichromate on addition of the stoichiometric quantity (or an excess) of pyridine [5]. Preparation by addition of a solution of hydrated $MnSO_4$ to a solution of CrO_3 and pyridine was reported by [4]. $Mnpy_4Cr_2O_7$ forms dark maroon, glittering, minute crystals [5], which smell strongly of pyridine and decompose on heating. The compound is only slightly soluble in water. It is insoluble in organic solvents such as alcohol, ether, benzene, and pyridine [4]. The aqueous solutions decompose rapidly [5].

$Mnpy_4[ReO_4]_2$ was obtained by precipitating it from a solution of $Mn[ReO_4]_2$ in a mixture of acetone and 2,2-dimethoxypropane (for dehydration) with excess pyridine. The IR spectrum recorded in the 1000 to 200 cm^{-1} range exhibits bands of the ReO_4^- ion and at 212 cm^{-1} the Mn-N stretching band. The magnetic moment (μ_{eff} = 6.02 μ_B at 21 to 23°C) and the IR data indicate a trans-octahedral structure [7].

References:

[1] H. Köhler, B. Seifert (Z. Anorg. Allgem. Chem. **379** [1970] 1/8, 3, 5, 8). – [2] E. Petrovici, I. Lupu (Rev. Roumaine Chim. **22** [1977] 719/24, 721). – [3] G. Spacu,

L. Caton (Bul. Soc. Stiinte Cluj **2** [1924/25] 332/53, 345; C.A. **1925** 1995/6). – [4] S. H. C. Briggs (Z. Anorg. Allgem. Chem. **56** [1908] 246/60, 255, 257). – [5] N. Parravano, A. Pasta (Gazz. Chim. Ital. **37** II [1907] 252/64, 260).

16.1.7.2.3 Tris(pyridine) Compounds

$R_4N[MnBrCl_2py_3]$ ($R=CH_3$, C_2H_5). The manganese(II) compounds were obtained by refluxing solutions of $[(CH_3)_4N]_2[MnBr_2Cl_2]$ in methanol or $[(C_2H_5)_4N]_2[MnBr_2Cl_2]$ in ethanol with the required amount of pyridine for 1 h. The light pink, crystalline compounds do not melt or decompose up to about 250°C. The effective magnetic moments are $\mu_{eff}=5.95$ for the tetramethylammonium salt and 5.89 μ_B for the tetraethyl compound. The molar electrical conductivities of 0.001M solutions in acetone, $\Lambda=90$ and 84 $\Omega^{-1}\cdot cm^2\cdot mol^{-1}$, respectively, are those of 1:1 electrolytes. The electronic spectrum of solutions of $(C_2H_5)_4N[MnBrCl_2py_3]$ in acetonitrile exhibits four absorption maxima, at 25000, 22370, 18520, and 17240 cm^{-1}, thus indicating an octahedral structure. Only significant IR absorption bands are reported for the region 5000 to 650 cm^{-1}. The shifts of the ν(CH) bands of pyridine at ~3000 cm^{-1} towards lower frequencies and of the ν(C⩴C) and ν(C⩴N) bands of pyridine in the 1600 to 1500 cm^{-1} region towards higher frequencies suggest a ligand to metal coordination through the N atom. The compounds are sparingly soluble in organic solvents [1].

$[py_2H][MnBr_3py_3]$. Metallic manganese flakes were dissolved in a solution of HBr in dry ether, excluding air. The solution was evaporated in vacuo until a solid of composition $HMnBr_3\cdot(C_2H_5)_2O$ formed. This solid was reacted with an excess of anhydrous pyridine, excluding moisture, in an exothermic reaction. The product which formed after 10 d in vacuo over H_2SO_4 has the composition $HMnBr_3\cdot5py$ (see "Mangan" C 5, 1978, p. 288) and melts at 112°C. The electronic spectrum of the molten complex exhibits absorption maxima at 208, 246, 252, 258, and 264 nm, which in ethanolic solution shift to somewhat longer wavelengths. The ^1H NMR spectrum suggests a weak py-H bonding. The IR spectrum shows an absorption band at 3200 cm^{-1}, which is assigned to ν(NH) vibrations. Bands at 1540, 1327, 1295, and 1250 cm^{-1} are due to the pyridinium ion. Maxima in the ranges 755 to 744 and 675 to 671 cm^{-1} indicate that pyridine is directly coordinated to manganese. Bands in the range between 287 and <200 cm^{-1} are assigned to ν(Mn-N) vibrations and those at 124 and 118 cm^{-1} to ν(Mn-Br) vibrations. The physical properties suggest the constitution $[py_2H][MnBr_3py_3]$, the anion $[MnBr_3py_3]^-$ probably having octahedral structure. The compound is strongly fluorescent in laser light. On further heating, the molten complex solidifies at 121°C, an indication of a chemical transition. Up to 126°C the complex loses 1.4% of its pyridine content, while above 300°C some 18% of the pyridine is released. If left in a vacuum desiccator for 5 months, the compound melts at 168°C [2].

References:

[1] A. K. Das, D. V. Ramana Rao (J. Indian Chem. Soc. **48** [1971] 823/7). – [2] E. Lehmann, J. Kouinis, A. Galinos (Monatsh. Chem. **106** [1975] 499/502).

16.1.7.2.4 Bis(pyridine) Compounds

Nitrates

$Mnpy_2(NO_3)_2\cdot2H_2O$ was obtained by adding pyridine to a mixture of concentrated $Mn(NO_3)_2$ [1, 2] and excess saturated NH_4NO_3 [2]. Crystallization in a desiccator afforded colorless crystals [2] or pinkish plates [1]. In the IR spectrum (Nujol) absorption bands at

627 and 417 cm^{-1} are due to pyridine ring vibrations (liquid pyridine, 602 and 404 cm^{-1}). Far-IR bands at 246 and 214 cm^{-1} are assigned to the $\nu(Mn\text{-}ONO_2)$ and $\nu(Mn\text{-}N_{py})$ frequencies, respectively. Strong bands in the 3600 to 3000 and 1670 to 1600 cm^{-1} regions indicate the presence of water of crystallization [3].

When heated quickly, the compound decomposes at 170°C with inflammation; heating it slowly drives off H_2O beginning at about 65°C. Pyridine is given off at about 115°C and NO_2 at about 160°C [2]. The DTA investigation of the (presumably anhydrous) compound in streaming N_2 (heating rate 56 K/min) showed two endothermic peaks, at 135 and 219°C [4]. $Mn\,py_2(NO_3)_2 \cdot 2H_2O$ is readily soluble in water [1, 2], slightly soluble in alcohol, and insoluble in ether or benzene [2]. The molar electrical conductivity of the aqueous solutions at 20°C increases from 231.6 $\Omega^{-1} \cdot cm^2 \cdot mol^{-1}$ at a dilution of 1:251.4 to 258.9 $\Omega^{-1} \cdot cm^2 \cdot mol^{-1}$ at 1:4022.4, thus indicating a 1:2 electrolyte [2].

References:

[1] H. Grossmann (Ber. Deut. Chem. Ges. **37** [1904] 1253/7, 1255). – [2] O. E. Zvyagintsev, O. Yu. Mamulaishvili, M. Chkoniya (Izv. Akad. Nauk SSSR Otd. Mat. Estestv. Nauk **1937** 1255/9; C.A. **1938** 2862). – [3] C. W. Frank, L. B. Rogers (Inorg. Chem. **5** [1966] 615/22, 616/9). – [4] P. B. Bowman, L. B. Rogers (J. Inorg. Nucl. Chem. **28** [1966] 2215/24, 2219/20).

Halides

$Mn\,py_2X_2$ (X = Cl, Br, I). The compounds with X = Cl and Br can be prepared by the following methods. (1) The manganese(II) halides and pyridine in stoichiometric amounts are reacted in ethanolic solution [12, 14, 15]. Similar preparations use hot ethanolic pyridine [7, 36], hot $MnCl_2$ solutions [36], absolute ethanol [4, 10, 13], and anhydrous manganese(II) chloride [10]. The first preparation of $Mn\,py_2Br_2$ by this method was reported by [19]. (2) Stoichiometric quantities of pyridine are added to a concentrated aqueous solution of hydrated $MnCl_2$ [6, 9] or $MnBr_2$ [8]. In this case, care must be taken to avoid oxidation of the solution [6]. For the formation of a contaminated product by this method, see [4, 11]. A pure product is said to result upon adding pyridine after saturating the $MnCl_2$ solution with NH_4Cl [5]. (3) $Mn\,py_2Cl_2$ and $Mn\,py_2Br_2$ are crystallized from solutions of anhydrous MnX_2 in dry pyridine [2]. An early preparation involved triturating $MnCl_2$ with excess pyridine [4]. For preparation by reacting $MnCl_2$ with pyridine in slight excess at 250°C in a sealed tube, see [3]. (4) Thermal degradation of $Mn\,py_4Cl_2$ [41] at 120°C or $Mn\,py_6Br_2$ at 140°C [1] is described on pp. 85 and 84.

$Mn\,py_2Cl_2$ [3, 6, 20] and $Mn\,py_2Br_2$ [8, 19] form pink needle-shaped monoclinic crystals. From X-ray investigations of single crystals the space group of $Mn\,py_2Cl_2$ was determined to be $P2/c\text{-}C^4_{2h}$ (No. 13). Lattice constants: a = 19.508, b = 3.738, c = 17.543 Å; β = 116.64° [6]. A previous work with single crystals resulted in $P2_1/n\text{-}C^5_{2h}$ (No. 14) and a = 17.40, b = 8.75, c = 3.76 Å, β = 91°; Z = 2 [3]. The first investigation yielded c = 3.73 Å [20]. A table of d values is given in [3]. The structure consists of polymeric chains as shown in **Fig. 10,** with $>Mn{<}^{Cl}_{Cl}{>}Mn<$ groups along the chains. Each Mn^{2+} ion is surrounded by an octahedron formed of four Cl atoms and two pyridine molecules in trans positions. The octahedral axis N-Mn-N is normal to the $\text{-}Mn\text{-}Cl_2\text{-}Mn\text{-}$ chains. The Mn-Mn distances are about 3.7 Å along and >7.5 Å between the chains [6]. For bond distances and angles see Fig. 10.

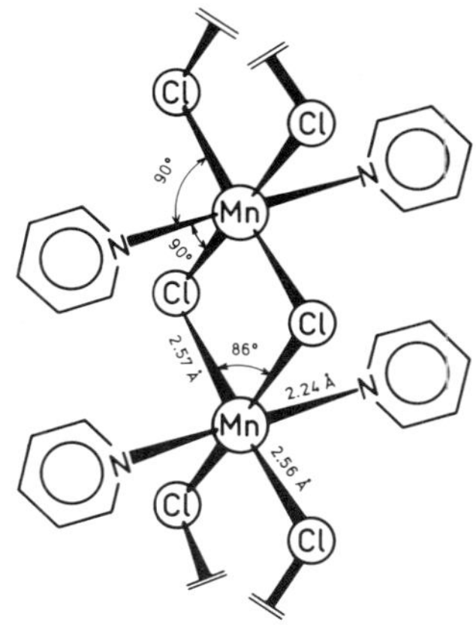

Fig. 10

Schematic representation of the linear chains in $Mnpy_2Cl_2$ with bond lengths and angles at the Mn atom.

Thermal analysis of $Mnpy_2Cl_2$ by differential scanning calorimetry in the 130 to 300 K temperature range reveals a polymorphic transition at 245 K with a hysteresis of 4 K and a transition enthalpy of $\Delta H = 78$ cal/mol, indicating a first-order reaction [50], whereas a transition at 230 K is revealed by Mössbauer spectroscopy (see p. 94) [36]. It is assumed that this transition involves a molecular displacement of the $Mnpy_2Cl_4$ octahedra along the polymeric linear chains [50].

The polymeric chain structure of $Mnpy_2Cl_2$ (and $Mnpy_2Br_2$) is also apparent from the far-IR spectrum [24], electronic spectrum [1, 11, 34], and magnetic susceptibility [6, 12, 14, 26]. Debye-Scherrer diagrams show that the compounds Mpy_2Cl_2 with M = Mn, Fe, Co (α-form), Ni [3, 12], Ca, Cd, Hg, Cu [3], and $Mnpy_2Br_2$ [12] belong to a series of similar monoclinic structures. According to recent high-resolution X-ray powder patterns, $Mnpy_2Cl_2$ is isomorphous with α-$Copy_2Cl_2$ [22], as stated earlier [20], but $Nipy_2Cl_2$ and $Cdpy_2Cl_2$ have different patterns. However, $Mnpy_2Br_2$ is isomorphous with $Nipy_2Br_2$ and $Cdpy_2Br_2$ [22]. The measured density is 1.63 g/cm³, and the calculated is 1.65 g/cm³ [3].

Experimental heat capacity (C_p) data for $Mnpy_2Cl_2$ and $Mnpy_2Br_2$ between 1.3 and 80 K are shown in **Fig. 11a** and **11b**, p. 94. The broad maxima in the curves are indicative of transitions to a short-range ordered state as expected for linear-chain exchange interactions. The superimposed sharp peaks are attributed to transitions to the antiferromagnetic state due to weak inter-chain interactions. However, the two maxima could be separated only for $Mnpy_2Cl_2$, for which the temperature for the transition to the short-range ordering was determined to be $T_{tr} = 2.25 \pm 0.05$ K. The Néel temperatures were found to be $T_N = 1.380 \pm 0.005$ K for $Mnpy_2Cl_2$ and 2.055 ± 0.005 K for $Mnpy_2Br_2$. The corresponding magnetic energies, $\Delta E(T_{tr})$ and $\Delta E(T_N)$, and entropies, $\Delta S(T_{tr})$ and $\Delta S(T_N)$ (for $Mnpy_2Br_2$ only an estimated T_{tr}) are given below:

compound	$\Delta E(T_{tr})$ in J/mol	$\Delta E(T_N)$ in J/mol	$\Delta S(T_{tr})$ in $J \cdot mol^{-1} \cdot K^{-1}$	$\Delta S(T_N)$ in $J \cdot mol^{-1} \cdot K^{-1}$
$Mnpy_2Cl_2$	73	2.7	15.0 ± 0.2	2.4
$Mnpy_2Br_2$	72	5.8	14.5 ± 0.3	3.1

The isotropic Heisenberg linear-chain model with uniaxial single-ion anisotropy for $S = 5/2$ gives the best fit with the intra-chain parameter $J/k = -0.59$ K for X = Cl and $J/k = -0.53$ K for X = Br [15].

Fig. 11

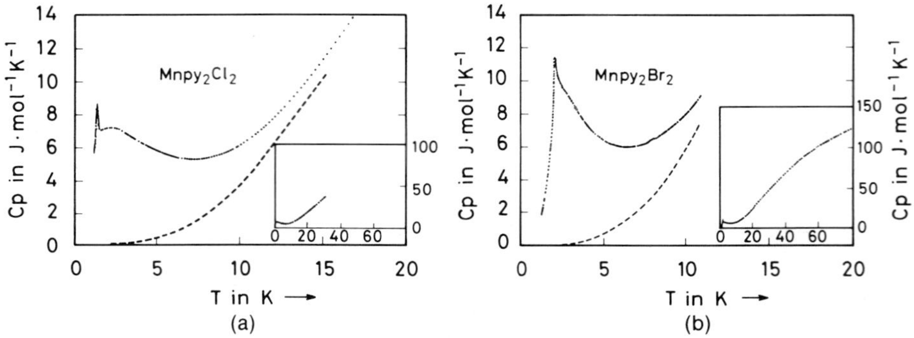

Heat capacity (C_p) of $Mnpy_2Cl_2$ (a) and $Mnpy_2Br_2$ (b) and estimated lattice contribution (dashed lines).

The molar magnetic susceptibility χ_{mol} of $Mnpy_2Cl_2$ measured by the Faraday method at 296 K and corrected for diamagnetism is $\chi_{mol} = 14750 \times 10^{-6}$ cm³/mol, corresponding to the magnetic moment $\mu_{eff} = 5.92$ μ_B. The magnetic moment μ_{eff} is almost independent of temperature, being 5.89, 5.88, 5.87 μ_B at T = 261, 172, and 83 K, respectively [37]. For older but similar values of μ_{eff}, see [10, 11, 26]. Measurements on $Mnpy_2Br_2$ at 18°C yielded $\mu_{eff} = 5.96$ μ_B [11], also see [38]. Magnetic susceptibility measurements on $Mnpy_2Cl_2$ and $Mnpy_2Br_2$ between 2 and 80 K show that both compounds obey the Curie-Weiss law in the high-temperature region with a negative paramagnetic Curie temperature Θ_p. At low temperatures the compounds show antiferromagnetic behavior. The experimental results are interpreted on the basis of the antiferromagnetic Heisenberg linear-chain model with spin $S = ^5/_2$. The following table lists the maximum susceptibility χ_{max} (in cm³/g) at T (χ_{max}), Θ_p in K, and the intra-chain exchange parameters J/k, calculated from the temperature of the susceptibility maximum χ_{max} or from the χ_{max} value itself, either by the scaling method (S) or an interpolation scheme (I) [14]:

compound	$10^2\chi_{max}$	T(χ_{max})	Θ_p	J/k (T at χ_{max})	J/k (χ_{max})
$Mnpy_2Cl_2$	26.3(3)	4.0(2)	−9(1)	−0.49(2)(S) −0.41(2)(I)	−0.58(1)(S) −0.54(1)(I)
$Mnpy_2Br_2$	20.3(2)	4.7(2)	−11(1)	−0.57(2)(S) −0.48(2)(I)	−0.74(1)(S) −0.70(1)(I)

The susceptibility measurements on $Mnpy_2Cl_2$ in the range 4 to 296 K show nearly linear dependence of $1/\chi$ on T. From these measurements, $J/k = -0.59 \pm 0.04$ K was calculated [6]. J/k was evaluated from ESR line-widths measurements at X- and Q-band frequencies from 110 to 296 K to be −0.75(5) at 110 K and −0.43(5) at 300 K for $Mnpy_2Cl_2$ and −0.7(1) at 110 K and −0.50(5) at 300 K for $Mnpy_2Br_2$ [14]. The ratio between inter- and intra-chain coupling has been estimated to be $|J'/J| = 1.3 \times 10^{-2}$ for $Mnpy_2Cl_2$ and 2×10^{-2} for $Mnpy_2Br_2$ [14]. The stronger inter-chain coupling for $Mnpy_2Br_2$ was also suggested by heat-capacity measurements [15].

The Mössbauer spectrum of ^{57}Fe-doped $Mnpy_2Cl_2$ was recorded at various temperatures between 5 and 295 K. Discontinuous changes of the quadrupole splitting value at 5 and 230 K indicate crystallographic phase transitions of the host lattice at these two temperatures [36].

For a study of the X-ray K-absorption spectrum of $Mnpy_2Cl_2$, see [35].

The electronic reflection spectrum of $Mnpy_2Cl_2$ recorded in the 12500 to 45000 cm^{-1} region exhibits bands at the following positions, which are listed with their assignment:

frequency	16670	19050	20400	23700	32800
transition $^6A_{1g}\rightarrow$	$^4T_{1g}$		$^4T_{2g}$	$^4A_{1g}+{}^4E_g$	charge transfer

A shoulder on the long-wave side of the band at 19050 cm^{-1} indicates a tetragonal splitting of the $^4T_{1g}$ state [34]. Similar data in the range 10000 to 27000 cm^{-1} are given in [11]. Three maxima in the range 38000 to 40000 cm^{-1} are assigned to transitions within the coordinated pyridine [34].

The IR spectra of $Mnpy_2Cl_2$ and $Mnpy_2Br_2$ (KBr discs [17], Nujol, or hexachlorobutadiene mulls [27]) show the low-lying pyridine ring vibration bands ν_{6a} and ν_{16b} at 625 and 420 cm^{-1}, respectively (free ligand bands, 601 and 403 cm^{-1}) [27], while ν_{6a} at 610 cm^{-1} (free ligand, 605 cm^{-1}) is reported by [17]. Shifts of +10 to 20 cm^{-1} on coordination were recorded for the following complex bands: 690 (ν_{11}), 1005 (ν_1) [17, 27], 1360 ($\nu_{6a}+\nu_{10b}$) [27], (ν_{14}) [17], 1390 (ν_{14}) [27], ($\nu_{6a}+\nu_{10b}$) [17], and 1600 cm^{-1} (ν_{8a}) [17, 27]. A new band appears at 1237 cm^{-1} [17, 27].

Calculations of the molecular force field of pyridine and its applications to $Mnpy_2Cl_2$ and other metal-pyridine complexes are given in [32]. Far-IR and Raman spectra are listed below:

compound	method	chain modes*) $a_{1g}+b_{2g}$	chain modes*) $b_{1u}+b_{2u}+b_{3u}$	ν_s(Mn-N) a_g	ν_{as}(Mn-N) b_{2u}	Ref. [24]
$Mnpy_2Cl_2$	IR		175s, 160s, 140sh		230m, 214m	[13, 24]
	IR		208**)	121**)	232**)	[32]
	IR		226, 212		167	[25]
	Raman	207m, 137s		160w, 157m		[24]
$Mnpy_2Br_2$	IR		135s, br	175m	212s	[24]
	Raman	140s, 108mw		159vw		[24]

*) The a_g, b_{2g}, b_{1u}, and b_{3u} species are in-plane deformations of the $[MnX_2]_n$ chains [24] and are referred as ν(Mn-X) modes by [13, 30, 32]. The b_{2u} species is an out-of-plane mode [24]. – **) Calculated values.

Far-IR spectra have been recorded from polyethylene [13, 25], wax discs [24], and Nujol mulls [25] in the frequency ranges 450 to 80 cm^{-1} [13], 200 to 20 cm^{-1} [24], and 650 to 150 cm^{-1} [25]. For data at −196°C, see [13, 30]. The spectra of the compounds with deuterated pyridine have been recorded and assigned based on a D_{2h} symmetry [24]. The IR bands observed in the 240 to 210 cm^{-1} region are assigned to ν_{as}(Mn-N) vibrations because they are shifted only slightly (<5 cm^{-1}) on deuteration or substitution of chlorine by bromine [13, 24, 29], and not to ν(Mn-X) vibrations [25]. For previous reports, see [21, 23, 28, 31, 33].

The Urey-Bradley valence force constants K calculated from the frequencies of ν_{as}(Mn-N) = 230 cm^{-1}, ν_s(Mn-N) = 144 cm^{-1}, and ν(Mn-Cl) = 208 cm^{-1} are $K_{Mn\text{-}N}=0.90$ and $K_{Mn\text{-}Cl}=0.87$ mdyn/Å [32].

The molar electrical conductivities of $Mnpy_2Cl_2$ and $Mnpy_2Br_2$ measured in nitrobenzene are $\Lambda = 0.4$ and 0.3 $\Omega^{-1} \cdot cm^2 \cdot mol^{-1}$, respectively, corresponding to those of nonelectrolytes. In water, the compounds behave as 1:2 electrolytes [39].

The chloro and bromo compounds are reported as stable in air [15]. But in an earlier work $Mnpy_2Cl_2$ was described as deliquescent needles [20]. On heating, the compounds lose one molecule of pyridine to give $MnpyX_2$. Thermoanalytic investigations showed the following temperatures of the maximum rate of decomposition (peak temperatures). For $Mnpy_2Cl_2$: 170°C (DTA, 5 K/min) [43], 180°C (DTA, TG, DTG, 3 K/min) [40], 187°C (DTA, 5.6 K/min) [42], 187°C (DSC, 16 K/min) [7], and 195°C (DTG, TG, 10 K/min) [41]. For $Mnpy_2Br_2$: 195°C [40], 210°C (TGA) [1]. For $Mnpy_2I_2$: 120°C [40]. According to [5] $Mnpy_2Cl_2$ starts to lose pyridine at 60°C.

The activation energy of the endothermic reaction $Mnpy_2Cl_2(cryst) \rightarrow MnpyCl_2(cryst) + py(g)$ (I) calculated from DTG and TG data by the Coats-Redfern method (heating rate 2°C/min) is $E_A = 25$ kcal/mol; the Freeman-Carrol method (10°C/min) gives $E_A = 23$ kcal/mol. The preexponential factor of the Arrhenius equation and the reaction order are $A = 1.67 \times 10^7$ $g^{0.5}/s$ and $n = 0.5$ by the first method and $A = 8.13 \times 10^7$ $g^{0.5}/s$ and $n = 0.5$ by the second method. The results show that the process is partially controlled by diffusion [41, 44]. The enthalpy of reaction $\Delta H = 14.7 \pm 0.3$ kcal/mol at 480 ± 4.0 K (210°C) results from DSC measurements [7]. $Mnpy_2Cl_2$ dissolves readily in water [4, 5]. $Mnpy_2Cl_2$ [9, 34, 39] and $Mnpy_2Br_2$ [39] are hydrolyzed completely to form free pyridine, $[Mn(H_2O)_6]^{2+}$, and the halogen ions as is shown by the electronic spectrum [34], the electrical conductivity of the aqueous solutions [9, 39], and cryoscopic measurements [9].

$Mnpy_2Cl_2$ dissolves in methanol with extensive solvolysis [34]. It is only very slightly soluble in ethanol and is insoluble in diethyl ether and benzene. It is decomposed by NH_3 and by warm alkali hydroxide solutions [5]. In aqueous HCl solutions the pyridinium salt $(pyH)_2MnCl_4$ is formed [9, 16]. The catalytic activity of $Mnpy_2Br_2$ in the oxidation of dodecylbenzene and 3,5-diethylbenzene is reported in [51]. The solubility of $Mnpy_2Cl_2$ in pyridine decreases with temperature. The solubility extrapolated from log S vs. 1/T plots (S = solubility) is 0.163 mol/kg at 20°C and 0.135 mol/kg at 30°C. Direct determination at 30°C yielded 0.133 mol/kg. The differential heat of dissolution in the saturated solution is $-13.5(\pm 0.4)$ kJ/mol [46]. The solubility of $Mnpy_2Cl_2$ in chlorobenzene is very low, $(1.2 \pm 0.2) \times 10^{-5}$ mol/L at 20 and 30°C, but increases in the presence of pyridine [45]. The solubility isotherms of $Mnpy_2Cl_2$ have been determined at 20 and 30°C in mixtures of pyridine with benzene [46], chlorobenzene [45], o-dichlorobenzene [46], ethyl acetate, and n-heptane [47] and at 25°C in mixtures with chloroform, dioxane, or toluene [52]. The solubility of $Mnpy_2Br_2$ in mixtures of pyridine and chlorobenzene rises from the origin to about 0.03 mol/kg at 23 mol% pyridine. At higher pyridine contents $Mnpy_6Br_2$ is formed [18]. For the complexes existing in the pyridine-solvent mixtures, see p. 83.

$Mnpy_2Cl_2 \cdot 2H_2O$. The dihydrate crystallized from diluted aqueous solutions of $MnCl_2$ and pyridine as elongated white nonhygroscopic crystals, which lose all H_2O on drying in vacuo and which dissolve readily in water [9].

$M_2[Mnpy_2Cl_4]$. A green compound which was initially formulated as $Mnpy_2Cl_4$ [4, 5] was found in a later work [48] to be the pyridinium salt $(C_5H_5NH)_2[MnCl_4]$, see "Mangan" C 5, 1978, p. 206. More recently, the compound was formulated as the dibasic complex acid $H_2Mnpy_2Cl_4$ [9]. This formulation has not been confirmed.

The pink compound $K_2[Mnpy_2Cl_4]$ was obtained on neutralizing a concentrated aqueous solution of the previous compound or the pyridinium salt $(C_5H_5NH)_2(MnCl_4)$ with concentrated KOH solution. It is fairly soluble in water [9].

The hexadecylpyridinium compound $(C_5H_5NC_{16}H_{33})_2[Mn\,py_2Cl_4]$ was prepared by cautiously heating $(C_5H_5NC_{16}H_{33})_2[MnCl_4]$ with excess pyridine until a clear solution formed. The light yellow complex crystallized on cooling. It melts at 73°C. The IR spectrum (in Nujol and hexachlorobutadiene, 5000 to 650 cm^{-1}, not reported) exhibits a sharp band at 3050 cm^{-1} assigned to the NH stretching vibrations. Two bands, at about 1600 and 1560 cm^{-1}, are due to $\nu(CN)+\nu(CC)$ vibrations of pyridine and suggest coordination of pyridine to Mn via the N atom. The electronic spectrum of solutions in acetone or alcohol shows three absorption bands, at 27000, 22900, and 17070 cm^{-1}, which indicate an octahedral structure for the $[Mn\,py_2Cl_4]^{2-}$ anion. An effective magnetic moment of $\mu_{eff}=5.93$ μ_B indicates a high-spin Mn^{II} complex. The molar electrical conductivity of solutions in nitrobenzene is $\Lambda=66.0$ $\Omega^{-1}\cdot cm^2\cdot mol^{-1}$, corresponding to that of a 2:1 electrolyte [49].

References:

[1] J. R. Allan, D. H. Brown, R. H. Nuttall, D. W. A. Sharp (J. Inorg. Nucl. Chem. **27** [1965] 1865/7). – [2] S. Akyüz, A. B. Dempster, J. E. Davies, K. T. Holmes (J. Chem. Soc. Dalton Trans. **1976** 1746/9). – [3] R. Zannetti, R. Serra (Gazz. Chim. Ital. **90** [1960] 328/36, 334; Structure Reports, Vol. 24, 1960/68, p. 690). – [4] F. Reitzenstein (Z. Anorg. Allgem. Chem. **18** [1898] 253/304, 290/3). – [5] O. E. Zvyagintsev, O. Yu. Mamulaishvili, M. Chkoniya (Izv. Akad. Nauk SSSR Otd. Mat. Estestv. Nauk **1937** 1255/9; C.A. **1938** 2862).

[6] P. M. Richards, R. K. Quinn, B. Morosin (J. Chem. Phys. **59** [1973] 4474/7). – [7] G. Beech, C. T. Mortimer, E. G. Tyler (J. Chem. Soc. A **1967** 1111/5, 1112). – [8] R. J. Meyer, H. Best (Z. Anorg. Allgem. Chem. **22** [1900] 169/91, 182). – [9] W. S. Fyfe (J. Chem. Soc. **1950** 790/3). – [10] R. L. Mital (Indian J. Chem. **3** [1965] 88).

[11] R. W. Asmussen, H. Soling (Acta Chem. Scand. **11** [1957] 1331/9, 1332, 1336). – [12] N. S. Gill, R. S. Nyholm, G. A. Barclay, T. I. Christie, P. J. Pauling (J. Inorg. Nucl. Chem. **18** [1961] 88/97, 96). – [13] M. Goldstein, W. D. Unsworth (Inorg. Chim. Acta **4** [1970] 342/6). – [14] H. T. Witteveen, B. Nieuwenhuijse, J. Reedijk (J. Inorg. Nucl. Chem. **36** [1974] 1535/41). – [15] F. W. Klaaijsen, H. Blöte, Z. Dokoupil (Physica B+C **81** [1976] 1/14, 2/7, 11/3).

[16] J. Burić, K. Nikolić (Fizika **6** [1974] 51/9, 52; C.A. **81** [1974] No. 129250). – [17] F. Watari, S. Kinumaki (Sci. Rept. Res. Inst. Tohoku Univ. A **16** [1964] 285/9). – [18] W. Libuś, M. Kluczkowski (J. Chem. Soc. Faraday Trans. I **73** [1977] 1870/81, 1872, 1876). – [19] H. Grossmann (Ber. Deut. Chem. Ges. **37** [1904] 559/69, 563/4). – [20] E. G. Cox, A. J. Shorter, W. Wardlaw, W. J. R. Way (J. Chem. Soc. **1937** 1556/9).

[21] R. J. H. Clark, C. S. Williams (Inorg. Chem. **4** [1965] 350/7, 352). – [22] A. A. McConnell, R. H. Nuttall (J. Mol. Struct. **49** [1978] 207/9). – [23] R. J. H. Clark, C. S. Williams (Chem. Ind. [London] **1964** 1317/8). – [24] M. Goldstein, W. D. Unsworth (Spectrochim. Acta A **28** [1972] 1107/11). – [25] J. E. Rüede, D. A. Thornton (J. Mol. Struct. **34** [1976] 75/81).

[26] D. P. Mellor, C. D. Coryell (J. Am. Chem. Soc. **60** [1938] 1786/7). – [27] N. S. Gill, R. H. Nuttall, D. E. Scaife, D. W. A. Sharp (J. Inorg. Nucl. Chem. **18** [1961] 79/87, 82). – [28] C. W. Frank, L. B. Rogers (Inorg. Chem. **5** [1966] 615/22, 616/9). – [29] M. Goldstein (Inorg. Chim. Acta **31** [1978] L425/L427). – [30] M. Goldstein, W. D. Unsworth (Inorg. Nucl. Chem. Letters **6** [1970] 25/8).

[31] R. H. Nuttall (Talanta **15** [1968] 157/69, 162). – [32] S. Suzuki, W. J. Orville-Thomas (J. Mol. Struct. **37** [1977] 321/7, 325). – [33] J. R. Allan, D. H. Brown, R. H. Nuttall, D. W. A. Sharp (J. Inorg. Nucl. Chem. **27** [1965] 1305/9, 1307). – [34] E. König, H. L. Schläfer

(Z. Physik Chem. [Frankfurt] **26** [1960] 371/403, 374, 389/91). – [35] O. Stelling (Z. Physik. Chem. B **24** [1934] 282/92, 284, 289).

[36] T. Yoshihashi, H. Sano (Chem. Phys. Letters **34** [1975] 289/91). – [37] V. Mincu, E. Segal (Rev. Roumaine Chim. **16** [1971] 1651/4). – [38] R. W. Asmussen (Proc. Symp. Coord. Chem., Copenhagen 1953 [1954], pp. 27/30; C.A. **1955** 15470). – [39] K. C. Dash, D. V. Ramana Rao (Indian J. Chem. **3** [1965] 514/6). – [40] G. Liptay, K. Burger, É. Mocsári-Fülöp, I. Porubszky (J. Therm. Anal. **2** [1970] 25/36, 30; Magy. Kem. Folyoirat **77** [1971] 84/90, 85; C.A. **74** [1971] No. 150550).

[41] D. Fătu, E. Segal (Rev. Roumaine Chim. **14** [1969] 709/15). – [42] P. B. Bowman, L. B. Rogers (J. Inorg. Nucl. Chem. **28** [1966] 2215/24, 2219/20). – [43] W. W. Wendlandt, S. Iftikhar Ali (Z. Anorg. Allgem. Chem. **337** [1965] 6/13, 10). – [44] I. G. Murgulescu, D. Fătu, E. Segal (J. Therm. Anal. **1** [1969] 97/106, 100/2). – [45] W. Libuś, I. Uruska (Inorg. Chem. **5** [1966] 256/64, 257, 260).

[46] W. Libuś, M. Kluczkowksi, A. Kłonkowski, W. Nierzwicki (J. Chem. Soc. Faraday Trans. I **70** [1974] 1057/66). – [47] W. Libuś, A. Kłonkowski, W. Nierzwicki (Bull. Acad. Polon. Sci. Ser. Sci. Chim. **21** [1973] 297/305, 300; C.A. **79** [1973] No. 70658). – [48] F. S. Taylor (J. Chem. Soc. **1934** 699/701). – [49] B. Pradhan, D. V. Ramana Rao (J. Indian Chem. Soc. **56** [1979] 455/7). – [50] G. A. Eisman, M. M. Reiff (Inorg. Chim. Acta **44** [1980] L171/L173).

[51] M. Hronec, S. Holotik, J. Ilavsky (Collection Czech. Chem. Commun. **45** [1980] 880/7, 885). – [52] W. Libuś, B. Janaszewski, R. Pastewski, H. Twardowska (J. Solution Chem. **9** [1980] 737/52, 741; C.A. **94** [1981] No. 146203).

Isothiocyanates and Isoselenocyanates

$Mnpy_2(NCS)_2$ was formed by thermal degradation of $Mnpy_4(NCS)_2$ [8]. It may also be prepared by prolonged boiling of an alcoholic solution of $Mn(NCS)_2$ with a stoichiometric amount of pyridine. On cooling the solution, yellow crystals appeared [5]. For analytical purposes $Mnpy_2(NCS)_2$ was precipitated by adding 0.2N thiocyanate solution and pyridine to a solution of manganese nitrate [6].

According to X-ray photographs, $Mnpy_2(NCS)_2$ is isomorphous with the analogous complexes of Fe, Co, and Ni [7]. The magnetic moment was determined to be $\mu_{eff} = 6.01\ \mu_B$ [8]. The electronic reflection spectrum of the colorless compound shows three bands, at about 26000, 22900, and 16900 cm^{-1}, and is thus similar to that of $Mnpy_4(NCS)_2$ [8]. The IR spectrum shows the prominent ligand frequencies as strong single bands: 1039 (ν_{12}, A_1), 1009 (ν_1, A_1), 627 (ν_{6a}, A_1), and 417 cm^{-1} (ν_{16b}, B_2) [9] or 623 and 417 cm^{-1} [10]. The bands are shifted by 10 to 25 cm^{-1} with respect to those of the free ligand, indicating coordination of the pyridine molecule via the nitrogen atom [9]. The characteristic far-IR bands were assigned as follows: 258 cm^{-1} to ν(Mn-NCS) (ν_3), 196 cm^{-1} to ν(Mn-NCS) combined with $\nu(Mn\text{-}N_{py})$ (ν_4), 164 cm^{-1} to $\nu(Mn\text{-}N_{py})$ (ν_5), 153 cm^{-1} to δ(N-Mn-N) (ν_6) [11]. The assignments [11] were made with the aid of ^{15}NCS labeling and pyridine deuteration. Similar far-IR data were given by [9, 10]. The band-for-band correspondence of the far-IR spectrum with that of the cobalt(II) complex, which was shown to be polymeric and octahedral, points to the same structure for the manganese complex [11]. The positions of the ν(CN) and ν(CS) anion bands (at 2090 and 789 cm^{-1}, respectively), [8, 9] as well as the splitting of the δ(NCS) band (at 470 and 476 cm^{-1}) [11] also suggest a polymeric octahedral structure with bridging NCS groups [8, 9, 11].

On heating, the compound decomposes: $Mnpy_2(NCS)_2(cryst) \rightarrow Mn(NCS)_2(cryst) + 2py(g)(l)$ [3]. The peak temperatures of DTA [1, 2, 4, 12, 13], DSC [3], DTG [1, 4, 12], and TG curves [1, 14] were found to be within the range 225°C (heating rate 4°C/min in air) [1] to 247°C (16°C/min in N_2) [3]. Decomposition below 200°C was observed by [13, 14]. The following stability sequences of the complexes $Mpy_2(NCS)_2$ were established: M = Ni > Zn > Co > Cd > Mn > Cu [1, 12], also see [13], and $X = NCS^- > Br^- > Cl^- > I^-$ [1]. The enthalpy of reaction $\Delta H = +33.7 \pm 1.0$ kcal/mol was evaluated from DSC measurements [3]. The activation energy and preexponential factor A of the Arrhenius equation for reaction I have been evaluated from a kinetic study with the Coats-Redfern method: $E_A = 19.60$ kcal/mol and $A = 90000\ g^{0.5}/s$. The observed reaction order n = 0.5 indicates a partially diffusion-controlled reaction [15], also see [16]. Different parameters were calculated from TG and DTG curves: $E_A = 160.8$ kJ/mol (= 38.43 kcal/mol), ΔS_A (activation entropy) = 21.20 $J \cdot mol^{-1} \cdot K^{-1}$, $A = 1.362 \times 10^{14}\ s^{-1}$ by the Coats-Redfern method with n = 1 (n was evaluated with a relation suggested by Horowitz and Metzger) [4].

$Mnpy_2(NCSe)_2$. The colorless compound was prepared by adding KSeCN in a 1:1.8 mole ratio to an aqueous solution containing $Mn(NO_3)_2$ and pyridine at 80°C and maintaining the mixture at 80°C for 30 min, a method given by [18] for $Nipy_2(NCS)_2$ [17]. In the IR spectrum, the ν(CN) band appears at 2100 cm^{-1} (KSeCN at 2050 cm^{-1}). The strong shift of the ν(CN) band with respect to that of KSeCN suggests a polymeric octahedral structure with bridging SeCN groups. The pyridine bands are only slightly shifted with respect to free pyridine. $Mnpy_2(NCSe)_2$ decomposes on heating to give $Mn(NCSe)_2$. Peak temperatures of DTA and DTG curves (3°C/min) were found to be at 220 and 230°C, respectively, values taken from a diagram which also contains a TG curve [17].

References:

[1] G. Liptay, K. Burger, E. Mocsári-Fülöp, I. Porubszky (J. Therm. Anal. **2** [1970] 25/36, 30; Magy. Kem. Folyoirat **77** [1971] 84/90, 85; C.A. **74** [1971] No. 150550), G. Liptay, E. Mocsári-Fülöp, I. Porubszky (Proc. 3rd Anal. Chem. Conf., Budapest 1970, Vol. 2, pp. 297/304, 299; C.A. **74** [1971] No. 27660). – [2] W. W. Wendlandt, S. Iftikhar Ali (Z. Anorg. Allgem. Chem. **337** [1965] 6/13, 10). – [3] G. Beech, G. B. Kauffman (Thermochim. Acta **1** [1970] 93/7, 95). – [4] C. G. R. Nair, P. Indrasenan (Indian J. Chem. **14** [1976] 402/5). – [5] H. Grossmann (Ber. Deut. Chem. Ges. **37** [1904] 559/69, 563/4).

[6] A. L. J. Rao, B. K. Puri (Z. Anal. Chem. **235** [1968] 176/8). – [7] T. A. Mal'kova (Issled. Khim. Khelatnykh Soedin. **1971** 37/42; C.A. **77** [1972] No. 83120). – [8] G. Liptay, K. Burger, E. Papp-Molnár, Sz. Szebeni, F. Ruff (J. Inorg. Nucl. Chem. **31** [1969] 2359/66, 2363/5). – [9] R. J. Clark, C. S. Williams (Spectrochim. Acta **22** [1966] 1081/90, 1084, 1089). – [10] C. W. Frank, L. B. Rogers (Inorg. Chem. **5** [1966] 615/22, 618).

[11] C. Engelter, D. A. Thornton (J. Mol. Struct. **42** [1977] 51/8). – [12] G. Liptay, E. Papp-Molnár, K. Burger (J. Inorg. Nucl. Chem. **31** [1969] 247/52, 248, 251). – [13] P. B. Bowman, L. B. Rogers (J. Inorg. Nucl. Chem. **28** [1966] 2215/24, 2219/20). – [14] T. Dupuis, J. Besson, C. Duval (Anal. Chim. Acta **3** [1949] 599/605, 603). – [15] I. G. Murgulescu, D. Fătu, E. Segal (J. Therm. Anal. **1** [1969] 97/106, 100/2).

[16] E. Segal, C. Vasile (Analele Univ. Bucuresti Ser. Stiint. Nat. Chim. **15** [1966] 17/24, 21, 24; C.A. **69** [1968] No. 90180). – [17] I. Porubszky, A. Nemeth, G. Liptay (Periodica Polytech. [Budapest] **17** [1973] 365/71, 367/8; C.A. **80** [1974] No. 140736). – [18] S. M. Nelson, T. M. Shepherd (Inorg. Chem. **4** [1965] 813/7, 814).

Other Bis(pyridine) Compounds

The complexes were prepared (A) by reacting the manganese(II) salt (or the starting materials for the manganese salt) and pyridine in a solvent, (B) by dissolving the manganese(II) salt in pyridine and crystallizing the complex, and (C) by heating the tetrakis (pyridine) compounds. The electronic absorption spectra [3, 6] and the magnetic moments [3, 4, 6] are those of high-spin complexes with octahedral symmetry. The electrical conductivities of the solutions indicate the compounds to be nonelectrolytes [2, 4]. The physical properties suggest that the complexes have a structure similar to that described for $Mn\,py_2Cl_2$, see p. 92.

The table below gives the method of preparation, the starting materials for the Mn^{II} salt (if the salt is not used), the solvent for method A, the temperature for method C, the general physical properties, the magnetic moment μ_{eff} in μ_B, the molar conductivity Λ in $\Omega^{-1} \cdot cm^2 \cdot mol^{-1}$, positions of IR bands in cm^{-1}, and some remarks, generally on the decomposition.

No.	compound	method	color, habit
1	$Mn\,py_2S_2O_4$	(A) $MnSO_4 \cdot H_2O$ and $Na_2S_2O_4$, in H_2O	tan needles
2	$Mn\,py_2[N(CN)_2]_2$	(A) $NaN(CN)_2$ and $Mn(NO_3)_2$ hydrate, in H_2O at 80°C	colorless, microcrystalline
3	$Mn\,py_2[C(CN)_3]_2$	(A) $Mn(NO_3)_2$ hydrate and $KC(CN)_3$, in H_2O	colorless
4	$Mn\,py_2(CF_3COO)_2$	(C) 100 to 140°C	—
5	$Mn\,py_2(CCl_3COO)_2$	(C) below 60°C	—
6	$Mn\,py_2(CNCH_2COO)_2$	(A) in 2,2′-dimethoxy-propane at 40°C under N_2 for 4 days	almost colorless
7	$Mn\,py_2(\text{-}C_6H_4COO)_2$	(B) stirring for several days	colorless
8	$Mn\,py_2[Hg(SCN)_4]$	(A) $Mn(NCS)_2$ and $Hg(SCN)_2$ in absolute C_2H_5OH or THF, 2 to 3 d	—
9	$Mn\,py_2(VO_3)_2 \cdot 3\,H_2O$	(A) $MnCl_2 \cdot 4\,H_2O$ and NH_4VO_3, in H_2O	brick yellow
10	$Mn\,py_2[Ni(CN)_4]$	(A) $Mn(CH_3COO)_2 \cdot 4\,H_2O$ and $K_2[Ni(CN)_4]$ in aqueous NH_3	microcrystalline

*) Further details for compounds No. 6, 7, 8, and 10 are given on pp. 101/2.

Mnpy$_2$(CNCH$_2$COO)$_2$. X-ray powder photographs indicate a crystal structure of low symmetry. A table of d values is given. Selected IR bands, especially those of the coordinated anion, are recorded. The IR spectum suggests that the carboxylate groups probably are O,O chelated and the CN group of the anion is not coordinated to Mn. The many bands observed below 300 cm^{-1} may indicate a favored cis-octahedral structure [4].

Mnpy$_2$(-C$_6$H$_4$COO)$_2$. A comparison of the IR bands of the complex with those of the free biphenyl-2,2′-dicarboxylic acid (=$C_{14}H_{10}O_4$=biphenic acid) reveals the disappearance of the bands of the OH group at 2760, 1320 to 1210, and 940 cm^{-1} upon complexation, indicating loss of the hydrogen atom of the carboxylic group. Only regions are given for the IR band positions of several transition metal complexes. The pyridine bands around 1600 cm^{-1} and 700 cm^{-1} are only slightly shifted with respect to the free ligand [5].

Mnpy$_2$[Hg(SCN)$_4$]. The compound was prepared by reacting the THF complex Mn[Hg(SCN)$_4$]·2THF with pyridine (2%) in absolute ethanol for 2 to 3 d. The low-lying

IR bands	remarks*)	Ref.
—	extremely air sensitive, it smells of SO_2 and py	[1]
—	loses pyridine in vacuo at 200°C, dissolves readily in C_2H_5OH; $\Lambda = 1.9$ at 25°C in pyridine	[2, p.246]
—	liberates pyridine in vacuo at 120°C; $\Lambda = 15.4$ in pyridine	[2, p.247]
—	$\mu_{eff} = 5.90\ \mu_B$	[3]
—	$\mu_{eff} = 5.90\ \mu_B$	[3]
6 bands between 286 and 186 cm^{-1} (ν(Mn-O) and ν(Mn-N))	dec. on heating; $\mu_{eff} = 5.87\ \mu_B$; $\Lambda = 15.1$ in CH_3CN	[4]
—	m.p. 300°C, dec.	[5]
210 ν(Mn-NCS) 196 ν(Hg-S) 159 $\delta(N_{py}\text{-}Mn\text{-}N_{py})$	dec. >250°C; $\mu_{eff} = 5.90\ \mu_B$	[6]
—	almost insoluble in H_2O and organic solvents, decomposed by NH_3 and strong acids	[7]
—	for further details on preparation see text	[8, 9]

pyridine-ring vibration bands appear at 625 and 420 cm^{-1}. The observed shifts (+21 and +15 cm^{-1}, respectively) with respect to the free ligand bands indicate an Mn-N_{py} coordination bond. The compound is insoluble in water and most organic solvents. It is slightly soluble in pyridine [6].

Mnpy$_2$[Ni(CN)$_4$]. The mixture given in the table on pp. 100/1, is placed in an atmosphere of pyridine. X-ray powder patterns show the compound to be monoclinic, space group C2/m-C^3_{2h} (No. 12), with the lattice constants a = 15.76, b = 7.46, c = 7.04 Å, and β = 104.7°; Z = 2. The structure consists of corrugated polymeric layers of $[MnNi(CN)_4]_\infty$, which are held together by the pyridine molecules with direct bonds to the Mn atoms above and below the layers. The successive layers are shifted against each other by an amount of a + b/2. The Mn atom is surrounded octahedrally by four N atoms of the CN groups and two N atoms of pyridine molecules. The coordination around the Ni atom is square planar. The proposed structure was derived from single-crystal data for the Cd compound. The powder patterns suggest isomorphous relations with the Fe, Co, Ni, Cu, Zn, and Cd compounds. The measured density is 1.64 g/cm^3; the calculated value is 1.56 g/cm^3 [8].

References:

[1] R. Scholder, G. Denk (Z. Anorg. Allgem. Chem. **222** [1935] 41/7, 46). – [2] H. Köhler (Z. Anorg. Allgem. Chem. **331** [1964] 237/48, 246/7). – [3] S. Amasa, D. H. Brown, D. W. A. Sharp (J. Chem. Soc. A **1969** 2892/4). – [4] M. Cartwright, D. A. Edwards, J. M. Potts (Inorg. Chim. Acta **34** [1979] 211/20, 212/3). – [5] A. Ara-Blesa (Ion [Madrid] **34** [1974] 685/9, 687; C.A. **82** [1975] No. 67573).

[6] R. Makhija, L. Pazdernik, R. Rivest (Can. J. Chem. **51** [1973] 2987/92). – [7] G. Spacu (Acad. Roumaine Bull. Sect. Sci. **22** [1939/40] 42/8, 47). – [8] R. L. Morehouse, K. Aytaç, D. Ülkü (Z. Krist. **145** [1977] 157/60).

16.1.7.2.5 Mono(pyridine) Compounds

MnpyX_2 (X = Cl, Br, I). The compounds were obtained by thermal degradation of the bis(pyridine) compounds Mnpy$_2$$X_2$ [1, 3, 5].

X-ray powder photographs indicate strict isomorphism of MnpyCl_2 with NipyBr_2 [2]. For MnpyCl_2 the effective magnetic moment μ_{eff} = 5.87 μ_B indicates a high-spin Mn^{II} d^5 complex [6].

The far-IR absorption bands (260 to 60 cm^{-1}, in polyethylene at about 300 K) and the Raman bands (260 to 30 cm^{-1}, at 300 K) of MnpyCl_2 and MnpyBr_2 are listed in the table together with assignments.

compound	method	ν(Mn-X)				ν(Mn-N)		
MnpyCl_2	IR	262	230	215	199	160	128	118
	Raman		248	201	177		119	
MnpyBr_2	IR	226	205	176	150	161	118	95, 88
	Raman	211		192	147		123	

The band positions of the compounds with pentadeuterated pyridine differ only slightly (-2 to -5 cm^{-1}) from those of the nondeuterated complexes [5]. The far-IR [2, 5], the electronic reflectance spectra, neither reported [2], and the magnetic moment of $MnpyCl_2$ are typical of Mn^{II} in an octahedral environment. The structure is polymeric, involving bridging halogen atoms, each Mn atom being surrounded by five halogens and one nitrogen. The MnX_5N octahedra must be linked together to form a double strand chain structure as shown in **Fig. 12** [5].

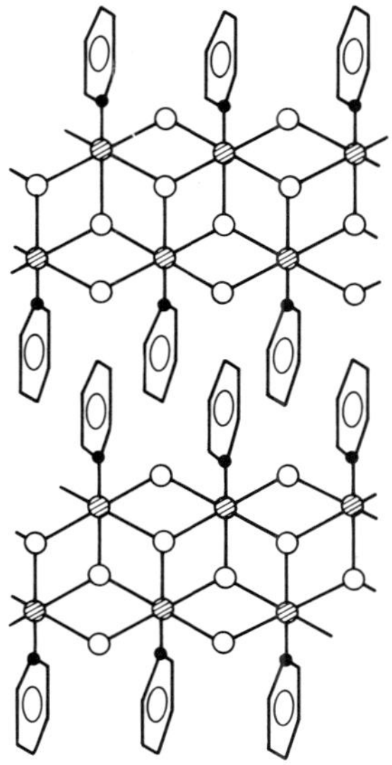

Fig. 12

Molecular structure of $[MnpyX_2]_n$ compounds.

Upon heating, the compounds $MnpyX_2$ lose pyridine in two steps. The peak temperatures of the first step (loss of $^1/_3$ mol of pyridine) are found to be 255, 265, and 200°C for X=Cl, Br, and I, respectively; the second step (loss of $^2/_3$ mol py) occurs at 315, 300, and 255°C [1]. The two degradation steps of $MnpyCl_2$ are also reported to be 290 and 354°C [3] or 267 and 327°C [4], also see [2]. For details on the methods used, see p. 96. The table on p. 104 lists the reaction order n, the enthalpies of reaction ΔH in kcal/mol [4], the energy of activation E_A in kcal/mol, and the preexponential factor of the Arrhenius equation A for the reactions (I) $MnpyCl_2(s) \rightarrow Mnpy_{2/3}Cl_2(s) + {}^1/_3$pyridine(g) and (II) $Mnpy_{2/3}Cl_2(s) \rightarrow MnCl_2(s) + {}^2/_3$pyridine(g). E_A and A were calculated from DTG data by the Coats-Redfern (CR) and Freeman-Carroll (FC) methods [3, 7].

reaction	n	ΔH	E_A(CR)	A(CR)	E_A(FC)	A(FC)
I	0.5	3.1±0.2	41.6	8.75×10^{13} $g^{0.5}/s$	42	3.16×10^{13} $g^{0.5}/s$
II	1	10.7±0.2	35	1.9×10^{9} s^{-1}	36	2.85×10^{10} s^{-1}

The reaction order of 0.5 indicates that reaction I is controlled partially by diffusion or other phenomena, while reaction II is not [7].

$(C_2H_5)_4N[MnBrCl_2py]$. The light green crystalline compound was prepared as described for $(C_2H_5)_4N[MnBrCl_2py_3]$ (p. 91) and has similar properties. It is assumed that the compound contains a 4-coordinate Mn atom [8].

[Mn(C(CH$_3$)$_3$COO)$_2$py]$_2$·2C(CH$_3$)$_3$COOH. The compound was prepared like the α-picoline complex [Mn(C(CH$_3$)$_3$COO)$_2$(α-pic)]$_2$ (p. 121), and it has a similar IR spectrum. Therefore, a dimeric structure is assumed for the pyridine complex. It is formulated LMn(C(CH$_3$)$_3$COO)$_4$MnL with L = C(CH$_3$)$_3$COOH·C$_5$H$_5$N. The complex is easily oxidized, forming a trinuclear oxo cluster, Mn$_3$O(C(CH$_3$)$_3$COO)$_7$ [9].

Mn py C$_2$O$_4$·H$_2$O. The compound was prepared by refluxing a suspension of MnC$_2$O$_4$·3H$_2$O (oxalate) in ethanol with pyridine for 4 to 6 d. The resulting product was washed with alcohol and ether and dried in a vacuum. Susceptibility measurements at 303 K (30°C) yield the magnetic moment μ_{eff} = 5.96 μ_B. The electronic reflectance spectrum recorded on Nujol mulls shows absorption maxima at 27100, 22640, and 16220 cm^{-1} due to d-d transitions. The IR spectrum presents absorption bands assigned to ν(CC) and ν(CN) vibrations at 1610 (broad), 1560 (sh), and 1490 cm^{-1}. A band assigned to ν(Mn-N$_{py}$) vibrations occurs at 245 cm^{-1}. The slightly raised frequencies of the ν(CC) and ν(CN) bands suggests coordination of the ligand to Mn through the N atom of the pyridine ring. The oxalato groups are coordinated to Mn as bridging bidentate ligands as indicated by the shift of ν_{as}(OCO) to higher frequency and the ν_s(OCO) bands to lower frequencies and the presence of a strong band at ~800 cm^{-1}. The coordination of H$_2$O is suggested by a band assigned to the rocking mode of coordinated H$_2$O at about 690 cm^{-1}. It is concluded from the magnetic and spectral data that the complex is polymeric octahedral and contains bridging oxalato groups. As shown by thermogravimetric analysis Mn py C$_2$O$_4$·H$_2$O loses its H$_2$O and the pyridine between 30 and 290°C to leave MnC$_2$O$_4$, which decomposes between 290 and 520°C. The complex is almost insoluble in organic solvents [10].

Mn py (H$_2$O)(C$_{11}$H$_7$O$_3$)$_2$ (C$_{11}$H$_8$O$_3$ = 2-hydroxynaphthoic acid = C$_{10}$H$_6$(OH)COOH). The compound was obtained by mixing aqueous solutions of MnCl$_2$·4H$_2$O (2 g), pyridine (3.5 mL), and 2-hydroxynaphthoic acid (3.7 g). The resulting precipitate had the consistency of a glue. It solidified within a few days. The yellowish brown compound dissolves only slightly in cold water but is soluble in aqueous solutions of (NH$_4$)$_2$C$_2$O$_4$, alcohol, ether, CHCl$_3$, and (more readily) in acetone. It is decomposed by acids and bases. On heating up to 160°C it evolves the water and pyridine [11].

Mn py (C$_{12}$H$_{12}$N$_2$)$_3$(NCS)$_2$. The compound crystallized from a solution of Mn py$_4$(NCS)$_2$ and benzidine (= C$_{12}$H$_{12}$N$_2$) in ethanol which has been heated at 60 to 70°C for about 4 min. It can be also obtained by refluxing a solution of benzidine in ethanol and Mn py$_4$(NCS)$_2$ on a water bath for 30 min. This procedure yielded big, well-formed, gray crystals [12].

References:

[1] G. Liptay, K. Burger, E. Mocsári-Fülöp, I. Porubszky (J. Therm. Anal. **2** [1970] 25/36, 30; Magy. Kem. Folyoirat **77** [1971] 84/90, 85; C.A. **74** [1971] No. 150550), G. Liptay, E. Mocsári-Fülöp, I. Porubszky (Proc. 3rd Anal. Chem. Conf., Budapest 1970, Vol. 2, pp. 297/304, 299; C.A. **74** [1971] No. 27660). – [2] J. R. Allan, D. H. Brown, R. H. Nuttall, D. W. A. Sharp (J. Inorg. Nucl. Chem. **27** [1965] 1865/7). – [3] D. Fătu, E. Segal (Rev. Roumaine Chim. **14** [1969] 709/15). – [4] G. Beech, C. T. Mortimer, E. G. Tyler (J. Chem. Soc. A **1967** 1111/5, 1112). – [5] M. Goldstein, R. J. Hughes (Inorg. Chim. Acta **37** [1979] 71/7).

[6] V. Mincu, E. Segal (Rev. Roumaine Chim. **16** [1971] 1651/4). – [7] I. G. Murgulescu, D. Fătu, E. Segal (J. Therm. Anal. **1** [1969] 97/106, 100/2). – [8] A. K. Das, D. V. Ramana

Rao (J. Indian Chem. Soc. **48** [1971] 823/7). – [9] A. A. Pasynskii, T. Ch. Idrisov, K. M. Suvorova, V. T. Kalinnikov (Koord. Khim. **2** [1976] 1060/8; Soviet J. Coord. Chem. **2** [1976] 813/9, 814/5, 818). – [10] D. Pandey, D. C. Rupainwar (Current Sci. [India] **49** [1980] 336/9).

[11] M. Poni, R. Cernătescu (Ann. Sci. Univ. Jassy I **30** [1948] 219/28, 221, 226; C.A. **1949** 2538). – [12] G. Spacu, T. I. Pirtea (Rev. Univ. C.I. Parhon Politehn. Bucuresti Ser. Stiint. Nat. **1** [1952] 104/14, 112; C.A. **1956** 16518).

16.1.7.2.6 Other Manganese(II) Complexes with Pyridine

$Mn(ClO_4)_2 \cdot 8py$ was prepared by dissolving 2 to 4 g $Mn(ClO_4)_2 \cdot 6H_2O$ in 20 to 30 mL of freshly distilled pyridine by warming. The lilac crystals which appeared on cooling were washed with cold pyridine and dried by pressing on filter paper. The compound is not deliquescent but is partially hydrolyzed by water [1].

$Mn(ClO_4)_2 \cdot 5py \cdot 3H_2O$. If 4.3 g (0.01 mol) of $Mn(ClO_4)_2 \cdot 10H_2O$ is treated with 7.9 g (0.1 mol) of pyridine, dissolution occurs with a considerable evolution of heat. Soon after cooling, the nearly colorless crystals of the adduct appear. They smell strongly of pyridine. The compound decomposes in air, turning brown. On heating, it explodes. It is soluble in water, alcohol, and pyridine, from which it may be recrystallized. On storing over H_2SO_4, the compound decomposes to form $Mnpy_4(ClO_4)_2$. The complex is formulated as $[Mnpy_5(H_2O)](ClO_4)_2 \cdot 2H_2O$ [2].

A compound of composition $(Cdpy_3)_2MnCl_6$ or $Cdpy_2Mnpy_4CdCl_6$ was obtained on dissolving $Cd_2MnCl_6 \cdot 12H_2O$ in a mixture of pyridine, ethanol, and redistilled gasoline; after the mixture had stood 48 h in a closed vessel, a white crystalline precipitate formed. It is insoluble in water at room temperature. Upon warming, it dissolves with decomposition (loss of pyridine). It is insoluble in cold alcohol and benzene, slightly soluble in boiling alcohol, and is decomposed by mineral acids [3].

The compounds $Mnpy_{2/3}X_2$ (with X = Cl, Br, I) were obtained by thermal degradation of $MnpyX_2$. $Mnpy_{2/3}Cl_2$ is strictly isomorphous with $Nipy_{2/3}Cl_2$, as shown by X-ray powder photographs [4]. Its magnetic moment, $\mu_{eff} = 5.91\ \mu_B$, indicates a spin-free Mn^{II} complex [5]. The thermal degradation of $Mnpy_{2/3}X_2$ is treated together with that of $MnpyX_2$ on p. 104.

$2MnF_2 \cdot 2NH_4F \cdot py$ was precipitated upon adding pyridine to an aqueous solution of MnF_2 and NH_4F. The fine colorless crystals dissolve readily in aqueous ammonia (with decomposition) but are insoluble in water, alcohol, ether, and benzene [6].

The aurintricarboxylato complex $Mn_5py_3(OH)(C_{22}H_{11}O_9)_3 \cdot 20H_2O$ was precipitated as a garnet-red solid on boiling the combined aqueous solutions of $MnCl_2 \cdot 4H_2O$ (5 g), triammonium aurintricarboxylate (2 g), and pyridine (3 mL). On heating, the compound loses 10 mol H_2O at 60°C and 13 mol H_2O at 70°C. The remaining 7 mol H_2O are split off on further heating together with the pyridine. It is assumed that three of the manganese atoms are coordinated by one pyridine molecule and one H_2O molecule and the two other manganese atoms are each coordinated by two H_2O molecules. A proposed structure of the complex is shown in the paper. The compound is decomposed by alkali hydroxides [7].

References:

[1] P. C. Sinha, R. C. Ray (J. Indian Chem. Soc. **20** [1943] 32/6, 34). – [2] R. Weinland, K. Effinger, V. Beck (Arch. Pharm. **265** [1927] 352/79, 357, 371). – [3] G. Spacu, L. Caton (Bul. Soc. Stiinte Cluj **2** [1924/25] 332/53, 342; C.A. **1925** 1995). – [4] J. R. Allan, D. H.

Brown, R. H. Nuttall, D. W. A. Sharp (J. Inorg. Nucl. Chem. **27** [1965] 1865/7). – [5] V. Mincu, E. Segal (Rev. Roumaine Chim. **16** [1971] 1651/4).

[6] O. E. Zvyagintsev, M. Chkoniya (Zh. Obshch. Khim. **10** [1940] 1647/52, 1650; C.A. **1941** 3187). – [7] R. Cernătescu, M. Poni, M. Bostan (Commun. Acad. Rep. Populare Romine **5** [1955] 159/65, 163; C.A. **1956** 11994).

16.1.7.3 Isolated Manganese(III) and Manganese(II,III) Complexes

$MnpyCl_3$ was synthesized by reacting the components in ether. The compound crystallizes in dark green glittering leaflets. It is stable only below −10°C [1].

$Mnpy_3Cl_3$ was obtained by reacting $MnpyCl_3$ with an ether solution of pyridine in a 1:2 mole ratio at 20°C. The brown leaflets are very sensitive to moisture. In the absence of moisture, the complex is stable up to 100°C [1].

$Mn_2py_2(SO_4)_3$. The chocolate-brown compound was formed on reacting $Mn_2(SO_4)_3$ with hot anhydrous pyridine for 2 h. It is hydrolyzed by water, only slightly soluble in liquid ammonia, and almost insoluble in pyridine [2].

$Mn_4py_4(RCOO)_m(OH)_nX_p$(?). On recrystallizing $[Mn_3O(CH_3COO)_6py_3]$ (see below) in a small amount of absolute ethanol, a complex of the composition $Mn_4py_4(CH_3COO)_8(OH)_4$ was obtained. Similar acetato pyridine complexes, but with additional NO_3^- or ClO_4^- ions, were prepared by reacting manganese(III) acetate with pyridine and $LiNO_3$ or $LiClO_4$ in absolute ethanol and/or glacial acetic acid. The compounds formed (four nitrates and four perchlorates) were formulated as $Mn_4py_4(CH_3COO)_m(OH)_nX_p$ with X = NO_3 or ClO_4, m varying between 7 and 10, n varying between 0 and 4, p = 1 or 2, and m + n + p = 12. The nitrates and perchlorates are brown, well-crystallized prismatic solids. They are stable if stored dry in a closed vessel. In air they decompose slowly, with a continuous slight odor of pyridine caused by hydrolysis. All perchlorates explode on heating. The compounds dissolve in water yielding brown solutions, which soon become turbid, especially on heating, to form a precipitate of manganic oxide hydrate. The compounds are soluble in methanol, ethanol, acetone, glacial acetic acid, and pyridine but insoluble in chloroform and benzene. They are decomposed by acidic and basic reagents [3, pp. 162/5, 170/6].

Similar benzoato complexes formulated $Mn_4py_4(C_6H_5COO)_8(OH)_4$ (gray-black) and $Mn_4py_2(C_6H_5COO)_7(OH)_5$ (dark brown) were synthesized by reacting a hot solution of manganese(III) benzoate (3.3 g) in absolute ethanol (50 mL) with benzoic acid (2 g) and pyridine (12 g for py_4 or 4 g for py_2) [3].

$[Mn_3O(CH_3COO)_6py_3]$, which had been formulated as $Mn_4py_5O_2(CH_3COO)_8$ [3, p. 175], was obtained by reacting hydrated manganese(III) acetate (2.7 g) with pyridine (4 mL) [2] and warming [3, 4]. The compound forms brown-black, glittering, octahedral crystals [3, p. 175]. According to an X-ray investigation these are trigonal, space group R32-D_3^7 (No. 155) with the lattice constants a = 10.794(4) Å and α = 108.94(5)° (hexagonal setting: a = 17.575 and c = 11.077 Å); Z = 3. The measured density is D = 1.48 g/cm³, corresponding to 1.4 mol of solvated pyridine per mole of complex. D = 1.30 g/cm³ was calculated for the unsolvated complex [4]. The structure is shown in **Fig. 13.** The molecule has a symmetry with three equivalent manganese atoms, each with an average oxidation state of 2.67. The distance Mn-N_{py} = 2.125 Å implies fairly strongly coordinated pyridine ligands. Solvated pyridine molecules occupy channels formed by the coordinated pyridine, thus causing a slight disordering. The solvated molecules are lost easily. The compound is isostructural with $[Fe_3O(CH_3COO)_6py_3]$. It is weakly antiferromagnetic as expected. Its magnetic moment per manganese atom drops from $\mu = 3.98\ \mu_B$ at 293 K to $\mu = 3.21\ \mu_B$ at 90 K. The reduction of the magnetic moment from the spin-only values of the Mn^{2+} and Mn^{3+} ions shows that this complex can be classified as a mixed valence compound in which the ions of formally different oxidation states interact strongly [4], also see [5]. The compound dissolves readily in methanol and ethanol, somewhat less well in pyridine and glacial acetic acid (brown solution), and smells of pyridine [3].

Fig. 13

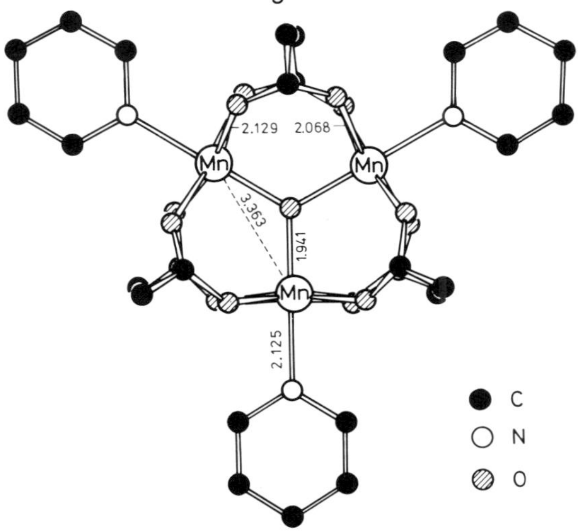

Molecular structure of $[Mn_3O(CH_3COO)_6py_3]$
with some interatomic distances in Å.

$[Mn_3O(CH_3COO)_6py_3]ClO_4$. The complex was synthesized by adding pyridine (0.5 g) and $NaClO_4 \cdot H_2O$ (0.2 g) to an ethanolic solution (10 mL) of the so-called manganese(III) acetate (0.3 g), which was obtained by oxidizing manganese(II) acetate with $KMnO_4$. The mixture was heated at 50°C for 2 min and allowed to stand for 30 min at room temperature to give a brown precipitate. As shown by X-ray powder diagrams $[Mn_3O(CH_3COO)_6py_3]ClO_4$ is isomorphous with the Cr, Fe, Ru, Rh, and Ir compounds. Measurements on solutions in CH_2Cl_2 yielded a molecular weight of 440 (872 calculated), indicating strong dissociation [6].

The complex melts at 218°C and has a magnetic moment of $\mu_{eff} = 4.25\ \mu_B$ per Mn atom at 298 K. The electronic spectrum of its solution in methanol exhibits an absorption maximum at 250 nm (molar extinction coefficient $\varepsilon = 18800\ L \cdot mol^{-1} \cdot cm^{-1}$) with shoulders at about 256, 440, and 550 nm. The IR spectrum (KBr discs) shows the band of the ClO_4^- ion at 1090 cm^{-1} and two bands of the bridging carboxylate groups at 1610 and 1395 cm^{-1} ($\nu_{as}(CO)$ and $\nu_s(CO)$, respectively). The complex thus probably has an oxo-centered triangular structure as described for $[Mn_3O(CH_3COO)_6py_3]$. Electrochemical studies show that $[Mn_3O(CH_3COO)_6py_3]ClO_4$ is reduced irreversibly at about +0.3 V versus SCE, presumably in a one-electron step. The reduced species could not be isolated. Reduction with H_2 (2 atm) in the presence of PtO_2 catalyst or $NaBH_4$ produces the simple hydrated Mn^{2+} ions. The compound has solubility similar to its Cr analogue, which is soluble in CH_2Cl_2, $CHCl_3$, acetone, acetonitrile, pyridine, dimethylformamide, and dimethyl sulfoxide; slightly soluble in alcohols, acetic acid, and tetrahydrofuran; and insoluble in water, ether, CCl_4, benzene, and petroleum ether [6].

$NH_4[Mn\,py_2(C_7H_4O_3)_2]\cdot py\cdot 5H_2O$. The salicylato complex crystallized from a solution of $NH_4[Mn(H_2O)_2(C_7H_4O_3)_2]\cdot 2H_2O$ (see "Mangan" D 2, 1980, p. 160) in a small amount of pyridine as green tabular crystals, which can be dried between sheets of filter paper. The compound loses one molecule of pyridine and turns brown on standing in the air, in vacuo, or also on being treated with chloroform. **$H[Mn\,py_2(C_7H_4O_3)_2]\cdot 2H_2O$** was obtained as a black lustrous solid by treating $NH_4[Mn\,py_2(C_7H_4O_3)_2]\cdot py\cdot 5H_2O$ with concentrated acetic acid. It is insoluble in water and does not smell of pyridine [7].

References:

[1] H. Funk, H. Kreis (Z. Anorg. Allgem. Chem. **349** [1967] 45/9, 47). – [2] K. W. Weber (Jahrb. Diss. Bonn Math. Nat. Fak. **2** [1952] 67; Diss. Univ. Bonn 1952, pp. 1/114, 12, 72/4). – [3] R. F. Weinland, G. Fischer (Z. Anorg. Allgem. Chem. **120** [1922] 161/80, 162/5, 168/76, 180). – [4] A. R. E. Baikie, M. B. Hursthouse, D. B. New, P. Thornton (J. Chem. Soc. Chem. Commun. **1978** 62/3). – [5] J. Catterick, P. Thornton (Advan. Inorg. Chem. Radiochem. **20** [1977] 291/362, 299, 320).

[6] S. Uemura, A. Spencer, G. Wilkinson (J. Chem. Soc. Dalton Trans. **1973** 2565/71, 2566/9). – [7] G. A. Barbieri (Ber. Deut. Chem. Ges. **60** [1927] 242/4).

16.1.8 With Halogenopyridines

Magnetic [1, 5], ESR [3], IR [5], and electrical-conductivity measurements [5] suggest the tetrakis(halogenopyridine) compounds to have a structure similar to that described for the tetrakis(pyridine) compounds. For the dihalogenobis(halogenopyridine) compounds $Mn(X\text{-}C_5H_4N)_2X_2$ magnetic and solubility properties indicate a polymeric octahedral structure as was shown for the bis(pyridine) compounds, whereas the complex $Mn(3\text{-}Br\text{-}C_5H_4N)_2(NCS)_2$ is considered to be monomeric tetrahedral [1].

$Mn(3\text{-}Cl\text{-}C_5H_4N)_4(NCS)_2$. The colorless crystalline compound was obtained by refluxing a mixture of $Mn(NCS)_2$ and the ligand in ethanol for 2 h and allowing the solution to stand for 5 or 6 d. It melts at 149°C; $\mu_{eff} = 5.91\ \mu_B$. A band at 2060 cm^{-1} in the IR spectrum is assigned to the terminal NCS group. The compound is freely soluble in the common organic solvents. Conductivity measurements indicate it to be a nonelectrolyte [1].

$Mn(3\text{-}Br\text{-}C_5H_4N)_4X_2$ (X = I, NCO). The complex $Mn(3\text{-}Br\text{-}C_5H_4N)_4I_2$ was obtained by refluxing MnI_2 with the ligand [3, 4]. Its ESR X-band spectrum shows two strong lines, at g = 6 and g = 2, indicating a tetragonal distortion of the octahedral ligand field (parameter D = 0.87 cm^{-1}) [3]. The isocyanato complex is prepared like the unsubstituted pyridine complex. It melts above 250°C; $\mu_{eff} = 6.01\ \mu_B$; $\Lambda = 11.5\ \Omega^{-1}\cdot cm^2\cdot mol^{-1}$ in dimethylformamide. The ν(CN) and ν(CO) bands of the terminal NCO groups appear in the IR spectrum at 2200 and 1325 cm^{-1}, respectively. The position of ν(CN) indicates coordination of NCO to Mn by the N atom [5].

$Mn(X\text{-}C_5H_4N)_2Cl_2$ and **$Mn(X\text{-}C_5H_4N)_2Br_2$** (X = 2-Cl, 3-Cl, and 3-Br). The complexes were precipitated by mixing ethanolic solutions of $MnCl_2$ or $MnBr_2$ and the ligands [2, 6]. The complexes with X = 3-Cl and 3-Br are colorless, amorphous powders, which do not melt or decompose up to 260°C.

The table below lists the effective magnetic moments μ_{eff} [1], the peak temperatures T_1 and T_2 in K (main reactions), and the enthalpies ΔH_1 and ΔH_2 (in kcal/mol) of the stepwise thermal degradation reactions (1) $MnL_2Cl_2(s) \rightarrow MnLCl_2(s) + L(g)$ and (2) $MnLCl_2(s) \rightarrow MnCl_2(s) + L(g)$ (L = halogenopyridine) obtained by DSC measurements (scanning rate 16 K/min) [6].

complex	μ_{eff}	T_1	ΔH_1	T_2	ΔH_2
$Mn(2\text{-}Cl\text{-}C_5H_4N)_2Cl_2$	—	490	15.6 ± 0.4	565	17.5 ± 0.4
$Mn(3\text{-}Cl\text{-}C_5H_4N)_2Cl_2$	5.60	485	16.0 ± 0.2	570	16.3 ± 0.3
$Mn(3\text{-}Br\text{-}C_5H_4N)_2Cl_2$	5.66	485	15.8 ± 0.3	585	16.5 ± 0.3
$Mn(3\text{-}Cl\text{-}C_5H_4N)_2Br_2$	6.03	—	—	—	—
$Mn(3\text{-}Br\text{-}C_5H_4N)_2Br_2$	5.98	—	—	—	—

The 3-halogenopyridine compounds are only sparingly soluble in common organic solvents. The properties suggest a polymeric octahedral structure [1] as described for the bis(pyridine) dihalogeno compounds on p. 92.

$Mn(3\text{-}Br\text{-}C_5H_4N)_2(NCS)_2$ was prepared like $Mn(3\text{-}Cl\text{-}C_5H_4N)_4(NCS)_2$ (p. 108). It is a crystalline light green substance melting at 220°C; $\mu_{eff} = 6.18\ \mu_B$. It is freely soluble in the common organic solvents. The IR spectrum shows a band at 2080 cm^{-1}, which is assigned to a terminal NCS group. The crystalline nature, the low melting point, and the solubility also rule out the presence of NCS bridges. The complex therefore must be considered to have a tetrahedral configuration [1].

$[Mn_3O(CH_3COO)_6(3\text{-}Cl\text{-}C_5H_4N)_3]$. The Mn^{III} compound formed on cooling a hot solution of $Mn(CH_3COO)_3 \cdot 2H_2O$ in 3-chloropyridine. The black crystals are monoclinic, space group $P2_1/a\text{-}C^5_{2h}$ (No. 14) with lattice constants a = 23.598(3), b = 8.477(1), c = 20.998(3) Å, and $\beta = 97.46(2)°$; Z = 4. The structure was determined in a single-crystal X-ray diffraction study by direct methods and refined by full-matrix least squares to reliability R = 0.06. The structure is shown in **Fig. 14**, p. 110, along with some of the important bond lengths (in Å). The isosceles triangle formed by the three Mn atoms clearly indicates that one of them (Mn 3) differs from the other two in that it is most probably in the +2 oxidation state. The other two are in the +3 state and have shorter Mn-O bond lengths. The measured density is 1.42 g/cm^3; the calculated value is 1.40 g/cm^3. Susceptibility measurements at 292 and 78 K yielded the magnetic moments $\mu = 4.59$ and 3.36 μ_B, respectively, which suggest antiferromagnetic properties similar to those of the related pyridine complex, see p. 105 [7].

References:

[1] A. K. Das, D. V. Ramana Rao (J. Inst. Chem. [India] **42** [1970] 127/30, 128; C.A. **74** [1971] No. 49141). – [2] S. Guru, K. C. Dash, D. V. Ramana Rao (Indian J. Chem. **6** [1968] 161/3). – [3] R. D. Dowsing, J. F. Gibson, M. Goodgame, P. J. Hayward (J. Chem. Soc. A **1969** 187/93, 189). – [4] J. R. Allan, D. H. Brown, R. H. Nuttall, D. W. A. Sharp (J. Inorg. Nucl. Chem. **27** [1965] 1865/7). – [5] A. K. Das, D. V. Ramana Rao (Current Sci. [India] **42** [1973] 56/7).

[6] C. T. Mortimer, J. L. McNaughton (Thermochim. Acta **10** [1974] 207/10). – [7] A. R. E. Baikie, M. B. Hursthouse, L. New, P. Thornton, R. G. White (J. Chem. Soc. Chem. Commun. **1980** 684/5).

Fig. 14

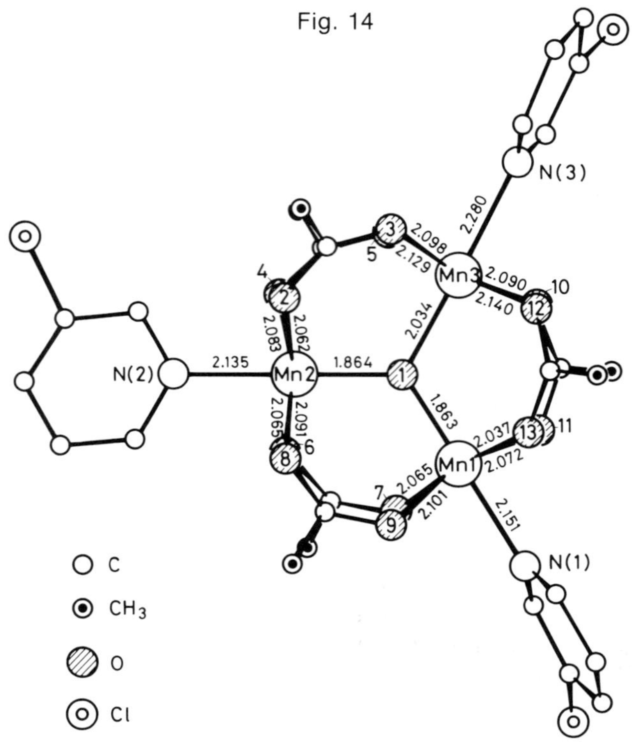

Molecular structure of $[Mn_3O(CH_3COO)_6(3\text{-}Cl\text{-}C_5H_4N)_3]$ with bond lengths (in Å) around the Mn atoms.

16.1.9 With Azido- and Cyanopyridines

16.1.9.1 With 2-Azidopyridine $N_3\text{-}C_5H_4N$ ($=C_5H_4N_4$)

A complex with 2-azidopyridine, which is known to exist mainly in its tetrazole form, is described together with manganese complexes with tetrazole compunds in „Manganese" D 4.

16.1.9.2 With 2-, 3-, and 4-Cyanopyridine $CN\text{-}C_5H_4N$ ($=C_6H_4N_2$)

$Mn(3\text{-}CN\text{-}C_5H_4N)_4(NCO)_2$ was prepared by dropwise addition of a cold aqueous solution of $Mn(NO_3)_2 \cdot 6H_2O$ (1 mmol) and NaOCN (2 mmol) to a solution of 3-cyanopyridine (5 mmol) in ether. On standing, colorless crystals appeared [1]. The magnetic moment and IR spectrum are reported and discussed together with those of the bis(cyanopyridine) complexes on p. 111.

MnL_2Cl_2. The complexes with L = 2-, 3-, 4-CN-C_5H_4N were prepared as described for $Ni(4\text{-}C_6H_4N_2)_2Cl_2$ [3] by shaking a solution of the hydrated metal chloride in a mixture of ethanol and 2,2-dimethoxypropane with an ethanolic solution of the ligand, both components in stoichiometric amounts. The thermal degradation of the chlorobis(cyanopyridine) complexes was studied by differential scanning calorimetry (DSC). The results – reaction type, temperature of the main peak (T_p in K), and the enthalpy of reaction ΔH in kJ/mol – are listed in the following table [2]:

ligand L	reaction	T_p	ΔH
2-CN-C_5H_4N	$MnL_2Cl_2 \rightarrow MnL_{1/3}Cl_2 + {}^5/_3 L$	485	99±3
	$MnL_{1/3}Cl_2 \rightarrow MnCl_2 + {}^1/_3 L$	570	6.0±0.5
3-CN-C_5H_4N	$MnL_2Cl_2 \rightarrow MnLCl_2 + L$	510	114±1
	$MnLCl_2 \rightarrow MnCl_2 + L$	650	26.9±0.9
4-CN-C_5H_4N	$MnL_2Cl_2 \rightarrow MnLCl_2 + L$	510	59±1
	$MnLCl_2 \rightarrow MnL_{1/3}Cl_2 + {}^2/_3 L$	575	39±2
	$MnL_{1/3}Cl_2 \rightarrow MnCl_2 + {}^1/_3 L$	635	25±1

An additional main peak at 555 K is observed on the DSC curve of Mn(3-CN-$C_5H_4N)_2Cl_2$. The stability and overall heat of thermal degradation of the bis(cyanopyridine) complexes decrease in the order L = 3-CN-C_5H_4N > 4-CN-C_5H_4N > 2-CN-C_5H_4N, and are lower than those of the corresponding cobalt and nickel complexes. The stoichiometries of the complexes and their decomposition products suggest structures similar to the pyridine analogues (p. 92), which all are polymeric octahedral [2].

MnL_2X_2 (L = 3-, 4-CN-C_5H_4N, X = NCO, N_3). The pale yellow complex Mn(3-CN-$C_5H_4N)_2$-$(NCO)_2$ was obtained by refluxing Mn(3-CN-$C_5H_4N)_4(NCO)_2$ in petroleum ether (b.p. 100 to 120°C) for 2 to 5 h. The yellow cyanatobis(4-cyanopyridine) complex formed on mixing $Mn(NCO)_2$ and ligand in any ratio. The azidobis(cyanopyridine) complexes can be prepared like Mn(3-CN-$C_5H_4N)_4(NCO)_2$ if NaN_3 is used instead of NaOCN and the $Mn(N_3)_2$ to ligand mole ratio is 1:2 [1].

The X-ray diffraction patterns of Mn(3-CN-$C_5H_4N)_2(NCO)_2$ and Mn(4-CN-$C_5H_4N)_2$-$(NCO)_2$ differ only slightly from those of the Ni, Co, and Fe^{II} complexes, which have six-coordinate metal ions; thus, the Mn^{II} compounds must have the same structure [1].

Below are listed the magnetic moments μ_{eff} in μ_B and paramagnetic Curie temperatures Θ_p (in K) from susceptibility measurements and the characteristic IR absorption bands of the complexes with their assignments [1]:

complex	μ_{eff} at 293.1 K	μ_{eff} at 173.1 K	μ_{eff} at 91.1 K	Θ_p
Mn(3-CN-$C_5H_4N)_4(NCO)_2$	5.97	6.00$^{1)}$	5.93	—
Mn(3-CN-$C_5H_4N)_2(NCO)_2$	5.77	5.44	4.89	−57
Mn(4-CN-$C_5H_4N)_2(NCO)_2$	5.54	5.36	4.93	−42
Mn(3-CN-$C_5H_4N)_2(N_3)_2$	—	—	—	—
Mn(4-CN-$C_5H_4N)_2(N_3)_2$	5.73	5.63	5.34$^{2)}$	−28

complex	IR $\nu(C\equiv N)$	$\nu_{as}(NCO)$ or $\nu_{as}(N_3)$	$\nu_s(NCO)$ or $\nu_s(N_3)$	$\delta(NCO)^{3)}$ or $\delta(N_3)$
Mn(3-CN-$C_5H_4N)_4(NCO)_2$	2235	2205	1334	628, 620
Mn(3-CN-$C_5H_4N)_2(NCO)_2$	2242, 2227	2185	1300	658, 629
Mn(4-CN-$C_5H_4N)_2(NCO)_2$	2240 sh	2200	1295	659, 630
Mn(3-CN-$C_5H_4N)_2(N_3)_2$	—	2090	1330	640, 603
Mn(4-CN-$C_5H_4N)_2(N_3)_2$	—	2090	1331	642, 610

$^{1)}$ At 193.1 K. – $^{2)}$ At 97.3 K. – $^{3)}$ Deformation mode.

In the IR spectrum the ν(C≡N) bands are shifted only slightly relative to those of the free ligands (3-cyanopyridine 2230 cm^{-1}, 4-cyanopyridine 2240 cm^{-1}), indicating that the cyano group is not coordinated to the manganese atom. The bis(cyanopyridine) complexes contain single-atom nitrogen bridges, Mn-N-Mn, rather than the three-atom anion bridges Mn-NCO-Mn or Mn-N-N-N-Mn. One indication is the splittings of the δ(NCO) and $\delta(N_3)$ bands, which are large compared with those of complexes containing terminally bonded NCO or N_3 groups. Another is the appearance of the $\nu_s(N_3)$ band as a medium strong band. (For silver azide, a three-atom bridged compound, $\nu_s(N_3)$ is so weak that it could not be detected.) Isomorphism relations between some cyanato and azido complexes show that the mode of bridging is the same in the two series of compounds

$Mn(3\text{-}CN\text{-}C_5H_4N_2)_4(NCO)_2$ has the regular and temperature-independent magnetic moment of a 6-coordinate high-spin Mn^{II} complex, whereas the moment of the bis(cyanopyridine) complexes is considerably lower and decreases distinctly at lower temperatures, obeying the Curie-Weiss law. These anomalously low magnetic moments suggest a weak antiferromagnetic exchange between the paramagnetic centers of the Mn-N-Mn bridged structures [1].

References:

[1] J. Nelson, S. M. Nelson (J. Chem. Soc. A **1969** 1597/603). – [2] G. Beech (Thermochim. Acta **3** [1972] 297/302, 298). – [3] R. A. Walton (J. Inorg. Nucl. Chem. **28** [1966] 2229/41, 2230).

16.1.10 With 2-, 3-, and 4-Methylpyridine $CH_3\text{-}C_5H_4N$
(= α-, β-, and γ-picoline = α-, β-, and γ-pic = C_6H_7N)

Complexes with methylpyridines and diketones are treated in "Mangan" D 1, 1979.

16.1.10.1 Complexes in Solution

Potentiometric measurements (Bjerrum's method) on aqueous solutions of $MnCl_2$ and HCl at 35°C and an ionic strength I = 1.4 (KCl) by titration with the ligands indicate the formation of 1:1 and 1:2 complexes with each picoline isomer. The stability constants K_1 and K_2 (in L/mol) and the calculated free enthalpies of formation ΔG_1 and ΔG_2 (in kcal/mol) are listed below [1]:

ligand	log K_1 (±0.05)	log K_2 (±0.05)	$-\Delta G_1$	$-\Delta G_2$
α-picoline	1.30	0.85	1.83	1.19
β-picoline	1.28	0.82	1.80	1.15
γ-picoline	1.32	1.17	1.86	1.64

The Mn^{II} complexes are less stable than the Cr^{III} complexes [1]. The Mn^{II} picoline complexes may be useful in the separation of single pyridine bases from mixtures [2].

References:

[1] P. S. Relan, P. K. Bhattacharya (J. Indian Chem. Soc. **46** [1969] 534/6). – [2] P. Tomasik, A. Woszczyk, F. Kret (Koks Smoła Gaz **21** [1976] 330/2; C.A. **87** [1977] No. 134941).

16.1.10.2 Isolated Complexes

16.1.10.2.1 Tetrakis(methylpyridine) Compounds

The compounds MnL_4X_2 (L = 3- and 4-methylpyridine) are prepared as described for the pyridine compounds on pp. 84/90. Magnetic measurements, IR spectra, and in some cases the electronic spectra show the complexes to be octahedral. ESR spectra indicate a tetragonal distortion from octahedral symmetry, this distortipn incrasing in the order X = NCS < Cl < Br < I. Comparison of the shift values of pyridine and 4-methylpyridine complexes suggests that back-donation from manganese to the ligand is unimportant in the 4-methylpyridine complexes, unlike in the pyridine complexes. The 3-and 4-methylpyridine compounds decompose on heating, as do the pyridine complexes. The selective formation of clathrates is observed for two 4-methylpyridine complexes: $Mn(\gamma\text{-pic})_4(NCS)_2$ with p-xylene and $Mn(\gamma\text{-pic})_4(HCOO)_2$ with o-xylene. These clathrates can be utilized for the separation of the xylene isomers.

Halides

$Mn(\beta\text{-pic})_4X_2$ (X = Cl, Br). The colorless or pinkish compounds with 3-methylpyridine were synthesized by refluxing the manganese halides with the ligand [1]. $Mn(\beta\text{-pic})_4Cl_2$ was also obtained by adding the ligand to a saturated aqueous solution of $MnCl_2$ [2], and $Mn(\beta\text{-pic})_4Br_2$ was obtained by adding β-picoline to an ethanolic solution of $MnBr_2$ [4].

The electronic reflectance spectrum of $Mn(\beta\text{-pic})_4Cl_2$ shows three spin-forbidden, thus very weak, bands at 550, 470, and 440 nm, which are assigned to d-d transitions from the ground state, $^6A_{1g}$, to the excited states $^4T_{1g}$, $^4T_{2g}$, and ^4E_g, $^4A_{1g}$, respectively. Far stronger are the charge-transfer bands, especially those of the bromo complex [4].

The far-IR spectrum of the chloro complex shows a band at about 235 cm^{-1}, assigned to the (Mn-Cl) stretching vibrations. A broad band at 230 cm^{-1} assigned to ν(Mn-N) vibrations is present in all complexes [1]. In the IR spectrum of $Mn(\beta\text{-pic})_4Br_2$ (Nujol or hexachlorobutadiene mulls) the following ligand bands were shifted by more than 10 cm^{-1} on coordination (free ligand bands in parentheses): ν_{13} at 1238 (1227), ν_{6b} at 646 (630), ν_{11} at 470 (454), and ν_{15} at 351 (337) cm^{-1}. A band at 222 cm^{-1} was assigned to ν(Mn-N) or ν(Mn-Cl) vibrations [4].

The ESR X-band spectra of $Mn(\beta\text{-pic})_4X_2$ recorded for powdered samples of the pure and diluted (in $Co(\beta\text{-pic})_4X_2$) compounds are interpreted in terms of the zero-field splitting parameters, D = 0.16 and λ = 0.10 cm^{-1} for X = Cl and D = 0.54 and λ = 0.01 cm^{-1} for X = Br. The parameters indicate a tetragonally distorted octahedral structure for both complexes (more clearly distorted for Br) [3].

The thermal degradation of $Mn(\beta\text{-pic})_4Cl_2$ and $Mn(\beta\text{-pic})_4Br_2$ with the loss of two molecules of β-picoline was studied by different methods (TG, DTG, T curves) [1, 2], heating rate 10 K/min [2]. The temperatures of the maximal decomposition rate were observed at 137°C [2] or at 160°C [1] for the chloro complex and at 180°C for the bromo complex [1]. The activation energy and the preexponential factor of the Arrhenius equation for the reaction $Mn(\beta\text{-pic})_4X_2(s) \rightarrow Mn(\beta\text{-pic})_2X_2(s) + 2\beta\text{-picoline}(g)$ were calculated from the DTG and TG data by the Coats-Redfern method to be E_A = 11.4 kcal/mol and A = 410 g/s. The Freeman-Carrol method yielded E_A = 10.5 kcal/mol and A = 646 g/s. The observed reaction order n = 0 indicates a diffusion-controlled reaction mechanism [2, 5].

$Mn(\gamma\text{-pic})_4X_2$ (X = Cl, Br, I). The pale pink chloro and bromo compounds with 4-methylpyridine are prepared in the same manner as the complexes with 3-methylpyridine [1, 2, 4]. $Mn(\gamma\text{-pic})_4I_2$ is obtained by adding a hot ethanolic solution of 4-methylpyridine to an ethanolic solution of MnI_2 in a 5:1 mole ratio; straw-colored needle-shaped crystals are washed with ether containing a little γ-picoline [6].

The IR spectra were obtained on samples dispersed in Nujol or hexachlorobutadiene mulls [4, 6], in discs, or in mulls [7]. The pertinent ligand bands were observed in the following frequency regions (in cm^{-1}), which are given together with the differing assignments of three authors:

frequency	3065	1615	1230	1220	1013	810,805	535
assignment [7] ..	ν(CH)	ν(ring)	δ(CH) i.p.	X-sens.	ν(ring)	X-sens.	X-sens.
	ν_1	ν_4	ν_6	ν_3	ν_9	ν_8	ν_{10}
assignment [4, 6]	—	ν_{8a}	ν_{9a}	ν_{13} [4], ν_{9a} [6]	ν_1	$\nu_{10b}+\nu_{12}$	ν_{6a}

The characteristic Raman bands appear at 1237 cm^{-1} (ν_6), 814 cm^{-1} (ν_8), and 538 cm^{-1} (ν_{10}) [7]. The IR bands are shifted by 8 to 25 cm^{-1}; the Raman bands, by 10 to 17 cm^{-1}, to higher wave numbers relative to the free ligand, the shifts indicating coordination of the ligand molecule via the nitrogen atom of the pyridine ring [4, 6, 7]. A band at 237 cm^{-1} for $Mn(\gamma\text{-pic})_4Br_2$ is assigned to Mn-N or Mn-Br vibrations [4]. The band shifts indicate that the metal-ligand bond strength in $Mn(\gamma\text{-pic})_4X_2$ is considerably weaker than that in the Fe, Co, and Ni complexes [7].

The ESR spectra of $Mn(\gamma\text{-pic})_4X_2$ are about the same as those of the β-picoline complexes, and the same distortion parameters have been evaluated for the chloro and the bromo complexes. For the iodo compound $D=0.87$ cm^{-1} and $\lambda=0.01$ were calculated. The parameters indicate an increasing tetragonal (axial) distortion of the octahedral complex on going from Cl^- to Br^- to I^-. Distortions from the axial symmetry are small as λ is very low [3].

During thermal analysis, $Mn(\gamma\text{-pic})_4Cl_2$ and $Mn(\gamma\text{-pic})_4Br_2$ release at 148°C [2] and 180°C [1], respectively, two molecules of γ-picoline. Decomposition of $Mn(\gamma\text{-pic})_4Cl_2$ to the mono(γ-picoline) complex, $Mn(\gamma\text{-pic})Cl_2$, at 240°C was observed by [1]. The activation energy and the preexponential factor of the Arrhenius equation for the degradation reaction calculated from the DTG and TG data by the Coats-Redfern method are $E_A=15.6$ kcal/mol and $A=8.6\times10^5$ $g^{0.5}/s$; the Freeman-Carrol method yielded $E_A=16.6$ kcal/mol and $A=2\times10^6$ $g^{0.5}/s$. The observed reaction order $n=0.5$ indicates a partially diffusion-controlled reaction [2, 5].

$(C_{16}H_{33}NC_5H_5)_2[MnCl_4]\cdot4L$ (L = 3- and 4-methylpyridine = β-pic, γ-pic). The hexadecylpyridinium compounds are prepared by dissolving $(C_{16}H_{33}NC_5H_5)_2MnCl_4$ in excess picoline while slowly heating. The light yellow crystals appear on cooling. The table lists the melting points (m.p. in °C), absorption maxima (ν_{max} in cm^{-1}) in the electronic spectra of solutions in acetone or alcohol, the effective magnetic moments μ_{eff} in μ_B, and the molar electrical conductivities Λ in $\Omega^{-1}\cdot cm^2\cdot mol^{-1}$ of solutions in nitrobenzene:

complex	m.p.	ν_{max}	μ_{eff}	Λ
$C_{16}H_{33}NC_5H_5[MnCl_4]\cdot4\beta$-pic	65	24390, 11220	5.86	66.3
$C_{16}H_{33}NC_5H_5[MnCl_4]\cdot4\gamma$-pic	70	25640, 11220	5.86	68.2

The electronic spectra differ from that of the tetrahedral complex ion $[MnCl_4]^{2-}$ (bands at 27000, 23810, 22470, and 20000 cm^{-1}) and that of the octahedral complex ion $[MnCl_4py_2]^{2-}$. Therefore it is assumed that the tetrakis(picoline) compounds have a different symmetry, possibly dodecahedral, with 8-coordinate manganese. They are formulated as $(C_{16}H_{33}NC_5H_5)_2[MnCl_4(pic)_4]$. The IR spectra (5000 to 650 cm^{-1}, not reported) show a sharp band at 3050 cm^{-1}, which is due to the NH stretching frequencies of the hexadecylpyridinium complexes. Two absorption bands at about 1600 and 1560 cm^{-1} are assigned to the CC and CN stretching vibrations of the pyridine rings and may indicate their coordination to Mn through the N atom. The complexes are 2:1 electrolytes in nitrobenzene, as shown by their molar conductivities. They lose two molecules of picoline upon heating at 100°C, leaving the bis(picoline) compounds [8].

References:

[1] J. R. Allan, D. H. Brown, R. H. Nuttall, D. W. A. Sharp (J. Inorg. Nucl. Chem. **27** [1965] 1865/7). – [2] D. Fătu, E. Segal (Rev. Roumaine Chim. **14** [1969] 709/15). – [3] R. D. Dowsing, J. F. Gibson, M. Goodgame, P. J. Hayward (J. Chem. Soc. A **1969** 187/93). – [4] N. S. Gill, H. J. Kingdon (Australian J. Chem. **19** [1966] 2197/211, 2201, 2202, 2208). – [5] I. G. Murgulescu, D. Fătu, E. Segal (J. Therm. Anal. **1** [1969] 97/106, 100/2).

[6] M. Goodgame, P. J. Hayward (J. Chem. Soc. A **1966** 632/4). – [7] S. Akyüz, J. E. D. Davies, K. T. Holmes (J. Mol. Struct. **42** [1977] 59/69, 64). – [8] B. Pradhan, D. V. Ramana Rao (J. Indian Chem. Soc. **56** [1979] 455/7).

Isocyanates, Isothiocyanates, and Isoselenocyanates

Mn(β-pic)$_4$(NCO)$_2$. The isocyanato complex with 3-methylpyridine was obtained by adding the ligand to an aqueous solution of $Mn(NO_3)_2 \cdot 6H_2O$ and KOCN. The compound melts at 150°C. Its IR spectrum in the 2200 cm^{-1} region indicates terminal NCO groups coordinated to Mn through their N atoms. The magnetic moment is $\mu_{eff} = 5.80\ \mu_B$. The molar electrical conductivity, $\Lambda = 6.4\ \Omega^{-1} \cdot cm^2 \cdot mol^{-1}$ in dimethylformamide, indicates the compound to be a nonelectrolyte [1].

Mn(β-pic)$_4$(NCS)$_2$. The isothiocyanato compound with 3-methylpyridine was obtained by reacting a manganese(II) salt with NH_4SCN and the ligand in ethanol [2]. The far-IR spectrum, which was recorded for Nujol mulls, shows two absorption bands, at 352 and 255 cm^{-1}, the latter assigned to the Mn-NCS stretching mode. The complex is stable under the usual conditions [2]. As shown by the TG curve (heating rate 3 K/min) the complex starts to lose β-picoline at about 80 to 90°C in an endothermic reaction, which on further heating proceeds continuously until all four moles of β-picoline are lost [3].

Mn(β-pic)$_4$(NCSe)$_2$ was prepared by reacting an ethanolic solution of $Mn(NCSe)_2$ (obtained from $MnCl_2$ and KSeCN) with β-picoline and (if required) refluxing the mixture for about 30 min. The pale pink crystalline compound melts at 145°C. Its IR spectrum shows two bands, one in the range between 2065 and 2040 cm^{-1}, the other between 788 and 738 cm^{-1}. The bands are assigned to the CN and CSe stretching vibrations of the SeCN group, respectively, indicating an Mn-NCSe coordination bond. The magnetic moment is $\mu_{eff} = 5.68\ \mu_B$. The molar electrical conductivity, $\Lambda = 9.5\ \Omega^{-1} \cdot cm^2 \cdot mol^{-1}$ in acetone, indicates the compound to be a nonelectrolyte. The complex turns red upon prolonged exposition to air [4].

Mn(γ-pic)$_4$(NCS)$_2$. The isothiocyanato compound with 4-methylpyridine was synthesized by reacting $MnCl_2 \cdot 4H_2O$, NH_4SCN or KSCN, and 4-methylpyridine (in a 1 to 2% excess) in aqueous solution [5]. The solid was dried in air at room temperature for 24 h [5, 6] and recrystallized from $CHCl_3$ [5, 6], acetone [6], or methanol [7]. The solvent was allowed to evaporate [5, 6]. Mn(γ-pic)$_4$(NCS)$_2$ crystallizes from methanol as colorless efflorescent quadrilateral prisms, which are best stored in their saturated methanolic solution. The density is $D_4^{20} = 1.371$ g/cm^3 (pycnometric). The refraction indices are $n_\alpha = 1.642$, $n_\beta = 1.558$, and $n_\gamma = 1.780$ [7].

The IR spectrum was obtained from KBr discs [9], discs, or mulls [8]; the Raman spectrum was obtained from solid samples [8]. Comparison of the IR and Raman spectra with those of the halogeno compounds show similar band positions in both cases. A larger shift, about +20 cm^{-1} (IR) and +30 cm^{-1} (Raman), is observed only for the δ(CH) out-of-plane band at 973 cm^{-1} for the free ligand. It is assumed that the isothiocyanato complexes are isomorphous with the halogeno complexes [8]. Assignments differing from those of [8] were given in [9]: bands at 1208 and 1228 cm^{-1} were assigned to δ(CH) out-of-plane, at 1020 cm^{-1} to ν(CC)+δ(CNC) and $\delta(CCH)_{CH_3}$, and at 495 cm^{-1} to a vibration of the NCS group. A comparison of the shift values of pyridine and 4-methylpyridine complexes suggests that back-donation from manganese to the ligand is much less important in the 4-methylpyridine complexes than in the pyridine complexes [8]. Anion bands at 2055 cm^{-1} (IR), 2060 cm^{-1} (Raman), and far-IR bands at 257 cm^{-1} [8], 248 [2] (IR), and 240 cm^{-1} (Raman) are assigned to the CN and Mn-NCS stretching vibrations. The ν(CN) band at 2055 cm^{-1} indicates terminal NCS groups which are bonded to Mn through their N atoms as in the related trans-octahedral Ni complex [8].

The ESR X-band spectrum of pure and diluted samples shows one line at $g_{eff}=2$. The zero-field splitting parameter $D=0.035$ cm^{-1} was calculated. λ is probably close to zero. The axial distortion of the O_h symmetry in $Mn(\gamma\text{-pic})_4(NCS)_2$ is much less than in the halogeno complexes [10], also see [11].

$Mn(\gamma\text{-pic})_4(NCS)_2$ on steady heating (3 K/min, under Ar) loses two molecules of 4-methylpyridine to give $Mn(\gamma\text{-pic})_2(NCS)_2$. The temperature of the maximum rate of decomposition was observed at 168°C [3]. In a previous work 220°C was found [7]. The solubility, in g/L, of $Mn(\gamma\text{-pic})_4(NCS)_2$ is found to be 7.8 at 20°C and 41.0 at 100°C in H_2O, 52.7 at 55°C in acetone, and 12.2 at 55°C in methanol. It is insoluble in CCl_4 and dichloroethane [7] but forms clathrates with aromatic hydrocarbons such as m-xylene [13], p-xylene [5, 9, 12, 13], naphthalene, and 1-bromonaphthalene [14].

Clathrates of $Mn(\gamma\text{-pic})_4(NCS)_2$. The clathrate with p-xylene was formed on shaking solid $Mn(\gamma\text{-pic})_4(NCS)_2$ with p-xylene and methanol, ethanol, or 2-propanol as diluents. It may be obtained also by adding p-xylene to a solution of the complex and evaporating the solvent ($CHCl_3$ or acetone) or crystallizing the clathrate (methyl cellosolve solvent). Single-crystal X-ray photographs yield a tetragonal lattice, space group $I4_1/a\text{-}C^6_{4h}$ (No. 88) with the lattice constants a=17.5 and c=26.0 Å [6]. The IR spectra of $Mn(\gamma\text{-pic})_4(NCS)_2$ and its p-xylene clathrate differ only slightly. Therefore it is assumed that the p-xylene molecules occupy interstitial positions in the $Mn(\gamma\text{-pic})_4(NCS)_2$ lattice and their bonding is very weak [13]. The selective forming of clathrates by $Mn(\gamma\text{-pic})_4(NCS)_2$ with m- and p-xylene enables their separation [15]. The Raman phonon spectra (at 50 K) of the clathrates with naphthalene and deuterated naphthalene and the electronic absorption spectrum of the clathrate with 1-bromonaphthalene (at 15 or 20 K) show very broad and inhomogenous absorption bands, which suggest a largely disordered structure [14]. The quenching of the fluorescence of anthracene in methanol and of the phosphorescence of biacetyl in H_2O by $Mn(\gamma\text{-pic})_4(NCS)_2$ was studied both in the liquid and glassy state of the solution (4×10^{-5} mol/L at 295 and 77 K). It was stated that $Mn(\gamma\text{-pic})_4(NCS)_2$ and anthracene or biacetyl, which form clathrates in the solid state with significant ground-state interaction, do not form these associations in solution. Thus such clathrates can not be invoked to explain the processes of luminescence decay in solution [16].

$Mn(\beta\text{-pic})_2(\gamma\text{-pic})_2(NCS)_2$. The compound was precipitated from an aqueous solution of $Mn(NCS)_2$ upon adding a mixture of 3- and 4-methylpyridine [17].

Mn(γ-pic)$_4$(NCSe)$_2$. The isoselenocyanato compound with 4-methylpyridine was prepared as described for the analogous β-picoline complex. The pale pink compound melts at 242°C. Its magnetic moment is $\mu_{eff} = 5.72\ \mu_B$. The molar electrical conductivity is $\Lambda = 8.6\ \Omega^{-1} \cdot cm^2 \cdot mol^{-1}$ in acetone, indicating a nonelectrolytic complex. The IR bands and chemical properties are similar to those of the β-picoline complex [4].

References:

[1] A. K. Das, D. V. Ramana Rao (Current Sci. [India] **42** [1973] 56/7). – [2] J. Burgess (Spectrochim. Acta A **24** [1968] 277/83, 278). – [3] T. A. Mal'kova (Issled. Khim. Khelatnykh Soedin. **1971** 37/42; C.A. **77** [1972] No. 83120). – [4] A. K. Das, D. V. Ramana Rao (Indian J. Chem. **12** [1974] 898/9). – [5] W. D. Schaeffer, W. S. Dorsey, D. A. Skinner, C. G. Christian (J. Am. Chem. Soc. **79** [1957] 5870/6, 5875).

[6] G. Gawalek, H. G. Könnecke (Rev. Chim. Acad. Rep. Populaire Roumaine **7** [1962] 875/99, 878, 897; C.A. **61** [1964] 5177). – [7] G. M. Panchenkov, A. I. Korosteleva (Zh. Neorgan. Khim. **15** [1970] 2573/5; Russ. J. Inorg. Chem. **15** [1970] 1333/4). – [8] S. Akyüz, J. E. D. Davies, K. T. Holmes (J. Mol. Struct. **42** [1977] 59/69, 64, 68), S. Akyüz (Hacettepe Bull. Natl. Sci. Eng. **4** [1975] 50/64; C.A. **86** [1977] No. 163066). – [9] G. M. Panchenkov, V. V. Basilevich, A. I. Korosteleva (Zh. Fiz. Khim. **46** [1972] 3084/7; Russ. J. Phys. Chem. **46** [1972] 1761/3). – [10] R. D. Dowsing, J. F. Gibson, M. Goodgame, P. J. Hayward (J. Chem. Soc. A **1969** 187/93, 188, 191).

[11] R. D. Dowsing, J. F. Gibson, D. M. L. Goodgame, M. Goodgame, P. J. Hayward (Nature **219** [1968] 1037/8). – [12] G. D. Jacobs, M. T. Lok (Inorg. Chem. **10** [1971] 1823/5). – [13] G. M. Panchenkov, V. V. Basilevich, A. I. Korosteleva (Tr. Mosk. Inst. Neftekhim. Gaz. Prom. No. **107** [1974] 140/2; C.A. **83** [1975] No. 177626). – [14] P. N. Prasad, J. C. Bellows, B. A. Bolton (J. Lumin. **18/19** [1979] 513/6). – [15] G. Gawalek, H.-G. Könnecke (Chem. Tech. [Leipzig] **15** [1963] 609/15, 614).

[16] A. Guarino, E. Possagno, R. Bassanelli (J. Chem. Soc. Faraday Trans. I **76** [1980] 2003/10, 2005/7). – [17] W. D. Schaeffer, A. C. McKinnis, W. S. Dorsey, Union Oil Co. of California (U.S. 2849511 [1953/58] 1/18, 11/3; C.A. **1959** 3145).

Other Tetrakis(methylpyridine) Compounds

Mn(γ-pic)$_4$(ClO$_4$)$_2$. The compound was obtained as described for the pyridine compound (p. 86) and has similar properties. The electronic reflectance spectrum shows bands at 23800 and 24700 cm^{-1}. The IR spectrum (mulls in Nujol and hexachlorobutadiene) shows ligand bands at 537 and 491 cm^{-1}, which were assigned to ligand ring vibrations. The free ligand bands were shifted on coordination towards higher frequencies by 24 and 8 cm^{-1}, respectively. Far-IR bands at 238 and 200 cm^{-1} were assigned to ν(Mn-N) vibrations. The coordination of the ClO_4^- ion as a unidentate ligand is suggested by a lowering of its symmetry from T_d to C_{3v} as shown by band splitting. The magnetic moment is $\mu_{eff} = 5.77\ \mu_B$. The corresponding tetrafluoroborate could not be isolated as a solid [1].

Mn(γ-pic)$_4$(HCOO)$_2$. To a solution of $MnCl_2 \cdot 4H_2O$ in warm water, formic acid and 4-methylpyridine are added. On cooling, a white solid precipitates. The compound forms a clathrate with o-xylene in preference to other xylene isomers [2].

Mn(γ-pic)$_4$(CX$_3$COO)$_2$ (X = F, Cl). The compounds were prepared as described for the pyridine compounds (p. 89). The magnetic moments are $\mu_{eff} = 5.90\ \mu_B$ for the compound with X = F and $\mu_{eff} = 5.94\ \mu_B$ for the compound with X = Cl [3].

References:

[1] D. H. Brown, R. H. Nuttall, J. McAvoy, D. W. A. Sharp (J. Chem. Soc. A **1966** 892/6). – [2] W. D. Schaeffer, Union Oil Co. of California (U.S. 2903456 [1956/59]; C. A. **1960** 25642). – [3] S. Amasa, D. H. Brown, D. W. A. Sharp (J. Chem. Soc. A **1969** 2892/4).

16.1.10.2.2 Tris(methylpyridine) Compounds

$R_4N[Mn(\gamma\text{-pic})_3BrCl_2]$ ($R = CH_3$, C_2H_5). The γ-picoline compounds were prepared in the same manner as the analogous pyridine complexes (p. 91) and have similar properties, A. K. Das, D. V. Ramana Rao (J. Indian Chem. Soc. **48** [1971] 823/7).

16.1.10.2.3 Bis(methylpyridine) Compounds

$Mn(\beta\text{-pic})_2X_2$ (X = Cl, Br). The compounds with 3-methylpyridine may be obtained by thermal degradation of $Mn(\beta\text{-pic})_4X_2$ [1, 5]. They can be synthesized also by reacting MnX_2 and 3-methylpyridine in ethanolic solution [2 to 4]. For $Mn(\beta\text{-pic})_2Cl_2$ a molecular weight of 690 (624 calculated for a dimer) was found by cryoscopic measurements in nitrobenzene [2]. The electronic reflectance spectrum of the pale pink complex is similar to that of $Mn(\beta\text{-pic})_4Cl_2$. It shows three weak bands, at 540, 490, and 425 nm, which were assigned to d-d transitions, and intense charge-transfer bands [1]. The IR spectra of both compounds show many ligand absorption bands, of which only those at about 1240, 1230, 650, 490, 470, and 350 cm^{-1} have been shifted by more than about 10 cm^{-1} on coordination [3]. The bands were assigned as for the tetrakis complexes, $Mn(\beta\text{-pic})_4X_2$ on p. 113. Somewhat different band positions are given elsewhere: 1612, 1590, 1210, 1068, 1055, 830, 818, 740, and 715 cm^{-1}. More ligand bands are split in the β-picoline compounds than in the γ-picoline compounds [2]. Two far-IR bands of $Mn(\beta\text{-pic})_2Cl_2$, at 222 and 204 cm^{-1}, can be assigned to the Mn-Cl or Mn-N stretching vibrations [3]. The magnetic moment of $Mn(\beta\text{-pic})_2Cl_2$ is $\mu_{eff} = 6.13\ \mu_B$ [2]. Results of conductivity measurements show the compounds to be nonelectrolytes [2].

The compounds $Mn(\beta\text{-pic})_2X_2$ decompose on steady heating according to the following scheme:

$$MnL_2X_2(s) \xrightarrow[\text{I}]{-L(g)} MnLX_2(s) \xrightarrow[\text{II}]{-\frac{1}{3}L(g)} MnL_{2/3}X_2(s) \xrightarrow[\text{III}]{-\frac{2}{3}L(g)} MnX_2(s)$$

The peak temperatures t_p (temperatures of the maximum rate of decomposition) corresponding to the single steps are listed below together with the methods of investigation:

compound	method	t_p(I)	t_p(II)	t_p(III)	Ref.
$Mn(\beta\text{-pic})_2Cl_2$	TG, DTG; 10°C/min	184	252	339	[5]
	DSC; 16°C/min	177	237	342	[4]
	TG	240	270	350	[1]
$Mn(\beta\text{-pic})_2Br_2$	TG	210	260	340	[1]

The table on p. 119 gives the enthalpy of decomposition ΔH in kcal/mol [4], the reaction order n, the activation energies E_A in kcal/mol, and the preexponential factor of the Arrhenius equation A (in g^{1-n}/s) for the degradation steps of $Mn(\beta\text{-pic})_2Cl_2$; n, E_A, and A were calculated by the Coats-Redfern (CR) and Freeman-Carrol (FC) methods [5, 6]:

reaction step	ΔH	n	E_A(CR)	E_A(FC)	A(CR)	A(FC)
I	13.6 ± 0.2	$^1/_3$	19.20	18.75	1.31×10^7	3.16×10^6
II	5.1 ± 0.1	$^1/_2$	33	34	2.3×10^{10}	2.5×10^9
III	13.4 ± 0.25	$^2/_3$	39.6	38.4	4.34×10^{11}	5.77×10^{10}

The observed reaction order n indicates a partially diffusion-controlled reaction [5, 6].

Mn(γ-pic)$_2$X$_2$ (X = Cl, Br). The pale pink [3], amorphous [2] compounds are prepared as described for the 3-methylpyridine compounds [1 to 3, 5, 7, 8], see p. 118. Mn(γ-pic)$_2$Br$_2$ is dimeric in nitrobenzene solutions as shown by cryoscopy [2].

The coordination-sensitive IR absorption bands are at 3070, 1620, 1232, 1217, 1208, 1018, 813, and 535 cm^{-1} [3, 7, 9]; the related Raman bands are at 1240, 1225, 820, and 542 [9]. The bands are assigned like those of the tetrakis compounds, Mn(γ-pic)$_4$X$_2$, on p. 114. They are shifted by 8 to 27 cm^{-1} with respect to the free ligand bands [3, 7, 9]. Further IR bands are given in [2]. The ν(Mn-N) bands are observed at 228 [9] or 230 cm^{-1} [8] for the complex with X = Cl and at 233 cm^{-1} for the complex with X = Br [9].

An interatomic distance Mn↔N of 2.27 ± 0.12 Å and a valence force constant of 0.80 were calculated from the ν(Mn-N) band of Mn(γ-pic)$_2$Cl$_2$ (230 cm^{-1}) [8]. The complexes have a polymeric, octahedral structure with bridging halogen atoms [2, 9].

The magnetic moment of Mn(γ-pic)$_2$Br$_2$ is μ_{eff} = 6.10 μ_B [2]. The complex behaves in conductivity experiments as a nonelectrolyte [2].

On heating, the compounds decompose like the 3-methylpyridine compounds. The following tables give the temperatures [1, 4, 5] and enthalpies of thermal decomposition [4] and kinetic values of the degradation steps for the 4-methylpyridine complexes [5, 6]:

compound	method	t_p(I)	t_p(II)	t_p(III)	Ref.
Mn(γ-pic)$_2$Cl$_2$	TG, DTG; 10°C/min	185	294	355	[5]
Mn(γ-pic)$_2$Cl$_2$	DSC; 16°C/min	167	267	342	[4]
Mn(γ-pic)$_2$Br$_2$	TG	240	280	340	[1]

reaction step*)	ΔH	n	E_A(CR)	E_A(FC)	A(CR)	A(FC)
I	15.1 ± 0.3	$^1/_2$	20.6	19.9	2.17×10^5	6.6×10^6
II	5.4 ± 0.15	$^1/_2$	41	42	8.47×10^{11}	5.5×10^{11}
III	14.0 ± 0.2	1	30	28.5	1.75×10^8	3.89×10^7

*) The reaction mechanisms resemble those of the β-picoline complexes described on p. 118 [5, 6].

(C$_{16}$H$_{33}$NC$_5$H$_5$)$_2$[MnCl$_4$L$_2$] (L = 2-, 3-, and 4-methylpyridine = α-pic, β-pic, and γ-pic). The hexadecylpyridinium compound with 2-methylpyridine was prepared by dissolving (C$_{16}$H$_{33}$NC$_5$H$_5$)$_2$MnCl$_4$ in excess ligand while heating slowly. The light yellow crystals appeared on cooling. The compound melts at 80°C. The electronic absorption spectrum of solutions in acetone or alcohol shows bands at 27000, 22730, and 16670 cm^{-1}, indicating an octahedral structure. The magnetic moment is μ_{eff} = 6.01 μ_B. The molar electrical conductivity of a solution in nitrobenzene is $\Lambda = 67.8\ \Omega^{-1} \cdot cm^2 \cdot mol^{-1}$. IR bands appear in the same regions as described for the corresponding tetrakis(pyridine) compound (p. 114) [12].

The complexes with 3- and 4-methylpyridine were obtained by heating the corresponding tetrakis complexes at 100°C to constant weight. They are intense yellow and their solutions in nitrobenzene show the molar conductivity $\Lambda = 63.9$ and 67.1 $\Omega^{-1} \cdot cm^2 \cdot mol^{-1}$, respectively [12].

$MnL_2(NCS)_2$ (L = 3- and 4-methylpyridine = β-pic and γ-pic). The compounds were prepared by heating the corresponding tetrakis compounds, p. 116, at 105°C to constant weight (3 h) [14]. $Mn(\gamma\text{-pic})_2(NCS)_2$ was also obtained by reacting $MnCl_2 \cdot 4H_2O$, NH_4SCN or KSCN, and 4-methylpyridine [13].

IR spectra and X-ray powder diagrams of the compounds $M(\beta\text{-pic})_2(NCS)_2$ with M = Mn, Co, and Ni are different, indicating different structures for these compounds [14]. The IR and Raman spectra of $Mn(\gamma\text{-pic})_2(NCS)_2$ were recorded and are discussed in [9]. The ligand band positions are similar to those of the halogeno complexes. In the IR spectrum the ν(CN), ν(CS), and ν(Mn-NCS) bands appear at 2102, 780, and 232 cm^{-1}, respectively, indicating N-coordinated bridging NCS groups [9].

$Mn(\gamma\text{-pic})_2(CX_3COO)_2$ (X = F, Cl). The trihalogenoacetato compounds were obtained by thermal degradation of $Mn(\gamma\text{-pic})_4(CX_3COO)_2$ at 100 to 140°C for X = F and very carefully below 60°C for X = Cl. The magnetic moments are $\mu_{eff} = 5.92$ and 5.90 μ_B, respectively, suggesting an octahedral configuration around the Mn atom with bidentate CX_3COO groups [10].

$Mn(\gamma\text{-pic})_2(C_6H_5COO)_2$. The benzoato compound was obtained by dissolving $Mn(C_6H_5COO)_2$ in γ-picoline while heating. The white crystalline solid appeared on standing. The IR spectrum shows two absorption bands, at 1625 and 1400 cm^{-1}, which are assigned to the antisymmetric and symmetric stretching vibrations of the bridging carboxylate groups. Susceptibility measurements yield the effective magnetic moments $\mu_{eff} = 5.44\ \mu_B$ at 293 K, 5.00 μ_B at 148 K, and 4.92 μ_B at 77 K [11].

References:

[1] J. R. Allan, D. H. Brown, R. H. Nuttall, D. W. A. Sharp (J. Inorg. Nucl. Chem. **27** [1965] 1865/7). – [2] S. Guru, K. C. Dash, D. V. Ramana Rao (Indian J. Chem. **6** [1968] 161/3). – [3] N. S. Gill, H. J. Kingdon (Australian J. Chem. **19** [1966] 2197/211, 2201, 2202, 2208). – [4] G. Beech, C. T. Mortimer, E. G. Tyler (J. Chem. Soc. A **1967** 1111/5, 1112). – [5] D. Fătu, E. Segal (Rev. Roumaine Chim. **14** [1969] 709/15).

[6] I. G. Murgulescu, D. Fătu, E. Segal (J. Therm. Anal. **1** [1969] 97/106, 100/2). – [7] M. Goodgame, P. J. Hayward (J. Chem. Soc. A **1966** 632/4). – [8] D. G. Brewer, P. T. T. Wong, M. C. Sears (Can. J. Chem. **46** [1968] 3137/41). – [9] S. Akyüz, J. E. D. Davies, K. T. Holmes (J. Mol. Struct. **42** [1977] 59/69, 62, 67). – [10] S. Amasa, D. H. Brown, D. W. A. Sharp (J. Chem. Soc. A **1969** 2892/4).

[11] A. A. Pasynskii, T. Ch. Idrisov, K. M. Suvorova et al. (Koord. Khim. **1** [1975] 799/803; C.A. **83** [1975] No. 125425). – [12] B. Pradhan, D. V. Ramana Rao (J. Indian Chem. Soc. **56** [1979] 455/7). – [13] W. D. Schaeffer, W. S. Dorsey, D. A. Skinner, C. G. Christian (J. Am. Chem. Soc. **79** [1957] 5870/7, 5875). – [14] T. A. Mal'kova (Issled. Khim. Khelatnykh Soedin. **1971** 37/42; C.A. **77** [1972] No. 83120).

16.1.10.2.4 Other Complexes with Methylpyridines

$MnLX_2$ and **$MnL_{2/3}X_2$** (L = β-pic, γ-pic; X = Cl, Br). The compounds were obtained by stepwise thermal degradation (during TG analysis) of the tetrakis- or bis(methylpyridine) compounds (see pp. 113, 114, 118, and 119) [1 to 3]. Mn(β-pic)Cl_2 was also prepared by isothermic heating of Mn(β-pic)$_2Cl_2$ at 95°C, and Mn(γ-pic)Cl_2 was also prepared by direct reaction of $MnCl_2$ with the ligand [2]. The compounds do not decompose up to 280°C and are almost insoluble in organic solvents [2].

The electronic reflectance spectra of both Mn(β-pic)Cl_2 and Mn_3(β-pic)$_2Cl_6$ show three absorption bands, at 540, 460, and 420 nm, which were assigned to d-d transitions and charge-transfer bands as reported for the tetrakis complex, Mn(β-pic)$_4Cl_2$ [1]. The IR spectra of the mono complexes are similar to those of the bis complexes Mn(β-pic)$_2X_2$ (p. 118) and Mn(γ-pic)$_2X_2$ (p. 119) [2, 4]. The molecular weights of Mn(β-pic)Cl_2, Mn(γ-pic)Cl_2, and Mn(γ-pic)Br_2 were determined (in nitrobenzene) to be 510, 480, and 640 (calculated for dimers: 438 and 616). The data indicate dimeric compounds in nitrobenzene [2]. The magnetic moments for these compounds are μ_{eff} = 6.10, 6.20, and 6.20 μ_B, respectively [2]. Conductivity measurements show the compounds to be nonelectrolytes [2]. The IR and electronic spectra favor six-coordinate manganese atoms in complexes with polymeric octahedral structures involving halogen bridging [1]. A polymeric structure is consistent with the other properties of the compounds [2]. The compounds decompose on heating to give, finally, anhydrous manganous salts. Temperatures, enthalpies, and the kinetics of the stepwise thermal decomposition of the compounds $MnLX_2$ are given on pp. 118 and 119. The decomposition of Mn(γ-pic)Cl_2 in one step to give $MnCl_2$ at 350°C (TG analysis in air) was observed by [1].

Mn(γ-pic)Cl_2·C_2H_5OH was synthesized by reacting anhydrous manganese(II) chloride and a slight excess of γ-picoline in absolute ethanol. The pale pink crystals were washed with alcohol and dried at a low temperature [6]. The magnetic moment is μ_{eff} = 5.88 μ_B at about 20°C [5]. The compound decomposes on heating or in boiling water, mineral acids, or alkalies (expelling the ligand) [6].

$(C_2H_5)_4N$[MnBrCl_2(γ-pic)]. The light green compound was prepared by refluxing a solution of $[(C_2H_5)_4N]_2[MnBr_2Cl_2]$ in ethanol with the required amount of pyridine for 1 h. It has properties like those of the mono(pyridine) complex $(C_2H_5)_4N$[MnBrCl_2py], p. 104 [7].

[Mn(C$(CH_3)_3$COO)$_2$(α-pic)]$_2$. The compound was prepared by adding α-picoline (1 mL) to a solution of Mn(C$(CH_3)_3$COO)$_2$ (0.009 mol) in benzene (30 mL) and concentrating the mixture at 50 to 60°C under 10 Torr of argon until colorless or yellowish green crystals appeared. The IR spectrum (KBr pellets) shows the antisymmetric and symmetric ν(OCO) vibration bands at 1625 and 1425 cm^{-1}, which indicate bridging carboxylato groups [8]. The magnetic moments from susceptibility measurements are 4.85, 4.34, 3.48, and 2.54 μ_B at 293, 221, 151, and 77 K, μ_{eff} decreasing monotonically for falling temperature. This behavior and the broad maxima of the χ vs. T curve characterize an antiferromagnetic dimer [9], also see [8]. The dependence of μ_{eff} on temperature can be described by the Heisenberg-Dirac-van Vleck model for a system with two exchange-coupled ions of spin $S = ^5/_2$ [8]. The isotropic g factor of 2.000 (±2.3% error) and isotropic exchange parameter J = 16 cm^{-1} were calculated from the magnetic data [9]. From the magnetic properties and the close coincidence of the IR spectrum with that of the similar Co compound, $C_7H_9NCo(C_6H_5COO)_4CoC_7H_9N$ (C_7H_9N = quinoline), the structure of which is known from X-ray investigation [10], it was concluded that the compound also has a dimeric structure of the so-called lantern type with four bridging carboxylic groups and two axial α-picoline molecules. Thus it was formulated $LMn(RCOO)_4MnL$ (L = α-picoline). It is easily oxidized to form a trinuclear oxo cluster, $[Mn_3O(C(CH_3)_3COO)_7]$ [8].

Mn(β-pic)$C_2O_4 \cdot H_2O$ and Mn(γ-pic)$C_2O_4 \cdot H_2O$. The complexes were prepared from $MnC_2O_4 \cdot 3H_2O$ and excess β- or γ-picoline in the same way as $Mn py C_2O_4 \cdot H_2O$, see p. 104. Susceptibility measurements at 303 K yielded the magnetic moment $\mu = 5.98\ \mu_B$ (β-pic) and 5.99 μ_B (γ-pic). The IR spectrum (Nujol mulls) shows bands assigned to ν(Mn-N) vibrations at 240 and 230 cm^{-1}, respectively. The slightly raised frequencies of the ν(CC) and ν(CN) vibration frequencies suggest coordination of the ligands by the N atom of the pyridine ring. The oxalate groups are coordinated to Mn in the same way as in $Mn py C_2O_4 \cdot H_2O$ to produce a polymeric octahedral structure. The complexes are almost insoluble in organic solvents. As shown by thermogravimetric analysis, Mn(γ-pic)$C_2O_4 \cdot H_2O$ loses its H_2O between 30 and 280°C and the ligand between 280 and 310°C to leave MnC_2O_4, which decomposes between 310 and 520°C [11].

$[Mn_3O(CH_3COO)_6(\beta\text{-pic})_3]ClO_4$. The Mn^{III} compound was synthesized like the pyridine complex (p. 106). A molecular weight of 440 was determined (calc. 914), which indicates strong dissociation. The complex melts at 193 to 195°C. The magnetic moment is $\mu_{eff} = 4.08\ \mu_B$ per Mn^{3+} ion at 298 K, indicating four unpaired electrons per metal ion. The electronic spectrum of the solution in methanol exhibits three absorption maxima, at 441, 260, and 255 nm (molar extinction coefficients ε = 740, 13000, and 12800 $L \cdot mol^{-1} \cdot cm^{-1}$), and a shoulder at 550 nm. The IR spectrum shows the expected ClO_4^- and bridging acetate bands as in $[Mn_3O(CH_3COO)_6py_3]ClO_4$, suggesting a trinuclear structure. The polarographic behavior and the solubilities are also much like those of the pyridine compound [12].

References:

[1] J. R. Allan, D. H. Brown, R. H. Nuttall, D. W. A. Sharp (J. Inorg. Nucl. Chem. **27** [1965] 1865/7). – [2] S. Guru, K. C. Dash, D. V. Ramana Rao (Indian J. Chem. **6** [1968] 161/3). – [3] D. Fătu, E. Segal (Rev. Roumaine Chim. **14** [1969] 709/15). – [4] M. Goodgame, P. J. Hayward (J. Chem. Soc. A **1966** 632/4). – [5] R. L. Mital (Indian J. Chem. **3** [1965] 88).

[6] S. Prasad, K. P. Kacker (J. Indian Chem. Soc. **35** [1958] 719/21). – [7] A. K. Das, D. V. Ramana Rao (J. Indian Chem. Soc. **48** [1971] 823/7). – [8] A. A. Pasynskii, T. C. Idrisov, K. M. Suvorova, V. T. Kalinnikov (Koord. Khim. **2** [1976] 1060/8; Soviet J. Coord. Chem. **2** [1976] 813/9, 815, 818). – [9] V. M. Novotortsev, Yu. V. Rakitin, A. A. Pasynskii, V. T. Kalinnikov (Dokl. Akad. Nauk SSSR **240** [1978] 355/7; Dokl. Chem. Proc. Acad. Sci. USSR **238/243** [1978] 220/2). – [10] J. Drew, M. B. Hursthouse, P. Thornton, A. J. Welch (J. Chem. Soc. Chem. Commun. **1973** 52/3).

[11] B. D. Pandey, D. C. Rupainwar (Current Sci. [India] **49** [1980] 336/9). – [12] S. Uemura, A. Spencer, G. Wilkinson (J. Chem. Soc. Dalton Trans. **1973** 2565/71, 2566, 2570).

16.1.11 With Dimethylpyridines $(CH_3)_2C_5H_3N$ (= C_7H_9N) and 2,4,6-Trimethylpyridine $(CH_3)_3C_5H_2N$ (= $C_8H_{11}N$)

L_1 = 2,4-dimethylpyridine (= α,γ-lutidine)
L_2 = 2,6-dimethylpyridine (= α,α′-lutidine)
L_3 = 3,5-dimethylpyridine (= β,β′-lutidine)
L_4 = 2,4,6-trimethylpyridine (= collidine)

16.1.11.1 Complexes in Solution

Potentiometric measurements (Bjerrum's method) on aqueous solutions of $MnCl_2$ (0.04 M) and HCl (0.09 M) at 35°C at ionic strength I = 1.4 (KCl) by titration with a 5 M ligand solution indicate the formation of 1:1 and 1:2 complexes with the ligands L_1, L_2, and L_4. The stability constants K_1 and K_2 (in L/mol) and the free enthalpies of formation ΔG_1 and ΔG_2 (in kcal/mol) are listed in the table below:

ligand	log K_1	log K_2	$-\Delta G_1$	$-\Delta G_2$
L_1	1.81	1.43	2.55	2.01
L_2	1.74	1.22	2.45	1.72
L_4	1.94	1.51	2.73	2.12

These lutidine and collidine complexes of manganese(II) are high-spin complexes [1]. The formation of manganese(II) complexes with these ligands may be used in the separation of single pyridine bases from pyridine-base mixtures [2].

16.1.11.2 Isolated Compounds

$Mn(L_3)_4X_2$ (X = Cl, Br, I). The complexes were prepared by refluxing MnX_2 in 3,5-dimethylpyridine. The resultant precipitates were washed with ether. The ESR X-band spectra resemble closely those of the corresponding complexes with β- or γ-picoline (pp. 113 and 114) and are also indicative of a tetragonally distorted octahedral structure [4].

$Mn(L_3)_4X_2$ (X = NCO, NCS, NCSe). The isocyanato and isoselenocyanato complexes were obtained as described for the corresponding pyridine complexes (p. 86) [5, 7]. The isothiocyanato complex was prepared as described for the 3-chloropyridine complex (p. 108) [6, 7]. The table below lists the colors, melting points m.p. (in °C), the magnetic moments μ_{eff} (in μ_B), and the molar electrical conductivities Λ (in $\Omega^{-1}\cdot cm^2\cdot mol^{-1}$) in the solvents given:

complex	color	m.p.	μ_{eff}	Λ	solvent	Ref.
$Mn(L_3)_4(NCO)_2$	colorless	175	5.65	8.5	DMF	[5]
$Mn(L_3)_4(NCS)_2$	colorless	215	5.83	18.6	aprotic solvent	[6]
$Mn(L_3)_4(NCSe)_2$	slight pinkish	230	5.65	11.6	acetone	[7]

The IR spectrum of the isocyanato complex shows two absorption bands, at 2190 and 1330 cm^{-1}, which are assigned to the CN and CO stretching vibrations of the NCO groups [5]. The corresponding bands of the isoselenocyanato complex were found at 2065 to 2040 cm^{-1} and 788 to 738 cm^{-1} [7]. The far-IR spectrum of the isothiocyanato complex shows a doublet at 275 and 244 cm^{-1}, which was assigned to ν(Mn-NCS) vibrations [8]. As shown by the physical data, all of the pseudohalogeno complexes contain 6-coordinate manganese(II) atoms with five unpaired electrons (d^5) in an octahedral environment. They are nonelectrolytes in DMF [5 to 7]. The pseudohalogenide ions are terminally coordinated to Mn through their nitrogen atoms [5, 7, 8]. The complexes are soluble in common organic solvents [7].

$R_4N[MnBrCl_2(L_3)_n]$ ($R = CH_3$, C_2H_5; n = 1, 3). The compounds were prepared by refluxing solutions of $[(CH_3)_4N]_2[MnBr_2Cl_2]$ in methanol or $[(C_2H_5)_4N]_2[MnBr_2Cl_2]$ in ethanol with the required amount of 3,5-dimethylpyridine for 1 h. The tris complexes are pink, while the mono complex is light green. All the compounds are crystalline and do not melt or decompose up to 250°C. The table below lists the magnetic moments μ_{eff} (in μ_B) and the molar electrical conductivities Λ (in $\Omega^{-1} \cdot cm^2 \cdot mol^{-1}$) in acetone.

complex	μ_{eff}	Λ
$(CH_3)_4N[MnBrCl_2(L_3)_3]$	6.01	121
$(C_2H_5)_4N[MnBrCl_2(L_3)_3]$	5.85	90
$(C_2H_5)_4N[MnBrCl_2(L_3)]$	5.80	111

The electronic spectrum of a $(C_2H_5)_4N[MnBrCl_2(L_3)]$ solution in acetonitrile shows four absorption maxima, at 28570, 25640, 22470, and 20640 cm^{-1}. The IR spectra have features similar to those of the pyridine compounds. On account of these similar properties the structures are assumed to also be similar [9].

$Mn(L_3)_2X_2$ (X = Cl, Br). The bis compounds were obtained by reacting stoichiometric amounts of MnX_2 and 3,5-dimethylpyridine in ethanol. Both compounds are colorless powders which do not melt or decompose up to 260°C. The magnetic moments are $\mu_{eff} = 5.59$ μ_B for the complex with Cl and 6.06 μ_B for that with Br. Conductivity measurements in aprotic media show the complexes to be nonelectrolytes [6].

$Mn(L_3)_2aca_2$ (acaH = 2,4-pentanedione). The compound is treated in “Mangan” D 1, 1979, p. 80.

$Mn(L_2)Cl_2$ and **$Mn(L_4)Cl_2$** were obtained by refluxing $MnCl_2$ in 2,6-dimethylpyridine or 2,4,6-trimethylpyridine. The resulting solids were washed with ether. The ultraviolet and visible spectra are similar to that reported for the β-picoline compound on p. 121; the IR spectra are similar to that reported for the pyridine compound on p. 95. On heating, $Mn(L_2)Cl_2$ and $Mn(L_4)Cl_2$ lose their organic component at 300 and 350°C, respectively, to leave $MnCl_2$ [3].

References:

[1] P. S. Relan, P. K. Bhattacharya (J. Indian Chem. Soc. **46** [1969] 534/6). – [2] P. Tomasik, A. Woszczyk, F. Kret (Koks Smoła Gaz **21** [1976] 330/1; C. A. **87** [1977] No. 134941). – [3] J. R. Allan, D. H. Brown, R. H. Nuttall, D. W. A. Sharp (J. Inorg. Nucl. Chem. **27** [1965] 1865/7). – [4] R. D. Dowsing, J. F. Gibson, M. Goodgame, P. J. Hayward (J. Chem. Soc. A **1969** 187/93, 188). – [5] A. K. Das, D. V. Ramana Rao (Current Sci. [India] **42** [1973] 56/7).

[6] A. K. Das, D. V. Ramana Rao (J. Inst. Chem. [India] **42** [1970] 127/30, 128; C. A. **74** [1971] No. 49141). – [7] A. K. Das, D. V. Ramana Rao (Indian J. Chem. **12** [1974] 898/9). – [8] J. Burgess (Spectrochim. Acta A **24** [1968] 277/83, 278). – [9] A. K. Das, D. V. Ramana Rao (J. Indian Chem. Soc. **48** [1971] 823/7).

16.1.12 With 3- and 4-Ethylpyridine (= $3\text{-}C_7H_9N$ or $4\text{-}C_7H_9N$)

$Mn(3\text{-}C_7H_9N)_4X_2$ (X = Cl, Br). The chloride and bromide were prepared by a method of [1], by refluxing MnX_2 in the ligand [2], also see [3]. The ESR X-band spectra resemble closely those of the corresponding complexes of MnX_2 with β- and γ-picoline (pp. 113 and 114) and may be interpreted likewise as indicative of a tetragonally distorted octahedral structure [2, 3]. The X-band spectrum of the compound $[Co,Mn(3\text{-}C_7H_9N)_4(H_2O)_2]Br_2$ was recorded and is discussed in [3].

$Mn(3\text{-}C_7H_9N)_4(NCO)_2$ was obtained by adding the stoichiometric amount of 3-ethylpyridine to an aqueous solution of $Mn(NO_3)_2 \cdot 6H_2O$ and KOCN. The dry compound melts at 108°C. Its IR spectrum shows absorption bands at 2195 (strong) and 1336 (medium) cm^{-1}, which are assigned to the CN and CO stretching vibrations of the NCO groups. The position of the ν(CN) band indicates terminal coordination of the NCO groups to Mn through their N atoms. The magnetic moment is $\mu_{eff} = 5.81\ \mu_B$, suggesting an octahedral high-spin complex of 6-coordinate divalent manganese. The molar conductivity, $\Lambda = 10.6\ \Omega^{-1} \cdot cm^2 \cdot mol^{-1}$ in dimethylformamide, indicates a nonelectrolyte [4].

$Mn(3\text{-}C_7H_9N)_4(NCS)_2$ was prepared like the β-picoline complex (p. 115). Their properties are also similar. The far-IR spectrum shows an absorption band at 255 cm^{-1}, which is assigned to the Mn-NCS stretching vibrations, and a weak unassigned band at 356 cm^{-1} [5].

$[Mn(3\text{-}C_7H_9N)_4(H_2O)_2]I_2$ was obtained by adding a large excess of 3-ethylpyridine to a hot solution of manganese(II) iodide in acetone. The IR spectrum shows a strong broad band at about 3300 cm^{-1} which was ascribed to the OH stretching vibrations of the H_2O molecule. The bands at 588 and 315 cm^{-1} were assigned to H_2O rocking and (tentatively) Mn-O stretching vibrations, respectively. The ESR X-band spectrum shows a multitude of bands, all at fairly low field (below 6000 G) and centered on $g_{eff} = 2$. The distortion parameters evaluated, $D = 0.057\ cm^{-1}$ and $\lambda (= E/D) = 0.033$, indicate a trans-octahedral structure for the dihydrate, with four 3-ethylpyridine and two H_2O molecules coordinated to the manganese atom [3].

$Mn(3\text{-}C_7H_9N)_2X_2$ (X = Cl, Br). The compounds were prepared according to [6] by reacting stoichiometric amounts of the manganese(II) halides and the ligand in ethanol. The precipitates were washed with ethanol and petroleum ether. The white amorphous compounds do not melt or decompose up to 260°C. Susceptibility measurements yielded the magnetic moment $\mu_{eff} = 5.50\ \mu_B$ for the complex with X = Cl and 6.00 μ_B for that with X = Br. Conductivity experiments indicated the compounds to be nonelectrolytes. The properties point to an octahedral, polymeric halogen-bridged structure [7].

$Mn(4\text{-}C_7H_9N)_4X_2$ (X = Cl, CN, NCO, NCS, HCOO). The chloro complex was prepared by reacting concentrated aqueous solutions of $MnCl_2$ and 4-ethylpyridine in a 1:4 mole ratio [8, 9]. The cyano, isocyanato, and isothiocyanato complexes were obtained by adding NaCN [9], NaOCN [8, 9], or $Na(K, NH_4)SCN$ [8, 10] to the $MnCl_2$ solution (mole ratio 2:1) and then slowly adding the organic ligand [8, 9]. The cyano complex may also be synthesized by reacting a solution of 4-ethylpyridine (0.4 mol) in aqueous ethanol (20 vol%) with aqueous solutions of KCN (0.2 mol) and $MnCl_2 \cdot 4H_2O$ (0.1 mol) [11]. $Mn(4\text{-}C_7H_9N)_4(NCS)_2$ may be recrystallized from chloroform or acetone. The solvent is removed by evaporation. The calorimetrically determined enthalpy of formation of $Mn(4\text{-}C_7H_9N)_4(NCS)_2$ from one mole of $Mn(NCS)_2$ and four moles of 4-ethylpyridine is $\Delta H = -18.760$ kcal/mol (±1.6%) [10]. $Mn(4\text{-}C_7H_9N)_4(HCOO)_2$ was obtained as described for $Mn(\gamma\text{-pic})_4(HCOO)_2$ on p. 117 [13].

The chloro complex is pink and melts at 164°C with decomposition [8, 9]. The tan cyano complex turns brown at 180°C and melts above 300°C [9, 11]. The tan isocyanato complex melts at 126°C [8, 9]. The formato [13], isothiocyanato [8], and isoselenocyanato [12] complexes are colorless [8, 12, 13] or light pink. The last is crystalline and melts at 165°C [12].

The far-IR spectrum of $Mn(4\text{-}C_7H_9N)_4(NCS)_2$ shows an absorption band at 259 cm^{-1}, which is assigned to the Mn-NCS stretching vibrations [13]. Susceptibility measurements at 20.2°C yield the magnetic moment $\mu_{eff} = 6.03\ \mu_B$ [10]. $Mn(4\text{-}C_7H_9N)_4(NCS)_2$ is stable under normal conditions [13]. The vapor pressure increases from about log (p/Torr) = 0.3 at 320 K to about 0.95 at 357 K, according to a graph [10, 14]. Solubility determinations in aromatic hydrocarbons at room temperature yielded the following contents in wt%: benzene 28.6, toluene 9.13, m-xylene 3.80, p-xylene 1.71, and ethylbenzene 4.78 [10].

The complexes $Mn(4\text{-}C_7H_9N)_4X_2$ with X = CN, NCO, and NCS are insoluble in water [8 to 11, 15, 16]; the complexes with X = Cl and HCOO are somewhat soluble [13]. All are soluble in chloroform [8 to 11, 13, 15, 16].

The complexes $Mn(4\text{-}C_7H_9N)_4X_2$ were found to form crystalline clathrate compounds with hydrocarbons such as the xylenes, cumenes, methylnaphthalenes, etc. In addition, they very selectively form clathrates with one xylene isomer, thus enabling a separation [8 to 11, 13 to 16]. The complexes with X = Cl, CN, NCO [8, 9, 11, 15, 16], and HCOO [13] selectively clathrate o-xylene, in preference to the other C_8 alkylbenzene isomers, while $Mn(4\text{-}C_7H_9N)_4(NCS)_2$ selectively clathrates p-xylene [8 to 10, 14 to 16].

Clathrates of $Mn(4\text{-}C_7H_9N)_4(NCS)_2$ with a xylene isomer or ethylbenzene were obtained by shaking or stirring the solid complex with the hydrocarbon in a diluent such as methanol, ethanol, or 2-propanol for 10 to 15 min [8]. The clathrates may also be prepared by adding the hydrocarbon to a solution of the complex and evaporating the solvent (chloroform and acetone solvents) or cooling the solution (methyl cellosolve solvent). The clathrates crystallize as octahedra and contain about one molecule of hydrocarbon per molecule of complex. In mixtures of different hydrocarbons the least soluble clathrate always forms first. The series of increasing solubility is p-xylene < m-xylene < ethylbenzene < o-xylene [10], which is about the same as the stability series obtained by calorimetric measurements. The table below lists the enthalpies ΔH of formation from $Mn(4\text{-}C_7H_9N)_4(NCS)_2$ and hydrocarbon, the volume contraction ΔV on clathrate formation [10, 14], the specific magnetic susceptibilities χ_g, and the magnetic moments μ_{eff} [10, 14]:

hydrocarbon	$-\Delta H$ in cal/mol	ΔV in %	$\chi_g \times 10^6$ at t in °C in cm^3/g		μ_{eff} in μ_B
p-xylene	5300	4.41	20.80	18.4	5.96
ethylbenzene	4205	3.14	21.05	17.6	5.99
m-xylene	3480	2.06	21.57	18.1	6.07
o-xylene	3090	1.63	21.80	19.0	6.11

All of the clathrates [10, 14] are rather unstable in solution [10]. In addition, vapor pressures, microscopic and kinetic observations, and separations of m- and p-xylene using various diluents are given in [10, 14]. The results suggest a weak bonding between complex and hydrocarbon. The selectivity of the clathrate formation may be due to the differing polarizabilities of the hydrocarbons.

$Mn(4\text{-}C_7H_9N)_4(NCSe)_2$ was prepared and shows properties similar to those described for $Mn(\beta\text{-pic})_4(NCSe)_2$ on p. 115. It is soluble in common organic solvents [12].

References:

[1] J. R. Allan, D. H. Brown, R. H. Nuttall, D. W. A. Sharp (J. Inorg. Nucl. Chem. **27** [1965] 1865/7). – [2] R. D. Dowsing, J. F. Gibson, M. Goodgame, P. J. Hayward (J. Chem. Soc. A **1969** 187/93, 188). – [3] M. Goodgame, P. J. Hayward (J. Chem. Soc. A **1971** 3406/9). – [4] A. K. Das, D. V. Ramana Rao (Current Sci. [India] **42** [1973] 56/7). – [5] J. Burgess (Spectrochim. Acta A **24** [1968] 277/83, 278).

[6] S. Guru, K. C. Dash, D. V. Ramana Rao (Indian J. Chem. **6** [1968] 161/3). – [7] A. K. Das, D. V. Ramana Rao (J. Inst. Chem. [India] **42** [1970] 127/30; C.A. **74** [1971] No. 49141). – [8] W. D. Schaeffer, W. S. Dorsey, D. A. Skinner, C. G. Christian (J. Am. Chem. Soc. **79** [1957] 5870/6, 5872). – [9] W. D. Schaeffer, Union Oil Co. of California (U.S. 2827463 [1954/58]; C.A. **1960** 21681). – [10] G. Gawalek, H. G. Könnecke (Rev. Chim. Acad. Rep. Populaire Roumaine **7** [1962] 875/99, 878/80, 884, 895/6; C.A. **61** [1964] 5177).

[11] W. D. Schaeffer, Union Oil Co. of California (U.S. 2876226 [1956/59]; C.A. **1959** 16164). – [12] A. K. Das, D. V. Ramana Rao (Indian J. Chem. **12** [1974] 898/9). – [13] W. D. Schaeffer, Union Oil Co. of California (U.S. 2903456 [1956/59]; C.A. **1960** 25642). – [14] G. Gawalek, H. G. Könnecke (Chem. Tech. [Leipzig] **15** [1963] 609/15). – [15] R. N. Fleck, C. G. Wight, Union Oil Co. of California (U.S. 2983767 [1959/61] 1/10, 7; C.A. **1961** 26572).

[16] W. D. Schaeffer, Union Oil Co. of California (U.S. 2798891 [1954/57] 1/28, 17/20; C.A. **1958** 2087).

16.1.13 With 4-Isopropylpyridine $(CH_3)_2CH\text{-}C_5H_4N$ $(=C_8H_{11}N)$

$Mn(4\text{-}C_8H_{11}N)_4(NCS)_2$. The isothiocyanate is prepared as described for the pyridine complex on p. 86. Its far-IR spectrum shows two absorption bands, at 319 and 259 cm^{-1}, the latter assigned to the Mn-NCS stretching mode, J. Burgess (Spectrochim. Acta A **24** [1968] 277/83, 278).

16.1.14 With Vinyl- and Polyvinylpyridine Compounds

16.1.14.1 With 4-Vinylpyridine $CH_2{=}CH\text{-}C_5H_4N$ $(=C_7H_7N=L)$

$MnL_4(NCS)_2$ and **MnL_nX_2** ($X=N_3$, Cl, Br, NCO; n=2, 1). The chloro and bromo complexes were prepared by reacting stoichiometric amounts of MnX_2 (X=Cl, Br) and 4-vinylpyridine in ethanolic solution [1, 2]. The azido, isocyanato, and isothiocyanato complexes were obtained from stoichiometric amounts of $Mn(NO_3)_2$, MX (M=K, Na, and $X=N_3$, OCN, SCN), and 4-vinylpyridine in aqueous ethanol [2].

The two tables list the colors, the magnetic moments μ_{eff} at 300 K (in μ_B), the absorption frequencies (in cm^{-1}) observed in the electronic spectra together with the corresponding excited states (ground state $^6A_{1g}$) and the frequencies of the principal pseudohalogen, Mn-X, and Mn-L vibrations in the IR spectrum:

complex	color	μ_{eff}	$^4T_{1g}(G)$	$^4T_{2g}(G)$	$^4A_{1g}(G)$, $^4E_g(G)$	Ref.
$MnL_4(NCS)_2$	cream	—	19450	—	—	[2]
MnL_2Cl_2	pale pink	6.10	18950	21500	23700	[2]
MnL_2Br_2	pink	6.18	18900	21600	23550	[2]
$MnL_2(N_3)_2$	pale yellow	5.64	16400, 19800	21800*)	23000*)	[2]
$MnL_2(NCO)_2$	cream	6.04	17500*)	—	—	[2]
$MnLCl_2$	light brown	5.98	—	—	—	[1]
$MnLBr_2$	light brown	6.00	—	—	—	[1]

*) at 90 K

complex	ν_{as}(X)	ν_s(X)	δ(X)	ν(Mn-X)	ν(Mn-L)	Ref.
$MnL_4(NCS)_2$	2050	813	485	— —	— —	[2]
MnL_2Cl_2	—	—	—	170, 160	244, 199	[2]
MnL_2Br_2	—	—	—	135, 118	— 191	[2]
$MnL_2(N_3)_2$	2100	1339, 1285	651, 615, 609	273 —	228, 202	[2]
$MnL_2(NCO)_2$	2200	1325, 1305	654, 633, 625	— —	— —	[2]

Further far-IR bands (unassigned) as well as electronic bands at 90 K are given in the paper. The pseudohalogen vibrations are similar to those of equivalent cobalt and nickel complexes, indicating N bonding to the metal [2]. The IR spectra of the halogenides were obtained as Nujol mulls. The coordination-sensitive ligand bands of the bis complexes MnL_2X_2 appear at the following frequencies (in cm^{-1}, free ligand bands in parentheses): 1615 (1600), 1440, 1425 (1422), 1025, 1010 (1010), 885, 855 (852), and 818 (808). For the mono complexes $MnLX_2$ far fewer bands were observed than for the bis complexes. Only those at about 1620 and 1030 cm^{-1} were shifted appreciably relative to the free ligand bands [1].

Low-temperature magnetic measurements yielded the magnetic moments μ_{eff} = 4.43 μ_B at 90 K and the paramagnetic Curie temperature Θ_p = 110 K for $MnL_2(N_3)_2$ and μ_{eff} = 5.38 μ_B at 90 K and Θ_p = 38 K for $MnL_2(NCO)_2$. The marked change in magnetic moment with temperature and the large paramagnetic Curie temperatures are indicative of antiferromagnetic exchange [2].

All of the halogenides do not melt or decompose up to 280°C. They are almost insoluble in organic solvents. Molecular-weight determinations show them to be dimeric in nitrobenzene. According to the magnetic moments they are high-spin complexes. The halogenides MnL_2X_2 are nonelectrolytes (solvent not given) [1].

On the basis of their properties all of the complexes are considered to be 6-coordinate halogen- or pseudohalogen-bridged polymers, except for $MnL_4(NCS)_2$, which is a 6-coordinate monomer [1, 2].

$R_4N[MnBrCl_2(C_7H_7N)_n]$ (R = CH_3, C_2H_5; n = 1, 3). The compounds are prepared in the same manner as described for the complexes with 3,5-dimethylpyridine on p. 123. The tris(4-vinylpyridine) compounds are pink. $(C_2H_5)_4N[MnBrCl_2(C_7H_7N)]$ is light green. All of them are crystalline and do not melt or decompose up to 250°C. The table below lists the magnetic moments μ_{eff} and the molar electrical conductivities Λ in acetone:

complex	μ_{eff} in μ_B	Λ in $\Omega^{-1}\cdot cm^2\cdot mol^{-1}$
$(CH_3)_4N[MnBrCl_2(C_7H_7N)_3]$	5.88	84
$(C_2H_5)_4N[MnBrCl_2(C_7H_7N)_3]$	6.00	88
$(C_2H_5)_4N[MnBrCl_2(C_7H_7N)]$	5.70	110

The electronic spectrum of a $(CH_3)_4N[MnBrCl_2(C_7H_7N)_3]$ solution in acetonitrile shows four absorption maxima, at 27000, 25000, 18520, and 16940 cm^{-1}. The IR spectrum shows the features described for the tris(pyridine) compound. All three compounds are high-spin $Mn^{II}(d^5)$ complexes and 1:1 electrolytes in acetone. The Mn atom is six-coordinate in $R_4N[MnBrCl_2(C_7H_7N)_3]$, as suggested also by the electronic spectrum, and probably four-coordinate in $R_4N[MnBrCl_2(C_7H_7N)]$ [3].

References:

[1] S. Guru, K. C. Dash, D. V. Ramana Rao (Indian J. Chem. **6** [1968] 161/3). – [2] R. J. Collin, L. F. Larkworthy (J. Inorg. Nucl. Chem. **37** [1975] 334/8). – [3] A. K. Das, D. V. Ramana Rao (J. Indian Chem. Soc. **48** [1971] 823/7).

16.1.14.2 With Polymerization Products of 4-Vinylpyridine and 2-Methyl-5-vinylpyridine

A complex with polymeric 4-vinylpyridine was obtained by dropping an ethanolic solution of the polymeric ligand (1.85 g) into a solution of $MnCl_2$ (0.3 g) in absolute ethanol under N_2. The light pink flocculent precipitate was washed with absolute ethanol and dried in a high vacuum. The IR spectrum shows a strong absorption band at 1606 cm^{-1}, which may be assigned to a CC or CN stretching mode. The manganese atom seems to be coordinated by two 4-vinylpyridine units and four chlorine atoms. The complex turns dark red brown on drying in air [1].

A polymeric product containing 5.61 mol% of $MnCl_2$ was obtained by reacting a solution of a copolymerization product of ethylene and 2-methyl-5-vinylpyridine (11.9 mol%, 1.5 parts) in xylene with an aqueous solution of $MnCl_2$ (0.45 parts) under Ar at 90°C for 1 h. The product melts above 250°C and is only slightly soluble in solvents for polyethylene [2].

References:

[1] H. G. Biedermann, P. Seidl (Makromol. Chem. **177** [1976] 631/43, 632, 639). – [2] K. Gehrke, J. Ulbricht, VEB-Kombinat Chem. Werke Buna Schkopau (Ger. [East] 130534 [1971/78] 1/7, 4; C.A. **90** [1979] No. 122286).

16.1.15 With 4-Phenylpyridine $C_6H_5\text{-}C_5H_4N$ (= $C_{11}H_9N$)

$Mn(C_{11}H_9N)_2Cl_2$ is prepared by adding a methanolic solution of 4-phenylpyridine (0.02 mol) to a warm solution of $MnCl_2 \cdot 4H_2O$ (0.01 mol) in a mixture of dry methanol and triethyl orthoformate (volume ratio 4:1). The precipitate was collected, washed with ether, air-dried under suction, and then recrystallized from methanol to give pale cream-colored needles. Susceptibility measurements at temperatures ranging from 1.6 to 70 K (at field strengths of 0 to 20 kOe) reveal a maximum magnetic susceptibility of about 270×10^{-3} cm^3/mol at 4.2 K. Above 17 K, the Curie-Weiss law is obeyed with the paramagnetic Curie temperature $\Theta_p = -11.1$ K and g = 2.03. Below 17 K the data increasingly deviate from that law, and a broad maximum occurs at 4.2 K. The susceptibility is independent of the magnetic field up to 10 kOe. The magnetic data may be fit by the scaled form of the infinite-spin-linear-chain equation modified by a molecular-field correction term with the parameters g = 1.98, J/k = −0.39 K, and the molecular-field term zJ′/k = −0.133 K. The magnetic data suggest a chain structure similar to those of $MnCl_2 \cdot 2H_2O$ and $Mnpy_2Cl_2$ (see p. 92) and with comparable interchain exchange effects, W. E. Estes, R. R. Weller, W. E. Hatfield (Inorg. Chem. **19** [1980] 26/31, 28).

16.1.16 With 2- and 4-Benzylpyridine $C_6H_5CH_2\text{-}C_5H_4N$ (= $C_{12}H_{11}N$)

$Mn(2\text{-}C_{12}H_{11}N)_2Cl_2$ and **$Mn(4\text{-}C_{12}H_{11}N)_4Cl_2$** were obtained by mixing ethanolic solutions of $MnCl_2$ and the benzylpyridine. The precipitates are washed with a mixture of ether and ethanol. The table on p. 130 lists the melting point (m.p. in °C), the principal absorption bands (frequencies in cm^{-1}) in the IR spectrum recorded for Nujol mulls in the

1700 to 200 cm^{-1} region with assignments and the shifts observed on complexation, and the magnetic moments μ_{eff} in μ_B.

compound	m.p.	$\nu(CC)+\nu(CN)$	ring vibrations	ν(Mn-Cl)	μ_{eff}
$Mn(2\text{-}C_{12}H_{11}N)_2Cl_2$	300	1618, 1575	1023, 627, 425	218	5.9
shift		+ 18, + 24	+ 36, + 25, + 22		
$Mn(4\text{-}C_{12}H_{11}N)_4Cl_2$	166	1618, 1568	1021, 625, 430	230	6.0
shift		+ 18, + 18	+ 25, + 20, + 26		

A polymeric trans-octahedral structure with bridging Cl atoms for the bis(2-benzylpyridine) compound and a monomeric trans-octahedral structure for the tetrakis(4-benzylpyridine) compound were concluded from the positions of the Mn-Cl stretching vibration. Both complexes are high spin and dissociate in dimethylformamide as shown by conductivity measurements, I. S. Ahuja, R. Singh, R. Sriramulu (Transition Metal Chem. **3** [1978] 185/90, 186).

16.1.17 With Hydroxyalkylpyridines

16.1.17.1 With 2-Pyridylmethanol $HOCH_2\text{-}C_5H_4N$ ($=C_6H_7NO$)

Complexes in Solution. Potentiometric pH measurements on aqueous solutions of $Mn(NO_3)_2$ and the ligand at 25±0.1°C and I=0.10 (KNO_3) (under N_2) reveal the presence of a 1:1 complex, which has the stability constant log $K_1\approx1$ (K in L/mol), Y. Murakami, M. Tagaki (Bull. Chem. Soc. Japan **38** [1965] 828/34, 831/2).

16.1.17.2 With Tris(2-pyridyl)methanol $HOC(C_5H_4N)_3$ ($=C_{16}H_{13}N_3O$)

$Mn(C_{16}H_{13}N_3O)_2(ClO_4)_2$. The complex was prepared by adding an ethanolic solution of the ligand (0.01 mol) to a solution of $Mn(ClO_4)_2\cdot6H_2O$ (0.005 mol) in a mixture of ethanol and dimethoxypropane, volume ratio 1:1. The mixture was heated gently for about ½ h, cooled, and diethyl ether was added to precipitate the complex, which was dried at 60°C in vacuo. Susceptibility measurements by the Faraday method yield the magnetic moment $\mu_{eff}=5.83\ \mu_B$. The IR spectrum (Nujol or hexachlorobutadiene mulls) shows absorption bands at 3374 and 659 cm^{-1} assigned to stretching and rocking vibrations of the carbinol OH group. Several bands in the 1600 to 720 cm^{-1} region are ascribed to vibrations of the pyridine rings, and three strong bands at 1110, 1095, and 617 cm^{-1} (with shoulder at 625 cm^{-1}) are due to vibrations of noncoordinated ClO_4^- ions although some splitting was noted. The shift of the ring stretching mode at 1595 cm^{-1} (+10) and that of the ring breathing mode at 1013 cm^{-1} (+22) indicate tridentate coordination of the ligands to Mn through the N atoms of all the pyridine rings, which are nearly equivalent in a symmetrical arrangement. The carbinol group apparently forms hydrogen bonds to the ClO_4^- ions, thus lowering the symmetry of the anions, as suggested by the observed band splitting. The molar electrical conductivity $\Lambda=298\ \Omega^{-1}\cdot cm^2\cdot mol^{-1}$ measured on 0.001M solutions of the complex in acetonitrile at 25.0°C is that of a 1:2 electrolyte, R. K. Boggess, S. J. Boberg (J. Inorg. Nucl. Chem. **42** [1980] 21/6).

16.1.17.3 With Other Pyridylalkanols

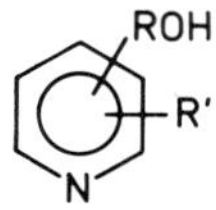

ligand	name	ROH	R′	formula
L_1	2-(4-pyridyl)propanol	4-$CH(CH_3)CH_2OH$	H	$C_8H_{11}NO$
L_2	2-(4-hydroxymethyl-3-pyridyl)-ethanol	3-CH_2CH_2OH	4-CH_2OH	$C_8H_{11}NO_2$
L_3	4-(4-pyridyl)butanol	4-$(CH_2)_3CH_2OH$	H	$C_9H_{13}NO$

$Mn(L_1)_4(NO_2)_2$, $Mn(L_2)_4(CN)_2$, and **$Mn(L_3)_4(NCS)_2$.** The compounds were precipitated by mixing concentrated aqueous solutions of $MnCl_2$ and $NaNO_2$, NaCN, or NaSCN with the ligands. The complexes are soluble in chloroform. They selectively form clathrates, thus separating mixtures of chemically and physically similar organic compounds such as the xylene isomers, W. D. Schaeffer, Union Oil Co. of California (U.S. 2798069 [1954/57]; C.A. **1958** 2088).

16.1.17.4 With Pyridoxine

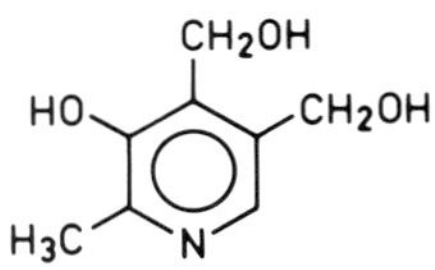

($= C_8H_{11}NO_3$)

Complex Formation in Solution. Potentiometric investigations (glass electrode) on aqueous solutions of $MnCl_2$ and pyridoxine at 25 ± 0.1°C and $I = 0.50$ (KCl) under N_2 revealed the formation of a 1:1 complex with the stability constant $\log K_1 = 1.70$. A precipitate is formed at pH ≈ 10. The stability constant is similar to that of the 1:1 pyridine complex [1].

$Mn(C_8H_{11}NO_3)_3Cl_2 \cdot 2H_2O$ was obtained by partial evaporation of an aqueous solution containing equimolar amounts of $MnCl_2$ and pyridoxine. The existence of the yellow complex is confirmed by X-ray powder diagrams. The IR spectrum shows features characteristic of a pyridinium ion (produced by migration of a proton from the phenolic OH group to the nitrogen of the pyridine ring). The bidentate chelating ligand is coordinated to the metal through the oxygens of the deprotonated phenol and the oxygens of a neighboring hydroxymethyl group [2].

References:

[1] M. S. El-Ezaby, F. R. El-Eziri (J. Inorg. Nucl. Chem. **38** [1976] 1901/5). – [2] A. T. Franklin, M. F. Richardson (Inorg. Chim. Acta **46** [1980] 191/7, 192).

16.1.18 With Pyridinones

16.1.18.1 With 2(1H)-Pyridinone and Its Derivatives

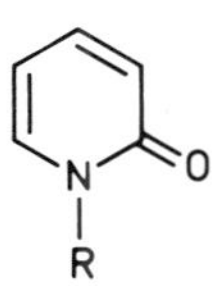

ligand	R	formula	name
I	H	C_5H_5NO	2(1H)-pyridinone = 2-pyridone ⇌ 2-pyridinol
II	CH_3	C_6H_7NO	1-methyl-2(1H)-pyridinone = N-methyl-2-pyridone
III	C_2H_5	C_7H_9NO	1-ethyl-2(1H)-pyridinone = N-ethyl-2-pyridone

2(1H)-Pyridinone is the stable tautomer of 2-pyridinol [5]. The tautomers are also discussed in [1, 6]. Crystalline 2(1H)-pyridinone is a hydrogen-bonded polymer in which the labile hydrogen is strongly bonded to the oxygen of a neighboring molecule [6]. The hydrogen-bonding strength is decreased on coordination as shown by the shift of the ν(NH) band in the IR spectrum to higher wavelengths [1, 4, 5].

$MnL_6(ClO_4)_2$ and $MnL_6(BF_4)_2$. The compounds were obtained by mixing solutions of $Mn(ClO_4)_2 \cdot 6H_2O$ or $Mn(BF_4)_2 \cdot 6H_2O$ in $HC(OC_2H_5)_3$ with solutions of excess ligand I or III in ethanol [1, 3] or II in $HC(OC_2H_5)_3$ [2]. They crystallized on standing [3] or on addition of dry ether [1, 3]. The table lists the melting points (in °C) and characteristic IR absorption bands (frequencies in cm^{-1}, sh = shoulder) of the white to buff complexes:

ligand	R	X	m.p.	IR anion bands ν_4	IR anion bands ν_3	ν(Mn-L)	Ref.
I	H	ClO_4	125 to 127	625	1085	270, 252(sh)	[1]
I	H	BF_4	105 to 115	518 to 522	—	—	[1]
II	CH_3	ClO_4	153 to 154	—	—	—	[2]
II	CH_3	BF_4	86 to 88	—	—	—	[2]
III	C_2H_5	ClO_4	100 to 103	624	1090	277, 242	[3]
III	C_2H_5	BF_4	122 to 124	520	—	264, 237	[3]

The X-ray powder patterns suggest isomorphism between the perchlorates and tetrafluoroborates for the compounds with ligand I and II [1, 2], but no isomorphism between the compounds with ligand III [3]. Isomorphism is also stated to exist between the ligand I compounds $M(C_5H_5NO)_6X_2$ (M = Mn, Fe, Co, Ni, Zn, and Cd) [1] and between the ligand III compounds $M(C_7H_9NO)_6(ClO_4)_2$ (M = Mn, Co) and $M(C_7H_9NO)_6(BF_4)_2$ (M = Mn, Co, and Cd) [3].

For $Mn(C_5H_5NO)_6(ClO_4)_2$ the magnetic moment was determined to be $\mu_{eff} = 6.02\ \mu_B$, which suggests a rather regular octahedral structure for $[Mn(C_5H_5NO)_6]^{2+}$ [1]. The octahedral geometry was confirmed by the far-IR spectrum (see table above) [1]. ESR spectra for the compounds with the ligands II and III (pure solids or doped in diamagnetic compounds) also indicate an octahedral geometry (A values of 95 G) [2, 3]. The octahedron is only slightly distorted, as expressed by the small zero-field splittings of the doped cadmium compounds ($D \approx 0.02\ cm^{-1}$ for $Cd(Mn)(C_7H_9NO)_6(ClO_4)_2$ and $D \approx 0.04\ cm^{-1}$ for $Cd(Mn)(C_7H_9NO)_6(BF_4)_2$) [3]. The g values are in the range 1.99 to 2.01 [2, 3].

The IR spectrum of $Mn(C_5H_5NO)_6(ClO_4)_2$ (in Nujol) shows a triplet at 1645, 1600, and 1592 cm^{-1}, which is assigned to the CO stretching modes of ligand I, with shifts of −12, −18, and −23 cm^{-1} on coordination. This suggests that the ligand is coordinated in the lactam form to Mn by its oxygen atom [1]. The ν(NH) absorption occurs between 2850 and 3160 cm^{-1} (solid 2-pyridone between 2830 and 3110 cm^{-1}). Two sharp unsplit bands at 1085 and 625 cm^{-1} are assigned to ν_3 and ν_4 vibration modes of the ClO_4^- ion, which apparently is not coordinated to Mn, but forms hydrogen bridges to the organic ligand as shown by the ν(NH) shift. The tetrafluoroborate complex shows an IR spectrum similar to the perchlorate complex, but the hydrogen bridging bonds between the BF_4^- ion and the organic ligand are weaker [1]. The IR and far-IR spectra of the compounds with ligands II and III also suggest 6-coordination of the organic ligands via the CO group to give the complexes octahedral structure [2, 3]. The compounds are described as air stable (ligand I) [1] or moderately stable in dry air (ligand II and III) [2, 3].

$Mn(C_5H_5NO)_4(NO_3)_2$. The complex with 2-pyridone was obtained by reacting ethanolic solutions of $Mn(NO_3)_2 \cdot 6H_2O$ and the ligand in stoichiometric quantities after having added $HC(OC_2H_5)_3$ for dehydration. The buff-colored compound melts at 157 to 158°C. According to X-ray powder photographs it seems to be isomorphous with the analogous compounds of Mg, Co, Ni, Zn, Cu, and Cd. The IR spectrum (not reported) suggests a coordination of Mn to ligand I via the oxygen atoms of its lactam form. Broad ν(NO) bands indicate a coordination of the NO_3^- ions. The ν(NH) bands, which occur at lower frequencies than in the perchlorate, suggest strong hydrogen bonds between the NH group of ligand I and the NO_3 groups. The ESR spectra of solid $Mn(C_5H_5NO)_4(NO_3)_2$ and Mn^{2+} ions doped in corresponding diamagnetic compounds suggest an octahedral symmetry with some distortion: $g_{iso} = 2.00$, $A_{\parallel} = 95$ G (only for doped compounds), $D = 0.05\ cm^{-1}$, $\lambda(=E/D) = 0.10$ [4].

$Mn(C_6H_7NO)_3(NO_3)_2$. The complex with N-methyl-2-pyridone was prepared by mixing solutions of $Mn(NO_3)_2 \cdot 6H_2O$ and the ligand in $HC(OC_2H_5)_3$ in a 1:4 mole ratio and precipitating the complex by addition of ether. The white compound melts at 87 to 88°C. The IR spectrum (two ν(NO) bands at about 1400 (split) and 1040 cm^{-1}) suggests coordination of the NO_3^- ions to Mn [2].

$Mn(C_5H_5NO)Cl_2$. The complex with 2-pyridone was obtained as a white precipitate on mixing ethanolic solutions of $MnCl_2$ and the ligand. It is isomorphous with the Co and Ni compounds, as shown by X-ray powder photographs, and has a structure similar to that of the Cu compound (space group Cc or C2/c for $Co(C_5H_5NO)Cl_2$ and $P\bar{1}$ for $Cu(C_5H_5NO)Cl_2$). The compounds contain a chlorine-bridged dimer as the basic structural unit. The dimers are loosely bound together into chains by long Cl-Mn interactions. The point symmetry at the metal atom is distorted octahedral. Susceptibility measurements at various temperatures between 80 and 298 K yield magnetic moments nearly independent of temperature. The magnetic moment at 298 K, $\mu_{eff} = 5.97\ \mu_B$, indicates high-spin Mn^{II}. The UV spectrum (200 to 350 nm) in ethanol shows two unshifted absorptions of the ligand molecule. The spectrum suggests considerable dissociation of the complex in polar solvents. The IR absorption spectrum shows the NH and CO stretching vibration bands of the ligand at 3450 and 1685 cm^{-1} (free ligand, at 3100 and 1685 cm^{-1}). In the range between 1250 and 1000 cm^{-1} there are four bands of 2-pyridone with very minor shifts on coordination. The compound is stable in moist air [5].

References:

[1] J. Reedijk (Rec. Trav. Chim. **88** [1969] 1139/55, 1142, 1148/53). – [2] J. Reedijk, H. Schrijver, J. A. Welleman (Rec. Trav. Chim. **94** [1975] 40/3). – [3] F. B. Hulsbergen, J. A. Welleman, J. Reedijk (Delft. Progr. Rept. A **1** No. 5 [1976] 137/42, 138, 141; C.A. **85** [1976] No. 55942). – [4] J. Reedijk, J. A. Smit (Rec. Trav. Chim. **91** [1972] 681/7, 682, 686). – [5] C. C. Houk, K. Emerson (J. Inorg. Nucl. Chem. **30** [1968] 1493/502).

[6] B. R. Penfold (Acta Cryst. **6** [1953] 591/600).

16.1.18.2 With 3-Hydroxy-2(1H)-pyridinone and Its Chloro Derivative

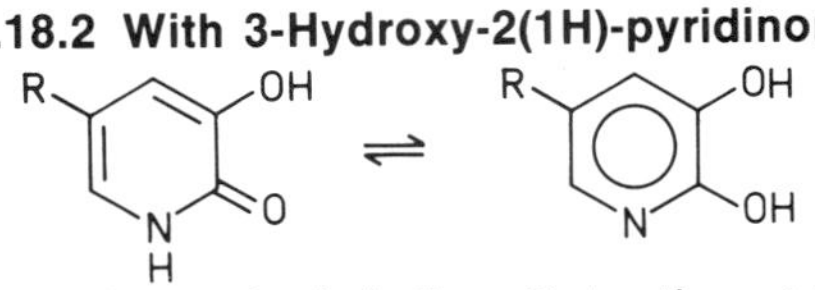

ligand I: R = H (= $C_5H_5NO_2$)
ligand II: R = Cl (= $C_5H_4ClNO_2$)

Complexes in Solution. Potentiometric pH investigations on solutions on a manganese(II) salt and the ligands in water [1] or aqueous dioxane (50 vol%) [1, 2] revealed the existence of 1:1 and 1:2 complexes with ligand I [1] and of a 1:1 complex with ligand II

[2]. The stepwise stability constants K_n (in L/mol) listed in the table below were obtained by glass-electrode measurements at 25°C for I = 0.1 ($NaClO_4$) [1].

ligand	solvent	log K_1	log K_2	Ref.
I	water	4.61	3.71	[1]
I	water-dioxane (50 vol%)	5.91	4.78	[1]
II	water-dioxane (50 vol%)	5.32	—	[2]

Ligand I mainly exists in its lactam (pyridone) form. The 3-OH group takes part in complex formation, the proton being replaced by the manganese(II) ion [1].

References:

[1] D. P. Goel, Y. Dutt, R. P. Singh (J. Inorg. Nucl. Chem. **32** [1970] 3119/22). – [2] V. Kushwaha, M. Katyal, R. P. Singh (Talanta **21** [1974] 763/9, 767).

16.1.19 With Pyridylamines and Pyridylalkylamines

16.1.19.1 With 2-Pyridylamine $H_2N\text{-}C_5H_4N$ (= $C_5H_6N_2$)

Complexes in Solution. Potentiometric pH measurements on aqueous solutions of $Mn(NO_3)_2$ [1] or $MnCl_2$ [2] and 2-pyridylamine revealed the 1:1 and 1:2 complexes [1, 2]. The stability constants (in L/mol) obtained at different temperatures and ionic strengths I are listed below:

t in °C	I	log K_1	log K_2	log β_2*)	Ref.
25	0.1 ($KNO_3 + HNO_3$)	2.19	—	(5.43)	[1]
35	0.1 ($KNO_3 + HNO_3$)	1.97	—	(5.22)	[1]
35	1.4 ($KCl + HCl$)	1.92	1.36	—	[2]

*) log β_2 values in parentheses were obtained by graphical extrapolation.

From the temperature slope of log K_1 and log K_2 between 25 and 40°C the enthalpies of formation $\Delta H_1 = -9.1$ and $\Delta H_2 = -10.7$ kcal/mol were calculated for the 1:1 and 1:2 complexes. The entropies of formation are $\Delta S_1 = -20.4$ and $\Delta S_2 = -10.9$ $cal \cdot mol^{-1} \cdot K^{-1}$ [1]. The free enthalpies of formation $\Delta G_1 = -2.70$ and $\Delta G_2 = -1.91$ kcal/mol were also calculated [2].

References:

[1] A. K. Basak, D. Banerjea (Indian J. Chem. A **14** [1976] 184/8, 186). – [2] P. S. Relan, P. K. Bhattacharya (J. Indian Chem. Soc. **46** [1969] 534/6).

16.1.19.2 With 2-Pyridylmethylamine $H_2NCH_2\text{-}C_5H_4N$ (= α-Picolylamine = $C_6H_8N_2$) and 6-Methyl-2-pyridylmethylamine $H_2NCH_2\text{-}C_5H_3N(CH_3)$ (= $C_7H_{10}N_2$)

Complex Formation in Solution

Potentiometric pH investigations on aqueous solutions of $Mn(NO_3)_2$ and the ligands revealed the existence of the 1:1 complexes. The stability constants measured at 20°C for an ionic strength of I = 0.1 ($NaNO_3$) are log K_1 = 2.66 for the 2-pyridylmethylamine compound and 1.95 for the complex with the 6-methyl derivative (K_1 in L/mol). Both ligands are bidentate, forming five-membered chelate rings [1].

In methanol and ethanol a colorless tris complex, $[Mn(C_6H_8N_2)_3]^{2+}$, is formed with excess ligand [3].

Isolated Compounds

$Mn(C_6H_8N_2)_3X_2$ ($X = ClO_4$, I). The perchlorate was obtained by mixing concentrated ethanolic solutions of $Mn(ClO_4)_2$ and 2-pyridylmethylamine in a 1:5 mole ratio [2] or by reacting $MnCl_2 \cdot 4H_2O$, $LiClO_4$, and the ligand in ethanol. Excess $LiClO_4$ was added to the $MnCl_2$ solution before addition of the ligand. The $MnCl_2$ to ligand ratio was 1:3. $Mn(C_6H_8N_2)_3I_2$ was prepared similarly, with LiI instead of $LiClO_4$ [3].

The tris chelates crystallize as cubes, which are almost colorless for $X = ClO_4$ and pale yellow for X = I. They melt at 277 and 211°C, respectively, the iodide with decomposition. The perchlorate decomposes at temperatures above 300°C [3].

The IR spectrum of $Mn(C_6H_8N_2)_3(ClO_4)_2$ shows one absorption band at 3350 cm^{-1} (free ligand at 3380 cm^{-1}), which is assigned to the NH stretching vibrations, and two bands, at 1609 and 1571 cm^{-1} (free ligand at 1589 and 1570 cm^{-1}), which are assigned to pyridine ring vibrations. The band shifts indicate coordination of the bidentate ligand via the pyridine nitrogen atom and the primary amino group to form three chelate rings. Two bands, at 1100 and 950 cm^{-1}, suggest the perchlorate group is ionic [2]. Bands occurring in the far-IR (680 to 150 cm^{-1}) region are assigned as follows: 637 and 563 to NH_2 rocking vibration, 625, 412 in-plane and out-of-plane ring deformation (412 coupled with Mn-N_{py} stretching vibration), 478 to skeletal deformation, 426 and 359 (or 435 and 370 [4]) to Mn-NH_2 stretching, 249 and 216 to Mn-N_{py} stretching, 188 and 174 to ligand-Mn-ligand deformation [5]. The far-IR spectrum is similar to those of the related Fe, Co, Ni, and Zn complexes. Therefore it is assumed that these compounds have the same pseudooctahedral configuration in which each amino-donor group is cis to the other two [4], see also [5].

The magnetic moments are $\mu = 5.85\ \mu_B$ for the perchlorate and 5.87 μ_B for the iodide at 20°C. The molar electrical conductivities of 0.001M solutions in nitromethane, $\Lambda = 128.4$ and 125.2 $\Omega^{-1} \cdot cm^2 \cdot mol^{-1}$ for $X = ClO_4$ and I, and in methanol, $\Lambda = 209.3$ and 201.5 $\Omega^{-1} \cdot cm^2 \cdot mol^{-1}$, respectively, are indicative of 1:2 electrolytes. The Mn atoms are shown by the magnetic moments to be high-spin [3].

$Mn(C_6H_8N_2)_2X_2$ (X = Cl, Br, I) and **$Mn(C_6H_8N_2)Cl_2$**. The chloro complexes with 2-pyridylmethylamine $Mn(C_6H_8N_2)_nCl_2$ (n = 1, 2), were obtained by mixing ethanolic solutions of $MnCl_2 \cdot 4H_2O$ and the ligand in a 1:n mole ratio. The bromo and iodo complexes are prepared similarly, except that LiBr or LiI is added in excess to the $MnCl_2 \cdot 4H_2O$ solution prior to the addition of the ligand. The iodide, which is obtained as a hydrate ($0.5H_2O$), may be dehydrated by heating at 110°C to give the anhydrous complex [3]. The bromide may also be prepared from $MnBr_2$ and the ligand in a 1:5 mole ratio, similar to $Mn(C_6H_8N_2)_3(ClO_4)_2$ [2]. The table below lists the colors, crystal shapes, melting points, magnetic moments μ_{eff} at 20°C, and molar electrical conductivities Λ (in $\Omega^{-1} \cdot cm^2 \cdot mol^{-1}$) of 0.001M solutions at 25°C [3]:

compound	color	crystal shape	m.p. in °C	μ_{eff} in μ_B	Λ (CH_3NO_2)	Λ (CH_3OH)
$Mn(C_6H_8N_2)_2Cl_2$	pale pink	rhombohedra	218(dec.)	5.91	19.9	183
$Mn(C_6H_8N_2)Cl_2$	colorless	prisms	270	5.97	3.9	163
$Mn(C_6H_8N_2)_2Br_2$*)	pale pink	rhombohedra	209(dec.)	5.93	20.5	180
$Mn(C_6H_8N_2)_2I_2 \cdot 0.5H_2O$	pale pink	micro-crystalline	203(dec.)**)	5.94	22.0**)	176**)

*) A yellow color and $\mu_{eff} = 5.9\ \mu_B$ were observed by [2]. – **) Values for the anhydrous complex.

The electronic absorption spectra of $Mn(C_6H_8N_2)_2X_2$ solutions in nitromethane show maxima at 266 nm for X=Cl, 268 nm for X=Br, and at 271 nm for X=I. The extinction coefficients (ε in $L \cdot mol^{-1} \cdot cm^{-1}$) are log ε≈3.7 [3]. The IR spectrum of $Mn(C_6H_8N_2)_2Br_2$ shows NH stretching and pyridine ring vibration bands which are similar to those of $Mn(C_6H_8N_2)_3(ClO_4)_2$ (p. 135) [2]. The physical properties suggest that the bis complexes are high-spin monomeric nonelectrolytes with a 6-coordinate structure containing two chelating pyridylmethylamine molecules and two coordinated anions. The structure of the mono chelate was not clarified by these investigations. It is somewhat soluble in nitromethane and lower alcohols. All of the complexes behave as 1:2 electrolytes in methanol and ethanol. Attempts to oxidize any of these Mn^{II} complexes give Mn_2O_3 [3].

References:

[1] G. Anderegg (Helv. Chim. Acta **54** [1971] 509/12, 511). – [2] B. Chiswell, E. J. O'Reilly (Inorg. Chim. Acta **7** [1973] 707/12, 710). – [3] G. I. Sutton (Australian J. Chem. **14** [1961] 550/4). – [4] M. L. Niven, G. C. Percy (Spectrosc. Letters **10** [1977] 519/25, 521). – [5] M. L. Niven, G. C. Percy, D. A. Thornton (J. Mol. Struct. **68** [1980] 73/80, 76).

16.1.19.3 With Pyridoxamine

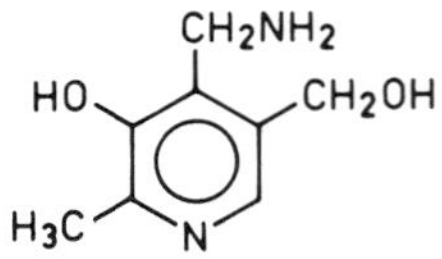

$(=C_8H_{12}N_2O_2)$

Complex Formation in Solution. Potentiometric investigations (glass electrode) in aqueous solutions of $MnCl_2$ and pyridoxamine at 25±0.1°C for I=0.50 (KCl) under N_2 yielded the formation constant log K_1=3.46 [1], log K_1=3.56 for I=0.1 (KNO_3) [2] (K in L/mol). A precipitate is formed at pH≈10. The possible coordination sites are discussed [1, 2].

References:

[1] M. S. El-Ezaby, F. R. El-Eziri (J. Inorg. Nucl. Chem. **38** [1976] 1901/5). – [2] R. L. Gustafson, A. E. Martell (Arch. Biochem. Biophys. **68** [1957] 485/98, 491).

16.1.19.4 With Bis(2-pyridyl)amine $HN(C_5H_4N)_2$ $(=C_{10}H_9N_3)$

Complexes in Solution

Potentiometric measurements on aqueous solutions of $Mn(NO_3)_2$ and the ligand revealed the 1:1 and 1:2 complexes [1]. The overall stability constants $\beta_1=K_1$ and $\beta_2=K_1K_2$ (in L^n/mol^n) obtained at temperature t for ionic strength I are listed below:

t in °C	I	log β_1	log β_2*)	Ref.
20	0.1 (KNO_3)	~2	—	[2]
25	0.1 (KNO_3+HNO_3)	1.78	5.94	[1]
35	0.1 (KNO_3+HNO_3)	1.50	5.86	[1]

*) Obtained by graphical extrapolation.

From the slope of the log K_1 and log K_2 versus temperature plots between 25 and 40°C, the enthalpies of formation $\Delta H_1=-10.9$ and $\Delta H_2=-3.3$ kcal/mol were calculated for the 1:1 and 1:2 complexes. The entropies of formation are $\Delta S_1=-28.3$ and $\Delta S_2=+16.2$ $cal \cdot mol^{-1} \cdot K^{-1}$. The coordination of the ligand to Mn apparently occurs through the N atoms of the pyridine rings as in the case of 2,2'-bipyridyl [1].

Isolated Compounds

$Mn(C_{10}H_9N_3)_2X_2$ (X = Cl, Br, I, NCS). The chelates were obtained by mixing hot ethanolic solutions of the manganese(II) salts and the ligand in stoichiometric quantities. They crystallize on cooling. The electronic and ESR X- and Q-band spectra are reported and discussed. They indicate considerable distortion of the octahedral complex symmetry, and the following distortion parameters D and λ (= E/D) were deduced:

X	Cl	Br	I	NCS
D in cm^{-1}	0.125	0.395	0.975	0.05
λ	0.267	0.267	0.300	0.333*)

*) Assumed.

Two Mn-N stretching bands in the far-IR spectra may also reflect low symmetry [3].

References:

[1] A. K. Basak, D. Banerjea (Indian J. Chem. A **14** [1976] 184/8, 186). – [2] G. Anderegg (Helv. Chim. Acta **54** [1971] 509/12, 511). – [3] R. D. Dowsing, J. F. Gibson, M. Goodgame, P. J. Hayward (J. Chem. Soc. A **1970** 1133/8).

16.1.19.5 With Bis(2-pyridylmethyl)amine $HN(CH_2\text{-}C_5H_4N)_2$ (= Di-α-picolylamine = $C_{12}H_{13}N_3$)

Complexes in Solution

Potentiometric investigations on aqueous solutions of $Mn(NO_3)_2$ and the ligand revealed the 1:1 and 1:2 complexes. In the following table the overall stability constants β_1 and β_2 (in L^n/mol^n) determined at 25°C for ionic strength I = 0.10 (KNO_3) with a glass electrode [1, 2] and thermodynamic data ΔH, ΔG (in kcal/mol), and ΔS (in cal·mol^{-1}·K^{-1}), ΔH obtained by titration calorimetry at 20°C for I = 0.1 [1, pp. 125/6] are listed:

$\log\beta_1$	$\log\beta_2$	$-\Delta G_1$	$-\Delta H_1$	ΔS_1	$-\Delta G_2$	$-\Delta H_2$	ΔS_2	Ref.
3.52	6.05	4.8	2.5	7.7	8.25	5	9	[1]
4.16	(7.07)*)	5.68	—	—	(9.66)*)	—	—	[2]

*) Calculated from the log K_2 = 2.91 and ΔG_2 = −3.98 kcal/mol given in [2].

It is assumed that in $[Mn(C_{12}H_{13}N_3)_2]^{2+}$ one molecule of the ligand is coordinated to the manganese atom as a tridentate and the other only as a monodentate ligand, the latter weakly bonded through the NH group [2]. The tridentate ligand is expelled by the hexadentate N,N,N′,N′-tetrakis(2-pyridylmethyl)ethylenediamine, which forms the far more stable 1:1 complex [1, p. 138].

References:

[1] G. Anderegg, E. Hubmann, N. G. Podder, F. Wenk (Helv. Chim. Acta **60** [1977] 123/40, 125/6, 138). – [2] J. K. Romary, J. D. Barger, J. E. Bunds (Inorg. Chem. **7** [1968] 1142/5).

16.1.19.6 With N-Phenyl-N-[2-(2-pyridyl)ethyl]amine $C_6H_5NH\text{-}CH_2CH_2\text{-}C_5H_4N$ (= $C_{13}H_{14}N_2$)

$Mn(C_{13}H_{13}N_2)_2$ was obtained by reacting $MnBr_2$·2THF with the lithium amide $LiC_{13}H_{13}N_2$ in tetrahydrofuran. The resulting deep yellow solution is concentrated cold.

By addition of twice its volume of ether, egg-yolk colored crystals of $Mn(C_{13}H_{13}N_2)_2$ are precipitated.

The IR spectrum does not show the NH stretching band in the 3400 cm^{-1} region. Therefore it is assumed that the NH group is deprotonated on complexation. Magnetic measurements at 20.5°C yield the magnetic moment $\mu_{eff}=5.74\ \mu_B$, indicating an Mn d^5 complex. A polymeric octahedral structure is assumed since the very air-sensitive compound is almost insoluble in ether or benzene but readily soluble in tetrahydrofuran. The polymeric structure may be formed by amide bridges, K. Schmiedeknecht (Z. Chem. [Leipzig] **15** [1975] 366/7).

16.1.19.7 With Bis[2-(2-pyridyl)ethyl]amine $HN(CH_2CH_2\text{-}C_5H_4N)_2$ (=$C_{14}H_{17}N_3$)

The compounds **$Mn(C_{14}H_{17}N_3)X_2$** (X = Cl, Br) crystallize on cooling a warm ethanolic solution containing MnX_2 and the ligand. They are recrystallized from ethanol. The following table lists the colors, molecular weights (mol. wt. determined by osmometry in 1,4-dichloroethane at 37°C, calculated values in parentheses), IR absorption bands (in cm^{-1}) assigned to the NH streching and pyridine ring vibrations (with shifts relative to the free ligand), magnetic moments μ_{eff} in μ_B, and molar conductivites Λ (in $\Omega^{-1}\cdot cm^2\cdot mol^{-1}$) of 0.001M solutions in nitromethane at 25°C:

X	color	mol. wt.	ν(NH)	pyridine ring vibrations	μ_{eff}	Λ
Cl	colorless	380(353)	3258(−49)	1605(+12), 1484(+12)	6.02	8.2
Br	white green	478(442)	3206(−101)	1605(+20), 1488(+16)	5.88	8.9

The physical data show that the complexes are monomeric and nonconducting in organic solvents. They contain high-spin $Mn^{II}\,d^5$ ions which are linked to the N atoms of the NH group and pyridine rings of the tridentate organic ligand and to the anions X. For the chloro complex, a six-coordinate structure, i.e., dimeric [1] or polymeric in the solid state, is suggested by the far-IR absorption (strong Mn-X frequencies at 250, 178, and 168 cm^{-1}) [2] and the diffuse reflectance spectrum, which shows a weak band at 23500 and a very weak band at 18000 to 17000 cm^{-1}. The X-ray powder diagram of the chloro complex suggests isomorphism with the related Cd compound. For the bromo complex the diffuse reflectance spectrum indicates a monomeric and 5-coordinate structure in the solid state; a band at 22700 cm^{-1}, both for Nujol mulls and solutions, is observed. The compounds are air stable and insoluble or decomposed in water. They are soluble to varying degrees in organic solvents [1].

References:

[1] D. P. Madden, S. M. Nelson (J. Chem. Soc. A **1968** 2342/8). – [2] D. Bryson, R. H. Nutall (Spectrochim. Acta A **26** [1970] 2275/80, 2278).

16.1.19.8 With N,N′-Bis(2-pyridylmethyl)alkylenediamines $C_5H_4N\text{-}NH\text{-}(CH_2)_n\text{-}NH\text{-}C_5H_4N$

Potentiometric investigations on aqueous solutions of a manganese(II) salt and the ligands revealed the 1:1 complexes in all cases. The stability constants K_1 (in L/mol) for ionic strength I = 0.10 (KNO_3) and the thermodynamic data ΔG, ΔH (in kcal/mol), and ΔS (in $cal\cdot mol^{-1}\cdot K^{-1}$), ΔH obtained by calorimetric titrations, are listed on p. 139:

ligand	n	formula	t in °C	log K_1	$-\Delta G$	$-\Delta H$	ΔS	Ref.
I	2	$C_{14}H_{18}N_4$	25	5.60, 5.9	7.6	4.5	10.4	[1 to 3]
II	3	$C_{15}H_{20}N_4$	25	4.45	6.1	3.2	9.7	[1]
III	4	$C_{16}H_{22}N_4$	25	2.27	3.6	0.2	11.4	[1]

It is assumed that the ligands are coordinated by all four nitrogen donor atoms [1].

References:

[1] G. Anderegg, N. G. Podder, P. Blaeuenstein, M. Hangartner, H. Stünzi (J. Coord. Chem. **4** [1975] 267/75, 268, 271; Proc. 16th Intern. Conf. Coord. Chem., Dublin 1974, Abstr. No. 2.30b, pp. 1/3; C.A. **85** [1976] No. 113325). – [2] R. G. Lacoste, G. V. Christoffers, A. E. Martell (J. Am. Chem. Soc. **87** [1965] 2385/8, 2387). – [3] R. G. Lacoste, A. E. Martell (Inorg. Chem. **3** [1964] 881/4, 883).

16.1.19.9 With N,N,N′,N′-Tetrakis(2-pyridylmethyl)ethylenediamine $(C_5H_4N\text{-}CH_2)_2\text{-}N\text{-}CH_2CH_2\text{-}N(CH_2\text{-}C_5H_4N)_2$ $(=C_{26}H_{28}N_6)$

The stability constant for the 1:1 complex, determined potentiometrically (glass electrode) in aqueous solution at 20°C for $I=0.10$ (KNO_3), and the thermodynamic data are $\log K_1=10.27$ (K in L/mol), $\Delta G=-13.77$ kcal/mol, $\Delta H=-11.43\pm0.70$ kcal/mol, $\Delta S=7.8\ cal\cdot mol^{-1}\cdot K^{-1}$ [1, 2].

$Mn(C_{26}H_{28}N_6)(ClO_4)_2\cdot H_2O$ was obtained by neutralizing a hot concentrated aqueous solution containing stoichiometric amounts of $MnCl_2$ (≈1M) and the perchlorate hydrate of the ligand with 2M NaOH solution. The complex, which crystallizes on cooling, is recrystallized from water. It forms yellowish needles. The IR spectrum (in KBr) shows three absorption bands, at 1655, 1607 (broad), and 1572 cm^{-1} (free ligand, 1590 and 1571 cm^{-1}), which are assigned to the coordination-sensitive pyridine-ring vibrations. The IR data suggest that the manganese(II) atom is coordinated only to the hexadentate ligand, while the ClO_4^- ions and the H_2O remain uncoordinated [1].

References:

[1] G. Anderegg, E. Hubmann, N. G. Podder, F. Wenk (Helv. Chim. Acta **60** [1977] 123/40, 127, 131, 137/9). – [2] F. Wenk, G. Anderegg (Chimia [Aarau] **24** [1970] 427/31).

16.1.19.10 With Tris(2-pyridyl)amine $N(C_5H_4N)_3$ $(=C_{15}H_{12}N_4)$

$Mn(C_{15}H_{12}N_4)Cl_2$ was obtained by reacting a solution of $MnCl_2$ in ethanol or acetone (or in a 1:1 mixture of one of these solvents with 2,2-dimethoxypropane) with an ethanolic solution of the ligand. The manganese chloride was dissolved under reflux, and the reaction mixture refluxed for a further 30 min and then cooled.

The far-IR spectrum shows two absorption bands, at 313 and 292 cm^{-1}, suggesting a tetrahedral structure with the bidentate ligand. In the diffuse reflectance spectrum, bands at 26000, 22990, and 22220 cm^{-1} were observed. The spectrum is typical of tetrahedral manganese(II) compounds. The complex fluoresces under UV light. Susceptibility measurements at 294 K yielded a magnetic moment of $\mu_{eff}=5.77\ \mu_B$. The molar electrical conductivities of 0.001M solutions in acetone or nitromethane are $\Lambda=2$ or $6\ \Omega^{-1}\cdot cm\cdot mol^{-1}$, respectively, thus indicating the complex to be a nonelectrolyte, G. C. Kulasingam, W. R. McWhinnie (J. Chem. Soc. A **1967** 1253/5).

16.1.19.11 With Tris(2-pyridylmethyl)amine $N(CH_2\text{-}C_5H_4N)_3$ (=Tri-α-picolylamine = $C_{18}H_{18}N_4$ = L_1) and Tris(6-methyl-2-pyridylmethyl)amine $N(CH_2\text{-}C_5H_3N\text{-}CH_3)_3$ (=Tri-α,α'-lutidylamine = $C_{21}H_{24}N_4$ = L_2)

Potentiometric investigations of aqueous solutions containing $Mn(NO_3)_2$ and the ligands indicated the presence of 1:1 complexes. The following stability constants (in L/mol) and free enthalpies of formation (in kcal/mol) were determined: log K_1 = 5.62 and $\Delta G_1 = -7.54$ for L_1 and log K_1 = 2.62 and $\Delta G_1 = -3.52$ for L_2 at 20°C and ionic strength I = 0.10 (KNO_3). The enthalpy and entropy of formation, $\Delta H_1 = -6.22 \pm 0.30$ and $\Delta S_1 = +4.5$ cal·mol^{-1}·K^{-1}, were obtained for the complex with L_1 by thermometric titration, also at 20°C and I = 0.10. It is assumed that this complex contains an additional coordinated H_2O molecule corresponding to a coordination number of seven as in the complex with ethylenediaminetetraacetic acid, which was studied by X-ray investigations. This assumption is supported by the value of the entropy of formation [1, 2].

References:

[1] G. Anderegg, E. Hubmann, N. G. Podder, F. Wenk (Helv. Chim. Acta **60** [1977] 123/40, 126, 137/8). – [2] F. Wenk, G. Anderegg (Chimia [Aarau] **24** [1970] 427/31).

16.1.20 With 2-Pyridylhydrazine $H_2N\text{-}NH\text{-}C_5H_4N$ (= $C_5H_7N_3$)

Complex Formation in Solution. Stability constant for the 1:1 complex, determined potentiometrically in aqueous solution at 20°C, I = 0.1 ($NaNO_3$): log K_1 = 2.64 (K in L/mol) [1].

$Mn(C_5H_7N_3)_3(ClO_4)_2$ and **$Mn(C_5H_7N_3)_2Br_2$** were obtained by mixing concentrated ethanol solutions of $Mn(ClO_4)_2$ or $MnBr_2$ and the ligand in a 1:5 mole ratio. Precipitation occurred toward the end of ligand addition. The table lists the color, IR absorption bands assigned to NH stretching and pyridine ring vibrations (frequencies in cm^{-1}), and the magnetic moment μ_{eff} (in μ_B):

compound	color	ν(NH)	ring vibrations	μ_{eff}
$Mn(C_5H_7N_3)_3(ClO_4)_2$	colorless	3345	1609, 1570	6.1
$Mn(C_5H_7N_3)_2Br_2$	pale yellow	3355	1609, 1571	—

$Mn(C_5H_7N_3)_3(ClO_4)_2$ has a molar electrical conductivity Λ between 160 and 180 Ω^{-1}·cm^2·mol^{-1} in nitromethane, corresponding to that of a 1:2 electrolyte. Thus the ClO_4^- groups are ionic, and the 2-pyridylhydrazine is coordinated as a bidentate ligand to form a high-spin 6-coordinate Mn^{II} complex. $Mn(C_5H_7N_3)_2Br_2$ is virtually nonconducting in nitromethane and may be considered as an octahedral complex with two Br and two bidentate ligand molecules coordinated to the manganese atom [2].

References:

[1] G. Anderegg (Helv. Chim. Acta **54** [1971] 509/12, 511). – [2] B. Chiswell, E. J. O'Reilly (Inorg. Chim. Acta **7** [1973] 707/12, 710).

16.1.21 With Pyridinecarbaldehydes

16.1.21.1 With 3-Pyridinecarbaldehyde $OHC\text{-}C_5H_4N$ (= C_6H_5NO)

$Mn(C_6H_5NO)_2Cl_2$ was obtained by reacting stoichiometric amounts of manganese(II) chloride and ligand in ethanolic solution. The IR spectrum in the 4000 to 400 cm^{-1} region showed two absorption bands, at 1707 and 1672 cm^{-1}, assigned to ν(CO) vibrations of the aldehyde group and several bands in the 1600 to 800 cm^{-1} region due to vibrations of the pyridine ring. As the position of the ν(CO) bands had not shifted to essentially lower frequencies and the position of some ring bands had shifted to somewhat higher frequencies, it was assumed that pyridine-3-carbaldehyde is coordinated to Mn only through the N atom of the pyridine ring. The finely crystalline compound is sparingly soluble in cold water, more upon heating, and readily soluble in ethanol, Yu. Ya. Kharitonov, G. V. Tsintsadze, T. G. Lomtadze, T. M. Chumakova (Zh. Neorgan. Khim. **25** [1980] 857/9; Russ. J. Inorg. Chem. **25** [1980] 478/9).

16.1.21.2 With Pyridoxal

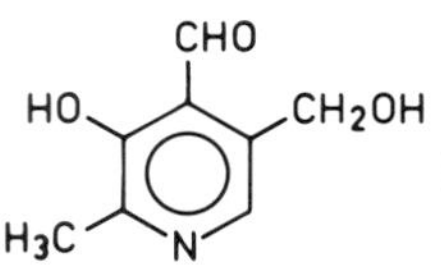

(= $C_8H_9NO_3$)

Potentiometric investigations (glass electrode) in aqueous solutions of $MnCl_2$ and pyridoxal at 25 ± 0.1°C for I = 0.50 (KCl) under N_2 revealed the formation of a 1:1 complex, which had the stability constant log K_1 = 1.70 (K_1 in L/mol). A precipitate is formed at pH ≈ 10. The stability constant is similar to that of the 1:1 pyridine complex, M. S. El-Ezaby, F. R. El-Eziri (J. Inorg. Nucl. Chem. **38** [1976] 1901/5).

16.1.22 With Pyridyl Ketones

16.1.22.1 With 4-Acetylpyridine CH_3CO-C_5H_4N (= C_7H_7NO)

Mn(C_7H_7NO)$_4$(NCS)$_2$ was obtained by mixing aqueous solutions of $MnCl_2$, NaSCN, and the ligand in a stoichiometric ratio as a white precipitate. The compound tends to form a selective clathrate with ethylbenzene, W. D. Schaeffer, Union Oil Co. of California (U.S. 2905684 [1957/59]; C.A. **1960** 3891).

16.1.22.2 With 2-Benzoylpyridine C_6H_5CO-C_5H_4N (= $C_{12}H_9NO$)

Mn($C_{12}H_9NO$)$_2$(H_2O)$_2$$Cl_2$ was obtained by refluxing a solution of $MnCl_2 \cdot 4H_2O$ and the ligand, mole ratio 1:3, in absolute ethanol for 3 h. The orange compound melts at 242°C with decomposition. The IR spectrum (in KBr) shows two strong bands, at 255 and 245 cm^{-1}, which are assigned to the Mn-Cl stretching modes and indicate a cis-octahedral structure. The positive shift of 25 cm^{-1} to 1020 cm^{-1} of the pyridine breathing motion and the positive shift of 30 cm^{-1} to 1600 cm^{-1} of the ν(CC) and ν(CN) pyridine ring band indicate coordination of the ligand to Mn through the N atom of the pyridine ring. The ν(CO) band at 1670 cm^{-1} is nearly unchanged, thus the carbonyl oxygen is not a binding site. The coordination of H_2O is indicated by the OH bending mode at 1630 cm^{-1}. The ν(OH) band appears at 3400 cm^{-1}. The Mn-N bond is confirmed by the electronic spectrum. A peak at 265.5 nm is due to a $\pi \rightarrow \pi^*$ transition. The magnetic moment of 5.30 μ_B indicates a high-spin manganese(II) complex. The conductivity in nitromethane is that of a nonelectrolyte [1].

Mn($C_{12}H_9NO$)$_2$$I_2$, Mn($C_{12}H_9NO$)$X_2$ (X = Cl, Br), and **Mn($C_{12}H_9NO$)(H_2O)$_2$$Cl_2$.** The anhydrous compounds crystallize on cooling hot ethanolic solutions which contain the manganese(II) halide and ligand in a 1:2 mole ratio [2]. Mn($C_{12}H_9NO$)(H_2O)$_2$$Cl_2$ was obtained by evaporating a 95% alcohol solution of $MnCl_2 \cdot 4H_2O$ and the ligand, mole ratio 4:5, in vacuo to dryness. It melts at 254°C with decomposition [3]. The table on p. 142 lists the colors, the ν(CO) and ν(Mn-X) bands (frequencies in cm^{-1}) in the IR spectra, the magnetic moments μ_{eff} (in μ_B), and the molar electrical conductivities Λ (in $\Omega^{-1}\cdot cm^2\cdot mol^{-1}$) of 0.001M solutions of the complexes in nitromethane at 20°C.

compound	color	ν(CO)*)	ν(Mn-X)	μ_{eff}	Λ	Ref.
$Mn(C_{12}H_9NO)_2I_2$	red brown	1620***)	—	5.63	34	[2]
$Mn(C_{12}H_9NO)Cl_2$	yellow	1630	250, 238**)	6.78	insoluble	[2]
$Mn(C_{12}H_9NO)(H_2O)_2Cl_2$	orange	1630	—	5.76	0.5	[3]
$Mn(C_{12}H_9NO)Br_2$	orange yellow	1620	—	6.64	20	[2]

*) Free ligand band at 1660 cm^{-1}. – **) The assignment of another band at 267 cm^{-1} is uncertain because of ligand absorption in this region. – ***) With shoulders at 1625 and 1632 cm^{-1}.

The unassigned electronic reflectance spectra of the anhydrous complexes are given in [2]. As $Mn(C_{12}H_9NO)X_2$ with X = Cl, Br do not fluoresce in UV light, and the position of the ν(Mn-Cl) bands is too low for a 4-coordinate species, a polymeric octahedral structure is favored. A polymeric octahedral structure of the chloride is consistent with the complexity of the ν(Mn-Cl) bands and the insolubility of $Mn(C_{12}H_9NO)Cl_2$ in organic solvents. $Mn(C_{12}H_9NO)Br_2$ is assumed to have a similar structure although its solubility is about 10^{-3} mol/L in nitromethane [2]. The IR and electronic spectra of $Mn(C_{12}H_9NO)(H_2O)_2Cl_2$ resemble those of $Mn(C_{12}H_9NO)_2(H_2O)_2Cl_2$ (p. 141). But the ν(CO) and δ(OH) bands occur at considerably lower frequencies (1630 and 1610 cm^{-1}, respectively), thus suggesting a bidentate ligand coordinated via the CO group and the N atom of the pyridine ring. The strong ν(OH) bands of coordinated H_2O appear at 3400 to 3320, and the ν(OD) bands of D_2O in $Mn(C_{12}H_9NO)(D_2O)_2Cl_2$ appear at 2500 to 2460 cm^{-1}. The magnetic moment of $Mn(C_{12}H_9NO)(H_2O)_2Cl_2$ indicates a monomeric high-spin Mn^{II} complex. The conductivity in nitromethane is that of a nonelectrolyte. The coordinated H_2O is not lost upon heating at 110°C. Thus an octahedral configuration with chelating ligand, and Cl and H_2O in cis-positions, has been proposed [3].

References:

[1] M. Plytzanopoulos, G. Pneumatikakis, N. Hadjiliadis, D. Katakis, V. Papadopoulos (Chem. Chron. **8** [1979] 109/17, 110, 116; C.A. **91** [1979] No. 150456). – [2] A. W. Downs, W. R. McWhinnie, B. G. Naik, R. R. Osbourne (J. Chem. Soc. A **1970** 2624/6). – [3] M. Plytzanopoulos, G. Pneumatikakis, N. Hadjiliadis, D. Katakis (J. Inorg. Nucl. Chem. **39** [1977] 965/72, 966, 970).

16.1.22.3 With 1-(2-Naphthyl)-2-(2-pyridyl)ethanone

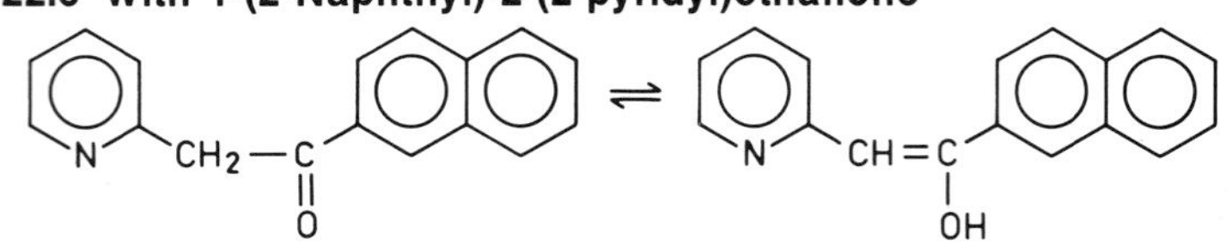

$(=C_{17}H_{13}NO)$

$Mn(C_{17}H_{12}NO)_3$. The manganese(III) compound is prepared by mixing a solution of 0.72 g $Mn(NO_3)_2\cdot6H_2O$ with a solution of 1.2 g of the ligand, both in 60 mL of hot methanol. Oxygen was bubbled through the mixture. The solution slowly turned brown. The complex was precipitated by addition of several milliliters of a saturated solution of sodium acetate in ethanol. It is a brown powder, which decomposes at 170(±3)°C. It dissolves in chloroform and benzene and is only slightly soluble in methanol, ether, and acetone. It is assumed that the ligand deprotonized in its enol form is coordinated to the manganese atom. The compound easily hydrolyzes, E. Uhlemann (J. Prakt. Chem. [4] **21** [1963] 277/85, 278, 284).

16.1.22.4 With Bis(2-pyridyl)ketone

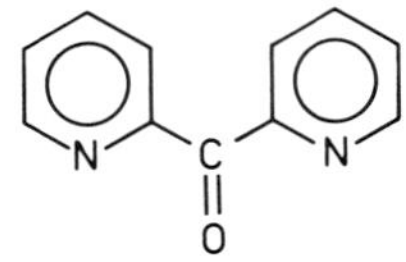

(= $C_{11}H_8N_2O$ = L)

Complexes in Solution. The electronic spectra of aqueous solutions containing Mn^{2+} and ligand L in various ratios (pH between 3.4 and 4.1) revealed a 1:1 complex with the stability constant log K = 1.03 ± 0.19 (K = $[MnL^{2+}]/[Mn^{2+}]\cdot[L]$) at 25°C and ionic strength I = 0.5 ($NaClO_4$). There is evidence from this and other investigations (e.g., see [2]) that at low pH coordination takes place via the two nitrogen atoms of the ligand and that water adds to the keto group to form a geminal diol. One of the OH groups is deprotonated in the pH range 5 to 8, leading to an N,O coordination of the Mn^{2+} ion. The acidity constant of the diol, pK($[H^+][Mn(L\cdot O^-)^+]/[MnL^{2+}]$) = 7.8 ± 0.3 was determined by potentiometric titrations at 25°C and I = 0.1. The overall stability constant log K = 6.8 (K = $[Mn(L\cdot O^-)^+]/[Mn^{2+}][L\cdot O^-]$) at I = 0.1 was calculated with the aid of the acidity constant of the free-ligand geminal diol [1].

$Mn(C_{11}H_8N_2O)_2Cl_2$. To prepare the complex, solutions of anhydrous $MnCl_2$ and ligand (mole ratio 1:3) in anhydrous dimethylformamide (DMF) were mixed and heated at 100°C under dry N_2 for 1 h. The magnetic moment is μ_{eff} = 6.1 μ_B. The diffuse reflectance spectrum shows a strong band at 28000 cm^{-1} and a shoulder at 22700 cm^{-1}, and the IR spectrum (KBr discs) shows two absorption bands at 1677 and 1585 cm^{-1}, which are assigned to the ν(CO) and pyridine ring stretching vibrations, respectively. The slightly greater frequency of the pyridine ring band (+6 cm^{-1}) suggests a bidentate coordination of the ligand to Mn through the N atoms. Although it was not possible to assign bands belonging to the ν(Mn-N) or ν(Mn-Cl) vibrations, a cis-octahedral structure (as in the Co and Ni complexes) may be assumed. According to conductivity measurements in DMF the molar conductivity is approximately half of that expected for a 1:1 electrolyte, thus suggesting partial dissociation. The compound is only slightly soluble in most organic solvents. In DMF the solubility is somewhat higher [2].

References:

[1] B. E. Fischer, H. Sigel (J. Inorg. Nucl. Chem. **37** [1975] 2127/32, 2129). – [2] M. C. Feller, R. Robson (Australian J. Chem. **21** [1968] 2919/27, 2921, 2926).

16.1.23 With Other Pyridyl Compounds

Complexes with 2,2′-bipyridyl, 4,4′-bipyridyl, and 2,2′,2″-terpyridyl and related compounds are described in Chapter 16.1.31 to 16.1.36, pp. 201/40; complexes with pyridylquinolines, in Chapter 16.1.25.10, p. 188.

16.1.23.1 With 2,2′- and 4,4′-(1,2-Ethenediyl)bispyridine (= $C_{12}H_{10}N_2$)

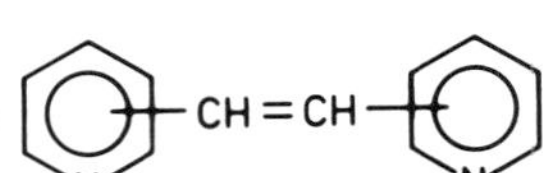

$(NO)_3Mn(2\text{-}C_{12}H_{10}N_2)Mn(NO)_3$ and **$(NO)_3Mn(4\text{-}C_{12}H_{10}N_2)Mn(NO)_3$**. To a solution of $Mn(NO)_3\cdot THF$ obtained by photo-induced nitrosylation of 1 mmol of $Mn_2(CO)_{10}$ in tetrahydrofuran, 1 mmol of the ligand was added under N_2, and the solution stirred overnight. The solvent is evaporated to leave a dark green residue, which is dissolved in benzene and purified by chromatography on silica gel and/or recrystallization. The yellow-green crystals melt at about 128°C (2-pyridyl compound) or 135°C (4-pyridyl compound under N_2) with decomposition. The IR spectrum (KBr discs) of each of the two compounds shows two absorption bands, at 1773 and 1667 cm^{-1}, which are assigned to

NO stretching vibrations, thus demonstrating the C_{3v} symmetry of the $Mn(NO)_3$ groups, M. Herberhold, A. Razavi (J. Organometal. Chem. **67** [1974] 81/6, 85).

16.1.23.2 With 4,4′-Trimethylenebispyridine $C_5H_4N\text{-}(CH_2)_3\text{-}C_5H_4N$ ($=C_{13}H_{14}N_2$)

$Mn(C_{13}H_{14}N_2)_2Cl_2$ was prepared by adding excess ligand to a hot ethanolic solution of manganese(II) chloride. The colorless complex has the magnetic moment $\mu_{eff}=5.9\ \mu_B$. The IR spectrum (Nujol) shows absorption bands which are assigned as follows: 1608 and 1556 cm^{-1}, ν(CC) (free ligand at 1650, 1600, and 1550 cm^{-1}); 1010 and 612 cm^{-1}, ring vibrations (free ligand at 992, 575, 402 cm^{-1}); and 235 cm^{-1}, ν(Mn-Cl). The IR spectrum clearly indicates the bidentate bridging nature of the ligand. It also gives evidence for the fact that the two pyridyl rings, which are nonplanar in the free ligand, become coplanar upon coordination. The position of the ν(Mn-Cl) band and the magnetic moment of the complex are consistent with terminal Cl atoms in a polymeric chain structure and an octahedral arrangement around the manganese atom by two Cl atoms and four N atoms of the ligand, I. S. Ahuja, R. Singh, C. P. Rai (Transition Metal Chem. [Weinheim] **3** [1978] 32/5, 33).

16.1.23.3 With Nicotine 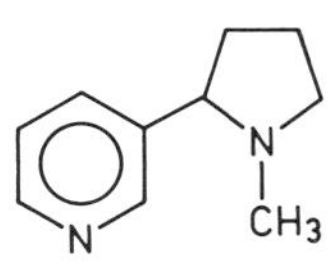 and Anabasine 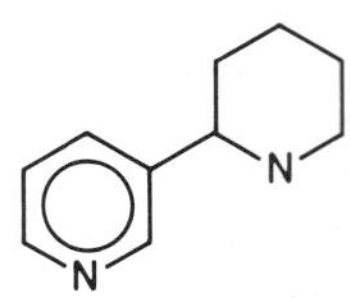

($=C_{10}H_{14}N_2$) ($=C_{10}H_{14}N_2$)

$Mn(C_{10}H_{14}N_2)_2X_2\cdot4H_2O$ with $X=(NO_2)_3C_6H_2O^-$ (picrate) or $(NO_2)C_6H_4COO^-$ (4-nitrobenzoate) and **$Mn(C_{10}H_{14}N_2)_2X_2\cdot6H_2O$** with $X=C_6H_5COC_6H_4COO^-$ (2-benzoylbenzoate) were obtained by reacting Mn^{II} acetate, the sodium or ammonium salt of the anion, and nicotine in water or aqueous ethanol. They crystallize well and are almost insoluble in water. The water of hydration is lost at 110°C [1].

A white instable compound, **$2MnCl_2\cdot C_{10}H_{14}N_2$**, is obtained by mixing alcoholic solutions of dehydrated $MnCl_2$ (slight excess) and nicotine. It turns black and viscous on exposure to air, decomposes on heating, and is hydrolyzed by boiling water, mineral acids, or soda lime [2].

$MnCl_2\cdot MnX\cdot3C_{10}H_{14}N_2$ and **$MnCl_2\cdot MnX\cdot3C_{10}H_{14}N_2\cdot6C_6H_6$** with $X={}^-OC_6H_4COO^-$ (salicylate) and **$MnCl_2\cdot2MnX\cdot2C_{10}H_{14}N_2$** with $X={}^-OC_{10}H_6COO^-$ (2-hydroxy-1-naphthoate) are reported in [3]. The products are believed to be bi- or trinuclear with bridging nicotine molecules between manganese atoms coordinated by the anions [3].

The salicylate **$Mn(C_{14}H_{10}N_2)_2(OH)(C_6H_4COO)_2$** was obtained as a colorless precipitate by reacting equal volumes of aqueous solutions of $MnCl_2$ (0.1M), anabasine (0.5M), and sodium salicylate (1M). It is soluble in $CHCl_3$ and can be extracted by this solvent at pH = 2.5 to 7 from aqueous solutions which are acidified with salicylic acid [4].

References:

[1] C. R. Smith (J. Am. Chem. Soc. **75** [1953] 2010/2). – [2] S. Prasad, K. P. Kacker (J. Indian Chem. Soc. **35** [1958] 719/21). – [3] A. Crăciun (Acad. Rep. Populare Romine Filiala Iasi Studii Cercetari Stiint. Chim. **8** No. 1 [1957] 101/26, 106, 112, 115/20; Ref. Zh. Khim. **1959** No. 18813). – [4] Kh. R. Rakimov, L. P. Nyrkova, K. I. Inoyantova (Nauchn. Tr. Tashkent. Gos. Univ. No. 257 [1964] 94/7, 95; C.A. **63** [1965] 2434).

16.1.24 With Pyridine N-Oxide and Its Derivatives

16.1.24.1 With Pyridine N-Oxide (= Pyridine 1-Oxide = C_5H_5NO = L)

A mixed-ligand complex with pyridine N-oxide and 2,4-pentanedione (acetylacetone) is described in "Mangan" D 1, 1979, p. 106.

16.1.24.1.1 MnL_6X_2 and MnL_6X_3 Complexes

$Mn(C_5H_5NO)_6(NO_3)_2$. The nitrate was obtained on mixing warm solutions of $Mn(NO_3)_2 \cdot 6H_2O$ and excess pyridine N-oxide in absolute ethanol and cooling. The crystals were filtered off with suction and dried at 60°C in a vacuum. The deep yellow complex melts at 159 to 161°C. Its IR spectrum (Nujol mulls) shows absorption frequencies of the uncoordinated NO_3^- ion at 1350 cm^{-1} (antisymmetric NO_2 stretching vibration), 1040 cm^{-1} (NO stretching vibration), 838, 825 cm^{-1} doublet (NO_2 deformation vibration), and 735 cm^{-1} (planar rocking vibration). The complex thus evidently has an octahedral structure with coordinated pyridine N-oxide molecules and uncoordinated NO_3^- ions [1].

$Mn(C_5H_5NO)_6(ClO_4)_2$. The perchlorate crystallized on cooling a warm methanolic solution containing $Mn(ClO_4)_2 \cdot 6H_2O$ and pyridine N-oxide in stoichiometric amounts [2 to 4]. A 1:7 mole ratio is recommended by [5], and absolute ethanol as the solvent and stirring are recommended by [6, 7]. The light yellow crystals were collected, washed with ethanol (99%) [4], ethanol, and ether, air dried, and recrystallized from ethanol [7] or dried in a vacuum at 60°C [2] or over P_2O_5 [4]. The compound should be stored in a cool inert atmosphere (N_2) because it is very sensitive toward air, light, and heat [3]. $Mn(C_5H_5NO)_6(ClO_4)_2$ has the same morphological structure as the compounds $M(C_5H_5NO)_6(ClO_4)_2$ with M = Co, Ni, and Zn [3]. It also shows an X-ray powder pattern much like those of the compounds with M = Mg, Fe, Co, Ni, and Zn and the tetrafluoroborates with M = Mn, Co, Ni, Cu, and Zn [8]. Thus these compounds seem to be isomorphous [3, 8] (for $Ni(C_5H_5NO)_6(BF_4)_2$ the space group R3-C_3^4 (No. 146) was found [3]). The electronic diffuse reflectance spectrum shows a charge-transfer band at 24330 cm^{-1} and a shoulder at 22570 cm^{-1} [9]. Solutions in acetonitrile show absorption maxima at 383.4 nm (26082 cm^{-1}) and 450 nm (22220 cm^{-1}), the first being very strong and causing the yellow color of the complex [2]. The charge-transfer bands were assigned as $e_g \rightarrow \pi^*$ transitions (Mn-to-ligand). The crystal-field parameters Dq = 850 cm^{-1} and B = 835 cm^{-1} were calculated. The spectral data were analyzed in terms of optical electronegativity [9].

The IR spectrum (Nujol [4, 10, 11] or between polyethylene sheets [8]) shows pertinent maxima (frequencies in cm^{-1}), which are given below with the shifts $\Delta\nu$ relative to the free ligand and the assignments:

band position	1220	919	776	620	551	316	223
$\Delta\nu$	−45	+32	+18	—	+10	—	—
assignment	ν(NO)	γ(C-H)*)	ligand	$\nu(ClO_4)$	**)	ν(Mn-O)	Mn-L
Ref.	[4,5,11]	[4]	[4]	[4,11]	[4,11]	[8]	[8]

*) Out-of-plane bending. – **) Pyridine ring vibrations.

Further bands in the range 4000 to 1300 cm^{-1} were assigned to ligand vibrations [4]; bands below 200 cm^{-1}, to Mn-L or ClO_4 vibrations [8, 10]. Assignments were based on band shifts induced by deuteration [10] and on the effects of metal [8, 10] and metal isotope substitution in isomorphous compounds together with a structure analysis by X-ray single crystal methods [8]. As shown by the considerable lowering of the ν(NO)-band frequency the ligands are coordinated to the manganese atom by their oxygen atoms [4], forming a regular octahedron of O_h site symmetry, which requires single unsplit ν(NO) and ν(Mn-O) bands [4, 10]. The ESR X-band spectra of $Mn(C_5H_5NO)_6(ClO_4)_2$ doped into the analogous complexes of Zn [3, 6, 11], Cd [6], and Hg [7] are reported and discussed. A g-factor of ~2.0, values for the hyperfine coupling constant of A = −86 to −94 Gauss, and values for the zero-field splitting parameter of D = +410 to +460 Gauss have been evaluated [3, 6, 7, 11].

The g-factor is isotropic (within experimental error) and close to the spin-only value (g = 2.0023). Possible reasons of the large deviations from cubic symmetry are discussed [3]. The values for the hyperfine coupling constant observed in the solid solutions suggest a rather low degree of covalency for the Mn-to-ligand bonding [6, 7] but one which is higher than that of the manganese(II) aquo complexes [6]. The covalency of the Mn-to-ligand bond does not depend on the size of the cation in the host lattice but is notably affected by substituents attached to the pyridine rings [7]. Susceptibility measurements at room temperature yield a magnetic moment of $\mu_{eff} = 5.88\ \mu_B$, which is consistent with a sextet ground state (d^5) for the Mn^{2+} ion [2].

If kept in air, the compound gradually turns greenish black and begins to smell of pyridine, a decomposition probably caused by O_2 and light [2], also see [3]. In the dark the compound is air stable. On irradiation with a high-pressure mercury lamp, the compound darkens and becomes black even in an inert atmosphere (N_2); pyridine was detected as a decomposition product by gas chromatography. The oxidation of Mn^{2+} to Mn^{3+} results from the electronic spectrum of a nitromethane solution of the previously irradiated solid complex. Light of wavelengths in the charge-transfer region at about 380 nm has proved to be most efficient. A sample which had been irradiated for 122 h contained about 10% of its Mn in the tervalent state. The reaction order is probably zero [12]. The compound is rapidly decomposed by water and is readily soluble in acetonitrile [2]. The solutions become dark brown on exposure to light and deposit some MnO_2 if the solvent is not dry. In anhydrous solutions (which also become brown on irradiation) iodides are not oxidized to I_2, and no precipitate is formed [12].

$Mn(C_5H_5NO)_6(BF_4)_2$. The tetrafluoroborate was obtained by mixing methanolic or ethanolic solutions of $Mn(BF_4)_2 \cdot 6H_2O$ and pyridine N-oxide in a 1:6 to 1:9 mole ratio [11, 14] and adding $HC(OC_2H_5)_3$ as a dehydrating agent [8]. The yellow crystals [8, 11, 14] which appeared upon standing [11] were filtered off (after 2 h), washed with benzene [14], with diethyl ether, and dried in a vacuum [11, 14] or recrystallized from ethanol and dried at 120°C. The complex melts at 240 ± 3°C [15] or 228 to 232°C [11, 14]. The specific heat, which has been measured in the temperature range 0.05 to 1.5 K, has a sharp Λ-shaped maximum at T = 0.158 ± 0.002 K (a figure is given in the paper), indicating antiferromagnetic behavior of the compound below this temperature. The magnetic exchange parameter $|J/k|$ is estimated to be about 7 mK. The total entropy change $\Delta S/R$ contained in the experimental curve supplemented by high- and low-temperature extrapolations is estimated to be 1.96 ± 0.1. The analysis of the specific heat curve is hampered by the similar magnitudes of exchange interaction, hyperfine splitting, and crystal-field splittings. The magnetic lattice formed by the Mn^{2+} spins is nearly cubic [16]. The electronic spectrum of $Mn(C_5H_5NO)_6(BF_4)_2$ dissolved in acetonitrile shows an absorption maximum at 378 nm (26400 cm^{-1}); in dimethylformamide the maximum is at 390 nm (25600 cm^{-1}) [14].

The IR spectrum of the solid was recorded for KBr discs in the 2000 to 400 cm^{-1} region. It shows the ν(NO) band at 1228 cm^{-1} (free ligand band at 1243 cm^{-1}). Further ligand bands were shifted only slightly relative to those of the uncoordinated ligand [14]. The far-IR spectrum, which was recorded in the 400 to 30 cm^{-1} region at 293 and 100 K is quite similar to that of the perchlorate. The ν(Mn-O) band appears at 317 cm^{-1} [8]. However, a band at 472 cm^{-1} was assigned to the ν(Mn-O) vibration by [14].

The thermal decomposition of $Mn(C_5H_5NO)_6(BF_4)_2$ starts at about 210°C and proceeds as a continuous reaction, which the DTA curve indicates to be complicated. The minimum observed between 230 and 250°C marks the melting of the compound, and some further peaks show the decomposition, which at 700°C yields Mn_2O_3 as the final product [14]. Exothermic decomposition at 301±3°C is indicated by an exothermic peak (initial point) of the DTA heating curve (heating rate 4, 8, or 12°C/min). Comparison with some other related metal complexes yields the thermal stability order: $Mn^{2+} > Ni^{2+} > Zn^{2+} > Co^{2+} >$ free ligand $> Cu^{2+} > Hg^{2+}$ [15].

$Mn(C_5H_5NO)_6(ClO_4)_3$. The manganese(III) compound is prepared by dissolving $Mn(CH_3COO)_3 \cdot 2H_2O$ (2.5 g) and pyridine N-oxide (6 g) in ice-cold methanol (25 mL) with stirring and adding an ice-cold methanolic (5 mL) solution of a stoichiometric amount of $HClO_4$. The red crystals which appear on standing are filtered off, washed several times with ether and with benzene, and dried in a vacuum over P_2O_5. The electronic spectrum of solid $Mn(C_5H_5NO)_6(ClO_4)_3$ was recorded in the 25000 to 10000 cm^{-1} region and shows two bands, at 19610 and 16000 cm^{-1}, which were assigned to the electron transitions $e_{1g} \rightarrow b_{1g}$ and $b_{2g} \rightarrow b_{1g}$, respectively. The presence of two bands indicates lowering of the complex symmetry from O_h to D_{4h} (tetragonal) due to a Jahn-Teller distortion. A third, low-energy band (at about 15000 cm^{-1}) required by D_{4h} symmetry was not observed. The crystal-field parameters $Dq = 1600$ cm^{-1} and $Dt = 838$ cm^{-1} were calculated, the latter by assuming $D_s = 2600$ cm^{-1}. The IR spectrum exhibits an absorption band at 1228 cm^{-1} assigned to the NO stretching mode of the ligand, which on complexation was shifted to a lower frequency (-37 cm^{-1}), and two bands of the ClO_4^- ion (symmetry T_d) at 1100 cm^{-1} (ν_3) and 624 cm^{-1} (ν_4). Susceptibility measurements yielded the magnetic moment $\mu_{eff} = 4.99$ μ_B, corresponding to four unpaired electrons. The molar electrical conductivity of a 0.001M solution in acetonitrile at 25°C, $\Lambda = 384$ $\Omega^{-1} \cdot cm^2 \cdot mol^{-1}$, indicates a 1:3 electrolyte [13].

References:

[1] R. L. Carlin, M. J. Baker (J. Chem. Soc. **1964** 5008/14, 5009/11). – [2] R. L. Carlin (J. Am. Chem. Soc. **83** [1961] 3773/5). – [3] C. J. O'Connor, R. L. Carlin (Inorg. Chem. **14** [1975] 291/6, 294). – [4] Y. Kakiuti, S. Kida, J. V. Quagliano (Spectrochim. Acta **19** [1963] 201/11, 202/4). – [5] L. C. Nathan, R. O. Ragsdale (Inorg. Chim. Acta **10** [1974] 177/81).

[6] G. M. Woltermann, J. R. Wasson (Inorg. Chem. **12** [1973] 2366/70, 2369). – [7] G. M. Woltermann, J. R. Wasson (J. Phys. Chem. **78** [1974] 45/9). – [8] A. D. van Ingen-Schenau, W. Groeneveldt, J. Reedijk (Spectrochim. Acta A **30** [1974] 213/21, 214, 227). – [9] W. Byers, B. Fa-Chun Chou, A. B. P. Lever, R. V. Parish (J. Am. Chem. Soc. **91** [1969] 1329/33, 1331). – [10] A. T. Hutton, D. A. Thornton (J. Mol. Struct. **39** [1977] 33/8, 36).

[11] J. Reedijk (Rec. Trav. Chim. **88** [1969] 499/512, 501/4, 511). – [12] L. C. Nathan, J. Cullen, R. O. Ragsdale (Inorg. Nucl. Chem. Letters **12** [1976] 137/40). – [13] C. P. Prabhackaran, C. C. Patel (J. Inorg. Nucl. Chem. **34** [1972] 2371/4). – [14] F. Kůtek, B. Dušek (Collection Czech. Chem. Commun. **35** [1970] 13/21, 14/6). – [15] A. J. Pappas, F. A. Osterman, H. B. Powell (Inorg. Chem. **9** [1970] 2695/8).

[16] H. A. Algra, L. J. de Jongh, R. L. Carlin (Physica B **92** [1977] 258/60).

16.1.24.1.2 $MnL_3X_2 \cdot H_2O$ and MnL_3X_3 Complexes

$Mn(C_5H_5NO)_3(NCS)_2 \cdot H_2O$ and **$Mn(C_5H_5NO)_3(NCSe)_2 \cdot H_2O$** were obtained on mixing ethanolic solutions of $Mn(NCS)_2$ or $Mn(NCSe)_2$ and pyridine N-oxide in a 1:6 mole ratio as yellow crystals. The compounds are isomorphous with the analogous complexes of Co and Ni as shown by their X-ray powder diffraction patterns. The electronic reflectance spectrum recorded on Nujol mulls shows the onset of a charge-transfer band at 21400 cm^{-1} for the isothiocyanate and at 21000 cm^{-1} for the isoselenocyanate. The IR spectra recorded on Nujol mulls in the 4000 to 200 cm^{-1} region show the notable absorption bands (frequencies in cm^{-1}) listed below together with their assignment:

$Mn(C_5H_5NO)_3(NCS)_2 \cdot H_2O$ ...	2078	1216, 1199	330 (broad)	230	—
$Mn(C_5H_5NO)_3(NCSe)_2 \cdot H_2O$ ...	2075	1218, 1200	330 (broad)	—	420
assignment	ν(CN)	ν(NO)	ν(Mn-O)	ν(Mn-N)	δ(NCSe)

The position of the ν(CN) band, which indicates terminal N-bonded NCS and NCSe groups, and the electronic (visible) spectrum are consistent with an octahedral coordination of the Mn^{2+} ions in these complexes. No data have been given for the anhydrous complexes [3].

$Mn(C_5H_5NO)_3Cl_3$. The manganese(III) compound was obtained on mixing solutions of $MnCl_3$ in ether (free from HCl and kept at −30°C) and a cold solution of pyridine N-oxide in ethanol, excluding moisture. The reddish black crystalline precipitate is washed with cold dry ethanol and dried in a vacuum. It is air stable at room temperature and melts at 139°C [1].

The electronic spectrum of the compound dissolved in $CHCl_3$ shows two absorption maxima, at 11490 cm^{-1} ($^5B_{1g} \rightarrow {}^5A_{1g}$) and 15870 cm^{-1} ($^5B_{1g} \rightarrow {}^5B_{2g}$), with molar extinction coefficients of $\varepsilon = 115$ and 262 $L \cdot mol^{-1} \cdot cm^{-1}$, respectively, which indicate a splitting of the E_g ground state of the Mn^{3+} ion due to a distortion of the surrounding octahedral ligand field. The IR spectrum (Nujol mulls) in the 4000 to 250 cm^{-1} region shows absorption bands at 1215 and 1185 cm^{-1}, assigned to the NO stretching vibrations, which are split and shifted towards lower frequencies from their free-ligand value (1228 cm^{-1}). Further bands at 420, 310, and 270 cm^{-1} are assigned to one ν(Mn-O) and two ν(Mn-Cl) vibrations [1].

The magnetic susceptibility, which has been measured at temperatures between 305 K ($10^6\ \chi_g = 27$ cm^3/g) and 70 K (125 cm^3/g), obeys the Curie-Weiss law with the paramagnetic Curie temperature $\Theta_p = -4.15$ K, yielding a magnetic moment of $\mu_{eff} = 5.45\ \mu_B$ [2]. The molar electrical conductivity of a freshly made 5×10^{-4}M solution in acetone at 20°C is $\Lambda = 33\ \Omega^{-1} \cdot cm^2 \cdot mol^{-1}$, that of a 3.01M solution in pyridine at 29°C is $\Lambda = 5\ \Omega^{-1} \cdot cm^2 \cdot mol^{-1}$, both indicating a nonelectrolyte [1].

The compound is instantly decomposed by water (precipitate of manganese oxides) and methanol (pink solutions of Mn^{2+}). It is slightly soluble in acetone, cold chloroform, and hot benzene, yielding green solutions; solutions in ethanol (slightly soluble) are grayish yellow. It is insoluble in CCl_4, ether, and n-hexane. It dissolves in pyridine, forming a maroon-red solution, which shows a single absorption band at 23810 cm^{-1} (electronic spectrum) probably due to a substitution of the ligand by the solvent [1, 2].

References:

[1] E. Contreras, V. Riera, R. Usón (Inorg. Nucl. Chem. Letters **8** [1972] 287/91). – [2] E. Contreras, V. Riera, R. Usón (Rev. Acad. Cienc. Exact. Fis. Quim. Nat. Zaragoza **28** [1973] 43/66, 46, 49, 52, 58, 61, 65; C.A. **80** [1974] No. 43582). – [3] G. B. Aitken, G. P. McQuillan (J. Chem. Soc. Dalton Trans. **1973** 2637/40).

16.1.24.1.3 MnL_2X_2 and MnL_2X_3 Complexes

$Mn(C_5H_5NO)_2(NO_3)_2$ was obtained by reacting $Mn(NO_3)_2 \cdot 6H_2O$ and pyridine N-oxide in absolute ethanol in a 1:2 mole ratio. The complex was precipitated by adding some ether, filtered off with suction, and dried in a vacuum at 60°C [1]. According to [6] an aqueous solution of manganese(II) nitrate (50 wt%) is mixed with butyl alcohol; 2.5 mole equivalents of pyridine N-oxide are added, and the solution is heated until all water has evaporated. Pale yellow crystals appear on cooling. These are washed with butyl alcohol and ether and dried in vacuo protected from sunlight. The crystals are photosensitive and slightly hygroscopic. The pale yellow complex melts at 204 to 206°C. The electronic reflectance spectrum of the solid complex shows one band at 375 nm with shoulders at 480 and 650 nm. The onset of charge transfer is located at 580 nm. The solution in acetonitrile shows an absorption maximum at 380 nm with a molar extinction coefficient of $\varepsilon_{max} = 190\ L \cdot mol^{-1} \cdot cm^{-1}$. The onset of charge transfer is located at 440 nm. The IR absorption spectrum recorded from Nujol mulls shows a strong band shoulder at 1499 cm^{-1}, a strong broad band at 1290 cm^{-1} with a strong shoulder at 1260 cm^{-1}, a sharp medium doublet at 1031 and 1024 cm^{-1}, and a strong shoulder at 782 cm^{-1}, which are assigned to the antisymmetric and symmetric NO_2 stretching, NO stretching, and out-of-plane NO_2 deformation modes. No ligand bands are given. It is assumed on the basis of the IR and electronic spectra that the NO_3 groups are coordinated to the manganese atom as bidentate chelating ligands via two oxygen atoms, thus completing an octahedral configuration. Susceptibility measurements at 297 K (Gouy balance, corrected for diamagnetism) yield the magnetic moment $\mu_{eff} = 5.86\ \mu_B$, which indicates a high-spin manganese(II) complex [1]. The product of thermal decomposition at 300 to 400°C is glassy, whereas that obtained by heating at 800 to 1000°C is a polycrystalline manganese oxide [6].

$Mn(C_5H_5NO)_2X_2$ (X = Br, I). The bromide and iodide were obtained by dissolving the hydrated metal halide in hot ethanol and adding a solution of the ligand in the same solvent. The precipitate was dried in a vacuum [2]. $Mn(C_5H_5NO)_2Br_2$ was made also by blending solutions of $MnBr_2$ (2 g) and the ligand (2 g), each in a mixture of 30 mL acetone and 5 mL ethanol. The precipitate was washed with acetone, recrystallized, and dried at 70°C in a vacuum. The yellow crystals darken in air and light. They melt at 232 to 235°C [3].

The electronic diffuse reflectance spectrum of $Mn(C_5H_5NO)_2Br_2$ shows a charge-transfer band at 26320 cm^{-1} and a shoulder at 24700 cm^{-1}. The bands were assigned as $e_g \rightarrow \pi^*$ transitions (Mn-to-ligand). The spectral data were analyzed in terms of the concept of optical electronegativity. The crystal-field parameter Dq = 850 cm^{-1} and the Racah parameter B = 835 cm^{-1} were calculated [4]. Measurements at 291 K yielded the specific magnetic susceptibility $\chi_g = 32.70 \times 10^{-6}\ cm^3/g$, which (corrected for diamagnetism) gave the effective magnetic moment $\mu_{eff} = 5.62\ \mu_B$. Conductivity measurements in acetonitrile at dilutions of 171 and 684 L/mol yielded the molar electrical conductivities $\Lambda = 21.3$ and $31.1\ \Omega^{-1} \cdot cm^2 \cdot mol^{-1}$, respectively, which indicate a nonelectrolyte, probably of tetrahedral configuration [3].

On heating to 240°C, $Mn(C_5H_5NO)_2Br_2$ forms a product of composition $Mn(C_5H_5NO)_{0.5}Br_2$ [2]. It is sensitive to light and air (the color turning dark). It is decomposed by water and ethanol, soluble in acetonitrile and phenol (decomposition), and insoluble in ether, benzene, and tetrahydrofuran [3].

$Mn(C_5H_5NO)_2Cl_3$. The compound was synthesized by reacting ethanolic solutions of the dioxane complex $Mn(C_4H_8O_2)_2Cl_3$ and pyridine N-oxide in a 1:3 mole ratio at −40°C in an N_2 atmosphere excluding daylight and moisture. After 1 h of stirring at −30°C an equal volume of diethyl ether was added at −40°C, and the precipitate was filtered off, washed with ether at −40°C, and dried in a vacuum over P_2O_5. The green powder is completely stable at room temperature. The molecular weight of 361 (calculated: 352) measured by cryoscopy in nitrobenzene indicates a monomeric compound. The complex melts at 140°C with decomposition.

The electronic reflectance spectrum and the absorption spectrum of the solution in acetonitrile show the maxima (frequencies in cm^{-1}) listed together with the molar extinction coefficients ε_{max} (in $L \cdot mol^{-1} \cdot cm^{-1}$) and the assignment:

solid	11500(sh)	15600	28600	—
solution	12000(sh)	16000	27000	29400
ε_{max}	355	835	5200	5400
assignment (transitions)	d-d	d-d	charge transfer	charge transfer

The IR spectrum (Nujol mulls) was recorded in the 4000 to 250 cm^{-1} region. It shows an absorption band at 1180 cm^{-1} due to NO stretching vibrations of the ligand, which was shifted towards lower frequencies by -65 cm^{-1} on complexation. Further bands in the 500 to 250 cm^{-1} region (at 423, 413sh, 327, 290, and 270 cm^{-1}), assigned to ν(Mn-O) and ν(Mn-Cl) vibrations, are compatible with a square-pyramidal or a trigonal-bipyramidal structure (5 possible isomers) of a monomeric 5-coordinate manganese complex. This is suggested also by the electronic spectra of both the solid and the solution. Susceptibility measurements at 302.27, 270.23, and 230.87 K (Faraday balance, $10^6 \cdot \chi_g = 27.49$, 30.77, and 36.05 cm^3/g, respectively) show that the magnetic susceptibility obeys the Curie-Weiss law with a paramagnetic Curie temperature of $\Theta_p = 1.69$ K. The magnetic moment $\mu_{eff} = 4.84(\pm 0.01)$ μ_B is independent of temperature and indicative of a high-spin manganese(III) d^4 complex. The molar electrical conductivity at 20°C of a 0.0005M solution in nitromethane is $\Lambda = 6.5$ $\Omega^{-1} \cdot cm^2 \cdot mol^{-1}$ and that of a 0.001M solution in nitrobenzene is $\Lambda = 0.10$ $\Omega^{-1} \cdot cm^2 \cdot mol^{-1}$, both values indicating a neutral nonelectrolyte [5].

References:

[1] R. L. Carlin, M. J. Baker (J. Chem. Soc. **1964** 5008/14, 5009/11). – [2] D. H. Brown, D. Kenyon, D. W. A. Sharp (J. Chem. Soc. A **1969** 1474/7). – [3] K. Issleib, A. Kreibich (Z. Anorg. Allgem. Chem. **313** [1961/62] 338/48, 342/6). – [4] W. Byers, B. Fa-Chun Chou, A. B. P. Lever, R. V. Parish (J. Am. Chem. Soc. **91** [1969] 1329/33, 1331). – [5] R. Usón, V. Riera, M. A. Ciriano, M. Valderrama (Transition Metal Chem. [Weinheim] [1976] 122/6).

[6] R. D. Richardson (J. Mater. Sci. **15** [1980] 2569/72, 2570).

16.1.24.1.4 $MnLX_2$ Complexes

$Mn(C_5H_5NO)Cl_2$. The anhydrous chloro compound can be obtained by heating $Mn(C_5H_5NO)Cl_2 \cdot H_2O$ to 130°C [1] or mixing ethanolic solutions of $MnCl_2 \cdot 4H_2O$ and pyridine N-oxide in equimolar quantities after dehydration with $HC(OC_2H_5)_3$ or 2,2′-dimethoxyacetone. The precipitate contained some water, which was removed by prolonged vacuum drying over P_2O_5. The cream-white complex is stable to atmospheric moisture [2], melting at 314°C [3]. The electronic spectrum of its saturated solution in a mixture of methanol and dichloroethane shows absorption maxima at $\lambda = 366$ and <300 nm. The IR spectrum recorded for Nujol mulls in the 1250 to 200 cm^{-1} region shows strong absorption bands (frequencies in cm^{-1}; t = terminal, b = bridging chlorine atom) which were assigned as follows:

band position	1190	449	395	370(t), 344(t)	313(b), 243(b)
assignment	ν(NO)	δ(NO)	ν(Mn-O)	ν(Mn-Cl)	ν(Mn-Cl)

A very strong maximum at 543 cm^{-1} was not assigned. The considerable lowering of the ν(NO) and δ(NO) frequencies (-53 and -13 cm^{-1}, respectively) clearly indicates coordination of the ligand via terminal NO groups. The complex probably has a binuclear chlorine-bridged structure with 4-coordinate Mn, as suggested also by the presence of five bands in the far-IR at <400 cm^{-1}. The symmetry of the $OMnCl_3$ units is most probably intermediate between distorted tetrahedral and square planar. Magnetic susceptibility measurements at 297 K yielded the magnetic moment $\mu_{eff} = 5.60\ \mu_B$, which suggests some possible spin-spin interaction. Nonbulky ligands such as water or methanol form adducts (1:1 with H_2O, see next compound) [2]. On heating to 210°C in a thermal balance, $Mn(C_5H_5NO)Cl_2$ loses 0.5 mol of the ligand, leaving $Mn(C_5H_5NO)_{0.5}Cl_2$ which has magnetic and spectral properties similar to those of $Mn(C_5H_5NO)Cl_2 \cdot H_2O$ [1]. A DTA curve shows two peaks at 238 and at 306°C, the latter being assigned to the exothermic decomposition [3].

$Mn(C_5H_5NO)Cl_2 \cdot H_2O$. The compound was prepared by mixing ethanolic solutions of $MnCl_2 \cdot 4H_2O$ and pyridine N-oxide; the precipitate was filtered off and dried in a vacuum [1] or recrystallized from alcohol and dried at 120°C [3]. The pink compound loses its H_2O in air on being heated to 130°C as shown by TG analysis [1] or at 177°C as shown by an endothermic peak in the DTA curve (heating rate 4, 8, and 12°C/min) [3]. The electronic reflectance spectrum of $Mn(C_5H_5NO)Cl_2 \cdot H_2O$ exhibits bands at 23810, 21510, and 17860 cm^{-1}, which were assigned to the electronic transitions from the $^6A_{1g}$ ground state to the excited states $^4T_{1g}(G)$, $^4T_{2g}(G)$, and $^4A_{1g}(G)$, respectively. Susceptibility measurements at 295 K yielded the effective magnetic moment $\mu_{eff} = 5.77\ \mu_B$. The magnetic and spectral data indicate an octahedral environment for the manganese atom in this complex for which a polymeric chlorine-bridged structure is assumed [1].

$Mn(C_5H_5NO)(CX_3COO)_2$ (X = F, Cl). The compounds were synthesized by adding a slight excess of pyridine N-oxide to a dry ethanolic solution of $Mn(CX_3COO)_2$ with stirring and warming. The precipitates were collected, washed with dry ethanol, and recrystallized from methanol. The IR spectrum (Nujol mulls) was recorded in the 4000 to 400 cm^{-1} region. Its most important absorption bands and their assignment (frequencies in cm^{-1}) and the magnetic moments μ_{eff} (in μ_B) obtained at 295 K are listed below:

compound/assignment	$\nu_{as}(COO)$	$\nu_s(COO)$	$(\nu_{as} - \nu_s)$	further bands	μ_{eff}
$Mn(C_5H_5NO)(CF_3COO)_2$	1700	1459	241	310/290; 240/220	5.92
$Mn(C_5H_5NO)(CCl_3COO)_2$	1680	1358	322	360/340; 340/320	5.90

The two bands observed in the far-IR at frequencies below 400 cm^{-1} are probably due to Mn-anion and Mn-ligand interactions, but detailed assignments could not be made. It is assumed that in these complexes the CX_3COO^- anions occupy more than one coordination site around the manganese atom [4].

References:

[1] D. H. Brown, D. Kenyon, D. W. A. Sharp (J. Chem. Soc. A **1969** 1474/7). – [2] N. M. Karayannis, C. M. Paleos, L. L. Pytlewski, M. M. Labes (Inorg. Chem. **8** [1969] 2559/62). – [3] A. J. Pappas, F. A. Osterman, H. B. Powell (Inorg. Chem. **9** [1970] 2695/8, 2696). – [4] S. Amasa, D. H. Brown, D. W. A. Sharp (J. Chem. Soc. A **1969** 2892/4).

16.1.24.2 With Monosubstituted Pyridine N-Oxides of the Type R-C_5H_4NO
R = Alkyl, Aryl, CN, or NO_2

This section treats complexes with the following substituted pyridine N-oxides RC_5H_4NO (= L), which generally act as monodentate ligands.

R	z-CH_3 (z = 2,3,4)	z-C_2H_5 (z = 2,3,4)	z-C_6H_5 (z = 2,4)	z-$CH_2C_6H_5$ (z = 2,4)
formula of L	C_6H_7NO	C_7H_9NO	$C_{11}H_9NO$	$C_{12}H_{11}NO$
R	4-$C(CH_3)_3$	2-CH_2Cl	4-CN	4-NO_2
formula of L	$C_9H_{13}NO$	C_6H_6ClNO	$C_6H_4N_2O$	$C_5H_4N_2O_3$

$Mn(RC_5H_4NO)_6(ClO_4)_2$. The compounds with R = 2-CH_3, 3-CH_3, 4-CH_3, and 2-C_2H_5 were obtained by refluxing $Mn(ClO_4)_2 \cdot 6H_2O$ with the appropriate ligand, mole ratio 1:7, in ethanol for a brief time and cooling the solution to −30°C, whereupon yellow crystals of the complexes appeared [1]. In another preparation hot ethanolic solutions of $Mn(ClO_4)_2 \cdot 6H_2O$ (1 g) and the ligands with R = 4-CH_3, 4-NO_2 (1 g) were mixed and cooled to 0°C for 24 h. The resulting solids were recrystallized from ethanol, washed with diethyl ether, and dried under reduced pressure [2]. The complex with R = 2-CH_2Cl was synthesized by adding the ligand slowly to an ethanolic solution of $Mn(ClO_4)_2 \cdot 6H_2O$. The yellow solid formed was washed with anhydrous ether and stored over $CaSO_4$ [13]. Doped complexes were obtained by mixing solutions of $Zn(ClO_4)_2$, $Cd(ClO_4)$ [3], or $Hg(ClO_4)_2$ [4] containing 1% [3] or 2% [4] by weight of $Mn(ClO_4)_2$ in absolute ethanol with an absolute ethanol solution of the ligand. Only complexes doped into those of Zn, Cd, or Hg were prepared for R = 4-CN [3, 4]. In the table below are listed the empirical formulas of the ligands L, the substituents R, the color, melting points (m.p. in °C), characteristic absorption bands in the electronic spectrum of 0.2 M solutions in nitromethane, and the important IR bands recorded for specimens in KI or suspensions in Nujol (frequencies in cm^{-1}; free ligand frequencies in parentheses):

ligand L	R	color	m.p.	charge transfer (Mn → L)	ν(NO)	ν(Mn-O)	Ref.
C_6H_7NO	2-CH_3	light yellow	142.5 to 144	26000	1218 (1252 L)	?	[1]
C_6H_7NO	3-CH_3	yellow	160 to 161.5	25300	1260 (1280 L)	258	[1]
C_6H_7NO	4-CH_3	golden yellow	200 to 202	25400	1212 (1255 L)	255	[1]
C_6H_7NO	4-CH_3	yellow [2]	—	24500 [5]	1208 (1232 L) [6]	—	[2, 5, 6]
C_7H_9NO	2-C_2H_5	light yellow	136 to 138	26100	1212 (1237 L) 1208	255	[1]
C_6H_6ClNO	2-CH_2Cl	yellow	—	25910	1220 (1248 L · HCl)	348	[13]
$C_5H_4N_2O_3$	4-NO_2	bright red [2]	—	20900 [5]	1235, 1222 (doublet) [2] (1228, 1208 L)	—	[2, 5]

The coordination of the ligands to Mn thus occurs through the N-bonded oxygen atoms as indicated by the lowering of the ν(NO) frequencies [1, 2, 6, 13] and appearance of ν(Mn-O) bands [1]. The charge-transfer bands are assigned to $e_g \rightarrow \pi^*$ electronic transitions. Besides the listed maxima, there are shoulders for the complexes with R = 4-CH_3 at 22300 cm^{-1} and for R = 4-NO_2 at 19310 cm^{-1}. Crystal-field parameters calculated from the spectral maxima are the same as for the unsubstituted pyridine N-oxide complex $Mn(C_5H_5NO)_6(ClO_4)_2$, see p. 145 [5]. Susceptibility measurements on the complexes with R = 4-CH_3 and R = 4-NO_2 at 25°C yielded the magnetic moments μ_{eff} = 5.93 and 5.92 μ_B, which indicate high-spin Mn^{II} d^5 complexes [2].

The ESR X-band spectra (9.4 GHz) of the hexakis manganese(II) perchlorate complexes with R = 2-CH_3, 3-CH_3, 4-CH_3, or 4-NO_2 doped into the corresponding solid complexes of Zn, Cd [3], or Hg [4] are reported and discussed in [3, 4]. The following parameters were found; A = hyperfine coupling constant, D and E = zero-field splitting parameters:

metal	ligand L	R	g (±0.002)	−A in G	D in G	E in G	Ref.
Hg	C_6H_7NO	2-CH_3	2.010	93	240	—	[4]
Zn	C_6H_7NO	3-CH_3	1.996	92	730	91	[3]
Cd	C_6H_7NO	3-CH_3	2.003	92	520	0	[3]
Hg	C_6H_7NO	3-CH_3	2.010	93	530	—	[4]
Zn	C_6H_7NO	4-CH_3	2.000	95	770	145	[3]
Cd	C_6H_7NO	4-CH_3	2.003	95	680	0	[3]
Hg	C_6H_7NO	4-CH_3	2.003	95	720	165	[4]
Zn	$C_5H_4N_2O_3$	4-NO_2	1.996	90	515	160	[3]
Cd	$C_5H_4N_2O_3$	4-NO_2	2.001	90	491	102	[3]
Hg	$C_5H_4N_2O_3$	4-NO_2	2.012	90	500	~20	[4]
Zn	$C_6H_4N_2O$	4-CN	2.001	90	189	0	[3]
Cd	$C_6H_4N_2O$	4-CN	2.006	90	224	<40	[3]
Hg	$C_6H_4N_2O$	4-CN	2.011	90	235	—	[4]

For A, D, and E in 10^6 cm^{-1} see original papers [3, 4]. The hyperfine coupling constants A show that the bonding of Mn to the nitro-, cyano-, and methylpyridine N-oxides is more covalent than the Mn-OH_2 bonding in the aquo complexes [3]. The complex with R = 2-CH_2Cl is sensitive to fluorescent and UV irradiation. After some months of exposure to such light, it changes to a green solid, which probably consists of an octahedral manganese(III) complex [13].

$Mn(RC_5H_4NO)_5(ClO_4)_2$ (R = 2-$CH_2C_6H_5$). The compound with 2-benzylpyridine N-oxide was prepared by blending solutions of $Mn(ClO_4)_2$ (1.35 mmol) and the ligand (5.4 mmol) in mixtures of ethanol and $HC(OC_2H_5)_3$ (3:2 v/v). The solution of the ligand in the solvent mixture had to be heated before to 50 or 60°C for 30 min. The pale yellow precipitate was washed with the solvent mixture, recrystallized from a mixture of CH_2Cl_2, ethanol, and $HC(OC_2H_5)_3$, and dried in a vacuum over anhydrous $CaSO_4$. The electronic reflectance spectrum (Nujol mulls) shows a very strong absorption at λ < 300 nm (charge transfer), a strong shoulder at λ = 318 nm, and a broad band at λ = 365 nm. The IR spectrum shows medium to strong absorption bands (ν in cm^{-1}), which were assigned as follows: 1219 ν(NO); 1090 $\nu_3(ClO_4)$; 1106, 1076, 1030 (ν_L, ν_3 region); 624 $\nu_4(ClO_4)$; 616 (ν_L, ν_4 region); 355 ν(Mn-O); 509, 460, and 303 (other ligand bands). Susceptibility measurements at 295 K yielded the magnetic moment $\mu_{eff}=5.96\ \mu_B$. The molar electrical conductivity $\Lambda=177\ \Omega^{-1}\cdot cm^2\cdot mol^{-1}$ measured at 25°C for a 10^{-3}M solution in nitromethane is that of 1:2 electrolyte. It appears that the manganese atom is coordinated only to the five organic ligands through the O atoms of the N-oxide function, as shown by the lowering of the ν(NO) frequency ($-30\ cm^{-1}$). The ClO_4^- ions are not coordinated and retain their tetrahedral T_d symmetry, as shown by the presence of the $\nu_3(ClO_4)$ and $\nu_4(ClO_4)$ single bands. The unusual 5-coordination of the manganese atom in this high-spin (d^5) complex evidently is due to steric hindrance by the bulky o-benzyl substituent [7].

$Mn(RC_5H_4NO)_4(ClO_4)_2$ (R = 4-C_6H_5). The compound with 4-phenylpyridine N-oxide was prepared by combining solutions of manganese(II) perchlorate (1 mmol) and the ligand (6 mmol) in mixtures of ethanol and $HC(OC_2H_5)_3$ (1:1 v/v) which had been heated previously at 50°C for 1 h. The yellow precipitate was dried in vacuo. The air-stable but photosensitive complex has the magnetic moment $\mu=5.68\ \mu_B$ (susceptibility measurement). Its electronic reflectance spectrum (Nujol mulls) shows strong or very strong absorption bands (sh = shoulder, b = broad) at λ = 207, 217, 252(sh), 271(sh), 283, 305(b), 360(b), and 442(b) nm, the last two bands due to charge-transfer absorption, which masks all d-d transitions. The IR spectrum shows medium to strong bands, which were assigned as follows: 1226 ν(NO); 1103, 1073(b), 1032(b) $\nu_3(ClO_4)$; 919 $\nu_1(ClO_4)$; 847 δ(NO); 632, 624, 611 $\nu_4(ClO_4)$; 479 $\nu_2(ClO_4)$; 445 $\nu(Mn\text{-}O_{ligand})$; and 319 $\nu(Mn\text{-}OClO_3)$. It can be concluded from the IR and electronic spectra that Mn is 5-coordinated to four oxygens of the organic ligand, as suggested by a shift of the ν(NO) band by 16 cm^{-1} toward lower frequencies, and one oxygen of one perchlorate group with C_{3v} symmetry (the other is ionic with T_d symmetry), as suggested by the triply split ν_3- and ν_4-$OClO_3$ bands. The molar electrical conductance $\Lambda=181\ \Omega^{-1}\cdot cm^2\cdot mol^{-1}$ of a 0.001M solution in nitromethane is that of a 1:2 electrolyte and indicates dissociation of the coordinated perchlorate group. The complex is soluble in polar organic solvents [14].

$Mn(RC_5H_4NO)_3Cl_3$ (R = 3-C_2H_5, 4-C_6H_5). The manganese(III) complexes with 3-ethyl- and 4-phenylpyridine N-oxide were prepared by reacting anhydrous ethanolic solutions of the complex with dioxane, $Mn(C_4H_8O_2)_2Cl_3$, and the ligands in stoichiometric quantities with continuous stirring (1 h at −30°C) and exclusion of light and air. Addition of ether cooled to −40°C forms green solids, which are kept at −30°C for four to seven days. They were washed with diethyl ether and dried in a vacuum at −30°C over P_2O_5. Both compounds are khaki green, stable at room temperature, and only slightly sensitive to moisture, but decomposed by water. The complex with R = 3-C_2H_5 melts at 70 to 71°C, and the complex with 4-C_6H_5 melts at 90 to 92°C (with decomposition).

The tables on p. 155 list the specific magnetic susceptibilities χ_g, the magnetic moments μ_{eff} (in μ_B), the paramagnetic Curie temperatures Θ_p (in K), the important IR bands recorded for Nujol mulls, the molar electrical conductivities Λ (in $\Omega^{-1}\cdot cm^2\cdot mol^{-1}$) of nitromethane solutions, and absorption bands in the electronic diffuse reflectance spectra of the solids and solutions in chloroform (molar extinction coefficients ε_{max} in $L\cdot mol^{-1}\cdot cm^{-1}$ in parentheses):

ligand	R	χ_g in $cm^3/g \cdot 10^{-6}$		μ_{eff}	Θ_p	$\nu(NO)$*)	ν(Mn-O) and ν(Mn-Cl)	Λ
C_7H_9NO	3-C_2H_5	23.91 (230.84 K)	18.83 (292.37 K)	4.83 (19°C)	−2.02	1250(vs)	475(m), 420(s), 370(m), 325(w), 270(s), 265(m)	12.80
$C_{11}H_9NO$	4-C_6H_5	19.90 (231.10 K)	15.61 (293.16 K)	4.97 (20°C)	−4.01	1215(vs)	460(m), 425(s), 375(m) 340(m), 270(m)	19.80

*) Free C_7H_9NO at 1270 cm^{-1}, free $C_{11}H_9NO$ at 1235 cm^{-1}.

R	state/ transition	d-d			charge transfer	charge transfer
3-C_2H_5	solid	10400	18200	22700		27000
4-C_6H_5	solid	11100	18000	22500(sh)		25000
3-C_2H_5	solution	12000(sh, 470)		15900(1000)	26300(6450)	29000(6900)
4-C_6H_5	solution	12200(sh, 430)		16100(1400)	22700(7050)	26000(9900)

Both complexes are nonelectrolytes in nitromethane, as shown by their low conductivity, and have a distorted octahedral structure. The organic ligand is bonded to the 6-coordinate Mn^{III} through the O atom of its N-oxide function. They are high-spin Mn^{III}-d^4 complexes. Their magnetic susceptibilities obey the Curie-Weiss law. Both complexes are soluble (with decomposition) in CH_2Cl_2 and $CHCl_3$, but insoluble in diethyl ether and n-hexane. The complex with 2-ethylpyridine N-oxide, $Mn(C_7H_9NO)_3Cl_3$, is also soluble in acetonitrile, slightly soluble in nitrobenzene, and insoluble in benzene. The complex with 4-phenylpyridine N-oxide, $Mn(C_{11}H_9NO)_3Cl_3$, is only slightly soluble in acetonitrile and benzene [8]. If the initial reaction mixture containing the complexes MnL_3Cl_3 in solution is allowed to warm to room temperature, a slow reduction, with formation of a complex of the type MnL_4Cl_2 in the case of R = 4-C_6H_5, occurs. In the cases R = 3-C_2H_5 (and 4-C_2H_5) compounds of the type $Mn_2L_4Cl_5 \cdot C_2H_5OH$ (p. 156) are obtained [10].

$Mn(RC_5H_4NO)_2X_2$ (X = Cl, R = 4-CH_3, 4-NO_2; X = Br, R = 4-CH_3; X = I, R = 2-CH_3, 3-CH_3, 4-CH_3). The complexes $Mn(RC_5H_4NO)_2I_2$ with R = 2-CH_3 and 3-CH_3 and $Mn(RC_5H_4NO)_2X_2$ with R = 4-CH_3 and X = Cl, Br, I were prepared by mixing ethanolic solutions of the hydrated manganese(II) halides with ethanolic solutions of 2-, 3-, or 4-methylpyridine N-oxide. The precipitate was filtered off and dried in a vacuum. Occasionally ether was added to facilitate precipitation [11]. $Mn(4\text{-}NO_2C_5H_4NO)_2Cl_2$ was obtained from $MnCl_2 \cdot nH_2O$ and 4-nitropyridine N-oxide by a way similar to that for the complex $Mn(4\text{-}NO_2C_5H_4NO)_6(ClO_4)_2$ [2, 5]. Susceptibility measurements on $Mn(4\text{-}CH_3C_5H_4NO)_2Br_2$ (pink) at 295 K yielded the effective magnetic moment $\mu_{eff} = 5.82\ \mu_B$, indicating a high-spin octahedral Mn^{II} complex. Its electronic reflectance spectrum shows two bands, at $\nu = 21370$ and 17860 cm^{-1}, which were assigned to the electronic transitions $^6A_{1g} \rightarrow {}^4T_{2g}(G)$ and $^6A_{1g} \rightarrow {}^4T_{1g}(G)$, respectively [11]. The electronic spectrum of $Mn(4\text{-}NO_2C_5H_4NO_2)_2Cl_2$ has a charge-transfer band at $\nu = 21800\ cm^{-1}$ ($e_g \rightarrow \pi^*$ transition) with a shoulder at 19800 cm^{-1}. A bridging function of the ligand 4-nitropyridine N-oxide is assumed in this complex [5]. The thermal degradation of the complexes MnL_2X_2 (L = 4-$CH_3C_5H_4NO$) takes place as follows: $MnL_2X_2 \rightarrow MnLX_2 + L$, at 260°C for X = Cl and at 240°C for X = Br and I [11].

$Mn(RC_5H_4NO)_2Cl_3$ (R = 2-CH_3, 3-CH_3, 2-C_2H_5, 4-$C(CH_3)_3$, 2-C_6H_5, and 4-$CH_2C_6H_5$). The manganese(III) complexes with the listed R were prepared much like the compound with unsubstituted pyridine N-oxide $Mn(C_5H_5NO)_2Cl_3$. The complexes with R = 2-C_2H_5, 2-C_6H_5, and 4-$CH_2C_6H_5$ were kept two days at −30°C prior to filtration. That with R = 4-$C(CH_3)_3$ was prepared without adding ether. They were stirred for 30 min at −30°C and then two days at room temperature. In the table below are listed the molecular weights M obtained by cryoscopic measurements in nitrobenzene, the melting points m.p. (in °C), the paramagnetic Curie temperatures Θ_p (in K), the IR bands assigned to the NO stretching vibrations, and the molar electrical conductivity Λ (in $\Omega^{-1}\cdot cm^2\cdot mol^{-1}$) of 5×10^{-4}M solutions in nitromethane at 20°C:

ligand L	R	M	m.p.	Θ_p	ν(NO)	$\nu(NO)_{ligand}$ **)	Λ
C_6H_7NO	2-CH_3	384(380)	153	−1.90	1185	1250	10.3
C_6H_7NO	3-CH_3	—(—)	139 to 140	+5.50	1245	1265	4.6
C_7H_9NO	2-C_2H_5	—(—)	156 to 158	−1.52	1185	1240	3.4
$C_9H_{13}NO$	4-$C(CH_3)_3$	466*)(464)	178*)	+0.07	1200	1241	8.5
$C_{11}H_9NO$	2-C_6H_5	516(504)	115 to 120*)	+0.85	1185	1255	1.3
$C_{12}H_{11}NO$	4-$CH_2C_6H_5$	554(532)	150 to 151*)	+3.28	1190	1230	7.9

*) Osmometric measurements in $CHCl_3$. – **) Free-ligand band.

In the IR spectrum, in the range 500 to 250 cm^{-1}, four bands were observed. These were assigned to the ν(Mn-O) and ν(Mn-Cl) vibrations. Other conductivity measurements in nitrobenzene (10^{-3}M) showed Λ to be <1 $\Omega\cdot cm^2\cdot mol^{-1}$. UV and visible spectra of the solids (diffuse reflectance) and of solutions in chloroform were recorded. Two absorptions, near 12000 and 16000 cm^{-1}, were assigned to d-d transitions. Two bands, near 25000 and 29000 cm^{-1}, must be due to charge transfer because of their intensity and location.

The compounds are nonconducting, neutral, and monomeric in solution. They exhibit magnetic moments μ_{eff} between 4.80 and 4.85 μ_B, indicating four unpaired electrons and thus high-spin Mn^{III}-d^4 complexes. The moments are nearly independent on temperature and obey the Curie-Weiss law as shown by measurements at about 300, 270, and 230 K. Based on the IR and electronic spectra and magnetic data, a distorted square pyramidal structure with a 5-coordinate manganese atom was proposed for these complexes. The compounds may be stored at room temperature in a vacuum desiccator [9].

$Mn_2(RC_5H_4NO)_4Cl_5\cdot C_2H_5OH$ (R = 3-C_2H_5, 4-C_2H_5). The complexes with 3- and 4-ethylpyridine N-oxide were obtained by reacting ethanolic solutions of the complex with dioxane, $Mn(C_4H_8O_2)_2Cl_3$, and the ligand in a 1:3 mole ratio at −40°C with stirring which was continued at room temperature. After ~24 h a reddish-brown solid precipitated. It was washed with ether and dried under N_2 and reduced pressure. The oxidation numbers of the complexes were found to be 2.49 for the complex with R = 3-C_2H_5 and 2.48 for the compound with R = 4-C_2H_5. They are stable at room temperature and melt at 93 to 94°C (R = 3-C_2H_5) and at 138 to 139°C (R = 4-C_2H_5). The electronic reflectance spectra of the solid complexes show four bands, at λ = 950, 530, 440(sh), and 360 nm, which suggest an octahedral ligand field around Mn^{III}. Solutions in chloroform or nitromethane show two absorption bands, at λ = 830(sh) and 630 nm, which were assigned to d-d transitions and two strong charge-transfer bands at λ = 380 and 350 nm. The solution spectra are similar to those of the 5-coordinate complexes MnL_2Cl_3 (see above) and suggest a distorted square-pyramidal structure in solution. In the IR spectra the bands assigned to the NO stretching vibrations are shifted towards lower frequencies, thus indicating the existence of a Mn-O bond. There are also bands indicating the presence of ethanol but no IR peaks were given. Susceptibility measurements at several temperatures (T in K) yielded the following effective magnetic moments μ_{eff} per 2 Mn, while conductivity measurements in nitromethane yielded the following electrical conductivities Λ (in $\Omega^{-1}\cdot cm^2\cdot mol^{-1}$):

ligand L	R		μ_{eff}/2 Mn in μ_B		Λ
C_7H_9NO	3-C_2H_5	7.75 (65.9 K)	7.71 (229.9 K)	7.67 (298.1 K)	40.0
C_7H_9NO	4-C_2H_5	—	7.25 (230.9 K)	7.27 (293.2 K)	43.2

The magnetic moments do not vary appreciably between 65 and 300 K in accordance with the Curie-Weiss law. They indicate the presence of both high-spin Mn^{3+}-d^4 ions and Mn^{2+}-d^5 ions (calculated magnetic moment, $\mu_{eff}=7.68\ \mu_B$). There are no oxygen bridges between the manganese atoms, but chlorine bridges are assumed in the solid complexes, which are probably binuclear or polymeric with 6-coordinate manganese atoms. In solutions the complexes are probably binuclear. They dissolve in nitromethane and acetonitrile with decomposition, as shown by the conductivity (which is low enough to exclude ions) and are insoluble in benzene and n-hexane [10].

$Mn(RC_5H_4NO)X_2\cdot nH_2O$ (R = 2-CH_3, 3-CH_3; X = Cl, Br; n = 0, 1) and **$Mn(RC_5H_4NO)X_2$** (R = 4-CH_3; X = Cl, Br, I). The hydrated complexes were synthesized by mixing ethanolic solutions of the hydrated manganese(II) halides and 2-methyl- or 3-methylpyridine N-oxide as described for the complexes $Mn(RC_5H_4NO)_2X_2$ on p. 155. Their thermal degradation occurs as follows:

$$\text{for R = 2-CH}_3\text{:}\quad MnLX_2\cdot H_2O \xrightarrow[(-H_2O)]{160°C\ (X=Cl),\ 130°C\ (X=Br)} MnLX_2 \xrightarrow[-0.5\,L]{240°C} MnL_{0.5}X_2$$

$$\text{for R = 3-CH}_3\text{:}\quad MnLX_2\cdot H_2O \xrightarrow[(-H_2O)]{120°C\ (Br,Cl)} MnLX_2$$

The compounds $Mn(RC_5H_4NO)X_2$ (R = 4-CH_3; X = Cl, Br, I) were obtained by heating the complexes $Mn(4\text{-}CH_3C_5H_4NO)_2X_2$ as indicated on p. 155. The complexes show similar visible reflectance spectra and magnetic moments (not reported) as the complex with unsubstituted pyridine N-oxide, $Mn(C_5H_5NO)Cl_2\cdot H_2O$, and the complex $Mn(RC_5H_4NO)_2Br_2$ with R = 4-CH_3 (pp. 151 and 155, respectively). The spectral and magnetic data suggest the manganese ions to be in a predominantly octahedral environment [11].

$Mn(4\text{-}CH_3C_5H_4NO)(CX_3COO)_2$ (X = F, Cl). The trifluoro- and trichloroacetato compounds were obtained by reacting $Mn(CX_3COO)_2$ and a slight excess of the ligand in absolute ethanol. They were recrystallized from methanol. Susceptibility measurements at 293 K yielded the magnetic moments $\mu_{eff}=5.92$ and 5.93 μ_B for the compounds with X = F and Cl, indicating high-spin octahedral manganese(II) complexes. In the IR spectra bands similar to those recorded for the acetato complex, $Mn(C_5H_5NO)(CH_3COO)_2$, were observed. The experimental data suggest coordination of the CX_3COO^- anions as bidentate ligands [12].

$Mn(2\text{-}CH_3C_5H_4NO)_{0.5}X_2$ (X = Cl, Br). The complexes were obtained by thermal degradation of $Mn(2\text{-}CH_3C_5H_4NO)X_2$ at 240°C. Spectral and magnetic data (only reported for some typical compounds) suggest that they have a structure similar to their precursors [11].

References:

[1] G. Schmauss, H. Specker (Z. Anorg. Allgem. Chem. **364** [1969] 1/19, 3, 7, 16, 18). – [2] W. Byers, A. B. P. Lever, R. V. Parish (Inorg. Chem. **7** [1968] 1835/40, 1838). – [3] G. M. Woltermann, J. R. Wasson (Inorg. Chem. **12** [1973] 2366/70, 2369). – [4] G. M. Wolter-

mann, J. R. Wasson (J. Phys. Chem. **78** [1974] 45/9, 48). – [5] W. Byers, B. Fa-Chun Chou, A. B. P. Lever, K. V. Parish (J. Am. Chem. Soc. **91** [1969] 1329/33, 1331).

[6] L. C. Nathan, R. O. Ragsdale (Inorg. Chim. Acta **10** [1974] 177/81, 178). – [7] A. N. Speca, L. S. Gelfand, F. J. Iaconianni, L. L. Pytlewski, C. M. Mikulski, N. M. Karayannis (J. Inorg. Nucl. Chem. **41** [1979] 283/6). – [8] R. Usón, V. Riera, M. A. Ciriano, M. Valderrama (Rev. Latinoam. Quim. **9** No. 2 [1978] 60/4; C.A. **89** [1978] No. 156706). – [9] R. Usón, V. Riera, M. A. Ciriano, M. Valderrama (Transition Metals Chem. **1** [1976] 122/6). – [10] M. Valderrama (Rev. Latinoam. Quim. **10** No. 2 [1979] 49/50; C.A. **91** [1979] No. 203492).

[11] D. H. Brown, D. Kenyon, D. W. A. Sharp (J. Chem. Soc. A **1969** 1474/7). – [12] S. Amasa, D. H. Brown, D. W. A. Sharp (J Chem. Soc. A **1969** 2892/4). – [13] D. X. West, T. P. Henning (J. Inorg. Nucl. Chem. **40** [1978] 915/7). – [14] A. N. Speca, F. J. Iaconianni, L. S. Gelfand, L. L. Pytlewski, C. M. Mikulski, N. M. Karayannis (J. Inorg. Nucl. Chem. **41** [1979] 957/61).

16.1.24.3 With 2,6-Dimethylpyridine N-Oxide (= C_7H_9NO)

$Mn(C_7H_9NO)_4(ClO_4)_2$. $Mn(ClO_4)_2 \cdot 6H_2O$ (1 mol) was mixed with a tenfold excess of $HC(OC_2H_5)_3$ for dehydration and a minimum of acetone to achieve complete dissolution. After 15 min of stirring at 40°C an excess of ligand was added (>6 mol). The precipitate was filtered off, washed with ether, and dried in a vacuum over P_2O_5 [1]. The very pale yellow [1, 2] compound is air stable. It melts at 200 to 202°C with decomposition [1]. The electronic spectrum recorded for Nujol mulls of the solid exhibits absorption bands at 25000 and >28000 cm^{-1}; the spectrum of 0.0014M solutions in nitromethane exhibits bands at 13620(sh), 17640(sh), and >27000 cm^{-1}. The latter is a charge-transfer band of high intensity [1]. In the IR spectrum, which was recorded for Nujol mulls in the 4000 to 200 cm^{-1} region, the significant bands (ν in cm^{-1}) were assigned as follows: 1203 ν(NO); 1076 ν_3 mode of ionic ClO_4 (T_d symmetry); 1133, 1024, and 920 ν_4, ν_1, and ν_2 modes of coordinated $OClO_3$ (C_{3v} symmetry) [1]; 772 δ(CH) out-of-plane; 338 $\nu(Mn\text{-}O_{ligand})$; and 260 $\nu(Mn\text{-}O_{OClO_3})$. Further bands in the range between 550 and 260 cm^{-1} were assigned to ligand vibrations [2]. The shift of the ν(NO) ligand band occurring on coordination, $\Delta\nu = -42$ cm^{-1}, indicates a coordination of the N-oxide to Mn through its oxygen atom, and the appearance of four $\nu(ClO_4)$ bands in the 1150 to 950 cm^{-1} region show the presence of both noncoordinated ClO_4^- ions and coordinated $OClO_3^-$ groups, which act as monodentate ligands. Thus the complex has to be formulated $[Mn(C_7H_9NO)_4OClO_3]ClO_4$, a structure of a symmetry close to square pyramidal with 5-coordinate manganese atoms [1], although a 6-coordinate structure, with bidentate ClO_4 groups, cannot be excluded [2]. Susceptibility measurements at 297 K yielded the magnetic moment $\mu_{eff} = 5.82$ μ_B, indicating a high-spin Mn^{II} complex. The molar electrical conductivity $\Lambda = 185.8$ $\Omega^{-1} \cdot cm^2 \cdot mol^{-1}$ measured at 25°C of a 0.001M solution in nitromethane indicates a 1:2 electrolyte formed by solvolysis of the complex cation $[Mn(C_7H_9NO)_4(OClO_3)]^+$ [1].

$Mn(C_7H_9NO)_2(NO_3)_2$. A solution of $Mn(NO_3)_2 \cdot 6H_2O$ (1 mol) in a mixture of $HC(OC_2H_5)_3$ (9 parts) and acetone (1 part) was warmed to 50 or 60°C with stirring, and excess ligand (~7 mol) was added. After 15 min of further stirring the solution was cooled to about 40°C, and ligroin (b.p. ~70°C) was added with further stirring for 20 to 30 min at room temperature. The pale yellow, air-stable precipitate was filtered off, washed with ligroin or ether, and dried in a vacuum over P_2O_5. The eletronic spectrum recorded for Nujol mulls of the solid shows a very strong absorption at $\lambda < 300$ nm ($\nu = 33000$ cm^{-1}), strong shoulders at 330 and 492 nm, and a weak shoulder at 602 nm. A 0.005M solution in nitromethane shows a strong absorption at $\lambda < 400$ nm and a shoulder at 510 nm (with

$\varepsilon = 20\ L\cdot mol^{-1}\cdot cm^{-1}$). The IR spectrum, which was recorded for Nujol mulls in the 1800 to 250 cm^{-1} region, shows the following significant absorption bands (in cm^{-1}; assignment in parenthesis): 994 ν_1 (NO_3); 819, 808 $\nu_2(NO_3)$; 1443(sh), 1304, 1299(sh) ν_3(NO_3 triplet); 740, 722 $\nu_4(NO_3)$; 1200 ν(NO) of the ligand L; 358 $\nu(Mn\text{-}O_{ligand})$; and 350, 257 $\nu(Mn\text{-}ONO_2)$. Four bands in the 1780 to 1720 range were assigned to ($\nu_1+\nu_4$) modes of the NO_3 group, and eight bands in the 830 to 280 range were assigned to ligand vibrations. The IR spectrum favors a structure involving two types of coordinated NO_3 groups, both monodentate and bidentate. The 2,6-dimethylpyridine N-oxide is coordinated to Mn through its oxygen atom ($\Delta\nu = -45\ cm^{-1}$). Susceptibility measurements at 300 K yielded the magnetic moment $\mu_{eff} = 6.13\ \mu_B$. The magnetic moment and the spectral data suggest a monomeric structure of the type $[MnL_2(ONO_2)(O_2NO)]$ for the complex with a 5-coordinate manganese atom and both mono- and bidentate nitrato ligands. The molar electrical conductivity of $\Lambda = 18\ \Omega^{-1}\cdot cm^2\cdot mol^{-1}$ measured on 0.001M solutions in nitromethane at 25°C indicates nonelectrolytic solutions [3].

$Mn(C_7H_9NO)_2X_2$ (X = Cl, Br, I). The chloro complex was obtained by heating a solution of $MnCl_2\cdot 4H_2O$ (1 mmol) and 2,6-dimethylpyridine N-oxide (3 mmol) in isopropyl alcohol (30 mL) to 60°C and then adding excess (30 to 50 mL) $HC(OC_2H_5)_3$ with stirring. The pale yellow precipitate was filtered off, washed with ligroin, and dried in a vacuum over P_2O_5 [4]. The bromo and iodo complexes were obtained by mixing ethanolic solutions of $MnX_2\cdot 4H_2O$ and the ligand. The precipitates were filtered off and dried in a vacuum [5]. The electronic spectrum of the chloro complex recorded for Nujol mulls shows absorption bands at $\lambda = 551$(m to w, sh), 453(m, b), 360(vs, b), 259(vs), 225(vvs), and 218(vvs) nm; the spectrum of a 0.005 solution in nitromethane has bands at $\lambda = 612$, 522(sh), and <400 nm. In the IR spectrum, which was recorded for Nujol mulls in the 1250 to 230 cm^{-1} region, the significant absorption bands (in cm^{-1}) were assigned as follows: 1191, 1159 ν(NO); 822 δ(NO); 773 γ(CH) (wagging); 376, 357 ν(Mn-O); 310, 300 $\nu(Mn\text{-}Cl_{termina})$; 263, 240 $\nu(Mn\text{-}Cl_{bridging})$. Other strong bands were assigned as ligand vibrations. The IR and electronic spectra suggest a binuclear complex containing a 5-coordinate manganese(II) atom with both bridging and terminal Cl^- ions: $[ClL_2MnCl_2MnL_2Cl]$. The 2,6-dimethylpyridine N-oxide is coordinated to the manganese atom through its oxygen atom. Susceptibility measurements at 294 K yielded the magnetic moment $\mu_{eff} = 5.84\ \mu_B$. The molar electrical conductivity of $\Lambda = 17\ \Omega^{-1}\cdot cm^2\cdot mol^{-1}$ indicates a nonelectrolyte. The compound is photosensitive (oxidation to trivalent Mn) and soluble in polar solvents such as nitromethane, acetonitrile, and alcohols [4]. $Mn(C_7H_9NO)_2Br_2$ and $Mn(C_7H_9NO)_2I_2$ each lose one molecule of the organic ligand on heating to 200°C (Br) or 210°C (I) [5].

$Mn(C_7H_9NO)Cl_2$ and **$Mn(C_7H_9NO)Cl_2\cdot C_2H_5OH$.** The ethanol adduct was obtained by mixing ethanolic solutions of $MnCl_2\cdot 4H_2O$ and 2,6-dimethylpyridine N-oxide. The precipitate was filtered off and dried in a vacuum. The compound loses its ethanol at 120°C to give $Mn(C_7H_9NO)Cl_2$ [5].

$Mn(C_7H_9NO)X_2$ (X = Br, I). The compounds were obtained by thermal degradation of the complexes $Mn(C_7H_9NO)_2X_2$ (X = Br, I) [5].

$Mn(C_7H_9NO)F(BF_4)\cdot CH_3OH$ and **$Mn(C_7H_9NO)F(BF_4)\cdot 4MnF_2$.** The methanol adduct was obtained on adding the stoichiometric amount of $HC(OC_2H_5)_3$ to a methanolic solution of $Mn(BF_4)_2\cdot 6H_2O$ and then adding a dry methanolic solution of the ligand L in a fourfold excess. The off-white crystals which appeared on keeping the solution overnight at room temperature were filtered off, washed with anhydrous diethyl ether, and dried in a vacuum over P_4O_{10}. $Mn(C_7H_9NO)F(BF_4)\cdot 4MnF_2$ was formed if a solution of $Mn(BF_4)_2\cdot 6H_2O$ in a mixture of alcohol and the stoichiometric amount of triethyl orthoformate were stirred for 2 h at 50°C and excess ligand (L:M mole ratio 6:1) was added. The pinkish-white precipitate was washed with the solvent and stored in a vacuum over P_4O_{10}. It is air stable when dry. In the tables below are listed the significant IR bands (in cm^{-1}) with their assignment, the main absorption bands in the electronic reflectance spectra (recorded for Nujol mulls), the magnetic moments μ_{eff} (in μ_B) from susceptibility measurements at 300 K, and the molar electrical conductivity Λ (in $\Omega^{-1} \cdot cm^2 \cdot mol^{-1}$) of 0.001M solutions in nitromethane measured at 25°C:

compound	ν(NO)	$\nu_3(BF_4)$	$\nu_1(BF_4)$	$\nu_4(BF_4)$	$\nu_2(BF_4)$	ν(Mn-O)	$\nu(Mn\text{-}O)_L$	ν(Mn-F)
$MnLF(BF_4) \cdot CH_3OH$	1209	1124 1063 1027	758	549 531 520	353	347*)	330 319	380
$MnLF(BF_4) \cdot 4MnF_2$	1202	1066	755	543	—	—	322	385

*) Band assigned to ν(Mn-O) vibrations of coordinated methanol.

compound	electronic bands, λ in nm				μ_{eff}	Λ
$MnLF(BF_4) \cdot CH_3OH$	393	261	—	209	5.79	29
$MnLF(BF_4) \cdot 4MnF_2$	384	264	253	208	5.31	insoluble

The methanol adduct shows also an IR band at 3420 cm^{-1}, which is assigned to ν(OH) vibrations of CH_3OH. It is assumed that the BF_4 ion was coordinated to Mn as a bidentate F_2BF_2 group because the ν_3 and ν_4 bands of the BF_4 group are split in the IR spectrum of the methanol adduct. In $Mn(C_7H_9NO)F(BF_4) \cdot 4MnF_2$ these bands are unsplit and indicate noncoordinated BF_4 ions. Both complexes are high spin, containing 6-coordinate Mn, as suggested by their magnetic data. The lower magnetic moment of $Mn(C_7H_9NO)F(BF_4) \cdot 4MnF_2$ may indicate a mixture of antiferromagnetic MnF_2 and $Mn(C_7H_9NO)F(BF_4)$ or MnF_2, $MnF(BF_4)$, and $Mn(C_7H_9NO)F_2$. For the methanol adduct a dimeric structure is proposed with Mn-F_2-Mn bridges, bidentate chelating F_2BF_2 ligands, and two neutral ligands per metal ion (one 2,6-dimethylpyridine N-oxide and one alcohol ligand). Reactions of the type $Mn(BF_4)_2 + nL \rightarrow MnF(BF_4)L_{n-1} + BF_3L$ or $Mn(BF_4)_2 + mL \rightarrow MnF_2L_{m-2} + 2BF_3L$ are assumed to take place during the formation of these complexes. The methanol adduct is soluble in nitromethane and other polar organic solvents with partial dissociation as shown by conductivity measurements, while the MnF_2 adduct is only sparingly soluble in these solvents [6].

References:

[1] N. M. Karayannis, L. L. Pytlewski, M. M. Labes (Inorg. Chim. Acta **3** [1969] 415/20, 416). – [2] N. M. Karayannis, C. M. Mikulski, M. J. Strocko, L. L. Pytlewski, M. M. Labes (J. Inorg. Nucl. Chem. **33** [1971] 3185/91, 3186). – [3] N. M. Karayannwas, C. M. Mikulski, L. L. Pytlewski, M. M. Labes (J. Inorg. Nucl. Chem. **34** [1972] 3139/52, 3140/2, 3144). – [4] C. M. Mikulski, L. S. Gelfand, L. L. Pytlewski, J. S. Skryantz, N. M. Karayannwas (Inorg. Chim. Acta **21** [1977] 9/14, 10). – [5] D. H. Brown, D. Kenyon, D. W. A. Sharp (J. Chem. Soc. A **1969** 1474/7).

[6] C. M. Mikulski, L. S. Gelfand, E. S. C. Schwartz, L. L. Pytlewski, N. M. Karayannis (Inorg. Chim. Acta **39** [1980] 143/51, 144/7).

16.1.24.4 With 2,4,6-Trimethylpyridine N-Oxide $(CH_3)_3C_5H_2NO$ (= $C_8H_{11}NO$)

$Mn(C_8H_{11}NO)_6(ClO_4)_2$. The complex was prepared from $Mn(ClO_4)_2 \cdot 6H_2O$ which had been heated in $HC(OC_2H_5)_3$ for dehydration and a solution of the ligand in absolute ethanol. The resulting beige precipitate was washed with diethyl ether and dried in vacuo over P_2O_5. The ν(NO) absorption band in the IR spectrum of the complex has a frequency of 1207 cm^{-1}, and thus is shifted by 33 cm^{-1} toward lower frequencies from the free-ligand value. Susceptibility measurements at 25°C yielded the magnetic moment $\mu_{eff} = 5.80$ μ_B, and conductivity measurements on solutions in dimethylformamide or nitrobenzene at 25°C yielded the molar electrical conductivities $\Lambda = 134.0$ and 48.8 $\Omega^{-1} \cdot cm^2 \cdot mol^{-1}$, respectively, both corresponding to a 1:2 electrolyte. The IR, magnetic, and conductivity data suggest an octahedral complex. It starts to decompose at 210°C and explodes at 240°C, B. Dušek, F. Kůtek, P. Bauer (Sb. Vys. Sk. Chem. Technol. Praze Anorg. Chem. Technol. B **24** [1980] 45/56, 49/52; C.A. **95** [1981] No. 53986).

16.1.24.5 With 2-Pyridinol N-Oxide $HO\text{-}C_5H_4NO$ (= $C_5H_5NO_2$)

Complexes in Solution. Potentiometric measurements (glass electrode) at 25(±0.1)°C and ionic strength I = 0.005 ($HClO_4$) for Mn-to-ligand ratios from 10:1 to 5:1 reveal the presence of 1:1 and 1:2 complexes with the stability constants log K_1 = 4.46 and log K_2 = 3.34 (K in L/mol), while spectrophotometric observations at λ = 215 nm, t = 25°C, and I = 0.1 for Mn-to-ligand ratios above 100:1 indicate the presence of only 1:1 complexes with log K_1 = 3.9, Peng-Joung Sun, Q. Fernando, H. Freiser (Anal. Chem. **36** [1964] 2485/8).

16.1.24.6 With 2-Pyridylmethanol N-Oxide $HOCH_2C_5H_4NO$ (= $C_6H_7NO_2$)

$Mn(C_6H_7NO_2)_4(ClO_4)_2$ and **$Mn(C_6H_7NO_2)_3(NO_3)_2$.** In order to prepare $Mn(C_6H_7NO_2)_4$-$(ClO_4)_2$ a solution of $Mn(ClO_4)_2 \cdot 6H_2O$ (1 mmol) in a mixture of methanol and 2,2-dimethoxypropane (1:1) was added dropwise to a warmed methanolic solution of ligand (6 mmol). The mixture was evaporated with gentle heating for 1 h, and crystallization was induced by adding acetone to the oily residue. The yellow crystals were washed with acetone and ether and dried over calcium sulfate [1]. $Mn(C_6H_7NO_2)_3(NO_3)_2$ was prepared similarly from $Mn(NO_3)_2 \cdot 6H_2O$ and the ligand in a 1:2 to 1:6 mole ratio to yield a solid on evaporation [2]. Below are listed the absorption maxima (in cm^{-1}) in the electronic spectrum assigned to charge-transfer transitions (Mn→L), the IR bands of the ligand sensitive to coordination (in cm^{-1}, with assignment) and the molar electrical conductivities Λ (in $\Omega^{-1} \cdot cm^2 \cdot mol^{-1}$) measured on solutions in acetonitrile:

compound	Mn→L	ν(NO)	ν(CO)	δ(NO)	ν(Mn-O)$_L$	Λ	Ref.
$Mn(C_6H_7NO_2)_4(ClO_4)_2$	26300	1212(b)	1070, 1028	858(sh), 846	375	361	[1]
$Mn(C_6H_7NO_2)_3(NO_3)_2$	27400	1207	1060, 1047(sh)	852, 838	—	36	[2]

The perchlorate has a magnetic moment of $\mu_{eff} = 5.80$ μ_B at room temperature, indicating a high-spin manganese(II) complex. As the perchlorate is a 1:2 electrolyte in acetonitrile it is assumed that ClO_4^- is ionic and not coordinated to Mn. A slight distortion of its symmetry from tetrahedral, which is suggested by the presence of a weak ν_1 and a split ν_4 band, is probably caused by a hydrogen-bridge bond between the ClO_4^- ion and the

CH_2OH group of the ligand. The manganese atom is probably 6-coordinate, the oxygen atoms of the NO groups and the CH_2OH groups of two bidentate ligands, being bonded to the metal atom, as in the octahedral nickel complex. Such a structure is suggested also by the positions of the ν(CO) and ν(NO) bands in the IR, which are shifted towards lower frequencies (or split) on coordination. The small shift of the ν(NO) band ($-7\ cm^{-1}$) is probably due to intramolecular hydrogen bridges present in the free ligand [1].

The low electrical conductivity of the nitrato complex in acetonitrile and its electronic and IR spectra suggest a structure containing 6-coordinate Mn^{II}, which is linked to the oxygen atoms of two monodentate nitrato groups and three neutral ligands, of which at least one is bidentate. The high conductivity, $\Lambda = 158\ \Omega^{-1}\cdot cm^2\cdot mol^{-1}$ in dimethylformamide, indicating a 1:1 electrolyte, is evidently due to a substitution of the nitrato groups by solvent molecules [2].

References:

[1] S. A. Boyd, R. E. Kohrman, D. X. West (J. Inorg. Nucl. Chem. **38** [1976] 1605/9). – [2] C. A. Frank, D. X. West (J. Inorg. Nucl. Chem. **40** [1978] 785/8).

16.1.24.7 With 4-Methoxypyridine N-Oxide $CH_3O\text{-}C_5H_4NO$ ($=C_6H_7NO_2$) and 4-Ethoxypyridine N-Oxide $C_2H_5O\text{-}C_5H_4NO$ ($=C_7H_9NO_2$)

$Mn(C_6H_7NO_2)_6(ClO_4)_2$ and $Mn(C_7H_9NO_2)_5(ClO_4)_2$. The complex with 4-methoxypyridine N-oxide, $Mn(C_6H_7NO_2)_6(ClO_4)_2$, was prepared by dehydrating $Mn(ClO_4)_2\cdot 6H_2O$ in a mixture of 2,2-dimethoxypropane and absolute ethanol at room temperature for 3 h [1] and then adding an ethanolic solution of the ligand in a 1:7 mole ratio [2]. The complex with 4-ethoxypyridine N-oxide, $Mn(C_7H_9NO_2)_5(ClO_4)_2$, was obtained by mixing ethanolic solutions of $Mn(ClO_4)_2\cdot 6H_2O$ and excess ligand. The yellow crystals which formed on cooling were washed with ether and dried in vacuo over $Mg(ClO_4)_2$. They melt at 177 to 179°C [3].

Magnetic measurements on the ethoxy compound, $Mn(C_7H_9NO_2)_5(ClO_4)_2$, at 297 K yielded the magnetic moment $\mu_{eff} = 6.00\ \mu_B$, which corresponds to a high-spin Mn^{II} complex. The electronic spectrum of its 0.72×10^{-3}M solution in acetonitrile shows two absorption maxima, at 20920 and 26950 cm^{-1}, with molar extinction coefficients of $\varepsilon_{max} = 6.9$ and 534.7 $L\cdot mol^{-1}\cdot cm^{-1}$, respectively, and a shoulder at 22940 cm^{-1}. There is an intense charge-transfer band in the UV region, which trails into the visible spectrum [3].

The IR spectra of $Mn(C_6H_7NO_2)_6(ClO_4)_2$ [2] and $Mn(C_7H_9NO_2)_5(ClO_4)_2$ [3] show absorption bands at 1207 or 1201 cm^{-1} with a shoulder at 1197 cm^{-1} [3], respectively, which were assigned to the NO stretching vibrations of the ligands. They were only slightly shifted upon coordination and indicate an essentially single-bond character of the NO bond [2, 3]. A band at 1076 cm^{-1} in $Mn(C_7H_9NO_2)_5(ClO_4)_2$ was assigned to ν_3 of the ionic ClO_4^- group [3]. Conductivity measurements on 5×10^{-5}M solutions of $Mn(C_7H_9NO_2)_5(ClO_4)_2$ in nitromethane at 21°C yielded the molar conductivity $\Lambda = 215.8\ \Omega^{-1}\cdot cm^2\cdot mol^{-1}$, which corresponds to a 1:2 electrolyte [3]. As shown by the electronic spectrum, IR spectrum, and electrical conductivity, $Mn(C_7H_9NO_2)_5(ClO_4)_2$ contains a 5-coordinate manganese(II) atom. Coordination of the ClO_4^- ions does not occur. A structural similarity with the analogous complexes of Zn, Co, and Ni is shown by the X-ray powder diffraction patterns. The compound darkens on aging. It could not be recrystallized from acetonitrile or other common solvents [3].

$Mn(C_6H_7NO_2)Cl_2$ and $Mn(C_7H_9NO_2)Cl_2$. The complexes with 4-methoxy- and 4-ethoxypyridine N-oxide were precipitated by adding triethyl orthoformate to solutions containing equimolar amounts of hydrated $MnCl_2$ and the ligand in a mixture of ethanol and acetone. The cream-white complexes contained some residual water, which was completely removed after prolonged vacuum desiccation over P_2O_5 [4]. $Mn(C_7H_9NO_2)Cl_2$ melts at 253 to 255°C [3]; $Mn(C_6H_7NO_2)Cl_2$ melts above 250°C [4], while exothermic decomposition at 245°C is reported by [5].

The electronic spectrum of a saturated solution of $Mn(C_7H_9NO_2)Cl_2$ in a mixture of dibromomethane and nitromethane shows two absorption maxima, at 22780 and 27800 cm^{-1} [3]. The following table lists the coordination-sensitive IR bands (in cm^{-1}; (b) = broad, (t) = terminal chlorine, (br) = bridging chlorine; shifts relative to the free ligand are in parentheses) and the magnetic moments μ_{eff} (in μ_B) at 297 K [4]:

compound	R	ν(NO)	δ(NO)	ν(Mn-O)	ν(Mn-Cl)	μ_{eff}
$Mn(C_6H_7NO_2)Cl_2$	4-OCH_3	1194 (−19)	438 (−25)	380(b)*)	380(b, t)*), 345(t), 298(br), 245(br)	5.40
$Mn(C_7H_9NO_2)Cl_2$	4-OC_2H_5	1198 (−11)	435 (−25)	390	386(t), 323**), 246(br)	5.54

*) Coupling of ν(Mn-O) and ν(Mn-Cl) (t). – **) May be assigned to ν(Mn-Cl) with terminal or bridging chlorine.

These chloro complexes do not fluoresce or triboluminesce. They are virtually anhydrous as they do not show IR absorption bands in the region of ν(OH) or δ(OH) vibrations [4], for $Mn(C_7H_9NO_2)Cl_2$ also see [3]. The IR spectra and magnetic data indicate a dinuclear structure containing 4-coordinate Mn ions, terminal and bridging chlorine atoms, and monodentate neutral ligands, the last coordinated to Mn through the oxygen of the NO unit. A symmetry around the manganese atom somewhere between distorted tetrahedral and square planar is suggested by the X-ray diffraction patterns (similar to those of the Ni complex) and the electronic spectra. The reduced magnetic moments of both complexes suggest a spin-spin interaction. The complexes are stable towards atmospheric moisture [4]. Both complexes are insoluble in the usual organic solvents, although sparingly soluble in dibromomethane and nitromethane. They are moderately soluble in methanol and its mixtures with halogenated hydrocarbons such as 1,2-dichloroethane [4].

$Mn(C_6H_7NO_2)Cl_2 \cdot H_2O$. The complex with 4-methoxypyridine N-oxide was obtained by mixing ethanolic solutions of $MnCl_2 \cdot 4H_2O$ and the ligand in equimolar amounts. The H_2O is lost at 118°C to give $Mn(C_6H_7NO_2)Cl_2$ [5].

References:

[1] R. Whyman, W. E. Hatfield, J. S. Paschal (Inorg. Chim. Acta **1** [1967] 113/9, 114). – [2] L. C. Nathan, R. O. Ragsdale (Inorg. Chim. Acta **10** [1974] 177/81, 178). – [3] N. M. Karayannis, J. V. Minkiewicz, L. L. Pytlewski, M. M. Labes (Inorg. Chim. Acta **3** [1969] 129/35, 130). – [4] N. M. Karayannis, C. M. Paleos, L. L. Pytlewski, M. M. Labes (Inorg. Chem. **8** [1969] 2559/62, 2560). – [5] A. J. Pappas, F. A. Osterman, H. B. Powell (Inorg. Chem. **9** [1970] 2695/8, 2696).

16.1.24.8 With 4-Benzyloxypyridine N-Oxide $C_6H_5CH_2O\text{-}C_5H_4NO$ (= $C_{12}H_{11}NO_2$)

$Mn(C_{12}H_{11}NO_2)_2Cl_3$ was prepared much like the compound with unsubstituted pyridine N-oxide, $Mn(C_5H_5NO)_2Cl_3$, but no ether was added. The reaction mixture was stirred for 30 min at −30°C and then for 2 d at room temperature. The green complex melts at 143°C with decomposition. The magnetic susceptibility between 65 and 295 K obeys the Curie-Weiss law with a paramagnetic Curie temperature of −0.36 K. The magnetic moment, μ_{eff} = 4.79 μ_B at 295 K, is that expected for a high-spin d^4 ion, without orbital contribution (4.9 μ_B). It is independent of temperature. In the IR spectrum, the ν(NO) band appears at 1194 cm^{-1} (1228 cm^{-1} for the free ligand). Four bands in the 450 to 300 cm^{-1} range were assigned to ν(Mn-O) and ν(Mn-Cl) vibrations. The UV and visible spectrum was recorded for solids (diffuse reflectance) and for solutions in chloroform. Absorptions at 28600 and 24300 cm^{-1} in the solution spectrum must be due to charge-transfer transitions because of their location and intensity. In the diffuse reflectance spectrum, only one band, at 22700 cm^{-1}, was observed in the 30000 to 20000 cm^{-1} region. Maxima near 16500 and 13000 cm^{-1} (shoulder), for both the solution and the solid, were assigned to d-d transitions. Measurements of the electric conductivity showed the compound to be a nonelectrolyte in nitromethane. On the basis of the physical data, a distorted square pyramidal structure was proposed for the complex, R. Uson, V. Riera, M. A. Ciriano, M. Valderrama (Transition Met. Chem. **1** [1976] 122/6).

16.1.24.9 With 2-Acetylpyridine N-Oxide $CH_3CO\text{-}C_5H_4NO$ (= $C_7H_7NO_2$)

$Mn(C_7H_7NO_2)_5(ClO_4)_2 \cdot H_2O$ was prepared from $Mn(ClO_4)_2 \cdot 6H_2O$ and the ligand (mole ratio 1:6) in the same way as the hydroxymethyl compound, $Mn(C_6H_7NO_2)_4(ClO_4)_2$ (see p. 161), but anhydrous ether was added under nitrogen to induce crystallization of the red-brown compound. In the electronic absorption spectrum, which was recorded from Nujol mulls, two weak maxima, at 15500 and 13000 cm^{-1}, were assigned to spin-forbidden d-d transitions. Peaks at 28900 and 25800 cm^{-1} correspond to charge-transfer transitions (manganese-to-ligand). Four maxima in the range between 45000 and 30000 cm^{-1} are presumably due to intra-ligand transitions. In the IR spectrum, the observed maxima are assigned as follows: 1690 ν(CO), 1217 ν(NO), 846 δ(NO), and 372 cm^{-1} ν($Mn\text{-}O_N$). A further strong band at 572 cm^{-1} is due to the CH_3CO group; other (partly split) bands in the range between 1150 and 450 cm^{-1} are due to the ClO_4^- group. According to the electronic spectrum the complex probably has an octahedral configuration with five ligand molecules coordinated to the manganese atom through the oxygen atoms of the NO groups, and one ligand molecule possibly coordinated as a bidentate ligand by the NO and the CO groups, although this has not been clearly shown by the IR studies ($\Delta\nu(NO) = -43$ cm^{-1}, $\Delta\nu(CO) = -5$ cm^{-1}). The ClO_4^- symmetry is apparently distorted from tetrahedral as its ν_3 band is strongly split and a sharp ν_4 band is present at 940 cm^{-1}. The distortion is probably caused by lattice requirements or by the uncoordinated H_2O molecules (formation of H bridges) rather than by a coordination to the manganese atom. The conductivity of the complex in acetonitrile, $\Lambda = 318$ $\Omega^{-1} \cdot cm^2 \cdot mol^{-1}$, corresponding to that of a 1:2 electrolyte, indicates noncoordinated ClO_4^- groups, V. S. Hagley, R. E. Kohrman, D. X. West (J. Inorg. Nucl. Chem. **39** [1977] 1959/63).

16.1.24.10 With 2-Pyridylamine 1-Oxides RC_5H_4NO

R = NH_2 (= $C_5H_6N_2O$)
R = $NHCH_3$ (= $C_6H_8N_2O$)
R = $N(CH_3)_2$ (= $C_7H_{10}N_2O$)

Complexes in Solution. From spectrophotometric investigations of the equilibrium $Mn^{2+} + C_5H_6N_2O \rightleftharpoons Mn(C_5H_6N_2O)^{2+}$ in aqueous dioxane (50 vol%) at 25°C, I = 0.5 ($NaClO_4$), the stability constant $K_1 = 0.75 \pm 0.06$ resulted as a mean value for various concentrations of Mn^{2+} [1].

$Mn(RC_5H_4NO)_6(ClO_4)_2$ (R = NH_2 or $NHCH_3$) and **$Mn(RC_5H_4NO)_5(ClO_4)_2 \cdot H_2O$** (R = $N(CH_3)_2$). The complex with R = NH_2 was obtained by mixing $Mn(ClO_4)_2 \cdot 6H_2O$ and the ligand, which were dissolved in 1:1 mixtures of methanol and acetone, in a 1:3 mole ratio

[1]. The complexes with R = $NHCH_3$ [2] and R = $N(CH_3)_2$ [3] were prepared by reacting $Mn(ClO_4)_2 \cdot 6H_2O$ and the ligands in a mixture of methanol and 2,2-dimethoxypropane, 2:1 v/v [2, 3].

The following table lists the color of the complexes, the ν(NO) frequencies (in cm^{-1}) of the complexes and of the free ligands L, and the molar electrical conductivities Λ (in $\Omega^{-1} \cdot cm^2 \cdot mol^{-1}$) measured on solutions in methanol:

R	empirical formula of ligand	complex	color	ν(NO)	ν(NO)(L)	Λ	Ref.
NH_2	$C_5H_6N_2O$	$MnL_6(ClO_4)_2$	yellow	1190	~1207	—	[1]
$NHCH_3$	$C_6H_8N_2O$	$MnL_6(ClO_4)_2$	tan	1202	1206	213	[2]
$N(CH_3)_2$	$C_7H_{10}N_2O$	$MnL_5(ClO_4)_2 \cdot H_2O$	buff	1210	1219	214	[3]

The complex with R = NH_2, $Mn(C_5H_6N_2)_6(ClO_4)_2$, in addition shows a band at 3400 cm^{-1}, which is assigned to NH vibrations not present in the solid ligand [1]. The comparatively slight lowering of the ν(NO) frequencies of the ligand with R = $NHCH_3$ observed on complexation is probably due to intramolecular hydrogen-bridge bonds in the free ligand. The positions of the δ(NCH_3) and ν(NCH_3) bands in the complex with this ligand do not suggest any coordination of the NH-CH_3 group to the manganese atom [2]. Coordination as unidentate ligand (via the oxygen atom of the NO group) is assumed also for the complex with R = NH_2 [1]. In the complex with R = $N(CH_3)_2$, $Mn(C_7H_{10}N_2O)_5(ClO_4)_2 \cdot H_2O$, the water molecule is not coordinated to the manganese atom, as shown by Karl-Fischer titrations and the absence of IR bands which could be assigned to coordinated H_2O. The unusual coordination of only five neutral ligands may be due to a steric hindrance which allows completion of the coordination sphere only by partially bidentate neutral ligands (as assumed for the Cu complex) or smaller ligands such as ClO_4^- ions. It cannot be decided from the IR spectra which of the two possibilities is realized in the complex [3]. UV absorption spectra were recorded for methanolic solutions of the complexes with R = $NHCH_3$ and R = $N(CH_3)_2$. Charge-transfer bands (Mn-to-ligand) were observed at 327, 249, and 229 nm for R = $NHCH_3$ and at 359, 265, 241 nm for R = $N(CH_3)_2$. The molar conductivities show that both complexes are 1:2 electrolytes in methanol [2, 3].

References:

[1] H. Sigel, H. Brintzinger (Helv. Chim. Acta **46** [1963] 701/12, 705/7). – [2] R. E. Kohrman, P. G. Phadtare, D. X. West (J. Inorg. Nucl. Chem. **37** [1975] 301/3). – [3] R. E. Kohrman, P. G. Phadtare, D. X. West (J. Inorg. Nucl. Chem. **37** [1975] 1296/8).

16.1.25 With Quinoline and Related Compounds

16.1.25.1 With Quinoline

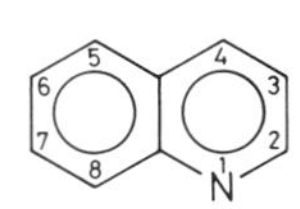

(= C_9H_7N)

Complexes with quinoline and diketones are described in "Mangan" D 1, 1979, pp. 70, 81, and 124. Mixed ligand complexes with quinoline and aminocarboxylic acids or N-heterocyclic carboxylic acids are described in "Manganese" D 4.

16.1.25.1.1 Manganese(II) Compounds

$Mn(C_9H_7N)_4(NCS)_2$ crystallizes from an ethanolic solution containing $Mn(NCS)_2$ and quinoline on prolonged standing as pale yellow plates which are readily soluble in water, alcohol, and ether [1].

$Mn(C_9H_7N)_2X_2$ ($X = Cl$, Br, ClO_4, BF_4). The chloro complex was obtained by refluxing $MnCl_2 \cdot 4H_2O$ with quinoline for several hours [2, 3] or by reacting anhydrous manganese(II) chloride and quinoline in absolute ethanol [4, 12]. The bromo complex was obtained by recrystallizing $MnBr_2$ from hot quinoline [5]. The perchlorato and tetrafluoroborato complexes were prepared by evaporating solutions of hydrated $Mn(ClO_4)_2$ or $Mn(BF_4)_2$ in absolute ethanol. The residue was redissolved in absolute ethanol and precipitated with anhydrous diethyl ether [6]. The chloro complex forms colorless crystals [2, 3, 12], which appear as elongated biaxial plates [3]. The perchlorato and tetrafluoroborato complexes are also colorless [6].

Magnetic measurements at room temperature show the magnetic moment $\mu_{eff} = 5.86\ \mu_B$ for $Mn(C_9H_7N)_2Cl_2$ [4], $\mu_{eff} = 5.91$ and $5.90\ \mu_B$ for the perchlorato and tetrafluoroborato complexes, respectively [6]. The electronic spectrum of solid $Mn(C_9H_7N)_2Cl_2$, which shows three bands (at 23800, 21280, and 18350 cm^{-1}) [2] and the spectra of $Mn(C_9H_7N)_2(ClO_4)_2$ and $Mn(C_9H_7N)_2(BF_4)_2$ (four bands in the same region) suggest octahedral structures. The anions ClO_4^- and BF_4^- are probably coordinated as bidentate ligands of symmetry C_{2v} as implied by the IR and magnetic data [6]. In the far-IR spectrum of the chloro complex, bands assigned to ν(Mn-N) and ν(Mn-Cl) vibrations occur at 196 and 246 cm^{-1}, respectively [7]. According to [8] a band at 241 cm^{-1} is ascribed to ν(Mn-N) vibrations. In the spectra of the perchlorato and tetrafluoroborato complexes a band at 200 cm^{-1} is assigned to ν(Mn-N) modes [6]. The ESR spectrum of $Mn(C_9H_7N)_2Cl_2$ shows a single resonance line at $g_{eff} = 2$, suggesting a polymeric structure, while $Mn(C_9H_7N)_2Br_2$ has its strongest line at $g = 4.3$, indicating a tetrahedral arrangement in the complex [9], which probably has C_{2v} symmetry because the observed X-band spectrum is consistent with the distortion parameters $D = 0.485\ cm^{-1}$ and $\lambda = 0.300$ [5]. The X-ray spectrum of $Mn(C_9H_7N)_2Cl_2$ was recorded and discussed by [10].

Investigation of $Mn(C_9H_7N)_2Cl_2$ by thermogravimetry and differential scanning calorimetry [11] (heating rate 8 or 16 K/min [11]) reveals successive thermal degradation to $Mn(C_9H_7N)Cl_2$ (first step), $Mn_3(C_9H_7N)_2Cl_6$ (second step), and $MnCl_2$ (third step) [2, 11]. The corresponding peak temperatures T_p (±30°C) and enthalpies ΔH (in kcal/mol) are reported as follows:

	first step	second step	third step
T_p in °C	154 [11], 190 [2]	243 [11], 220 [2]	331 [11], 340 [2]
ΔH	14.16±0.02 [11]	5.14±0.10 [11]	14.21±0.07 [11]

$Mn(C_9H_7N)_2Cl_2$ is insoluble in cold water and decomposed by hot water. It is soluble in cold alcohol [12].

$Mn(C_9H_7N)_2Cl_2 \cdot 3H_2O$ was obtained as a white amorphous precipitate upon mixing equivalent amounts of $MnCl_2$ and quinoline in concentrated alcoholic solutions. It is hygroscopic, readily soluble in water, and moderately soluble in alcohol. It easily loses quinoline, even on gentle heating. Upon drying at 100°C the three molecules of H_2O and one molecule of quinoline are lost. The compound adds two molecules of HCl, yielding yellowish-green crystals of $(C_9H_7NH)_2MnCl_4$ [13], see "Mangan" C 5, 1978, p. 208.

$Mn(C_9H_7N)Cl_2$ and **$Mn_3(C_9H_7N)_2Cl_6$** were obtained by thermal degradation of $Mn(C_9H_7N)_2Cl_2$, see above [2, 8, 11]. $Mn(C_9H_7N)Cl_2$ may be prepared also by refluxing stoichiometric amounts of manganese(II) chloride and quinoline for 2 h and cooling [14] or by mixing ethanolic solutions of anhydrous manganese(II) chloride and quinoline, followed by slow evaporation until crystallization occurs [15]. The compounds are white [2, 14]. They are clearly distinguible by their X-ray powder patterns [2]. The electronic reflectance spectra of both compounds, which show three bands in the 18000 to 24000 cm^{-1} region, are very similar to those of $Mn(C_9H_7N)_2Cl_2$ (see p. 166) and $Mnpy_2Cl_2$ (see p. 95), suggesting Mn atoms in octahedral coordination spheres. The bands corresponding to the transitions $^6A_{1g}(S) \rightarrow {}^4T_{1g}(G)$, $^6A_{1g}(S) \rightarrow {}^4T_{2g}(G)$ show a decrease in ligand field strength in the order $n = 2 > 1 > 0 > ^2/_3$, where n = number of ligand molecules. It is assumed that $Mn(C_9H_7N)Cl_2$ has a simple polymeric chain structure, while the structure of $Mn_3(C_9H_7N)_2Cl_6$ is more complex and probably implies cross-linked chains with Cl atoms acting as bridges between three Mn atoms [2]. The far-IR spectrum of the complexes shows two absorption bands, at 240 and 230 cm^{-1}, for $Mn(C_9H_7N)Cl_2$ and one band, at 262 cm^{-1}, for $Mn_3(C_9H_7N)_2Cl_6$; all are assigned to ν(Mn-N) vibrations [8]. For $Mn(C_9H_7N)Cl_2$ the magnetic moment is $\mu_{eff} = 5.93$ μ_B [14] or 5.90 μ_B [4]. The molar electrical conductivity, $\Lambda = 43.79$ $\Omega^{-1} \cdot cm^2 \cdot mol^{-1}$ in acetone, indicates a nonelectrolyte [14]. $Mn(C_9H_7N)Cl_2$ is readily hydrolyzed on boiling with water. It is decomposed by mineral acids and soda lime with the separation of quinoline [15]. It is very sparingly soluble in acetone or other organic solvents, probably due to its polymeric nature [14]. The thermal behavior is reported on p. 166.

$[Mn(C(CH_3)_3COO)_2(C_9H_7N)]_2$. The complex was prepared in the same way as the α-picoline complex (p. 121), and it has an IR spectrum and other properties similar to that complex. Therefore a dimeric structure of the type $LMn(RCOO)_4MnL$ is assumed also for the quinoline complex. Susceptibility measurements at 298, 188, and 80 K yield the effective magnetic moments $\mu_{eff} = 4.54$, 3.40, and 2.21 μ_B, respectively. The temperature dependence of μ_{eff} is best described by the Heisenberg-Dirac-van Vleck model for a system of two exchange-coupled cations with spins $S = ^5/_2$ with $g \approx 2.00$ and $-2J \approx 30 \pm 2$ cm^{-1} [17].

$Mn(C_9H_7N)C_2O_4 \cdot H_2O$. The compound was prepared by stirring and refluxing a suspension of $MnC_2O_4 \cdot 3H_2O$ and excess quinoline in methanol for 4 to 6 d. The resulting product was washed with alcohol and ether and dried in vacuo. Susceptibility measurements at 303 K yielded the magnetic moment $\mu_{eff} = 5.93$ μ_B. The electronic reflectance spectrum (recorded for Nujol mulls) shows absorption maxima at 26800, 21850, and 16190 cm^{-1} due to d-d transitions. The IR spectrum (Nujol mulls) shows absorption bands assigned to ν(CC) and ν(CN) vibrations at 1645(b), 1600(b), and 1590(sh) cm^{-1}; a band assigned to ν(Mn-ligand) vibrations occurs at 270 cm^{-1}. The slightly raised frequency of the ν(CC) and ν(CN) ligand bands in the complex suggests coordination of the ligand to Mn through the N atom of the quinoline. The oxalate groups, as bridging ligands, and water are coordinated to Mn in the same way as in $MnpyC_2O_4 \cdot H_2O$. It is concluded from the magnetic and spectral data and the near insolubility in organic solvents that the complex is octahedral and polymeric with bridging oxalate groups [16].

$[Mn(RCOO)_2(C_9H_7N)]_2$ ($R = C_6H_5$, C_4H_3O) and **$[Mn(C_4H_3OCOO)_2(C_9H_7N)]_n$.** The dibenzoato and the di(furan-2-carboxylato) complexes were prepared by adding quinoline to $Mn(C_6H_5COO)_2$ or $Mn(C_4H_3OCOO)_2$ in boiling tetrahydrofuran and concentrating the mixture in vacuo until crystallization occurred. The crystals were collected after cooling overnight. The polymeric compound, $[Mn(C_4H_3OCOO)_2(C_9H_7N)]_n$, was obtained by dissolving $Mn(C_4H_3OCOO)_2$ in quinoline with heating. Upon addition of hexane to the orange-brown solution, a dark, oily layer separated. It was treated with benzene to yield a beige solid [18]. In the following table are listed the color [18] and magnetic moments μ_{eff}

at temperatures between 77 and 300 K (T in parentheses) together with the isotropic g-factors and exchange parameters J [19]:

compound	color	μ_{eff} in μ_B (at T in K)	g value	J in cm^{-1}
$[Mn(C_6H_5COO)_2(C_9H_7N)]_2$	pale green	4.80(287), 4.24(184), 3.06(77)	2.001	13
$[Mn(C_4H_3OCOO)_2(C_9H_7N)]_2$	yellow green	4.90(289), 4.21(187), 2.83(77)	2.003	13
$[Mn(C_4H_3OCOO)_2(C_9H_7N)]_n$ *)	beige	5.34(294), 5.17(103), 4.77(77)	—	—

*) μ_{eff} from [18]

The IR spectra of the compounds (ν_s(OCO) at ~1420 cm^{-1} and ν_{as}(OCO) at ~1620 cm^{-1}) indicate the presence of bridging carboxylate groups. A dimeric dinuclear structure of the type $LMn(RCOO)_4MnL$ (compare the α-picoline complex, p. 121) is concluded from a comparison of the experimental dependence of μ_{eff} on temperature with that calculated for two exchanging paramagnetic ions of spin $S = {}^5/_2$ [18]. The antiferromagnetic character of these compounds is deduced from the monotonic decrease in μ_{eff} on lowering the temperature and a broad maximum in the molar susceptibility versus temperature plot [19]. As the decrease in μ_{eff} of $[Mn(C_4H_3OCOO)_2(C_9H_7N)]_n$ at lower temperatures is about the same as that of $Mn(C_4H_3OCOO)_2$ (see "Mangan" D 2, 1980, p. 85) and much less than that of the dimeric compound, a polymeric structure containing quinoline solvate molecules may be assumed in analogy with known polymeric Co adducts [18].

$Mn_5(C_9H_7N)_2(OH)(C_{22}H_{11}O_9)_3 \cdot 18H_2O$ ($C_{22}H_{14}O_9$ = aurintricarboxylic acid). The compound was obtained by boiling an alcoholic solution containing ammonium aurintricarboxylate (1 g), quinoline (2.4 g), and some water after addition of $MnCl_2 \cdot 4H_2O$ (2.5 g) in alcohol. The red-brown precipitate was washed with alcohol and air dried [20].

References:

[1] H. Grossmann, F. Hünseler (Z. Anorg. Allgem. Chem. **46** [1905] 361/405, 383). – [2] D. H. Brown, R. H. Nuttall, D. W. A. Sharp (J. Inorg. Nucl. Chem. **26** [1964] 1151/6, 1152). – [3] J. M. Mutchler, H. B. Bradley (Anal. Chem. **30** [1958] 1371/4, 1372). – [4] R. L. Mital (Indian J. Chem. **3** [1965] 88). – [5] R. D. Dowsing, J. F. Gibson, D. M. L. Goodgame, M. Goodgame, P. J. Hayward (J. Chem. Soc. A **1969** 1242/8, 1245).

[6] D. H. Brown, R. H. Nuttall, J. McAvoy, D. W. A. Sharp (J. Chem. Soc. A **1966** 892/6, 894). – [7] C. W. Frank, L. B. Rogers (Inorg. Chem. **5** [1966] 615/22, 616/9). – [8] J. R. Allan, D. H. Brown, R. H. Nuttall, D. W. A. Sharp (J. Inorg. Nucl. Chem. **27** [1965] 1305/9, 1307). – [9] R. D. Dowsing, J. F. Gibson, D. M. L. Goodgame, M. Goodgame, P. J. Hayward (Nature **219** [1968] 1037/8). – [10] D. Stelling (Z. Physik Chem. B **24** [1934] 282/92, 284, 289).

[11] G. Beech, C. T. Mortimer, E. G. Tyler (J. Chem. Soc. A **1969** 512/5). – [12] F. Reitzenstein (Z. Anorg. Allgem. Chem. **18** [1898] 253/304, 295). – [13] J. V. Dubský, V. Dostál (Spisy Vydávané Přirodvěd. Fak. Massarykovy Univ. No. 196 [1934] 17/23, 18; C. **1935** II 1123; C.A. **1935** 2109). – [14] K. C. Dash, D. V. Ramana Rao (Indian J. Chem. **5** [1967] 333/4). – [15] S. Prasad, K. P. Kacker (J. Indian Chem. Soc. **35** [1958] 719/21).

[16] B. D. Pandey, D. Rupainwar (Current Sci. [India] **49** [1980] 336/9). – [17] A. A. Pasynskii, T. Ch. Idrisov, K. M. Suvorova, V. T. Kalinnikov (Koord. Khim. **2** [1976] 1060/8; Soviet J. Coord. Chem. **2** [1976] 813/9, 814, 818). – [18] A. A. Pasynskii, T. Ch. Idrisov, K. M. Suvorova, V. M. Novotortsev, V. T. Kalinnikov (Koord. Khim. **1** [1975] 799/803; Soviet

J. Coord. Chem. **1** [1975] 673/7, 674). – [19] V. M. Novotortsev, Yu. V. Rakitin, A. A. Pasynskii, V. T. Kalinnikov (Dokl. Akad. Nauk SSSR **240** [1978] 355/7; Dokl. Chem. Proc. Acad. Sci. USSR **238/243** [1978] 220/2). – [20] R. Cernătescu, M. Poni, M. Bostan (Commun. Acad. Rep. Populare Romine **5** [1955] 159/65, 164; C.A. **1956** 11994).

16.1.25.1.2 Manganese(III) Compounds

$Mn(C_9H_7N)_3Cl_3$. The complex was formed on treating $Mn(C_9H_7N)Cl_3$ with excess quinoline. It is a brown hygroscopic compound, which is stable under dry N_2 up to 100°C.

$Mn(C_9H_7N)Cl_3$ was obtained as green crystals on storing a solution containing $MnCl_3$ and quinoline in ether at −10°C for several hours. It decomposes at temperatures above −10°C with loss of color. With excess quinoline $Mn(C_9H_7N)_3Cl_3$ is formed.

Reference:

H. Funk, H. Kreis (Z. Anorg. Allgem. Chem. **349** [1967] 45/9, 48).

16.1.25.2 With 2-Methylquinoline CH_3-C_9H_6N (=Quinaldine=$C_{10}H_9N$)

$Mn(C_{10}H_9N)_nX_2$ (X=Cl, Br, I; n=1,2,3, see table below). The compounds $Mn(C_{10}H_9N)_2Cl_2$, $Mn(C_{10}H_9N)_2Br_2$, and $Mn(C_{10}H_9N)_3I_2$ were precipitated by adding 2-methylquinoline in a 2:1 mole ratio to ethanolic solutions of MnX_2 (X=Cl, Br, I) at room temperature. If the reactant mixtures were refluxed for 3 h and cooled, compounds of composition $Mn(C_{10}H_9N)_2X_2$ resulted [1]. According to [2] $Mn(C_{10}H_9N)_2Cl_2$ is said to form after mixing ethanolic solutions of equivalent quantities of the reactants, but no analysis is given. The compounds $Mn(C_{10}H_9N)X_2$ (X=Cl, Br) were obtained by heating $Mn(C_{10}H_9N)_2X_2$ at 130°C. They are dusty, white, amorphous, and do not melt up to 280°C. $Mn(C_{10}H_9N)_3I_2$ is stable at 130°C [1]. Below are listed the magnetic moments of the complexes $Mn(C_{10}H_9N)_nX_2$ at 25°C and the molar electrical conductivity Λ of the solutions in acetone:

compound (L=$C_{10}H_9N$)	MnL_2Cl_2	$MnLCl_2$	MnL_2Br_2	$MnLBr_2$	MnL_3I_2	MnL_2I_2
μ_{eff} in μ_B	6.1	6.0	6.1	5.8	5.8	5.9
Λ in $\Omega^{-1}\cdot cm^2\cdot mol^{-1}$	26.7	25.0	28.6	27.5	30.4	33.0

The IR spectra of the complexes in the range between 1630 and 720 cm^{-1} are recorded, but not assigned. As shown by the magnetic and conductivity data all compounds are high-spin Mn(II)-d^5 complexes and nonelectrolytes in acetone. As they are high melting and only slightly soluble in organic solvents, a polymeric tetrahedral structure is assumed for the complexes of composition $MnLX_2$, and a polymeric octahedral structure is assumed for those of composition MnL_2X_2 and MnL_3X_2 [1]. $Mn(C_{10}H_9N)_2Cl_2$ dissolves in aqueous HCl to form the compound $(C_{10}H_9N\cdot H)_2MnCl_4$ [2], see "Mangan" C 5, 1978, p. 208.

$[Mn(C(CH_3)_3COO)_2(C_{10}H_9N)]_2$. The compound is prepared as described for the α-picoline complex (p. 121) and has an IR spectrum and other properties similar to that complex. A dimeric, so-called lantern structure is therefore assumed for the 2-methylquinoline complex [3].

References:

[1] S. Guru, D. V. Ramana Rao (Current Sci. [India] **37** [1968] 225/6). – [2] I. Burić, K. Nicolić (Fizika **6** [1974] 51/9, 52; C.A. **81** [1974] No. 129250). – [3] A. A. Pasynskii, T. Ch. Idrisov, K. M. Suvorova, V. T. Kallinnikov (Koord. Khim. **2** [1976] 1060/8; Soviet J. Coord. Chem. **2** [1976] 813/9, 815).

16.1.25.3 With 2,6-Dimethylquinoline $(CH_3)_2C_9H_5N$ $(=C_{11}H_{11}N)$

$\mathbf{Mn(C_{11}H_{11}N)_2Cl_2}$ was prepared by mixing ethanolic solutions containing equivalent quantities of $MnCl_2$ and the ligand (no analysis given), I. Burić, K. Nicolić (Fizika **6** [1974] 51/9, 52; C.A. **81** [1974] No. 129250).

16.1.25.4 With 8-Hydroxyquinoline 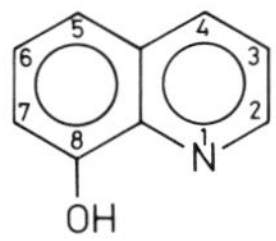 (=8-Quinolinol = Oxine = C_9H_7NO = HL)

16.1.25.4.1 Complexes in Solution

Stability constants for the manganese(II) chelate complexes MnL^+ and MnL_2 were determined under various experimental conditions. Values for log K_1 and log K_2 are listed below:

method	solvent	ionic strength	t in °C	log K_1	log K_2	Ref.
glass electrode	H_2O	I = 0.01	20	6.8	5.8	[5, 22]
spectrophotometry	H_2O	I = 0.1 (KNO_3)	16	6.24	—	[6]
kinetic data	H_2O	I = 0.1 (KNO_3)	25	5.84	—	[7]
glass electrode	50% dioxane/H_2O	I = 0.1 ($NaClO_4$)	25	7.30	6.19	[1]
glass electrode	50% dioxane/H_2O	I = 0.1 ($NaClO_4$)	25	7.62	6.70	[3]
glass electrode	75% dioxane/H_2O	—	30	10.8	9.6	[4]
kinetic data	dimethyl sulphoxide	I = 0.2 ($NaClO_4$)	19.0	2.63	—	[8]

Earlier results (in 50 vol% aqueous dioxane) are given by [2]. Thermodynamic data for solutions in 50 vol% aqueous dioxane: $\Delta H_1 = -3.5$ kcal/mol, $\Delta S_1 = 22$ cal·mol^{-1}·K^{-1}, $\Delta H_{\beta 2} = -10.5$ kcal/mol, and $\Delta S_{\beta 2} = 27$ cal·mol^{-1}·K^{-1}, ΔH values from calorimetric measurements [9]. Potentiometric measurements on aqueous dioxane solutions show that the average formation constant K_{av} increases with the dioxane content [10, 11]. The stability constants of the manganese complex are compared with those of other hydroxyquinolinates in [2, 12]. For the reaction $Mn^{2+}_{aq} + HL_{aq} \xrightleftharpoons{k_f} [MnHL]^{2+}_{aq}$ at 25°C (I = 0.1), the values of the rate constant log $k_f = 5.69$ (k in mol^{-1}·s^{-1}), the activation enthalpy $\Delta H^{\ddagger}_f = 9.1$ kcal/mol, and the entropy $\Delta S^{\ddagger}_f = -2$ cal·mol^{-1}·K^{-1} are reported for 25°C [7], also see [6]. The reaction rate constants k of the reaction $Mn^{2+}_{aq} + L^-_{aq} \xrightleftharpoons[k_d]{k_f} MnL^+_{aq}$ obtained by the temperature-jump-relaxation [7] or stopped-flow method [8, 13] are listed in the following table:

method	medium	t in °C	log k_f	log k_d	Ref.
relaxation	H_2O, I = 0.1 (KNO_3)	25	8.52	2.68	[7]
relaxation	H_2O, I = 0.1 (KNO_3)	16	8.04	2.15	[6]
stopped-flow	methanol, I = 0.2 ($NaClO_4$)	−46.7	2.92	—	[13]
stopped-flow	DMSO, I = 0.2 ($NaClO_4$)	19.0	2.30	—	[8]

The activation enthalpies $\Delta H^{\ddagger}_f = 8.8$ and $\Delta H^{\ddagger}_d = 7.0$ kcal/mol and entropies $\Delta S^{\ddagger}_f = +10$ and $\Delta S^{\ddagger}_d = -22$ cal·mol^{-1}·K^{-1} were calculated [7]. It is assumed that on complex formation in aqueous medium two H_2O molecules of $[Mn(H_2O)_6]^{2+}$ are replaced by the bidentate 8-hydroxyquinoline to yield $[Mn(H_2O)_4(HL)]^{2+}$ or $[Mn(H_2O)_4L]^+$ [7].

The extraction of manganese complexed with 8-hydroxyquinoline from aqueous solution has been investigated for analytical purposes. Distribution measurements with $CHCl_3$ solvent, 10^{-4} M Mn^{2+}, and 0.1 M 8-hydroxyquinoline at I = 0.1 ($KClO_4$) and 20°C for a shaking time of several minutes show complete extraction from pH 6.5 to 11. At higher pH values NH_2OH had to be added to prevent oxidation. From the data, the extraction constant $\log K_0 (= [MnL_2]_{org}[H^+]^2/[Mn^{2+}][HL]^2_{org}) = -9.32 \pm 0.03$ was deduced, with $MnL_2 = Mn(C_9H_6NO)_2$ assumed to be the extracted species [14]. Higher pH values were found for 10^{-3} mol/L hydroxyquinoline in the presence of a reducing agent (0.1N sulfite) [15] and in the presence of additional complexing agents [14]. Values of half-extraction under various conditions [14, 16] and also values for the back-extraction of Mn^{2+} into the aqueous phase [16] are given. The extracted species was found to contain an Mn-to-ligand ratio of 1:3 and thus is formulated $H[Mn(C_9H_6NO)_3]$ [15]. In the exchange extraction of Mn^{II} with 8-hydroxyquinoline, the distribution of manganese between $CHCl_3$ and the aqueous phase (acetate buffer, pH 6.10 to 7.78, I = 0.1 ($NaClO_4$)) gave a distribution constant for MnL_2 of $\log K_D = 1.94$, which was used to calculate the two-phase stability constant of $\log \beta_2(K_D) = 15.42$ [23]. In the presence of butylamine and in the absence of oxidizing agents, extraction of the complex $C_4H_9NH_3[Mn(C_9H_6NO)_3]$ is assumed [15]. Complex solutions which were obtained by extracting aqueous solutions containing 10^{-3} mol of Mn^{2+} ions per liter and the same or higher concentrations of hydroxyquinoline in $CHCl_3$ or $CHBr_3$ at pH 11 were studied spectrophotometrically. The intensity of the absorption spectra ($\lambda = \sim 400$ nm, $\varepsilon = \sim 10500$ $L \cdot mol^{-1} \cdot cm^2$) is roughly twice that of the complex $Mn(C_9H_6N)_2$ in pyridine ($\lambda = \sim 400$ nm, $\varepsilon = 5300$ $L \cdot mol^{-1} \cdot cm^2$), thus demonstrating that the solvate hydroxyquinoline molecules in $MnL_2 \cdot 2HL$ are bonded to manganese in the same manner as in MnL_2. The complex is assumed to contain 8-coordinate manganese. The extra 8-hydroxyquinoline molecules are assumed to be in the quinone form [18]. Synergistic extractions of Mn^{2+} from aqueous solution (pH 6, potassium phthalate/KOH buffer) into $CHCl_3$ by 8-hydroxyquinoline (= HL) and N-heterocyclic bases B (e.g., pyridine, 2,4-dimethylpyridine (= 2,4-C_7H_9N), 2,6-dimethylpyridine (= 2,6-C_7H_9N), and 2,4,6-trimethylpyridine (= 2,4,6-$C_8H_{11}N$)) were studied by using a ^{54}Mn radiotracer. Equilibrium constants were calculated and the formation constants of the extractable 1:1 adducts $K = [MnL_2 \cdot B]/[MnL_2] \cdot [B]$ were determined at 30°C and I = 0.25 ($NaClO_4$) [25]:

base	pyridine	2,4-C_7H_9N	2,6-C_7H_9N	2,4,6-$C_8H_{11}N$
log K	1.78	1.75	1.41	1.50

A manganese complex is also extractable with 1,2-dichloroethane. In the presence of tetradecyldimethylbenzylammonium chloride (zephiramine) a 1:3 complex containing additional organic base is extracted [17]. Quantitative extraction of manganese(II) by 0.01M solutions of HL in isopentyl alcohol occurs at about pH 8.3 [19]. Quantitative extraction of manganese(III) (10^{-3}M) in the presence of 2×10^{-3}M $K_3[Fe(CN)_6]$ and 0.1 equiv./L tartrate with 0.01M 8-hydroxyquinoline in $CHCl_3$ as $Mn(C_9H_6NO)_3$ occurs about pH 8 to 10 [15].

Manganese(III) complexes are formed by oxidation of $Mn(C_9H_6NO)_2$ in 1×10^{-5}M aqueous solutions with O_2, H_2O_2, $KMnO_4$, or $K_3Fe(CN)_6$ (in the presence of additional ligand). Spectrophotometric and pH-potentiometric measurements at 22°C reveal the formation of the complexes $[Mn(C_9H_6NO)_3]$, $[Mn(C_9H_6NO)_2OH]$, and possibly $[Mn(C_9H_6NO)_2(OH)_2]^-$, which are stable only in the narrow pH range between 9 and 10. The oxidation by O_2 takes place at pH > 9.5 and is complete at pH 10 within 10 min. The oxidation by H_2O_2 (1×10^{-5}M) at pH 9 requires about 3 min. The oxidation by $K_3Fe(CN)_6$ proceeds rapidly without significantly affecting the ligand. The oxidant $KMnO_4$, however, must be added in small portions to prevent such oxidation. The manganese(III) complexes decompose at pH > 10 to give manganese(IV) oxide and 8-hydroxyquinoline. At pH < 9 the oxidation of manganese(II) by H_2O_2 or $K_3Fe(CN)_6$ is incomplete, and $KMnO_4$ affects the ligand. If solutions of manganese(III) hydroxyquinolinates are acidified to pH = 8, the ligand again decomposes. The complexes are more stable to dissociation than $Mn(C_9H_6NO)_2$, but on extreme dilution ($\leqslant 10^{-5}$ M) they also will dissociate [24].

The complex species existing in DMSO solutions of $Mn^{II}(C_9H_6NO)_2 \cdot 2H_2O$ and $Mn_2^{III}O(C_9H_6NO)_4(C_9H_7NO)_2 \cdot 2CH_3OH$ (see pp. 175 and 178, respectively) and the species formed on redox reactions of those compounds are discussed in [20, 21].

References:

[1] R. L. Stevenson, H. Freiser (Anal. Chem. **39** [1967] 1354/8, 1355). – [2] W. D. Johnston, H. Freiser (J. Am. Chem. Soc. **74** [1952] 5239/42, 5240), H. Freiser (Analyst **77** [1952] 830/45, 835). – [3] H. F. Steger, A. Corsini (J. Inorg. Nucl. Chem. **35** [1973] 1621/36, 1625/7). – [4] L. G. Van Uitert, W. C. Fernelius (J. Am. Chem. Soc. **76** [1954] 375/9, 376). – [5] A. Albert (Biochem. J. **54** [1953] 646/54, 650).

[6] D. N. Hague, M. S. Zetter (Trans. Faraday Soc. **66** [1970] 1176/84, 1181). – [7] D. N. Hague, S. R. Martin, M. S. Zetter (J. Chem. Soc. Faraday Trans. I **68** [1972] 37/46, 43). – [8] D. M. W. Buck, P. Moore (J. Chem. Soc. Dalton Trans. **1976** 638/42, 640). – [9] G. Gutnikov, H. Freiser (Anal. Chem. **40** [1968] 39/44, 42). – [10] H. F. Steger, A. Corsini (J. Inorg. Nucl. Chem. **35** [1973] 1637/43, 1641).

[11] H. F. Steger, A. Corsini (J. Inorg. Nucl. Chem. **35** [1973] 3277/85, 3279, 3284). – [12] R. D. Gupta, G. S. Manku, A. N. Bhat, B. D. Jain (Australian J. Chem. **23** [1970] 1387/95, 1392). – [13] D. J. Benton, P. Moore (J. Chem. Soc. Chem. Commun. **1972** 717/8). – [14] J. Starý (Anal. Chim. Acta **28** [1963] 132/49, 138, 143). – [15] F. Umland (Z. Anal. Chem. **190** [1962] 186/208, 198).

[16] T. Honjo, K. Ameno, M. Horiuchi, T. Kiba (Bull. Chem. Soc. Japan **47** [1974] 1799/800). – [17] S. Noriki, M. Nishimura (Anal. Chim. Acta **94** [1977] 57/62, 59). – [18] Yu. A. Bankovskii, A. F. Ievin'sh, M. R. Buka, E. A. Luksha (Zh. Neorgan. Khim. **8** [1963] 110/8; Russ. J. Inorg. Chem. **8** [1963] 56/60). – [19] Yu. A. Zolotov, N. M. Kuz'min, V. G. Lambrev (Tr. Komis. Analit. Khim. Akad. Nauk SSSR Inst. Geokhim. Analit. Khim. **15** [1965] 51/8, 55; C.A. **63** [1965] 3672). – [20] J. K. Howie, D. T. Sawyer (J. Am. Chem. Soc. **98** [1976] 6698/700).

[21] J. K. Howie, M. M. Morrison, D. T. Sawyer (ACS Symp. Ser. No. 38 [1977] 97/112, 102; C.A. **86** [1977] No. 152029). – [22] A. Albert, M. I. Gibson, S. D. Rubbo (Brit. J. Exptl. Pathol. **34** [1953] 119/30, 126). – [23] D. Vančo, F. Macášek (Anal. Chim. Acta **131** [1981] 323/6). – [24] I. E. Kalinichenko, E. Ya. Matveeva, A. T. Pilipenko (Zh. Neorgan. Khim. **26** [1981] 115/9; Russ. J. Inorg. Chem. **26** [1981] 60/3). – [25] A. T. Rane (J. Inorg. Nucl. Chem. **42** [1980] 1520/1).

16.1.25.4.2 Isolated Manganese(II) Complexes

Solid complexes with 8-hydroxyquinoline and diketones are described in "Mangan" D 1, 1979, pp. 106, 117, and 118.

$\mathbf{Mn(C_9H_6NO)_2}$ was obtained by heating the dihydrate (see p. 174) at 155°C to constant weight [1, 2] or in vacuo at 150°C [6]. The compound was purified by recrystallization from a mixture of ethanol and acetone and dried at 105°C for 2 h. Susceptibility measurements yield the magnetic moment $\mu = 5.85\ \mu_B$. The ESR spectrum gives a signal of line width $\Delta H = 400$ mT and $g_{eff} \approx 2$ at 295 K [3].

The X-ray photoelectron spectrum of the solid indicates an N(1s) binding energy of 399.4 eV. From the shift relative to the free ligand (+0.3 eV) a 43% ionic character is calculated for the Mn-N bond, also see [36].

Strong IR bands at 1326 [12] and 1107 cm^{-1} [12, 13] are assigned to the antisymmetric and symmetric CO stretching modes, respectively; no other ligand bands are given. The first band does not occur in the ligand spectrum [12]. Bands at 724, 645, 596, 575, 505, and 492(s) for KBr discs are shifted by +10 to 30 cm^{-1} relative to the ligand. The first three are assigned to in-plane ring-deformation modes, the next two to ring-distortion modes, and the peak at 492 cm^{-1} to the in-plane CO bending vibration. Four bands in the range 740 to 820 cm^{-1} are shifted only slightly (5 to 10 cm^{-1}) [11]. The main peaks between 730 and 1110 cm^{-1} are also reported by [13]. The absorbancies of the ν(CO) band at 1111 cm^{-1} and of the ν(Mn-O) band at 495 cm^{-1} are decreased after γ-irradiation. The linear rate of decrease suggests a first-order degradation mechanism, which is also obtained for other 8-hydroxyquinolinates [37]. As the IR spectrum resembles closely that of the copper complex, which is square planar according to X-ray studies, it is assumed that $Mn(C_9H_6NO)_2$ is also square planar with the two chelate-bonded ligand molecules arranged equatorially [11].

Solutions of $Mn(C_9H_6NO)_2$ fluoresce at the temperature of liquid N_2. A 0.25×10^{-3}M solution of $Mn(C_9H_6NO)_2$ in a mixture of ethanol (95%), ether, and isopentane shows a weak emission maximum at 415 nm (365 or 315 nm exciting wavelength) [7]. Solutions in ethanol (95%) show an emission maximum at 20200 cm^{-1} (495 nm) at 80 K with a quantum yield of $\Phi = 0.015$ (relative to the $\Phi = 0.55$ of quinine bisulphate; 366 nm exciting wavelength) [8]. Absorption maxima of solutions of $Mn(C_9H_6NO)_2$ or the dihydrate in ethanol are observed at 377 nm [7] or at 382, 320, 261, and 242 nm with log ε = 3.20, 3.52, 4.40, and 4.42 $L \cdot mol^{-1} \cdot cm^{-1}$, respectively, [29] and in chloroform at 382 nm [7], 396 nm [9], or 395 nm with log ε = 3.93 [34]. The longest wavelength band is assigned to a $\pi \rightarrow \pi^*$ transition in the ligand [7].

The mass spectrum of solid $Mn(C_9H_6NO)_2$, recorded at 360°C, shows three peaks which indicate the presence of the ions $Mn_2(C_9H_6NO)_3^+$, $Mn(C_9H_6NO)_3^+$, and $Mn(C_9H_6NO)_2^+$ and suggest some polymerization in the solid state [15]. Further investigations at 250°C suggest the presence of one molecular ion and two fragment ions, which are, however, not clearly designated [16].

Differential thermal analyses, heating rate 10 K/min [17] or 12 K/min [18], in an argon atmosphere show a small exothermic effect at 250°C, which is not reversible and probably indicates crystallization of the initially amorphous dihydrate complex [18]. The large endothermic effect at 490°C [17] or 502°C [18] is assigned to decomposition [17] or, more exactly, melting followed immediately by thermal decomposition [18]. On heating in vacuo almost complete sublimation of the unaltered complex occurs above 400°C [19], while in air gradual decomposition is observed above 320°C, which leads to formation of Mn_3O_4 at about 700°C [4]. The volatile pyrolysis products obtained by heating 4 h at 500 and 700°C under argon consist of H_2, CH_4, CO_2, CO, and 8-hydroxyquinoline (HL). The ratios of product to complex are given. The nonvolatile product resulting at 500°C consists mainly of MnO [6].

The oxidation of $Mn(C_9H_6NO)_2$ in diluted aqueous solution by various oxidants (O_2, H_2O_2, $K_3Fe(CN)_6$, $KMnO_4$) was studied spectrophotometrically and potentiometrically. The reaction products are $Mn(C_9H_6NO)_3$, $Mn(C_9H_6NO)_2OH$, and possibly $[Mn(C_9H_6NO)_2(OH)_2]^-$. The complexes are stable in aqueous solutions at pH 9 to 10. Partial oxidation of the ligand is also observed [38]. $Mn(C_9H_6NO)_2$ in acetone or acetonitrile reacts with a solution of 0.1M $HClO_4$ in glacial acetic acid containing 20% acetic anhydride according to $Mn(C_9H_6NO)_2 + 4HClO_4 \rightarrow Mn(ClO_4)_2 + 2C_9H_7NO \cdot HClO_4$ [20]. The compound is soluble in hexamethylphosphoric triamide [21]. In ethanolic solution $Mn(C_9H_6NO)_2$ reacts with pyridine 2,6-dicarboxylic acid to give $Mn(C_7H_3NO_4)$ by ligand exchange [39].

$Mn(C_9H_6NO)_2 \cdot 2H_2O$ can be prepared by reacting $Mn(CH_3COO)_2 \cdot 4H_2O$ and 8-hydroxyquinoline in a deaerated 1:1 mixture of water and methanol. The dihydrate was formed if solutions of manganese(II) chloride in water and the ligand in ethanol [24] or acetone [40] are mixed in a 1:1.05 mole ratio and adjusted to pH 6.4 to 7.0 by adding aqueous ammonia [24] and tartaric acid [40]. A similar preparation is reported by [25]. Earlier preparations are given by [4, 6, 13]. Precipitations in gravimetric analysis are described by [3, 27]. The precipitate is filtered off under argon, washed with deaerated water and methanol, and dried 2 h in vacuo [22]. The precipitation of the complex from aqueous solutions with buffered ammonium acetate containing various Mn^{2+}, ligand, and H^+ concentrations was studied by optical microscopy [26]. For the reaction $Mn^{2+}(aq) + 2C_9H_6NO^-(aq) \rightarrow Mn(C_9H_6NO)_2 \cdot xH_2O(s) + yH_2O$, the enthalpy $\Delta H = -16.4$ kcal/mol at ~25°C was determined by calorimetric measurements, $\Delta G = -29.6$ kcal/mol was estimated, and $\Delta S = 44$ cal·mol^{-1}·K^{-1} was calculated [28]. $Mn(C_9H_6NO)_2 \cdot 2H_2O$ forms yellow [22], dull yellow [3, 27], orange-yellow [4] prisms, which may grow up to 10^{-9} mm in length as shown by studies with a polarizing microscope [26].

The IR spectrum of $Mn(C_9H_6NO)_2 \cdot 2H_2O$ was recorded in the 650 to 100 cm^{-1} region (Nujol mulls) [24], in the 5000 to 665 cm^{-1} region [30], and in the 5000 to 250 cm^{-1} region usually without assignment [10, 31]. There is a broad absorption band in the 3300 cm^{-1} region, which is assigned to H_2O of hydration. For the other bands, only regions are given for a number of metal chelates [30]. Two bands, at 1325 and 1106 cm^{-1}, are assigned to the antisymmetric and symmetric CO stretching modes [12]. Strong bands in the 650 to 120 cm^{-1} region are assigned as follows):

cm^{-1}	647	630, 604	501	373
assignment	δ(ligand)	H_2O rocking or wagging + δ(ligand)	δ(CO)	δ(chelate ring)
cm^{-1}	279	(200 to 190)*)	200 to 130	
assignment	ν(Mn-O)	ν(Mn-N)	ligand vibrations	

*) Overlaps with ν($Mn-OH_2$) and strong ligand vibration.

Assignments were made with the aid of isotopic shifts. No shifts relative to the free ligand or free ligand bands are given [24]. It is assumed that $Mn(C_9H_6NO)_2 \cdot 2H_2O$ has nearly the same trans-planar structure as $Co(C_9H_6NO)_2 \cdot 2H_2O$, which had been studied by X-ray analyses, since the far-IR spectra of both compounds are closely related. The H_2O molecules must occupy axial positions to produce C_{2v} symmetry [24].

The electronic absorption spectrum is reported together with that of the anhydrous complex on p. 173. Thermogravimetric measurements show that $Mn(C_9H_6NO)_2 \cdot 2H_2O$ starts to lose its H_2O at 117°C. Constant weight is attained at 188°C [4]. On the DTA curve (heating rate 10 K/min) the dehydration peak is observed at 185°C [17]. Appreciable dehydration is observed during isothermal heating above 110°C [27, 35]. All H_2O is lost on heating 3 h at 155°C [3], also see [2], or in vacuo at 150°C [6]. Differential scanning calorimetric measurements yield a dehydration enthalpy of $\Delta H = 28.5 \pm 1.4$ kcal/mol. The activation energy of dehydration, $E_A = 25 \pm 3$ kcal/mol, was calculated from thermogravimetric data. The value may indicate coordinated H_2O. The decrease of reflectance intensity observed on heating may be due to a change in coordination number from six to four [25].

The solubility product K_S of $Mn(C_9H_6NO)_2 \cdot 2H_2O$ in aqueous solution was determined to be $K_S = 1.6 \times 10^{-18}$ [32]. Solubility measurements in aqueous 1M acetic acid yielded $K_S = 4.75 \times 10^{-20}$ [33]. $Mn(C_9H_6NO)_2 \cdot 2H_2O$ is soluble in $CHCl_3$. The solubility product is $K_S = 0.0646$ [35].

Ethanolic solutions of $Mn(C_9H_6NO)_2 \cdot 2H_2O$ are decomposed by acids and alkalies, as shown by the electronic spectra [29].

Cyclic voltammetry and conductance measurements on 0.001M solutions in dimethyl sulfoxide (DMSO) suggest that $Mn(C_9H_6NO)_2 \cdot 2H_2O$ dissociates to about 10% into a 1:1 electrolyte consisting of the species $[Mn(C_9H_6NO)]^+$ and $[Mn(C_9H_6NO)_3]^-$. As further shown by cyclic voltammetry in DMSO containing 0.1M $(C_2H_5)_4NClO_4$, the complex is anodically oxidized at $E = +0.16$ V (versus standard calomel electrode) in a reversible one-electron step to $[Mn^{III}(C_9H_6NO)_2]^+$, which hydrolyzes, forming $Mn(C_9H_6NO)_2OH$ and then, in the presence of free ligand, $[Mn_2^{III}O(C_9H_6NO)_4(C_9H_7NO)_2]$, see p. 178. Controlled potential coulometry at 0.25 V, however, indicates that the total anodic oxidation of $Mn(C_9H_6NO)_2 \cdot 2H_2O$ in DMSO is only a 0.5-electron process: $2Mn^{II}(C_9H_6NO)_2 + {}^1/_2H_2O \rightarrow {}^1/_2Mn_2^{III}O(C_9H_6NO)_4(C_9H_7NO)_2 + [Mn^{II}(C_9H_6NO)]^+ + 1e^-$ [22, 23].

Cryoscopic molecular-weight determinations in dimethyl sulfoxide (DMSO) indicate that the compound is 80% dissociated into $Mn(C_9H_6NO)_2$ and H_2O [22, 23]. Solutions of $Mn(C_9H_6NO)_2 \cdot 2H_2O$ in DMSO are easily hydrolyzed. They are oxidized slowly by both O_2 and H_2O_2, forming $Mn_2^{III}O(C_9H_6NO)_4(C_9H_7NO)_2$ and $Mn^{II}(C_9H_6NO)^+$. They catalyze the disproportionation of O_2^- ions into O_2 and H_2O_2. This reaction was investigated by electrochemical methods to elucidate biological oxidation-reduction processes. The manganese 8-quinolinol complexes were used as model compounds. Solutions in pyridine react with H_2O_2 and O_2^- ions like those in DMSO. The reaction with O_2, however, occurs very rapidly, yielding a solution which contains no electrochemically active species [22, 23].

$Mn(C_9H_7NO)_2Cl_2$ and $Mn(C_9H_7NO)_2Cl_2 \cdot 2py$. The chloride $Mn(C_9H_7NO)_2Cl_2$ was prepared by reacting anhydrous manganese(II) chloride with 8-hydroxyquinoline in absolute ethanol. The golden-yellow compound is moderately stable in dry air and insoluble in cold water. It is hydrolyzed by hot water and NaOH solutions; it is soluble in mineral acids. The 8-hydroxyquinoline is assumed to be coordinated to manganese only through the N atoms [41]. Preparation and properties of $Mn(C_9H_7NO)_2Cl_2$ and its pyridine adduct $Mn(C_9H_7NO)_2Cl_2 \cdot 2py$ (X-ray diffraction, IR spectra, electrical conductivity, and thermal decomposition) are reported in [42].

Carboxylato Compounds. Mixed ligand complexes with 8-hydroxyquinoline and halogenoacetic acids (RCOOH with $R = CF_3$, CCl_3, $CHCl_2$, CH_2Cl) were obtained by adding $Mn(C_9H_6NO)_2$ to solutions of the acids in benzene and refluxing the mixtures for 1 h. Different types of buff-yellow products were isolated: a complex of composition $Mn(C_9H_6NO)(Cl_2HCOO) \cdot C_9H_7NO$, a 2:3:1 Mn-hydroxyquinoline-CF_3COOH compound containing H_2O, and a complex of composition $Mn_2(C_9H_6NO)_2(CCl_3COO)_2 \cdot C_9H_7NO$. All complexes contain bidentate O,N-bonded 8-hydroxyquinolato ions and bidentate O,O-bonded halogenoacetato ions. The coordination of the monodentate 8-hydroxyquinoline zwitterion to the manganese ion through its phenolate O atom and the presence of the system $^+N\text{-}H \cdots O$ are indicated by typical IR bands [43].

References:

[1] T. Dupuis, J. Besson, C. Duval (Anal. Chim. Acta **3** [1949] 599/605, 603). – [2] F. J. Welcher (Organic Analytical Reagents, Vol. I, New York – Toronto – London 1955, pp. 264/323, 309). – [3] R. Stoesser, A. Rerichs, G. Roebisch (Z. Chem. [Leipzig] **21** [1981] 332/3). – [4] M. Borrel, R. Pâris (Anal. Chim. Acta **4** [1950] 267/85, 269/72). – [5] Sh. Taguchi, M. Taga, H. Yoshida, S. Hikime (Bunseki Kagaku **20** [1971] 570/4, 574; C.A. **75** [1971] No. 71018).

[6] R. G. Charles, A. Perrotto, M. A. Dolan (J. Inorg. Nucl. Chem. **25** [1963] 45/55, 51). – [7] W. E. Ohnesorge, L. B. Rogers (Spectrochim. Acta **15** [1959] 27/40, 30/2). – [8] D. C. Bhatnagar, L. S. Forster (Spectrochim. Acta **21** [1965] 1803/7, 1805). – [9] K. Sone (J. Am. Chem. Soc. **75** [1953] 5207/11, 5208). – [10] R. J. Magee, L. Gordon (Talanta **10** [1963] 851/9, 853).

[11] J. E. Tackett, D. T. Sawyer (Inorg. Chem. **3** [1964] 692/6). – [12] B. A. Kulkarni, A. J. Mukhedkar (J. Univ. Poona Sci. Technol. **1969** No. 36, pp. 21/6, 25; C.A. **71** [1969] No. 26169). – [13] R. G. Charles, H. Freiser, R. Friedel, L. E. Hilliard, W. D. Johnston (Spectrochim. Acta **8** [1956] 1/8, 3). – [14] T. Yoshida, Sh. Sawada (Bull. Chem. Soc. Japan **47** [1974] 1024/5). – [15] A. E. Jenkins, J. R. Majer, M. J. A. Reade (Talanta **14** [1967] 1213/20, 1218).

[16] Y. Kidani, Sh. Naga, H. Koike (Bull. Chem. Soc. Japan **46** [1973] 2105/7; Chem. Letters **1972** 507/10). – [17] W. W. Wendlandt, G. R. Horton (Anal. Chem. **34** [1962] 1098/101). – [18] R. G. Charles (Anal. Chim. Acta **27** [1962] 474/9, 478). – [19] R. G. Charles, A. Langer (J. Phys. Chem. **63** [1959] 603/5). – [20] V. K. Akimov, A. J. Busev, I. A. Emel'yanova (Zh. Obshch. Khim. **40** [1970] 2706/8; J. Gen. Chem. [USSR] **40** [1970] 2698/700).

[21] T. Takeuchi, Y. Kosugi, K. Itoh (Z. Anal. Chem. **294** [1979] 275/7). – [22] J. K. Howie, D. T. Sawyer (J. Am. Chem. Soc. **98** [1976] 6698/700). – [23] J. K. Howie, M. M. Morrison, D. T. Sawyer (ACS Symp. Ser. No. 38 [1977] 97/112, 99/103; C.A. **86** [1977] No. 152029). – [24] N. Ohkaku, K. Nakamoto (Inorg. Chem. **10** [1971] 798/805, 803). – [25] R. H. Gore, W. W. Wendlandt (Anal. Chim. Acta **52** [1970] 83/9, 85).

[26] A. Packter, P. Chauhan (Krist. Tech. **10** [1975] 995/1002, 997). – [27] R. Berg (Z. Anal. Chem. **76** [1929] 191/204, 194). – [28] R. G. Charles (J. Phys. Chem. **61** [1957] 1640/3). – [29] J. Császár (Acta Chim. Acad. Sci.Hung. **32** [1962] 437/49, 440, 445/6). – [30] J. P. Philips, J. F. Deye (Anal. Chim. Acta **17** [1957] 231/3).

[31] R. J. Magee, L. Gordon (Talanta **10** [1963] 961/6, 963). – [32] M. Borrel, R. A. Pâris (Anal. Chim. Acta **6** [1952] 389/99, 393, 397). – [33] J. P. Philips, H. P. Price (J. Am. Chem. Soc. **73** [1951] 4414/5). – [34] F. Umland (Z. Anal. Chem. **190** [1962] 186/208, 198). – [35] A. Ya. Polukhina (V Sb. Vopr. Khim. Biokhim. Sist. Soderzh. Marganets Polifenoly Chelyabinsk **1975** No. 3, pp. 33/7; C.A. **86** [1977] No. 64805).

[36] T. Yoshida (Bull. Chem. Soc. Japan **53** [1980] 1327/30, 1328). – [37] A. M. El-Agrami, F. M. Abd el Kerim, H. A. Ali, M. H. Atya (Isotopenpraxis **15** [1979] 108/9). – [38] I. E. Kalinichenko, E. Ya. Matveeva, A. T. Pilipenko (Zh. Neorgan. Khim. **26** [1981] 115/9; Russ. J. Inorg. Chem. **26** [1981] 60/3). – [39] M. M. Aly (Inorg. Nucl. Chem. Letters **15** [1979] 161/5, 163). – [40] A. I. Busev, N. V. Kobrya, P. P. Korostelev, G. K. Mikhailov (Izv. Akad. Nauk SSSR Neorgan. Materialy **16** [1980] 1842/4; Inorg. Materials [USSR] **16** [1980] 1259/61).

[41] S. Prasad, P. J. Sarma (J. Proc. Inst. Chemists [India] **37** [1965] 246/8; C.A. **64** [1966] 10735). – [42] L. E. Pogosyan, S. N. Avakyan (Mater. 5th Resp. Soveshch. Neorgan. Khim., Yerevan 1976 [1977], pp. 124/7 from C.A. **91** [1979] No. 133336). – [43] M. M. Aly, A. M. Shalaby (Monatsh. Chem. **111** [1980] 935/43).

16.1.25.4.3 Isolated Manganese(III) Complexes

$Mn(C_9H_6NO)_3$ was prepared by mixing stoichiometric amounts of manganese(III) acetate dihydrate and 8-hydroxyquinoline in isobutyl alcohol or heating the ethanol solvate 1 h at 180°C [1]. Solutions of $Mn(acac)_3$ (acacH = 2,4-pentanedione = $C_5H_8O_2$) and ligand in dry ethanol were mixed in a 1:3 mole ratio. The crystals which had appeared after $^1/_2$ h of standing at room temperature were collected, washed with dry ethanol, and dried in vacuo over fused $CaCl_2$. The crystals are shiny, almost black needles. An analysis is given for the unsolvated complex [2]. By repeating the preparation of [2], Burns et al. [1] have only been able to obtain an ethanol adduct. To an aqueous solution containing equimolar amounts of $K[Mn(C_3H_2O_4)_2(H_2O)_2] \cdot 2H_2O$ and malonic acid, an acetate-buffered (pH 4) solution of the ligand was added dropwise at about +5°C. The precipitate was washed with water, ethanol (96%), and ether and dried in vacuo over P_2O_5. The complex is a dark green powder. Differences between the analysis and calculated values are attributed to incomplete removal of 8-hydroxyquinoline [3].

The electronic spectrum of a 10^{-4}M solution of the complex in $CHCl_3$ shows three absorption bands, at 19500 (shoulder), 25500, and 30100 cm^{-1}, with molar extinction coefficients log ε = 2.80, 4.01, and 3.76 $L \cdot mol^{-1} \cdot cm^{-1}$, respectively. The broad band at 25500 cm^{-1} is assigned to an internal transition in the ligand anion (free ligand band at 28200 cm^{-1}) [2]. This band (λ = 396 nm, in $CHCl_3$) was also observed by [4]. The mean absorption maximum of a solution in $CHCl_3$ is observed at 420 nm (~23800 cm^{-1}) by [3]. Burns et al. [1] did not observe the shoulder at 19500 cm^{-1}, which was assigned to the $^5E_g \rightarrow {}^5T_{2g}$ transition by [2]. They did observe a band in the region about 11000 cm^{-1}, which is assigned to the $^5B_{1g} \rightarrow {}^5A_{1g}$ transition associated with a tetragonally distorted octahedral field [1]. The crystal-field parameter Dq = 1950 cm^{-1} was calculated, indicating a ligand-field strength roughly equivalent to that in $[Mn(H_2O)_6]^{3+}$. The magnetic moment was determined to be μ_{eff} = 4.48 μ_B [2] or 5.0 μ_B at 25°C [1]. The magnetic moments and the UV and visible spectra of the unsolvated and solvated complexes correspond to high-spin octahedral Mn^{III} complexes [1]. A 0.1M solution in nitrobenzene is virtually nonconducting [2].

The manganese(III) complex is indefinitely stable in the solid state [2]. It is practically insoluble in water [2, 3] and ethanol, and it is surprisingly stable towards hydrolysis in aqueous suspension [2]. It is moderately soluble in $CHCl_3$, acetone, benzene [2, 3], nitromethane, nitrobenzene [2], dioxane, and long-chain alcohols [3]. In dimethyl sulfoxide, solid $Mn(C_9H_6NO)_3$ reacts with water to form $Mn_2O(C_9H_6NO)_4(C_9H_7NO)_2$ [5]. Bromination in $CHCl_3$ by Br_2 probably yields $Mn(C_9H_4Br_2NO)_3$ contaminated with HBr [2]. $Mn(C_9H_6NO)_3$, its more soluble mono-ethanol solvate (see below), and Mn^{III} complexes of 5-nitro-, 5-chloro-, and 5-cyano-hydroxyquinoline are reduced by the dimethylformamide species $Fe(DMF)_6^{2+}$ in acetonitrile or propanediol-1,2-carbonate. The kinetics and mechanism of the reduction were investigated spectrophotometrically. The electron transfer takes place in a dinuclear intermediate. The direction of the substituent effect indicates that the formation of the intermediate is rate determining. Subsequent ligand exchange yields the final products, $Fe(C_9H_6NO)_3$ and Mn^{2+} [9].

Alcohol Solvates. The mono-ethanol solvate can be prepared by mixing dry ethanolic solutions of manganese(III) acetate dihydrate or manganese(II) acetate tetrahydrate and 8-hydroxyquinoline in a 1:3 mole ratio. In the second case, oxidation and crystallization occurred on standing overnight. The shiny black crystals were washed with dry ethanol and dried at 25°C in vacuo. The adducts with other alcohols were obtained from the manganese(III) salt in the same way. They are listed in the table below, together with the solvent-to-complex mole ratio and the thermogravimetric decomposition temperature t:

alcohol	methyl	ethyl	1-propyl	1-butyl	1-pentyl	1-hexyl	1,2-ethanediol
ratio	1.03	1.03	0.55	0.56	*)	0.54	0.89
t in °C	160	160	160	210	210	210	210

*) Complex decomposes slowly at the temperature required to remove the solvent [1].

The solvates have UV and visible spectra similar to those of the unsolvated complex, indicating tetragonally distorted octahedral structures. The alcohol molecules are trapped in cage-like cavities in the structures formed by the ligand molecules. No solvates are formed with the branched-chain isopropyl or isobutyl alcohols [1].

$Mn(C_9H_6NO)_3 \cdot CH_3OH$ and $Mn(C_9H_6NO)_3 \cdot 0.5\,n\text{-}C_6H_{13}OH$ are isomorphous according to Weissenberg photographs. They are monoclinic; space group $P2_1/n\text{-}C_{2h}^5$ (No. 14). The lattice constants are a = 10.847(3), b = 13.201(4), c = 17.285(5) Å, and β = 97.56(5)° for the methanol solvate and a = 11.201(4), b = 13.342(4), c = 17.200(5) Å, and β = 95.09(6)° for the n-hexanol solvate; Z = 4. In both adducts the bond lengths Mn-N vary between 2.057 and 2.266, while the bond lengths Mn-O vary between 1.905 and 1.924. The bond angles N-Mn-N, N-Mn-O, and O-Mn-O vary between 90.2° and 95.5° in both adducts. The Mn atom of the tris(quinolato) species is coordinated to the oxygen and nitrogen atoms of each ligand to form a distorted octahedron (see **Fig. 15**). The complex configuration is meridional. The alcohol molecules are hydrogen bonded to a meridional oxygen atom, the distances $O\text{-}OH_{alcohol}$ being 2.79 and 2.91 Å for the methanol and 1-hexanol adduct, respectively. The methanol solvate structure contains one molecule of alcohol per asymmetric unit, whereas in the hexanol adduct structure there is only half a molecule of alcohol. Thus the hexanol molecule is disordered, and the center of the central C-C bond lies at the crystallographic center of symmetry. In the methanol adduct, which is most likely isomorphous with the related Cr^{III} compound, two methanol molecules lying in cage-like cavities of the asymmetric units are related by the crystallographic center of symmetry at a distance C-C = 4.93 Å. The compound is a racemate [6, 7]. The densities of $Mn(C_9H_6NO)_3 \cdot CH_3OH$ and $Mn(C_9H_6NO)_3 \cdot 0.5\,n\text{-}C_6H_{13}OH$ obtained by flotation are D = 1.38(3) and 1.39(3) g/cm^3; densities calculated from X-ray data are 1.406 and 1.402 g/cm^3, respectively [7].

$Mn_2O(C_9H_6NO)_4(C_9H_7NO)_2 \cdot 2CH_3OH$ (or $Mn(C_9H_6NO)_3 \cdot 0.5H_2O \cdot CH_3OH$ from the magnetic data) was obtained by reacting a solution of $Mn(C_9H_6NO)_2 \cdot 2H_2O$ (see p. 175) in a 1:1 mixture of water and methanol with air. The crystals are black. The susceptibility of the solid yields the magnetic moment μ_{eff} = 5.02 μ_B [5], corresponding to high-spin Mn^{III}-d^4 ions. Similar values are found in acetonitrile (4.8 μ_B) and pyridine (4.91 μ_B), but in dimethyl sulfoxide (DMSO) a considerably lower value (4.56 μ_B) was obtained. This lower value may be due to some antiferromagnetic spin-spin coupling, while such coupling does not occur in the other two solvents and a monomeric structure is assumed [8]. However, NMR studies on solutions in dimethyl sulfoxide reveal the diamagnetic nature of the complex in this solvent, which implies strong antiferromagnetic coupling. Presumably its molecular form in DMSO is $[(C_9H_7NO)Mn^{III}(C_9H_6NO)_2OMn^{III}(C_9H_6NO)_2(C_9H_7NO)]$ [5].

Cyclic voltammetric investigations on the complex in DMSO (0.1M $(C_2H_5)_4NClO_4$) show reversible reduction to $[Mn(C_9H_6NO)_3]^-$ at E = −0.31 V (versus standard calomel electrode). Controlled potential coulometry at −0.50 V indicates a one-electron reaction per Mn atom. A large peak at E = +0.75 V is due to a one electron per Mn atom oxidation, which is irreversible in DMSO and pyridine but quasireversible in acetonitrile and may produce an (unstable) dinuclear Mn^{IV} complex. The complex reacts with O_2^- to form O_2 and H_2O but at a slower rate than $Mn(C_9H_6NO)_2 \cdot 2H_2O$ (see p. 175) [5, 8].

Fig. 15

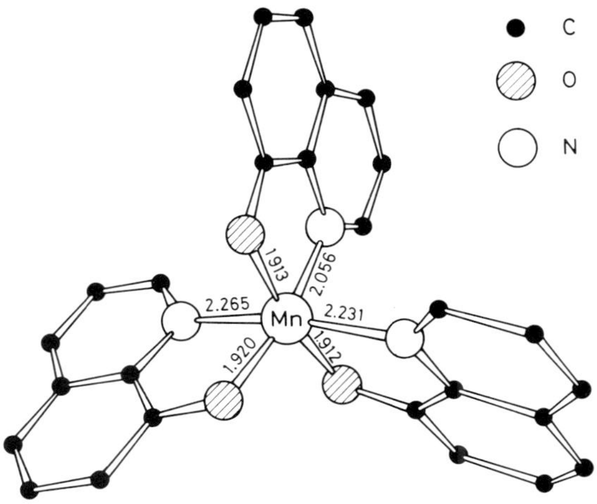

Perspective view of $Mn(C_9H_6NO)_3$. Bond lengths refer to the mean value for $Mn(C_9H_6NO)_3 \cdot CH_3OH$ and $Mn(C_9H_6NO)_3 \cdot 0.5\,n\text{-}C_6H_{13}OH$.

References:

[1] A. R. Burns, T. J. Cardwell, R. W. Cattrall (Australian J. Chem. **24** [1971] 661/3). – [2] M. M. Ray, J. N. Adhya, D. Biswas, S. N. Poddar (Australian J. Chem. **19** [1966] 1737/40). – [3] A. Ya. Polukhina (V Sb. Vopr. Khim. Biokhim. Sist. Soderzh. Marganets Polifenoly Chelyabinsk **1975** No. 3, pp. 33/7; C.A. **86** [1977] No. 64805). – [4] F. Umland (Z. Anal. Chem. **190** [1962] 186/208, 198). – [5] J. K. Howie, D. T. Sawyer (J. Am. Chem. Soc. **98** [1976] 6698/700).

[6] R. Hems, T. J. Cardwell, M. F. Mackay (Australian J. Chem. **28** [1975] 443/5). – [7] R. Hens, M. F. Mackay (J. Cryst. Mol. Struct. **5** [1975] 227/45, 229, 231/2, 235). – [8] J. K. Howie, D. T. Sawyer (ACS Symp. No. 38 [1977] 97/112, 99/102, 105; C.A. **86** [1977] No. 152029). – [9] R. W. Soukup, V. N. Sapunov, R. Schmid, B. Gutmann (Z. Phys. Chem. [Wiesbaden] **119** [1980] 147/58, 150, 157).

16.1.25.5 With Derivatives of 8-Hydroxyquinoline

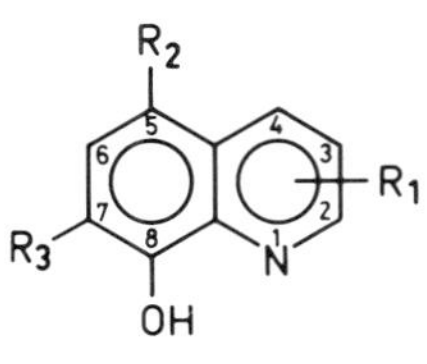

Complexes with nitroso derivatives of 8-hydroxyquinoline are described in the Chapter "Complexes with Oximes and Nitroso Compounds" in "Manganese" D 4.

Complexes in Solution

There are generally 1:1 and 1:2 complexes. The following stability constants K_1 and K_2 (L/mol) of the manganese(II) complexes were obtained by potentiometric (glass electrode) measurements in aqueous dioxane (50 vol% [1 to 3, 7, 22], 60 vol% [8], 75 vol% [19]) at 25°C [1 to 3, 7, 8, 22] or 35°C [19] and ionic strength I:

No.	R_1	R_2	R_3	formula	log K_1	log K_2	I	Ref.
I	2-CH_3	H	H	$C_{10}H_9NO$	6.81	6.29	0.1 ($NaClO_4$)	[1]
II	4-CH_3	H	H	$C_{10}H_9NO$	7.74	7.07	0.1 ($NaClO_4$)	[2]
III	2-CH_2OH	H	H	$C_{10}H_9NO_2$	7.5	—	0.1 ($NaClO_4$)	[1]
IV	2-CH_2NH_2	H	H	$C_{10}H_{10}N_2O$	8.32	7.52	0.1 ($NaClO_4$)	[1]
V	2-NH-NH_2	H	H	$C_9H_9N_3O$	8.81	7.65	0.1 ($NaClO_4$)	[22]
VI	2-C_6H_5	H	H	$C_{15}H_{11}NO$	6.22*⁾	—	—	[3]
VII	H	CHO	H	$C_{10}H_7NO_2$	5.73	4.97	0.3 (NaCl)	[7]
VIII	H	$COCH_3$	H	$C_{11}H_9NO_2$	6.78	6.30	0.1 ($NaClO_4$)	[8]
IX	H	CH_2OCH_3	H	$C_{11}H_{11}NO_2$	see text below			[10]
X	H	Cl	H	C_9H_6ClNO	7.13	6.43	0.1 ($NaClO_4$)	[8]
XI	H	Cl	Cl	$C_9H_5Cl_2NO$	5.90	4.63	0.2 ($NaClO_4$)	[19]
XII	H	Br	Br	$C_9H_5Br_2NO$	6.00	4.73	0.2 ($NaClO_4$)	[19]
XIII	2-CH_3	Cl	Cl	$C_{10}H_7Cl_2NO$	6.84	5.96	—	[6]
XIV	2-CH_3	Br	Br	$C_{10}H_7Br_2NO$	7.15	6.33	—	[6]
XV	2-CH_3	I	I	$C_{10}H_7I_2NO$	8.32	7.13	—	[6]
XVI	H	NO_2	H	$C_9H_6N_2O_3$	6.15	5.11	0.1 ($NaClO_4$)	[8]
XVII	H	NO_2	NO_2	$C_9H_5N_3O_5$	4.10	2.94	0.1 ($NaClO_4$)	[8]

*⁾ Extrapolated.

Further data for the complexes with the ligands I and II at 0.7 and 40°C were given by [3]; for earlier data also see [4, 5]. On comparing the stability of the complexes of substituted 8-hydroxyquinolines compounds, the following substituent series 4-CH_3 > H > 5-SO_3H > 2-CH_3 results [2]. Ligands III and IV are considered to be tridentate in their manganese chelates [1]. For the complex with 5-formyl-8-hydroxyquinoline (ligand VII) a 1:3 complex was found: log $K_3 = 4.58$ [7].

A complex of composition $[Mn(C_9H_4Cl_2NO)_3]^-$ is probably formed from Mn^{2+} ions and ligand XI (mole ratio 1:3) in aqueous solution with pH 7.2 to 8.0 in the presence of Rhodamine 6 G. It may be extracted with benzene, and the electronic absorption spectrum of such solution exhibits a maximum at $\lambda = 540$ nm with the molar extinction coefficient $\varepsilon_{max} = 70000$ L·mol^{-1}·cm^{-1} [4].

For the complexes with the ligands I and II the following thermodynamic data of formation were obtained by differential calorimetry [2] or calculated from the slope of log β_2 vs. temperature (0.7, 25, and 40°C) [3]:

ligand	$-\Delta H_1$	$-\Delta H_2$	$-\Delta G_1$	$-\Delta G_2$ (at 25°C)	ΔS_1	ΔS_2
I	3.3	6.3 (6.6)	9.29	17.87 (19.10)	20	39 (42)
II	4.1	6.4 (11.6)	10.56	20.20 (21.15)	22	46 (32)
Ref.	[2]	[2] ([3])	[2]	[2] ([3])	[2]	[2] ([3])

Polarography of solutions containing 0.001 mol of Mn^{III} salt, 0.05 to 0.15 mol of ligand IX (= HL) per liter of aqueous dimethylformamide (50%) at 25 ± 0.2°C, pH = 4.6, and I = 1 ($NaClO_4$) reveals the formation of a 1:1 complex $[MnL]^{2+}$ with the instability constant $K = (4.4 \pm 0.4) \times 10^{-40}$ L/mol. This complex is formed also at pH = 10. It is electrochemically reduced to form $[MnL]^+$ in alkaline media or Mn^{2+} + HL in acidic media, and the ratio of the instability constants K of the oxidized and reduced forms of the complex is $K_{ox}/K_{red}=(1.7\pm0.1)\times10^{-38}$ [10].

Synergistic extractions of Mn^{2+} from aqueous solutions (pH 6, potassium phthalate/KOH buffer) into $CHCl_3$ by the 2-methyl or 4-methyl derivative of 8-hydroxyquinoline (=HL) and N-heterocyclic bases B, e.g., pyridine (=py), 2-methylpyridine (=2-C_6H_7N), 4-methylpyridine (=4-C_6H_7N), 2,4-dimethylpyridine (=2,4-C_7H_9N), 2,6-dimethylpyridine (=2,6-C_7H_9N), and 2,4,6-trimethylpyridine (=2,4,6-$C_8H_{11}N$), were studied by using a ^{54}Mn radiotracer. Equilibrium constants were calculated and the formation constants of the extractable 1:1 adducts $K=[MnL_2\cdot B]/[MnL_2]\cdot[B]$ were determined at 30°C and I=0.25 ($NaClO_4$) [26]:

chelating ligand HL	B	log K	B	log K	B	log K
2-methyl-8-hydroxy-	py	1.85	2-C_6H_7N	1.71	4-C_6H_7N	1.97
quinoline	2,4-C_7H_9N	1.88	2,6-C_7H_9N	1.43	2,4,6-$C_8H_{11}N$	1.78
4-methyl-8-hydroxy-	py	1.53	2-C_6H_7N	1.30	4-C_6H_7N	1.67
quinoline	2,4-C_7H_9N	1.47	2,6-C_7H_9N	1.40	2,4,6-$C_8H_{11}N$	1.40

Isolated Compounds

$Mn(C_{10}H_8NO)_2$. The manganese(II) complexes with the 2- and the 4-methyl derivative of quinolinol (ligand I and II) were prepared by adding a small excess of the ligands (1% solution in ethanol) to an aqueous solution of $Mn(NO_3)_2$ or $Mn(ClO_4)_2$ (0.0025 M) which had been heated to 60°C. The mixture is made slightly alkaline with aqueous ammonia and cooled [11]. The complex with ligand I was obtained also by precipitating it from a solution of $MnSO_4$ and the ligand, mole ratio 1:2, in diluted hydrochloric acid with aqueous NaOH at 20°C; it is brown yellow [12]. The fluorescence spectra of the complexes being excited with Hg light (366 nm) at 80 K show maxima at 20100 cm^{-1} (quantum yield $\Phi=0.020$) for the 2-methyl derivative and at 20500 cm^{-1} ($\Phi=0.035$) for the 4-methyl derivative [13]. Their IR spectra show absorption bands at 1100 and 1153 cm^{-1}, respectively, which may be assigned to the CO stretching vibrations of the ligands; further principal absorption peaks are reported but not assigned [11].

The solubility product S of $Mn(C_{10}H_8NO)_2$ (derived from the 2-methylquinolinol) was determined by pH determination in aqueous solution to be $S=4.5\times10^{-19}$ (log S=−18.35) [14]. In another work $S=1.66\times10^{-18}$ (log S=−17.78) was found at 20°C. Precipitation occurs at pH=4.8. The solubility of the complex in $CHCl_3$ is found by colorimetry to be 8.08×10^{-4} mol/L at 20°C. The extraction from the aqueous phase by $CHCl_3$ is exhaustive at pH=5 to 10. At 20°C the partition coefficient is P=1270 (log P=3.10), and the calculated extraction coefficient is $K_{ex}<1.02\times10^{-11}$ (log $K_{ex}<-10.99$) [12].

$Mn(C_{10}H_8NO)_3$. The manganese(III) complex with 8-hydroxy-2-methylquinoline was prepared by mixing stoichiometric amounts of manganese(III) acetate dihydrate and the ligand in ethanolic solutions. Small black shiny crystals were obtained after 2 to 3 days. The electronic spectrum shows a band shoulder in the 14500 cm^{-1} region, which was assigned to a low energy $^5A_{1g}\leftarrow{}^5B_{1g}$ transition associated with a tetragonally distorted octahedral ligand field. The effective magnetic moment $\mu_{eff}=5.0\ \mu_B$ indicates high-spin manganese(III) complex. Thermogravimetric studies suggest that this complex is less stable than the 8-hydroxyquinoline complex [15].

$Mn(C_{13}H_{14}NO)_2$ was obtained by reacting a slightly alkaline aqueous solution of a manganese salt with a methanolic solution of 2-butyl-8-hydroxyquinoline (=$C_{13}H_{15}NO$).

The mass spectrum, which was recorded at temperatures between 280 and 320°C, shows the molecular ion and fragments $Mn(C_{13}H_{14}NO)_2^+$ (= MnL_2^+; m/e = 455) and $[MnL_2-C_2H_5]^+$ (m/e = 426) as the main components, together with $[MnL_2-200]^+$ (m/e = 255, corresponding to the loss of one ligand molecule), $[MnL_2-230]^+$ (m/e = 225), and other fragments as minor components. The complex splits off first ethyl radicals, then propene, one ligand molecule, and hydrogen. Deuterated derivatives were also studied [16].

$Mn(C_{15}H_{10}NO)_2 \cdot H_2O$. The manganese(II) complex with 8-hydroxy-2-phenylquinoline (= $C_{15}H_{11}NO$) was prepared as follows: To a mixture containing equal volumes of the ligand (0.05 M) in acidified aqueous ethanol and manganese(II) salt (0.025 M) in water, alcohol and ether were added until the mixture contained 30 and 15%, respectively, of these solvents. The light brown precipitate appeared on neutralization with NaOH; the pH of formation is 7.55 to 8.30. Thermogravimetric studies (heating rate 3 K/min up to 700°C) show that the compound loses one molecule of H_2O at 60°C and decomposes above 315°C [17].

$Mn(C_{22}H_{17}N_2O)_2$. The complex with the 7-(α-anilinobenzyl) derivative of 8-hydroxyquinoline R-C_9H_6NO (R = $CH(C_6H_5)NHC_6H_5$; HL = $C_{22}H_{18}N_2O$) was prepared as follows: An aqueous 0.01 M solution of a manganese salt in dilute acid was combined with the stoichiometric amount of the ligand dissolved in ethanol and aqueous ammonia. Sodium or ammonium acetate was added to the mixture until precipitation was complete [18].

$Mn(C_9H_5ClNO)_2$ was prepared by reacting 5-chloro-8-hydroxyquinoline (= C_9H_6ClNO) and manganese(II) perchlorate, likely in ethanolic solution. The kinetics of its bromination in a two-phase system consisting of a 10^{-4} M chelate in $CHCl_3$ and a phosphate-buffered aqueous solution was studied. The reaction rate is first-order in the concentration of bromine in the organic phase and of a fractional order in the chelate concentration [21].

$Mn(C_9H_5ClNO)_3$. The manganese(III) complex with 5-chloro-8-hydroxyquinoline was obtained by reacting stoichiometric amounts of manganese(III) acetate dihydrate and the ligand in ethanol. After recrystallization from benzene, the product was dried for 8 h at 70°C in a high vacuum. The compound is reduced by $Fe(DMF)_6^{2+}$ in the same way as described for $Mn(C_9H_6NO)_3$ on p. 177. The kinetic data obtained by spectroscopy of solutions in acetonitrile or propanediol-1,2-carbonate are reported together with those of the corresponding complexes of 5-nitro- or 5-cyano derivatives. The reduction is faster in acetonitrile than in propanediol-1,2-carbonate. The steps are essentially the same as those suggested for $Mn(C_9H_6NO)_3$ on p. 177, but the ligand exchange in propanediol-1,2-carbonate occurs far more slowly. The electronic effect of the substituents follows a linear free energy relationship. The electron transfer takes place in a dinuclear intermediate. The direction of the substituent effect indicates that the formation of the intermediate is rate determining [23].

$Mn(C_9H_4Br_2NO)_2$ and **$Mn(C_9H_4I_2NO)_2$.** The fluorescence spectra of the complexes with 5,7-dibromo-8-hydroxyquinoline (= $C_9H_5Br_2NO$) and 5,7-diiodo-8-hydroxyquinoline (= $C_9H_5I_2NO$) (wavelength of excitation λ = 366.0 nm) at 80 K exhibit maxima at 20500 (quantum yield Φ = 0.010) and 21400 (Φ = 0.005) cm^{-1}, respectively; the quantum yields are much lower than those of the corresponding free ligands [13].

$Mn(C_9H_5N_2O_3)_3$ was prepared by reacting Mn^{III} acetate dihydrate and 8-hydroxy-5-nitroquinoline (= $C_9H_6N_2O_3$) in ethanol. It was recrystallized from a mixture of CH_2Cl_2 and pentane and dried for 5 h at 120°C in a high vacuum. The kinetic data of its reduction by $Fe(DMF)_6^{2+}$ species are reported together with those of $Mn(C_9H_6NO)_3$ (see p. 177) and complexes of other substituted 8-hydroxyquinoline derivatives [23].

$Mn(C_{10}H_5N_2O)_3$ was prepared from stoichiometric amounts of Mn^{III} acetate and 5-cyano-8-hydroxyquinoline ($=C_{10}H_6N_2O$) in isobutyl alcohol and was recrystallized from a mixture of benzene and hexane. The black crystals were dried at 100°C in a high vacuum. The kinetics of its reduction by $Fe(DMF)_6^{2+}$ in acetonitrile has been studied spectrophotometrically [24]. The reduction in propanediol-1,2-carbonate was investigated by means of the stopped-flow method [25]. The kinetics and mechanisms of the reduction of manganese(III) complexes with 8-hydroxyquinoline and its substituted derivatives were compared in [23].

References:

[1] R. L. Stevenson, H. Freiser (Anal. Chem. **39** [1967] 1354/8, 1355). – [2] G. Gutnikov, H. Freiser (Anal. Chem. **40** [1968] 39/44, 42). – [3] W. D. Johnston, H. Freiser (Anal. Chim. Acta **11** [1954] 201/13, 203/6). – [4] H. Freiser (Analyst **77** [1952] 830/45, 835). – [5] W. D. Johnston, H. Freiser (J. Am. Chem. Soc. **74** [1952] 5239/42, 5240).

[6] Tiao-Hsu Chang (J. Chinese Chem. Soc. [Taipei] **13** [2] [1966] 143/9, 146). – [7] J. C. Tomkinson, R. J. P. Williams (J. Chem. Soc. **1958** 1153/8, 1157). – [8] H. F. Steger, A. Corsini (J. Inorg. Nucl. Chem. **35** [1973] 1621/36, 1625). – [9] H. F. Steger, A. Corsini (J. Inorg. Nucl. Chem. **35** [1973] 1637/43, 1640). – [10] E. G. Chikryzova, Sh. E. Vassershtein, I. V. Aronina (Zh. Analit. Khim. **25** [1970] 1568/74; J. Anal. Chem. [USSR] **25** [1970] 1353/8, 1356).

[11] R. C. Charles, H. Freiser, R. Friedel, L. E. Hilliard, W. D. Johnston (Spectrochim. Acta **8** [1956] 1/8, 3). – [12] L. Serrano Berges (Anales Real Soc. Espan. Fis. Quim. [Madrid] B **49** [1953] 417/26, 419, 422, 425). – [13] D. C. Bhatnagar, L. S. Forster (Spectrochim. Acta **21** [1965] 1803/7, 1805). – [14] M. Borrel, R. A. Pâris (Anal. Chim. Acta **6** [1952] 389/99, 397). – [15] A. R. Burns, T. J. Cardwell, R. W. Cattrall (Australian J. Chem. **24** [1971] 661/3).

[16] H. Budzikiewicz, E. Plöger (Org. Mass Spectrom. **3** [1970] 709/20, 711). – [17] G. Bocquet, A. Pâris (Anal. Chim. Acta **14** [1956] 201/8, 202/5). – [18] J. P. Phillips, A. L. Duckwall (J. Am. Chem. Soc. **77** [1955] 5504/6). – [19] R. D. Gupta, G. S. Manku, A. N. Bhat, B. D. Jain (Z. Anorg. Allgem. Chem. **379** [1970] 312/9, 316). – [20] J. Minczewski, J. Chwastowska, E. Lachowicz (Chem. Anal. [Warsaw] **18** [1973] 199/206, 201/3; C.A. **79** [1973] No. 73293).

[21] R. K. Baker, Q. Fernando (Anal. Chem. **43** [1971] 1561/5). – [22] A. Corsini, R. M. Cassidy (Talanta **26** [1979] 297/301, 299). – [23] R. W. Soukup, V. N. Sapunov, R. Schmid, V. Gutmann (Z. Physik. Chem. [Wiesbaden] **119** [1980] 147/58, 148, 150, 157). – [24] R. W. Soukup, V. N. Sapunov, R. Schmid, V. Gutmann (Z. Physik. Chem. [Wiesbaden] **119** [1980] 129/37, 130). – [25] R. W. Soukup, V. N. Sapunov, R. Schmid, V. Gutmann (Z. Physik. Chem. [Wiesbaden] **119** [1980] 139/46).

[26] A. T. Rane (J. Inorg. Nucl. Chem. **42** [1980] 1520/1).

16.1.25.6 With 8-Hydroxy-5-quinolinesulfonic Acid and Its Derivatives

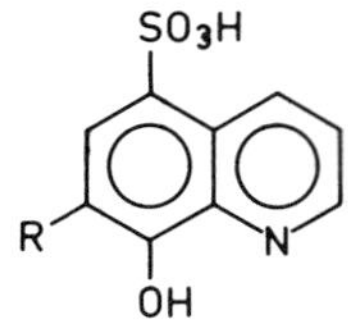

$(=H_2L)$

Complexes in Solution

Formation constants of 1:1 and 1:2 complexes (K in L/mol), determined potentiometrically (glass electrode) in H_2O [1, 2, 5, 6, 8, 9] or aqueous dioxane (50% v/v [4] or 60% v/v [3]) at temperatures t in °C and ionic strengths I are listed below:

No.	R	ligand	I	t	log K_1	log K_2	Ref.
1	H	$C_9H_7NO_4S$	0.1 (KNO_3)	25	5.67	5.05	[1]
1	H	$C_9H_7NO_4S$	1.0 ($NaClO_4$)	25	5.47	4.89	[2]
1	H	$C_9H_7NO_4S$	0.1 ($NaClO_4$)	25	7.43	6.31	[3]
1	H	$C_9H_7NO_4S$	0.1 ($NaClO_4$)	25	7.05	6.13	[4]
2	I	$C_9H_6INO_4S$	0.03 (NaCl, HCl)	20	5.25*)	—	[5]
2	I	$C_9H_6INO_4S$	0.1 (KCl)	25	5.3	4.3	[6]
3	NO_2	$C_9H_6N_2O_6S$	0.2 (KCl)	40	3.85	3.05	[8]
3	NO_2	$C_9H_6N_2O_6S$	$I \to 0$	25	4.76	3.04	[9]

*) The solubility method yields log $K_1 = 5.42$ [5].

Ligand 1 also forms a 1:3 complex, $Mn(C_9H_5NO_4S)_3^{4-}$, with log $\beta_3 = 14.30$ [2]. Further data are given for ligand 1 by [10, 11] and for ligand 2 by [7, 12]. The average stability constant of the manganese complex with ligand 1 increases slightly as the dioxane content of the solvent is increased [13]. Differential calorimetry of the complexing reaction with ligand 1 yields the thermodynamic formation data $\Delta H_1 = -3.2$, $\Delta H_{\beta 2} = -6.7$ kcal/mol, $\Delta G_1 = -9.62$, $\Delta G_{\beta 2} = -17.97$ kcal/mol, $\Delta S = 22$, and $\Delta S_{\beta 2} = 46$ $cal \cdot mol^{-1} \cdot K^{-1}$ for 50% v/v aqueous dioxane at 25°C and $I = 0.1$ [4]. The formation of the 1:1 and 1:2 complexes with ligand 1 in a replacement titration with the fluorescent complex ZnL_2^{2-}, the end point being the appearance of fluorescence, was studied in aqueous solution. The exchange constants log $K' = -2.2$ ($K' = [MnL] \cdot [ZnL]/[Mn^{2+}] \cdot [ZnL_2^{2-}]$) and log $K'' = -3.5$ ($K'' = [MnL_2^{2-}] \cdot [Zn^{2+}]/[MnL] \cdot [ZnL]$) were determined [16]. The iodinations of the ligand-1 complex and the free ligand were studied in [17].

Isolated Compounds

Complexes of composition $MnL \cdot nH_2O$ (n = 0, 2, 6) were precipitated from aqueous solutions containing Mn^{2+} and the ligands 1 to 3 (in the table above). $Mn(C_9H_5NO_4S)$ was precipitated at pH ~4.7 to give a yellowish solid [14], $Mn(C_9H_4INO_4S) \cdot 6H_2O$ at pH 2 to 3 [5]. The complex with ligand 3 formed at pH 3 to 3.7 (no analysis given) [15].

The electronic spectrum of $Mn(C_9H_5NO_4S)$ in aqueous solution (0.01M, pH 8.2) exhibits an absorption maximum at about 362 nm with a molar extinction coefficient $\varepsilon = 4000$ $L \cdot mol^{-1} \cdot cm^{-1}$ [14]; the spectrum of $Mn(C_9H_4INO_4S)$ is probably closely related [18]. $Mn(C_9H_4N_2O_6S)$ has three peaks, at about 400, 342, and 290 nm, with $\varepsilon \sim 7000$, ~6000, and ~16500 $L \cdot mol^{-1} \cdot cm^{-1}$, respectively [15].

The thermodynamic solubilities have been determined by potentiometric and spectrophotometric measurements on aqueous solutions of the 1:1 complexes with the ligands listed below. The measurements were made at 25°C [14, 15, 18] or 20°C [5], ionic strengths I, and various pH and reported as p[MnL] and pS (S = $[Mn^{2+}][L^{2-}]$).

No.	ligand H_2L	R	added electrolyte	ionic strength I	p[MnL]	pS	Ref.
1	$C_9H_7NO_4S$	H	$MnCl_2$	→0	—	9.79	[14]
2	$C_9H_6INO_4S$	I	KCl	→0	3.66	9.71	[18]
2	$C_9H_6INO_4S$	I	(Na, H)Cl	0.03	—	9.14	[5]
3	$C_9H_6N_2O_6S$	NO_2	KCl	→0	3.35	8.11	[15]

Manganese(II) complexes derived from 8-hydroxyquinoline-5-sulfonic acid and additional ligands, such as 5-sulfosalicylic acid or pyrocatechol-3,5-disulfonic acid (Na salt = Tiron) are reported by [19]. The over-all stability constants of the species formed in aqueous solutions were determined potentiometrically at 25°C, pH > 7, and I = 1.0 ($NaClO_4$). The mixed-ligand complexes exhibit in all cases greater stability than the simple ones [19].

References:

[1] C. F. Richard, R. L. Gustafson, A. E. Martell (J. Am. Chem. Soc. **81** [1959] 1033/40, 1037, 1039). – [2] J. P. Scharff, R. Genin (Anal. Chim. Acta **78** [1975] 201/9, 207). – [3] H. F. Steger, A. Corsini (J. Inorg. Nucl. Chem. **35** [1973] 1621/36, 1625). – [4] G. Gutnikov, H. Freiser (Anal. Chem. **40** [1968] 39/44, 42). – [5] I. V. Kotlyarova, N. I. Chukhrova, N. A. Skorik (Zh. Neorgan. Khim. **22** [1977] 1482/5; Russ. J. Inorg. Chem. **22** [1977] 807/9).

[6] Å. R. Storm (Acta Chem. Scand. **17** [1963] 667/72, 671). – [7] P. Lingaiah, J. Mohan Rao, U. V. Seshaiah (Current Sci. [India] **36** [1967] 197/9). – [8] S. K. Patel, K. P. Soni, I. M. Bhatt (Indian J. Chem. **11** [1973] 1324/5). – [9] R. Näsänen, E. Uusitalo (Suomen Kemistilehti B **28** [1955] 17/23, 18, 21; C.A. **1955** 8671). – [10] R. Näsänen, E. Uusitalo (Acta Chem. Scand. **8** [1954] 112/8, 115/6).

[11] A. Albert (Biochem. J. **54** [1953] 646/54, 650). – [12] A. Ekman (Suomen Kemistilehti B **35** [1962] 111/3; C.A. **58** [1963] 971). – [13] H. F. Steger, A. Corsini (J. Inorg. Nucl. Chem. **35** [1973] 1637/43, 1640). – [14] R. Näsänen, E. Uusitalo (Acta Chem. Scand. **8** [1954] 835/41, 837/9). – [15] R. Näsänen, E. Uusitalo (Acta Chem. Scand. **9** [1955] 1222/9, 1227).

[16] J. A. Bishop (Anal. Chim. Acta **35** [1966] 224/30, 228). – [17] C. Bostic, Q. Fernando, H. Freiser (Inorg. Chem. **2** [1963] 232/3, **4** [1965] 602/3). – [18] A. Ekman (Suomen Kemistilehti B **35** [1962] 166/8; C.A. **58** [1963] 971). – [19] J. P. Scharff, R. Genin (Anal. Chim. Acta **78** [1975] 211/8, 213/5).

16.1.25.7 With N-(p-Tolyl)-8-hydroxy-5-quinolinesulfonamide and Its Derivatives (= H_2L)

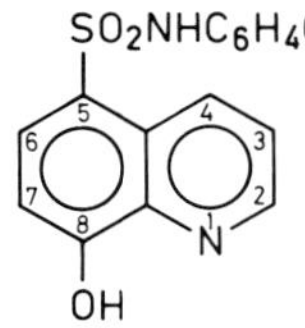

ligand I (= $C_{16}H_{14}N_2O_3S$)

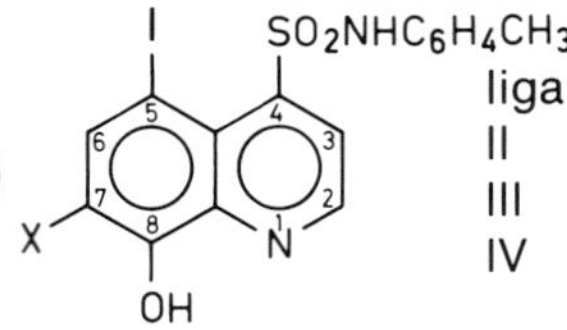

ligand	X	formula
II	H	$C_{16}H_{13}IN_2O_3S$
III	Cl	$C_{16}H_{12}ClIN_2O_3S$
IV	I	$C_{16}H_{12}I_2N_2O_3S$

The complexes of general composition MnL_2 were prepared by adding the appropriate amount of $MnCl_2$ in aqueous solution to aqueous solutions of the ligands I to IV at pH 8 to 9. The resulting precipitates were collected, washed with warm water, and dried at 120°C. $Mn(C_{16}H_{12}N_2O_3S)$ is gray, $Mn(C_{16}H_{11}IN_2O_3S)$ and $Mn(C_{16}H_{10}ClIN_2O_3S)$ are brown, and $Mn(C_{16}H_{10}I_2N_2O_3S)$ is grayish brown to blackish brown. The complexes melt above 300°C. They are insoluble in common organic solvents, but soluble in 50% acetic acid. Intramolecular hydrogen-bonded chelate structures have been proposed on the basis of the IR spectra, which show an absorption band at 3350 cm^{-1}, possibly due to hydrogen bonding. The compounds are bactericides, G. D. Tenari, M. N. Mishra, R. K. Mishra (Sci. Cult. **46** [1980] 331/3).

16.1.25.8 With 8-Vinyloxyquinoline

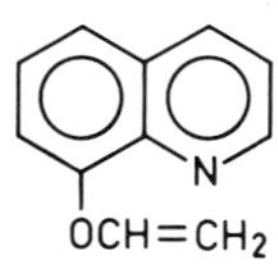

(= $C_{11}H_9NO$)

$Mn(C_{11}H_9NO)_2Cl_2$ was prepared by reacting $MnCl_2$ and the ligand in ethanol or ether. It is light yellow and melts at 190°C. The IR spectrum (recorded for KBr pellets in the 3700 to 400 cm^{-1} region) shows medium to strong absorption bands at 1642 $\nu(CH{=}CH_2)$, 1600 and 1512 ν(ring), 1400 and 1380 ν(skeleton), 890 and 876 cm^{-1} δ(CH). The $\nu(CH{=}CH_2)$ band remains unchanged upon coordination, while the bands of the stretching vibrations of the heterocyclic ring and the δ(CH) band are shifted by +10 to +35 cm^{-1} relative to the free ligand, but no ultimate interpretation was made. The complex dissociates completely in ethanol as shown by the UV spectrum, N. N. Chipanina, D. G. Kim, M. A. Andriyankov, D. D. Taryashinova, G. G. Skvortsova (Zh. Obshch. Khim. **46** [1976] 1118/23; J. Gen. Chem. [USSR] **46** [1976] 1108/11).

16.1.25.9 With 8-Quinolylamine

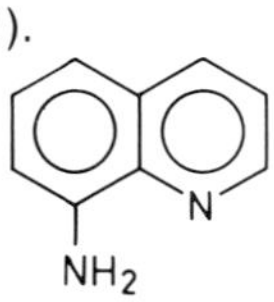

(= $C_9H_8N_2$)

$Mn(C_9H_8N_2)_n(ClO_4)_2$ (n = 3 and 4). $Mn(C_9H_8N_2)_4(ClO_4)_2$ was prepared by mixing concentrated ethanolic solutions of $Mn(ClO_4)_2$ and the ligand in a 1:5 mole ratio. $Mn(C_9H_8N_2)_3(ClO_4)_2$ was obtained by dissolving $Mn(C_9H_8N_2)_4(ClO_4)_2$ in acetone or nitromethane, adding a drop of $HClO_4$, and slowly evaporating the solution. Both compounds are colorless and have the magnetic moment $\mu_{eff} = 6.0\ \mu_B$.

The IR spectra (in Nujol or KBr) exhibit the following characteristic absorption bands:

complex (L = $C_9H_8N_2$)	ν(NH)	heterocyclic ring vibrations	ν(CN)
$MnL_4(ClO_4)_2$	3335	1611, 1570	1250
$MnL_3(ClO_4)_2$	3350	1610, 1570	1250
free ligand	3390	1591, 1570	1273

Bands in the perchlorate-band region, a broad band at about 1100 cm^{-1} with a smaller peak at about 950 cm^{-1}, show that both complexes contain ionic ClO_4^-. This is supported by the molar electrical conductivity of the tris complex, $\Lambda = 160$ to 180 $\Omega^{-1} \cdot cm^2 \cdot mol^{-1}$ in nitromethane, indicating the tris complex to be a 1:2 electrolyte. The existence of the tetrakis complex in nitromethane solutions is doubtful as only the tris complex could be isolated from such solutions. The shifts of the heterocyclic ring, ν(NH), and ν(CN) bands of the ligand suggest its bidentate coordination to Mn through the N atom of the heterocyclic residue and the N atom of the NH_2 group in both compounds. Thus the manganese atom is 6-coordinate in $MnL_3(ClO_4)_2$, and a structure with an 8-coordinate central atom is proposed for $MnL_4(ClO_4)_2$ [1].

$Mn(C_9H_8N_2)_2X_2$ (X = Cl, Br, I, NCS). The complexes with X = Cl or Br were prepared by reacting concentrated ethanolic solutions of MnX_2 and the ligand, mole ratio 1:5 [1]. $Mn(C_9H_8N_2)I_2$ was obtained by dissolving 0.179 g manganese(II) nitrate, 0.498 g KI, and 0.288 g ligand in 75 mL acetone and adding a minimum amount of water required to achieve dissolution of all solids. The mixture was refluxed for 2 h and evaporated almost to dryness. $Mn(C_9H_8N_2)_2(NCS)_2$ was prepared by reacting an aqueous solution of a manganese(II) salt with the stoichiometric amount of the ligand in ethanol and a slight excess of a saturated solution of NH_4SCN. The mixture was heated for 1 h [2]. $Mn(C_9H_8N_2)_2Cl_2$ and $Mn(C_9H_8N_2)_2Br_2$ are yellow [1]; $Mn(C_9H_8N_2)_2I_2$ is yellowish green [2]. $Mn(C_9H_8N_2)_2Br_2$ has the magnetic moment $\mu_{eff} = 6.0\ \mu_B$ [1].

The IR spectra of the bromide, iodide, and isothiocyanate (in Nujol or KBr) exhibit the following characteristic absorption bands (ν = stretching, δ = deformation, ω = wagging mode):

X	ν(NH)	ν(C-C) + ν(C-N) pyridyl ring	δ(NH_2)	ν(C-N)	ω(NH_2)	ν(Mn-NH_2)	Ref.
Br	3360	1611, 1570	—	1255	—	—	[1]
I	3270	1620, 1560	1590	*)	1010	495	[2]
NCS	3230	1620, 1570	1590, 1585	*)	1010	500	[2]

*) Not assigned.

Only the highest frequency bands are given for ν(NH). The ν_{as}(NH) and ν_s(NH) bands of the free ligand are observed at 3340 and 3440 cm^{-1}, respectively. No further free ligand bands are given by [2]. The isothiocyanato complex also shows bands at 2075, 2057, 830, and 490 cm^{-1}, which are assigned to the vibrations of the NCS group. For further IR bands of the complexes see the tables in [2].

$Mn(C_9H_8N_2)_2Br_2$ is nonconducting in nitromethane. The shifts of the pertinent IR bands (free ligand bands are given on p. 186) and the conductivity show $Mn(C_9H_8N_2)_2Br_2$ to be an octahedral complex containing bidentate 8-aminoquinoline and coordinated Br^- ions [1]. $Mn(C_9H_8N_2)_2I_2$ consists probably of a tetrahedral complex cation $[MnL_2]^{2+}$ of symmetry T_d and two noncoordinated I^- anions since the molar electrical conductivity is $\Lambda = 141.5\ \Omega^{-1} \cdot cm^{-2} \cdot mol^{-1}$ for a 0.001M solution in dimethylformamide at 25°C. The IR spectrum suggests that $Mn(C_9H_8N_2)_2(NCS)_2$ has an octahedral structure of ideal symmetry O_h with the thiocyanate groups coordinated to Mn via the N atoms. A cis configuration is concluded from a twofold splitting of the ν(CN) band in the 2050 to 2080 region.

The isothiocyanato complex is quite insoluble in the ordinary solvents. In dimethyl sulfoxide and dimethylformamide it dissolves with decomposition. $Mn(C_9H_8N_2)_2I_2$ is fairly soluble in acetone but more soluble in dimethylformamide, where no decomposition is noted [2].

$Mn(C_9H_8N_2)Cl_2$ was prepared by refluxing equimolar quantities of manganese(II) chloride and the ligand in ethanol for $^1/_2$ h. The crystals are pale yellow. The magnetic moment, $\mu_{eff}=5.90$ μ_B at 305 K, indicates an octahedral high-spin complex. The compound is probably polymeric since it has a high melting point and is only slightly soluble in organic solvents. It is stable and does not decompose or oxidize even on heating at 120°C for several hours [3].

References:

[1] B. Chiswell, E. J. O'Reilly (Inorg. Chim. Acta **7** [1973] 707/12, 710). – [2] J. Casabo, J. Ribas, J. Bartroli (Rev. Chem. Minerale **13** [1976] 149/56). – [3] A. Syamal (J. Indian Chem. Soc. **44** [1967] 989/90).

16.1.25.10 With 2-Pyridylquinoline and Related Compounds

16.1.25.10.1 With 2-(2-Pyridyl)quinoline

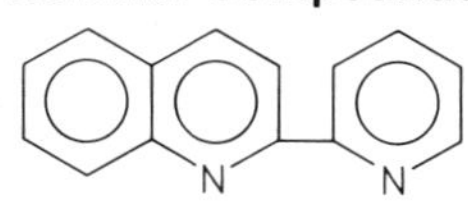

$(=C_{14}H_{10}N_2)$

$Mn(C_{14}H_{10}N_2)(H_2O)Br_2$. A solution of $MnBr_2 \cdot 4H_2O$ in methanol was added to the equimolar amount of the ligand dissolved in acetone. The precipitate was recrystallized from nitrobenzene. According to X-ray studies the resulting small yellow crystals are monoclinic, space group $P2_1/c-C^5_{2h}$ (No. 14) with lattice constants a = 8.172(3), b = 11.868(4), c = 16.456(5) Å, and β = 98.83(4)°; Z = 4. The unit cell volume is V = 1537 Å^3. The structure has been determined by the heavy atom technique (three-dimensional Patterson functions) and refined by the method of least squares to a reliability R = 8.0%. It consists of well isolated, distorted trigonal bipyramidal molecules containing 5-coordinate Mn^{II} as the central atom, which is surrounded by two Br, one O (from H_2O), and two N atoms of the bidentate ligand. The bond lengths are shown in **Fig. 16.** A very weak intermolecular hydrogen bond may exist between the water hydrogens and Br atoms of neighboring molecules. The apices of the pentagonal bipyramid are occupied by O of H_2O and N(2) of the ligand. The axis O-Mn-N(2) is bent away from linearity such as to minimize the electronic interaction of O with Br(1) and Br(2). A dimerization of the

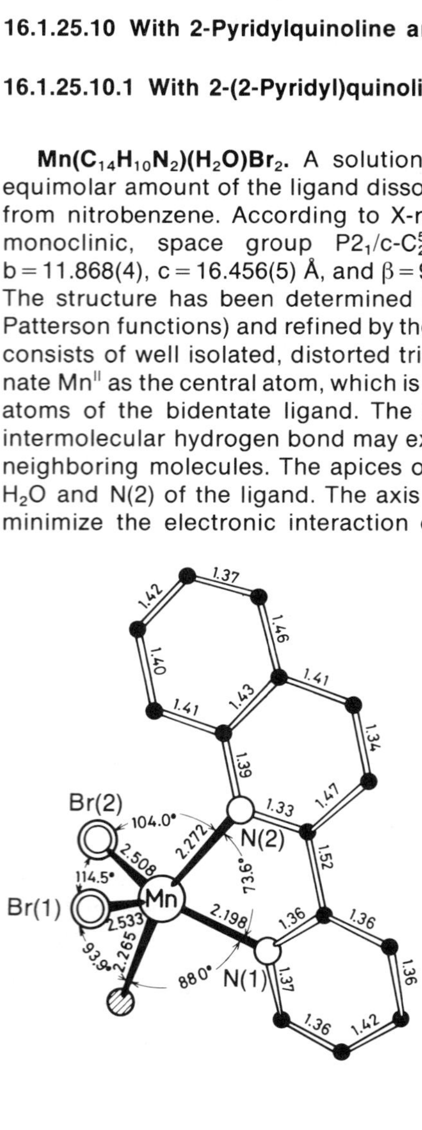

Fig. 16

Molecular structure of $Mn(C_{14}H_{10}N_2)(H_2O)Br_2$ with bond lengths (in Å) and some bonding angles around the Mn atom.

◎ Br

○ N

⊘ O(H_2O)

● C

complex via bromide bridges, similar to the biquinolyl complex $Mn_2(C_{18}H_{12}N_2)_2Cl_4$, is prevented by the firmly bound H_2O molecule. Measured density D = 1.87 g/cm³, calculated density from X-ray data D = 1.88 g/cm³, R. J. Butcher, E. Sinn (J. Chem. Soc. Dalton Trans. **1976** 1186/9).

16.1.25.10.2 With 8-Hydroxy-2-(2-pyridyl)quinoline

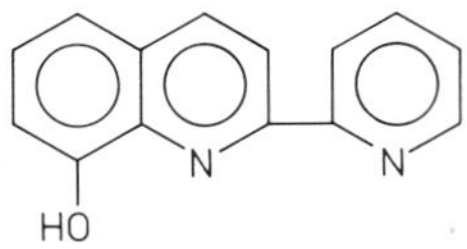

(= $C_{14}H_{10}N_2O$ = HL)

Complexes in Solution. Potentiometric titrations (glass electrode) of a manganese(II) salt and the ligand HL in aqueous dioxane (50 vol%) at 25°C for an ionic strength I = 0.1 M revealed the formation of the deprotonated complexes $[MnL]^+$ and $[MnL_2]$, with stability constants log K_1 = 8.44 and log β_2 = 16.50 (β in L^n/mol^n), respectively. It is assumed that 8-hydroxy-2-(2-pyridyl)quinoline is coordinated to Mn as a chelating tridentate ligand, A. Corsini, R. M. Cassidy (Talanta **26** [1979] 297/301, 299).

16.1.25.10.3 With 2,2′-(2,6-Pyridinediyl)bisquinoline

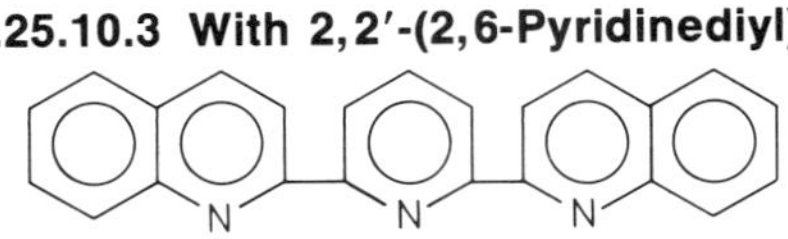

(= $C_{23}H_{15}N_3$)

$Mn(C_{23}H_{15}N_3)_2(ClO_4)_2$. A solution of $Mn(ClO_4)_2 \cdot 6H_2O$ in acetone was added to a solution of the ligand in hot benzene in a 1:2 mole ratio. The precipitate is decanted and treated with acetone. The resulting solution was filtered and concentrated in a vacuum rotating evaporator at room temperature until crystallization occurs. The yellow crystals were collected and dried in vacuo at 30°C [1].

The electronic spectra of the solid complex and its solution in acetonitrile show a group of five absorption bands in the 500 to 330 nm region. There are three shoulders on the very intense band in the 360 nm region, at 485, 450, and 420 nm. The spectrum is similar to those of other Mn^{II} complexes with an octahedral environment. The IR spectrum, which was recorded for Nujol and hexachlorobutadiene mulls in the 4000 to 625 cm^{-1} range, shows three absorption bands, at 1115, 1090, and 625 cm^{-1}, which are assigned to vibrations of the tetrahedral (T_d) ClO_4^- ion. Many other bands in the 1620 to 745 cm^{-1} region belong to ligand stretching, deformation, or skeletal vibrations. Some of them are strongly split upon complex formation (especially the 1505 cm^{-1} band, which shifts to 1510 and 1485 cm^{-1}) [1].

Conductivity measurements on solutions in acetonitrile (0.001 to 0.002 M) at 25°C yielded a molar electrical conductivity $\Lambda = 285.7\ \Omega^{-1} \cdot cm^2 \cdot mol^{-1}$, corresponding to a 1:2 electrolyte [1]. Polarographic analysis (dropping Hg cathode) on solutions in acetonitrile containing 0.05 M $(C_2H_5)_4NClO_4$ at 25°C revealed a five-step reduction wave at half-wave potentials $E_{1/2}$ = −0.875, −1.090, −1.650, −1.848, and −2.330 V (versus saturated calomel, SCE), agreeing fairly well with the results obtained by cyclic voltammetry. The first three steps correspond to diffusion-controlled reversible one-electron reactions leading to the species $[Mn(C_{23}H_{15}N_3)_2]^+$, $[Mn(C_{23}H_{15}N_3)_2]$, and $[Mn(C_{23}H_{15}N_3)_2]^-$, respectively. The last two steps are related to the discharge of the ligand liberated from the complex. The complex ions probably have a distorted octahedral structure (symmetry D_{2d}) similar to that of the complexes with terpyridyl, $[Mn(C_{15}H_{11}N_3)_2]^{n+}$, but the electrochemical behavior corresponds far more to that of $[Mn bpy_3]^{n+}$ (which also is reduced to $[Mn bpy_3]^-$) [2].

The air-stable complex decomposes in air at 270°C as shown by thermogravimetry (heating rate 2 K/min). It is quite soluble in acetone and acetonitrile, but only sparingly soluble in methanol and ethanol [1].

References:

[1] R. P. Bonomo, S. Musumeci, E. Rizzarelli, A. Seminara (Z. Anorg. Allgem. Chem. **414** [1975] 185/92, 186/8). – [2] R. P. Bonomo, S. Musumeci, E. Rizzarelli, S. Sammartano (J. Electroanal. Chem. Interfacial Electrochem. **56** [1974] 435/42).

16.1.25.11 With 2,2′-Biquinoline 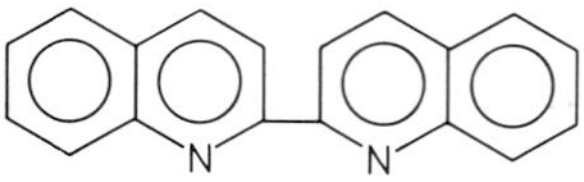 (= $C_{18}H_{12}N_2$)

A mixed-ligand complex with 2,2′-biquinoline and acetylacetone is described in "Mangan" D 1, 1979, p. 81.

$Mn_2(C_{18}H_{12}N_2)_2Cl_4$. An equimolar amount of the ligand dissolved in benzene was added to a methanolic solution of $MnCl_2$. The precipitate was recrystallized from nitrobenzene.

According to X-ray studies the resulting very small pale yellow crystals are triclinic, space group $P\bar{1}$-C_i^1 (No. 2), with lattice constants a = 11.275(3), b = 10.115(6), c = 14.681(8) Å, α = 95.71(7)°, β = 133.49(4)°, and γ = 118.50(4)°; Z = 1. The unit cell volume is V = 791 Å^3. The structure has been determined by the heavy atom technique and refined by the least squares method and introduction of anisotropic thermal parameters to R = 2.7%. It consists of well separated, neutral binuclear complex molecules with a pair of Cl atoms bridging the Mn atoms in a central Mn_2Cl_2 plane as shown in **Fig. 17.** The bridging Mn-Cl bonds are slightly elongated, and three stronger Mn-N and Mn-Cl bonds hold the bidentate ligands (via the N atoms) and the remaining Cl atoms to each 5-coordinate Mn atom. The bulky ligands are especially effective in isolating the Mn_2Cl_4 units and no significant intermolecular interactions likely exist. The observed distortion of the ligands corresponds to a marked curving of the ligand into a banana shape, with

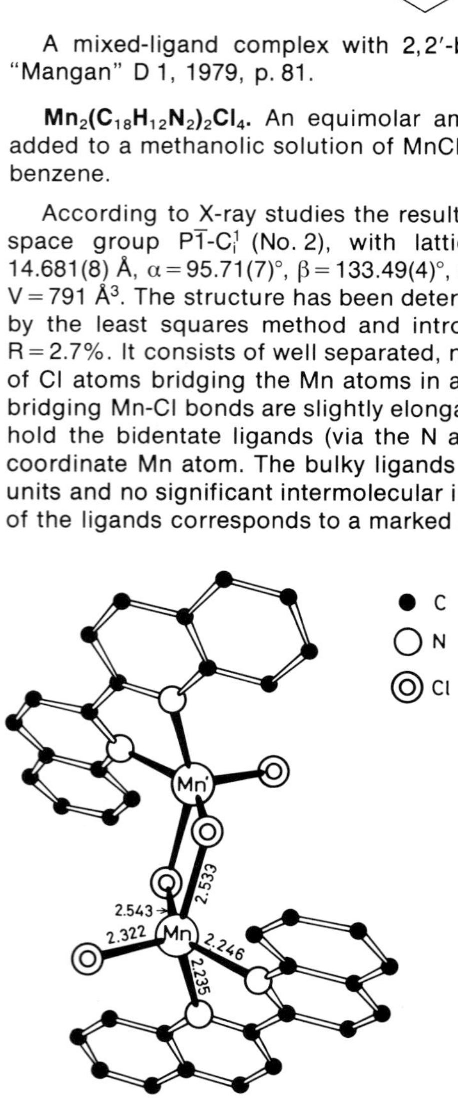

Fig. 17

Molecular structure of $Mn_2(C_{18}H_{12}N_2)_2Cl_4$ with bond lengths (in Å) involving the Mn atom.

curvatures in the quinolyl rings as well as in the central single bond, but there is no significant twisting of the planes of the aromatic rings around the single bond, probably due to the metal chelation. The Mn-Cl_2-Mn bridging plane forms an angle of 80° with the ligand. The bridging Mn-Cl bonds apparently are sufficiently strong to permit significant magnetic superexchange interaction via the ligand bridges. A direct Mn-Mn interaction seems to be ruled out by the large distance (3.887 Å) between the Mn atoms. Measured density D = 1.58 g/cm³, calculated density D = 1.61 g/cm³ (from X-ray results). The compound is slightly air sensitive, E. Sinn (J. Chem. Soc. Dalton Trans. **1976** 162/5).

16.1.25.12 With 8,8′-Dihydroxy-5,5′-biquinoline

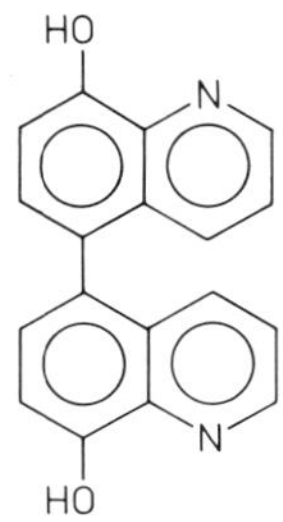

(= $C_{18}H_{12}N_2O_2$)

$(MnC_{18}H_{10}N_2O_2 \cdot H_2O)_n$. The polymeric compound was prepared by refluxing equimolar amounts of manganese(II) acetate and ligand in dimethylformamide at 130 to 140°C under N_2 for 5 to 6 h. The hot reaction mixture is then filtered and the precipitate is treated with boiling water and alcohol, then washed with diethyl ether, and dried in vacuo over boiling acetone for 12 h. The yellow to dull green powdery complex is crystalline as shown by its X-ray powder diagram. As the compound contains somewhat less Mn than required for a 1:1 complex, a polymerization coefficient n between 4 and 10 is assumed. The IR absorption spectrum shows a strong band at about 1125 cm^{-1} assigned to vibrations of the Mn-OC bond. The CN vibration of the quinoline ring is found in the 1565 to 1575 cm^{-1} region (free ligand 1580 to 1600 cm^{-1}). The band positions suggest chelation of the ligand. Upon heating the compound turns black between 250 and 300°C, losing 14 to 15% by weight in the air at 300°C (1 h). The compound is insoluble in common organic solvents, E. W. Berg, A. Alam (Anal. Chim. Acta **27** [1962] 454/9, 456).

16.1.25.13 With 5,5′-Methylenebis(8-hydroxyquinoline)

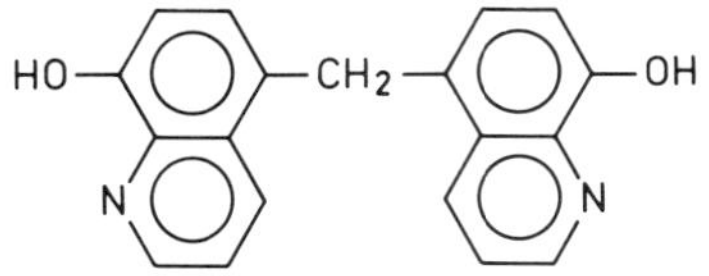

(= $C_{19}H_{14}N_2O_2$)

$[Mn(C_{19}H_{12}N_2O_2) \cdot x H_2O]_n$ (x = 1 and 0.5). The polymeric compound $[Mn(C_{19}H_{12}N_2O_2) \cdot H_2O]_n$ was prepared by heating stoichiometric amounts of manganese(II) acetylacetonate and the ligand in vacuo at 290°C [1]. In another procedure equimolar amounts of manganese(II) acetylacetonate and the ligand were reacted in dimethylformamide at 120 to 140°C for 4 to 6 h under a stream of N_2 [2].

The X-ray data for the dull green complex suggest a chain structure, and a true lattice formed by the manganese atoms may be considered [2]. Susceptibility measurements at 24, 83, and 160°C yield the specific magnetic susceptibility $10^6\ \chi_g$ = 40.0, 33.2, and 27.4 cm³/g, respectively, corresponding to a temperature-independent magnetic moment of $\mu_{eff} = 5.8 \pm 0.1\ \mu_B$ and a paramagnetic Curie temperature of $\Theta_p = 0 \pm 1$ K. Thus a tetrahedral configuration can be discussed [3].

In the IR spectrum, a free-ligand band occurring at about 3340 cm^{-1}, the region of the OH vibration frequency, does not appear in the spectrum of the complex. This is an indication of coordination. A band at 1573 to 1575 cm^{-1} (1580 cm^{-1} for the free ligand) is ascribed to the CN vibrations of the quinoline ring. An absorption band at 1089 cm^{-1} may be assigned to the CO stretching vibrations at the CO-Mn site. The data suggest a polymeric complex with a quadridentate ligand [1].

The thermal degradation of the complex occurs gradually at a heating rate of 2.5°C/min in vacuo, as shown by thermogravimetric analysis. A definite acceleration of the decomposition is noted above ~400°C, and manganese oxide is produced after 8 h of pyrolysis at 1000°C (weight loss 83.0%) [1]. On heating in a capillary tube, darkening of the complex is observed at 290°C [2]. On isothermic heating at 250°C in a vacuum of 10^{-5} Torr, no weight loss was observed [5]. The principal mechanism of the degradation seems to involve an initial rupturing of the metal-to-ligand bonds [1, 4]. The complex is infusible. It is not soluble in water, ethanol, methanol, ether [2], dimethylformamide [1], or the other usual solvents. It is not soluble in diluted acetic acid, sulfuric acids, or alkalies, but it dissolves in glacial acetic acid, formic acid, or, with apparent decomposition, in concentrated sulfuric acid [2]. The catalytic activity of the complex (which was formulated with one molecule of H_2O) in the decomposition of N_2O (to $2N_2+O_2$) at 270°C was studied by [5].

References:

[1] E. Horowitz, T. P. Perros (J. Inorg. Nucl. Chem. **26** [1964] 139/59, 141/2, 151, 155, 158). – [2] V. V. Korshak, S. V. Vinogradova, T. M. Babchinitser (Vysokomol. Soedin. **2** [1960] 498/507; Polymer Sci. [USSR] **2** [1961] 344/54, 346, 350, 354). – [3] V. V. Korshak, A. A. Slinkin, S. V. Vinogradova, T. M. Babchinitser (Vysokomol. Soedin. **3** [1961] 1624/32, 1630; C.A. **56** [1962] 14457). – [4] E. Horowitz (Mod. Plastics **45** No. 3 [1967] 146/8).

16.1.26 With Quinoline N-Oxide and Its Derivatives

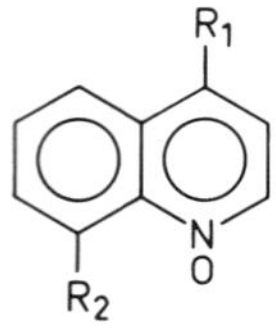

No.	R_1	R_2	formula(L)
1	H	H	C_9H_7NO
2	Cl	H	C_9H_6ClNO
3	CH_3	H	$C_{10}H_9NO$
4	OCH_3	H	$C_{10}H_9NO_2$
5	H	OH	$C_9H_7NO_2$

16.1.26.1 Complexes in Solution

$MnCl_2$ and ligand 1 form 1:1 and 1:2 complexes in ethanol, but only a 1:1 complex in isopentyl alcohol and a 1:2 complex in dimethylformamide. The following instability constants K_{inst} were determined from spectrophotometric data by graphical methods (no frequencies or graphs given): $K_{inst}=3.60$ (ethyl alcohol), 4.1 (isopentyl alcohol) for the 1:1 complex and $K_{inst}=0.4$ or 2.3 (ethyl alcohol), 0.045 (dimethylformamide) for the 1:2 complex. The two different values were found by two graphical methods. The two values for the other complexes agreed, T. T. Tuichiboev (Nauchn. Tr. Tashkentsk. Gos. Univ. No. 379 [1970] 148/51; C.A. **76** [1972] No. 118034; C.A. **81** [1974] No. 98754).

16.1.26.2 Isolated Manganese(II) Compounds

$MnL_6(ClO_4)_2 \cdot nH_2O$ (L = ligand 1 to 4 given in the table on p. 192, n = 0, 2, 7). The complexes were prepared by adding an ethanolic solution of the required ligand to a solution of $Mn(ClO_4)_2 \cdot 6H_2O$ in a mixture of ethanol and triethyl orthoformate or 2,2'-dimethoxypropane which had been stirred for 15 min to completely dehydrate the metal salt. A 7:1 ligand-to-Mn salt mole ratio was used in each case. The resulting precipitates were collected, washed with anhydrous ether, and dried in vacuo over P_2O_5. The dried complexes are bright orange to deep red [1]; for L = C_9H_7NO also see [5]. In the following two tables are listed the decomposition points (dec.), the magnetic moments μ_{eff} in μ_B, molar conductivities Λ (in $\Omega^{-1} \cdot cm^2 \cdot mol^{-1}$) of 0.002M solutions in nitromethane, principal IR bands (in cm^{-1}), and electronic absorption maxima (in Nujol, cm^{-1}) together with the transitions from the ground state $^6A_{1g}(S)$ to some excited states, and the ligand field parameters Dq (in cm^{-1}) and β:

ligand	R_1	n*)	dec. at °C	μ_{eff}	Λ	ν(NO)	ν(Mn-ON)	Ref.
1	H	0	174 to 182	6.12	177	1228	338	[1]
1	H	0	—	5.86	297**)	~1230	—	[5]
2	Cl	0	169 to 170	6.10	180	1302, 1256	365	[1]
3	CH_3	2	193 to 200	6.28	172	1212	387	[1]
4	OCH_3	7	200 to 201	6.00	179	1289	414, 389	[1]

*) The water content of the complexes is not considered in the physical data [1]. – **) In acetonitrile at 30°C, Λ = 59.6 in nitrobenzene at 30°C [5].

No.	$\rightarrow {}^4T_{1g}(G)$	$\rightarrow {}^4T_{2g}(G)$	$\rightarrow {}^4E_g$, ${}^4A_{1g}(G)$	$\rightarrow {}^4T_{2g}(D)$	$\rightarrow {}^4E_g(D)$	Dq	β	Ref.
1	18200	22000	24400	~27600	~29200	703	0.77	[1]
1	18200	20000	24700	27800	29400	—	—	[5]
2	18900	—	25200	~27500	~29200	690	0.78	[1]
3	18200	22000	25100	~28100	~29200	747	0.78	[1]
4	18200	21800	—	~28200	~29400	710	0.76	[1]

Unassigned bands in the 19700 to 20800 cm^{-1} region are possibly due to a metal-to-ligand charge transfer, which is thought to cause also the intensive color of the complexes. The magnetic moments and molar conductivities indicate that all of the compounds are high-spin octahedral complexes and 1:2 electrolytes in nitromethane [1] or acetonitrile [5]. A metal-to-ligand π-bonding effect for the 4-substituted complexes was concluded from the IR spectra [1]. The ClO_4^- ions show the T_d symmetry of non-coordinated perchlorate as they give sharp, unsplit ν_3 and ν_4 bands [5].

In the dark the complexes are stable in air. By irradiation with light of a mercury lamp, solid samples were decomposed, even under N_2, to give darkened or black decomposition products. Chemical and spectrophotometric studies of solutions of the decomposition products showed that the manganese(II) had been oxidized partially (10%) to manganese(III). The wavelengths which cause decomposition lie in the charge-transfer region at 500 nm. The decomposition exhibits zero-order kinetics during the first few hours of irradiation [2].

$[Mn(C_9H_7NO)_6]MnBr_4$ and **$Mn(C_9H_7NO)_2X_2$** (X = Cl, I). The complexes were obtained on mixing solutions of MnX_2 (X = Cl, Br, I) and ligand 1 in ethanol, 2-propanol, or 1-butanol in a 1:6 mole ratio (with a slight excess of ligand) followed by partial evaporation of the solvent under reduced pressure and cooling if required. The crystallization of the brown bromo and iodo complexes was induced by addition of $HC(OC_2H_5)_3$ and rapid stirring. Magnetic measurements yield $\mu_{eff}=5.85$ μ_B for the red-brown chloro complex, $\mu_{eff}=5.92$ μ_B for the bromo complex, and $\mu_{eff}=5.8$ μ_B for the iodo complex. Below are listed the molar electrical conductivities Λ of solutions in acetonitrile and nitrobenzene at 30°C, absorption bands in the electronic spectrum with the assigned excited states (ground state $^6A_{1g}(S)$), and probable symmetries of the complexes:

complex $L=C_9H_7NO$	Λ in CH_3CN	Λ in $C_6H_5NO_2$	assignment, $^6A_{1g}(S)$ to $^4T_{1g}(G)$	$^4T_{2g}(G)$	^4E_g, $^4A_{1g}(G)$	probable symmetry
$[MnL_6]MnBr_4$	117.7	27.64	515(sh)	450	316	cation octahedral, anion tetrahedral
MnL_2Cl_2	—	0.4	630(b)	540(sh)	490	pseudo-tetrahedral
MnL_2I_2	165.8	—	570(b)	360	—	tetrahedral (T_d)

Additional bands are observed for the chloro compound at 470 and 360 nm, corresponding to transitions to $^4T_{2g}(D)$ and $^4E_g(D)$. $Mn(C_9H_7NO)_2Cl_2$ behaves as a nonelectrolyte in acetonitrile, while $[Mn(C_9H_7NO)_6][MnBr_4]$ and $Mn(C_9H_7NO)_2I_2$ are 1:1 electrolytes in acetonitrile, the latter apparently due to a solvolysis equilibrium: $Mn(C_9H_7NO)_2I_2+CH_3CN \rightleftharpoons [Mn(C_9H_7NO)_2I\,CH_3CN]+I^-$. The bromo complex may be formulated as $[Mn(C_9H_7NO)_6]^{2+}[MnBr_4]^{2-}$ on the basis of its molar conductivity and molecular weight in nitrobenzene [5].

$Mn(C_9H_7NO)_2(NCS)_2$ and **$Mn(C_9H_7NO)_3(NCSe)$** were prepared by mixing ethanolic solutions of $Mn(NCS)_2$ or $Mn(NCSe)_2$ and ligand 1 in a 1:6 mole ratio. $Mn(C_9H_7NO)_2(NCS)_2$ exists in two forms. Form I was obtained by slow crystallization (3 d); form II, by rapid removal of the ethanol on a rotatory evaporator [3].

$Mn(C_9H_7NO)_3(NCSe)_2$ is isomorphous with the corresponding complex of Co as shown by X-ray powder diagrams. The electronic spectrum of $Mn(C_9H_7NO)_2(NCS)_2$ shows four absorption bands, at 18800(sh), 20200, 21500, and 22600 (form I) or 22400 (form II) cm^{-1}, while the spectrum of $Mn(C_9H_7NO)_3(NCSe)_2$ shows bands at 18600, 20200(sh), and 22400 cm^{-1}. The maxima in the 22500 cm^{-1} region may be associated with charge-transfer absorption which also causes the orange color of the compounds.

The following table lists the melting points and characteristic absorption bands in the IR spectrum of the complexes recorded for mineral oil mulls in the 4000 to 200 cm^{-1} region:

complex		m.p. in °C	ν(CN)	ν(Mn-O)	ν(Mn-N)
$Mn(C_9H_7NO)_2(NCS)_2$	I	222	2100, 2055	365, 330	260
	II	131	— 2045	328	—
$Mn(C_9H_7NO)_3(NCSe)_2$		175	— 2062	345	245

ν(NO) bands were not observed. For form I of $Mn(C_9H_7NO)_2(NCS)_2$, a tetrahedral configuration may be deduced from the position of the ν(Mn-N) band. Form II probably has an ionic structure as no far-IR band could be assigned to a ν(Mn-N) vibration and the single ν(CN) band at 2045 cm^{-1} is typical of free SCN^- ions. Thus it is suggested that the manganese atom is tetrahedrally coordinated by bridging quinoline N-oxide ligands. Both forms are soluble in nitromethane, yielding colorless solutions as a possible consequence of dissociation. The IR spectrum of $Mn(C_9H_7NO)_3(NCSe)_2$ is consistent with a complex structure containing the ionic species $[Mn(C_9H_7NO)_6]^{2+}$ and $[Mn(NCSe)_4]^{2-}$ [3].

$Mn(C_9H_6NO_2)_2$. The complex with 8-hydroxyquinoline N-oxide was obtained by mixing hot solutions of a manganese(II) salt and ligand 5, mole ratio 1:2, in aqueous ethanol (1:1) and adjusting to pH ~ 6.5 with dilute aqueous ammonia or saturated CH_3COONa solutions. The compound melts at 233°C with decomposition. Its magnetic moment is $\mu_{eff} = 3.95\ \mu_B$ at 300 K. The electronic spectrum shows a broad charge-transfer band at 350 nm [4].

The IR spectrum (Nujol mull) shows characteristic absorption bands at about 1290 and 1040 to 1020 cm^{-1}, which are assigned to stretching vibrations of coordinated NO and CO groups, respectively, of the deprotonated ligand. A band in the 460 to 440 cm^{-1} region is assigned to Mn-O stretching vibrations involving the oxygen of the NO group. The bidentate nature of the ligand, which is indicated by the IR spectrum, and the magnetic moment of the complex suggest a square-planar structure [4].

$[Mn(C_9H_7NO_2)_4(H_2O)_2](ClO_4)_2 \cdot 2H_2O$. Solutions of manganese(II) perchlorate in ethyl acetate (0.5 M) and of ligand 5 in chloroform (both dried over molecular sieve 4 A) were filtered and mixed in a 1:3 mole ratio. The resulting precipitate was collected, washed with chloroform and anhydrous ether, and stored in vacuo over anhydrous $CaSO_4$. The olive-yellow compound has a high-spin magnetic moment of $\mu_{eff} = 6.07\ \mu_B$ at 294 K (from susceptibility measurements corrected for diamagnetism). The IR absorption spectrum shows bands of coordinated H_2O at 3420 to 3300 cm^{-1} and of lattice H_2O in the 3600 to 3100 cm^{-1} region (extremely broad). The single bands at 1100 cm^{-1} ($\nu_3 ClO_4$) and 625 cm^{-1} ($\nu_4 ClO_4$) indicate the presence of ionic perchlorate, and three bands, at 1334, 1310, 1275 cm^{-1}, with shoulders at 1284 and 1260 cm^{-1}, are assigned to ν(NO) vibrations, including a δ(OH) contribution (sharp band at 1275 cm^{-1}). Metal-sensitive bands appear at 390 and 277 cm^{-1}; they are tentatively assigned to ν(Mn-O) vibration of coordinated H_2O and of the phenolic oxygen in the ligand, respectively, whereas the NO group apparently is noncoordinated. The molar electrical conductivity, $\Lambda = 160\ \Omega^{-1} \cdot cm^2 \cdot mol^{-1}$, of a 0.001 M solution in nitromethane at 25°C is that of a 1:2 electrolyte, thus confirming the proposed six-coordinate structure $[Mn(C_9H_7NO_2)_4(H_2O)_2](ClO_4)_2 \cdot 2H_2O$ [6].

References:

[1] L. C. Nathan, R. O. Ragsdale (Inorg. Chim. Acta **3** [1969] 473/7). – [2] L. C. Nathan, J. Cullen, R. O. Ragsdale (Inorg. Nucl. Chem. Letters **12** [1976] 137/40). – [3] G. B. Aitken, G. P. McQuillan (J. Chem. Soc. Dalton Trans. **1973** 2637/40). – [4] K. D. Ghuge, P. Umapathy, D. N. Sen (J. Indian Chem. Soc. **55** [1978] 864/8). – [5] K. Durairaj, S. J. Arulraj, C. C. Patel (Indian J. Chem. A **17** [1979] 465/8).

[6] F. J. Iaconianni, L. L. Pytlewski, C. M. Mikulski (Inorg. Chim. Acta **53** [1981] L21/L22).

16.1.26.3 Isolated Manganese(III) Compounds

$Mn(C_9H_7NO)_6X_3$ ($X = ClO_4$, I) and **$Mn(C_9H_7NO)_3X_3$** (X = Cl, Br). The complexes were prepared from manganese(III) acetate and quinoline N-oxide in the presence of ammonium halogenide or perchlorate in 1-butanol. Their physical data (color, magnetic moment μ_{eff} in μ_B, molecular conductance Λ in $\Omega^{-1} \cdot cm^2 \cdot mol^{-1}$ at 30°C, and electronic spectra) are listed below. They all have distorted octahedral structures, C_{3v} or D_{4h} symmetry, as suggested by their magnetic moments and electronic spectra. The chloro and bromo complexes are nonelectrolytes (coordinated halogen), whereas the iodide and perchlorate are 1:3 electrolytes with ionic I^- or ClO_4^-.

complex	color	μ_{eff}	Λ in CH_3CN	Λ in $C_6H_5NO_2$	electronic transitions (at λ in nm)			probable symmetry
MnL_3Cl_3	dark red	4.88	14.39	—	1210 $^5A_1 \rightarrow {}^5B_1$	1030 $^5B_2 \rightarrow {}^5B_1$	950 $^5E_2 \rightarrow {}^5B_1$	(C_{3v})
MnL_3Br_3	red brown	4.79	65.8	—	570 $^5A_{2g} \rightarrow {}^5B_{1g}$	465 $^5B_2 \rightarrow {}^5B_{1g}$	420 $^5E_g \rightarrow {}^5B_{1g}$	(D_{4h})
MnL_6I_3	dark green	4.82	377.2	71.1	690 $^5A_{1g} \rightarrow {}^5B_{1g}$	605 $^5B_{2g} \rightarrow {}^5B_{1g}$	500 $^5E_{1g} \rightarrow {}^5B_{1g}$	(D_{4h})
$MnL_6(ClO_4)_3$	dark red	4.95	365.4	—	700 $^5A_{1g} \rightarrow {}^5B_{1g}$	620 $^5B_{2g} \rightarrow {}^5B_{1g}$	510(sh) $^5E_{1g} \rightarrow {}^5B_{1g}$	(D_{4h})

Reference:

K. Durairaj, S. J. Arulraj, C. C. Patel (Indian J. Chem. A **17** [1979] 465/8).

16.1.27 With Isoquinoline 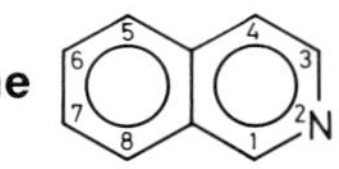 (= C_9H_7N = L)

Complexes with isoquinoline and diketones are described in "Mangan" D 1, 1979, pp. 70, 81, and 132.

Halides

The existence of $Mn(C_9H_7N)_2Cl_2$, $Mn(C_9H_7N)_5Cl_2$, and a solid of even higher isoquinoline content is demonstrated by the solubility isotherms of $MnCl_2$ in mixtures of isoquinoline and chlorobenzene at 27 and 36.8°C. $Mn(C_9H_7N)_2Cl_2$ appears for isoquinoline contents up to about 20 mol%, $Mn(C_9H_7N)_5Cl_2$ appears between 20 and about 75 mol%, and the isoquinoline-rich phase occurs above about 75 mol%. The electronic spectra show that the solutions contain mainly the octahedral complex $[Mn(C_9H_7N)_4Cl_2]$ while the tetrahedral $[Mn(C_9H_7N)_2Cl_2]$ is not present in appreciable amounts. The activity coefficient of the complex $[Mn(C_9H_7N)_4Cl_2]$ as a function of solvent composition was determined at 27 and 36.8°C. For the equilibrium $[Mn(C_9H_7N)_4Cl_2] \rightleftharpoons [Mn(C_9H_7N)_2Cl_2] + 2C_9H_7N$ the stability constants $K \approx 10^{-4}$ and $\approx 10^{-3}$ were determined at 27°C by spectrophotometric and solubility methods [1].

$Mn(C_9H_7N)_5Cl_2$. The yellow compound apparently was prepared by the action of excess isoquinoline on solid $Mn(C_9H_7N)_2Cl_2$ [1].

$Mn(C_9H_7N)_4I_2$ and **$Mn(C_9H_7N)_nX_2$** (n = 2, 1; X = Cl, Br). $Mn(C_9H_7N)_4I_2$ and the chloro and bromo complexes with n = 2 were prepared by reacting stoichiometric amounts of MnX_2 (X = Cl, Br, I) and isoquinoline in ethanolic solutions. The mono complexes were obtained by heating $Mn(C_9H_7N)_2X_2$ at 125 to 130°C for several hours to constant weight. With the exception of $Mn(C_9H_7N)_4I_2$, which is a pale yellow powder melting at 210°C, all the compounds are white amorphous solids melting above 280°C. The effective magnetic moments μ_{eff} in μ_B, the molar electrical conductivities Λ in $\Omega^{-1}\cdot cm^2\cdot mol^{-1}$ in acetone, and the solubilities in acetone are listed below ($L = C_9H_7N$):

complex	MnL_4I_2	MnL_2Cl_2	MnL_2Br_2	$MnLCl_2$	$MnLBr_2$
μ_{eff}	5.78	6.23	6.00	5.98	6.45
Λ	7.32	17.13	15.79	—	—
solubility in acetone	soluble	sparingly soluble	sparingly soluble	insoluble	insoluble

The low solubilities in organic solvents and the high melting points of $Mn(C_9H_7N)_2X_2$ and $Mn(C_9H_7N)X_2$ (X = Cl, Br) suggest a polymeric 6-coordinate halogen-bridged structure. However, $Mn(C_9H_7N)_4I_2$ is low melting and fairly soluble in acetone (as a nonelectrolyte), such properties indicating a monomeric octahedral complex [2]. The structures proposed by [2] are supported for $Mn(C_9H_7N)_4I_2$ and $Mn(C_9H_7N)_2X_2$ by the X-band frequencies of the ESR spectra. The spectra show lines at $g_{eff} = 6$ and 2. The zero-field splitting parameters are D = 0.87 for X = I and D = 0.54 for X = Br, with $\lambda = 0.01$ for both complexes, indicating a strong tetragonal distortion from O_h symmetry [16].

$(C_5H_5NC_{16}H_{33})_2[Mn(C_9H_7N)_4Cl_4]$. The compound was obtained by cautious heating of $(C_5H_5NC_{16}H_{33})_2MnCl_4$ with excess isoquinoline until a clear solution results. The light yellow crystals appear on cooling. They melt at 68°C. The electronic spectrum and the IR spectrum show bands similar to those of the complexes with 3- and 4-methylpyridine. The effective magnetic moment is $\mu_{eff} = 5.56$ μ_B. The molar electrical conductivity, $\Lambda = 65.9$ $\Omega^{-1}\cdot cm^2\cdot mol^{-1}$, is that of a nonelectrolyte. These properties indicate a structure with an 8-coordinate manganese atom much like that assumed for the methylpyridine complexes. If the complex is heated at 100°C to constant weight, two molecules of isoquinoline split off to leave the bis(isoquinoline) complex [3].

$(C_5H_5NC_{16}H_{33})_2[Mn(C_9H_7N)_2Cl_4]$ is formed from the tetrakis complex as described above. Its molar electrical conductivity is $\Lambda = 61.2$ $\Omega^{-1}\cdot cm^2\cdot mol^{-1}$ in nitrobenzene [3].

Cyanides and Chalcogenoisocyanates

$Mn(C_9H_7N)_4(CN)_2$. A solution of $Mn(C_9H_7N)_4(CN)_2$ was prepared by reacting isoquinoline with a hot solution of $Mn(CN)_2$ in a primary solvent consisting of 2-amino-2-methyl-1-propanol, $(CH_3)_2C(NH_2)CH_2OH$, 2-amino-2-methyl-1-propanol hydrocyanide, $(CH_3)_2C(NH_3CN)CH_2OH$, and H_2O [4]. Precipitation of the complex also occurred on reacting the components in this solvent at room temperature [15].

$Mn(C_9H_7N)_4(CN)_2$ is capable of forming clathrates with aromatic hydrocarbons, and this capability is used for the separation of hydrocarbons [4, 15]. For example, a clathrate with mesitylene is obtained by contacting a hot 35% complex solution with mesitylene at a weight ratio 10:1 and adding water until the precipitation is complete [4].

$Mn(C_9H_7N)_4(NCX)_2$ (X = O, S, Se). $Mn(C_9H_7N)_4(NCO)_2$ was obtained by mixing aqueous solutions of $Mn(NO_3)_2\cdot 6H_2O$, KOCN, and isoquinoline in a 1:2:4 mole ratio [5]. $Mn(C_9H_7N)_4(NCS)_2$ was prepared similarly from $MnCl_2\cdot 4H_2O$, NH_4SCN, and isoquinoline, the last added in slight excess (1 to 2%) [6, 7] or from an ethanolic solution of $Mn(NCS)_2$ and isoquinoline in a stoichiometric ratio [2]. $Mn(C_9H_7N)_4(NCSe)_2$ was prepared by reacting an ethanolic solution of $Mn(NCSe)_2$ (obtained from $MnCl_2$ and KSeCN by removing the precipitated KCl) and isoquinoline. The precipitate formed immediately or after refluxing the mixture for 30 min and allowing it to stand for 6 or 7 d [11]. In the following table are listed the color, melting point (m.p.), IR absorption bands (in cm^{-1}), magnetic moments μ_{eff} in μ_B, and molar electrical conductivities Λ (in $\Omega^{-1}\cdot cm^2\cdot mol^{-1}$) in acetone of $Mn(C_9H_7N)_4(NCX)_2$:

X	color	m.p. in °C	ν(CN)	ν(CX)	μ_{eff}	Λ	Ref.
O	—	225	2200	1350	5.85	12.0*)	[5]
S	pale yellow	186	2100 [8]	860 [8]	5.89	21.78	[2]
Se	colorless	165	2065 to 2040	788 to 738	5.75	8.5	[11]

*) Measured in dimethylformamide [5].

The positions of the ν(NC) and ν(CX) bands in $Mn(C_9H_7N)_4(NCX)_2$ indicate terminal NCX groups coordinated to Mn through their N atoms [2, 8, 11]. The magnetic moments and molar conductivities are indicative of high-spin octahedral Mn^{II} complexes (five unpaired d electrons) which are nonelectrolytes in organic solvents [2, 5, 11] such as dimethylformamide [5] or acetone [2, 11].

The ESR spectrum (X-band frequencies) of the isothiocyanato complex shows only one line, at $g_{eff}=2$. The zero-field splitting parameter $D=0.040\ cm^{-1}$ was evaluated, indicating a tetragonal distortion from O_h symmetry [16].

As shown by TG and DTG analyses, the isothiocyanato complex starts decomposing at 100°C. It loses one CN group on heating to 180°C. An NCS^- ion is lost at 210°C and subsequently four isoquinoline molecules are lost to form MnS. The decomposition is complete at 390°C [9]. The complex is fairly soluble in organic solvents like acetone [2], chloroform, and benzene [7]; it is insoluble in water [6]. It selectively forms crystalline clathrates with certain aromatic hydrocarbons, for example, o-xylene and m-xylene, in preference to p-xylene and ethylbenzene, or o-cymene, in preference to m-cymene and p-cymene. This capability enables the separation of the hydrocarbon isomers [7], see also [6, 10]. Crystalline clathrates with o- and m-xylene are obtained by mixing a saturated solution of the complex in chloroform with the hydrocarbons and allowing the mixture to stand 15 min at room temperature. The other clathrates can also be prepared in this way [7].

The isoselenocyanate complex slowly turns red on prolonged exposure to air. It is soluble in common organic solvents [11].

Other Compounds

$Mn(C_9H_7N)_4Cr_2O_7$ was obtained by mixing aqueous solutions of isoquinoline (or its HCl salt), a manganese(II) salt, and dichromate (the last is added dropwise) and heating the mixture to about 80°C, as described by [12] for related complexes. The reddish-brown precipitate [13] was collected and dried at 100°C [12]. Its decomposition starts, as shown by TG studies, at 120°C and is complete at 215°C with a weight loss of 47.8% (49.2% calculated for a residue of $MnCr_2O_7$) [13].

$Mn(C_9H_7N)_4(\text{-}C_6H_4COO)_2$. The 2,2′-biphenyldicarboxylate was obtained as a colorless microcrystalline precipitate by heating a solution of $Mn(\text{-}C_6H_4COO)_2$ in molten isoquinoline in a closed glass vessel at about 80°C. Or, an aqueous solution of isoquinoline and $Mn(\text{-}C_6H_4COO)_2$ was refluxed for several days. The two layers which formed were separated, and the lower layer, containing the manganese salt in isoquinoline, was heated at 80°C. In the IR spectrum (for KBr tablets) the ν(CH) vibrations and the stretching vibrations of the isoquinoline core appear in the ranges 3060 and 3025 cm^{-1} (ligand 3095 to 3012 cm^{-1}) and 1500 to 1365 cm^{-1} (ligand 1500 to 1392 cm^{-1}), respectively. Strong bands of the carboxylate group are observed in the range 1610 to 1560 cm^{-1} (anion 1650 cm^{-1}), 1495 to 1365 cm^{-1} (anion 1450 cm^{-1}), and 1390 to 1260 cm^{-1} (anion 1410 cm^{-1}). Only these ranges are given for the complexes of several metals. The band shifts indicate coordination of the isoquinoline to the manganese atom. On heating, the complex decomposes at 250 to 300°C without melting [14].

References:

[1] I. Uruska, M. Szpakowska (J. Chem. Soc. Faraday Trans. I **72** [1976] 2381/9, 2382/5, 2388). – [2] K. C. Dash, D. V. Ramana Rao (Indian J. Chem. **5** [1967] 333/4). – [3] B. Pradhan, D. V. Ramana Rao (J. Indian Chem. Soc. **56** [1979] 455/7). – [4] W. D. Schaeffer, Union Oil Co. of California (U.S. 3197525 [1961/65] 1/22, 17; C.A. **63** [1965] 14754). – [5] A. K. Das, D. V. Ramana Rao (Current Sci. [India] **42** [1973] 56/7).

[6] W. D. Schaeffer, W. S. Dorsey, D. A. Skinner, C. G. Christian (J. Am. Chem. Soc. **79** [1957] 5870/6, 5875). – [7] W. D. Schaeffer, Union Oil Co. of California (U.S. 2876227 [1956/59]; C.A. **1959** 16166). – [8] K. C. Dash, D. V. Ramana Rao (Current Sci. [India] **35** [1966] 462). – [9] B. Lóránt, M. Bojtor (J. Therm. Anal. **5** [1973] 379/87, 384). – [10] W. D. Schaeffer, Union Oil Co. of California (U.S. 3049575 [1960/62] 1/18, 8; C.A. **59** [1963] 1530).

[11] A. K. Das, D. V. Ramana Rao (Indian J. Chem. **12** [1974] 898/9). – [12] B. Lóránt (Thermochim. Acta **6** [1973] 205/14, 206). – [13] B. Lóránt (Thermochim. Acta **9** [1974] 67/75, 68). – [14] A. Ara Blesa (Ion [Madrid] **36** [1976] 295/9, 297). – [15] W. D. Schaeffer, Union Oil Co. of California (U.S. 3161694 [1961/64] 1/24, 15; C.A. **62** [1965] 16116).

[16] R. D. Dowsing, J. F. Gibson, M. Goodgame, P. J. Hayward (J. Chem. Soc. A **1969** 187/93).

16.1.28 With 3-Methylisoquinoline

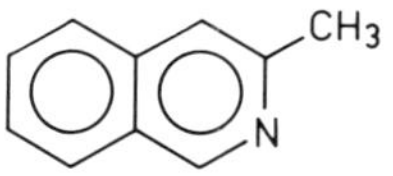

(= $C_{10}H_9N$)

$Mn(C_{10}H_9N)_2Br_2$ was prepared by mixing hot acetone solutions of manganese bromide and of excess ligand and allowing the mixture to cool. Bright yellow crystals precipitated. The ESR spectrum of the complex was reported at X-band and Q-band frequencies, together with that of the complex diluted in the analogous zinc complex. The main feature of the spectra is a strong line near $g_{eff}=4.3$ (7500 G). The distortion parameters $D=0.365$ cm^{-1} and $\lambda(=E/D)=0.267$ were deduced. The parameters indicate a distorted tetrahedral symmetry (C_{2v}), R. D. Dowsing, J. F. Gibson, D. M. L. Goodgame, M. Goodgame, P. J. Hayward (J. Chem. Soc. A **1969** 1242/8, 1245/8).

16.1.29 With 3-Methylisoquinoline N-Oxide (= $C_{10}H_9NO$)

$Mn(C_{10}H_9NO)_5(ClO_4)_2$. Solutions of $Mn(ClO_4)_2 \cdot 6H_2O$ (1.5 mmol) and the ligand (6.3 mmol) in a mixture of ethanol and triethyl orthoformate, volume ratio of 3:2, were warmed separately at 50°C for 30 min and combined by adding the ligand solution slowly to the manganese(II) perchlorate solution. The complex can be recrystallized from CH_2Cl_2 by adding n-hexane until the hot solution becomes slightly cloudy and cooling to 5°C overnight.

The yellow-orange compound has an effective magnetic moment of $\mu_{eff}=6.04$ μ_B at 295 K. Its electronic reflectance spectrum (Nujol mull) shows two strong maxima, at 338 (very broad) and 371 nm, probably due to Mn-to-ligand charge transfer. The IR spectrum shows seven medium-strong bands in the region of the ν(NO) vibrations between 1280 and 1146 cm^{-1} (free ligand between 1276 and 1169 cm^{-1}) and one band in the δ(NO) region at 835 to 840 cm^{-1} (free ligand at 830 cm^{-1}). The bands in the ν(NO) region are shifted partially to higher and partially to lower wavenumbers. The shifts of the NO bands indicate coordination via the oxygen atom of the NO group. Four strong bands, in the ranges 1150 to 1020 cm^{-1} and 640 to 610 cm^{-1}, are assigned to the ν_3 and ν_4 vibrations of the perchlorate group. The strong splitting of the ν_3 and ν_4 bands indicates the presence of the perchlorate group as both coordinated $OClO_3$ groups of symmetry C_{3v} and uncoordinated ClO_4^- ions of symmetry T_d. The medium-strong bands at 336 and 284 cm^{-1} are tentatively assigned to ν(Mn-O) vibrations of the coordinated ligand and $OClO_3$ groups, respectively. It is concluded from the physical data that the Mn complex has an octahedral configuration and may be formulated as $[Mn(C_{10}H_9NO)_5OClO_3][ClO_4]$. The molar electrical conductivity $\Lambda = 176\ \Omega^{-1} \cdot cm^2 \cdot mol^{-1}$ of a 0.001M solution in nitromethane measured at 25°C is that of a 1:2 electrolyte and indicates replacement of the coordinated perchlorate by nitromethane, A. N. Speca, L. S. Gelfand, F. J. Iaconianni, L. L. Pytlewski, C. M. Mikulski, N. M. Karayannis (Inorg. Chim. Acta **33** [1979] 195/203).

16.1.30 With Acridine N-Oxide

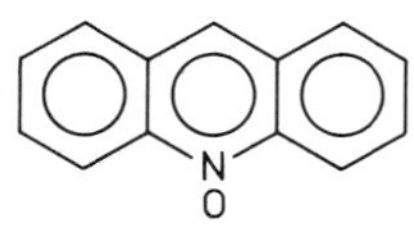

(= $C_{13}H_9NO$ = L)

$Mn(C_{13}H_9NO)_2X_2$ (X = NO_3, NCS) and **$Mn(C_{13}H_9NO)Cl_2$.** The acridine N-oxide complexes were prepared by boiling ethanolic solutions containing hydrated manganese(II) chloride, nitrate, or thiocyanate and the ligand in a 1:2 mole ratio for 2 to 3 min. The green complexes will crystallize upon standing for at least 1 h. Below are listed the melting points (m.p.) and principal absorption bands (in cm^{-1}) of the complexes in the IR spectrum (Nujol mull):

compound L = $C_{13}H_9NO$	m.p. in °C	ν(NO)		ν(Mn-O)	ν(Mn-X)	ligand mode	anion mode
$MnL_2(NO_3)_2$	188	1325, 1318	1080, 1060	—	320, 280sh, 270, 252	360b	1475, 1305, 1030, 815
$MnL_2(NCS)_2$	256	1327 —	1085, 1069	—	290, 270, 250	360	2060 ν(CN)
$MnLCl_2$	270	1326, 1319sh	1071 —	260	—	351	—

The ligand is apparently coordinated to the manganese atom through the oxygen atom of the ligand as suggested by the negative shift of the lower frequency ν(NO) band (at 1085 cm^{-1} in the free ligand). The relatively low metal-to-ligand frequencies suggest the complexes to be 6-coordinate with the acridine N-oxide ligands functioning as bridging elements. The nitrato and thiocyanato complexes contain coordinated monodentate anions, whereas in the chloro compounds both ligands and anions act as bridging elements, G. B. Aitken, G. P. McQuillan (Inorg. Chim. Acta **15** [1975] 221/5).

16.1.31 With 2,2'-Bipyridyl 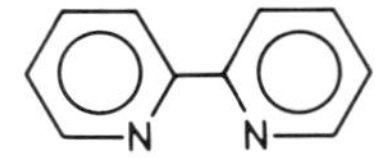 ($=C_{10}H_8N_2=$ bpy)

16.1.31.1 [Mnbpy$_3$] and Related Compounds

[Mnbpy$_3$] was obtained by reacting the manganese(II) complex Mnbpy$_3$Br$_2$, p. 206, with excess Li$_2$bpy in tetrahydrofuran [1] under N_2 [2]. The black-brown glittering needles which crystallized from the deep brown solution were washed and recrystallized from benzene [1]. On the basis of the electronic spectrum it is assumed that [Mnbpy$_3$] is an ionic complex consisting of Mn^{2+} cations and bpy$^-$ anions. The actual charge on Mn cannot be zero [3]. The IR spectrum (recorded for Nujol mulls in the 4000 to 100 cm^{-1} range) shows characteristic ligand absorption bands at 1600, 1573, 958, ~415, 354, 258, 218, and 188 cm^{-1}; the two bands at 258 (overlapped) and 227 cm^{-1} are assigned to Mn-N$_{bpy}$ stretching vibrations. The medium intensity bands at 1600 and 1573 cm^{-1} are assigned to the ν(CC) or ν(CN) vibrations of bpy. The strong band at 958 cm^{-1}, which appears only in low-valency complexes of manganese may indicate the negative charge on bpy in such complexes, which apparently is due to a transfer of electrons from Mn to bpy orbitals [2]. The nature of the bonding in [Mnbpy$_3$] was discussed qualitatively on the basis of LCAO-MO theory and the available magnetic results [5]. $\mu_{eff}=4.10\pm0.03$ μ_B at room temperature was reported in [1] Susceptibility measurements at temperatures between 413.4 and 81.2 K yield the following effective magnetic moments μ_{eff} in μ_B [4]:

T in K	413.4	293	81.2
μ_{eff}	4.31	4.09	3.81

A Curie temperature $\Theta_p=-282$ K was calculated from the slope of $1/\chi$ vs. T curve, which follows the Curie-Weiss law above 150 K. Below 150 K it obeys the Curie law, yielding a moment μ_{eff} corresponding to three unpaired electrons located on the molecular orbitals $(3e)^4(a_1)^1(e)^2$. Above 150 K, the magnetic moment rises clearly, apparently due to a decoupling of two other electrons which occupy closely adjacent orbitals, thus producing a five unpaired electron state. The ESR spectrum yields a factor g = 2.000, which is somewhat lower than a spin-only value, but the absorption lines are very broad ($\Delta H=240$ Oe) and may suggest a covalent Mn-to-ligand bond [4].

[Mnbpy$_3$] is soluble in ether, benzene, tetrahydrofuran, dioxane, and pyridine, forming reddish-brown solutions. It decomposes instantly on exposure to air, evolving heat. Attempted sublimation in a high vacuum at 180°C also caused decomposition. It is decomposed by water and alcohols and is oxidized by I_2 to yield the manganese(II) complex [1].

[Mnbpy$_3$]·2THF is obtained by crystallizing [Mnbpy$_3$] from tetrahydrofuran. It forms deep brown needles. Its effective magnetic moment $\mu_{eff}=4.1_3$ μ_B at 290 K (from susceptibility measurements between 1.6 and 320 K) is quite close to that of the unsolvated compound and follows the Curie-Weiss law between 20 and 160 K with $\Theta_p=-4$ K. A notable deviation of the susceptibility from this law was noted for temperatures above 160 K. The susceptibility of [Mnbpy$_3$]·$2C_4H_8O$ may be explained by a radical-ion model (similar to Li[Mnbpy$_3$]·$4C_4H_8O$) where two ·bpy$^-$ radical anions and a neutral bpy molecule are coordinated to an Mn^{2+} ion as bidentate ligands. The presence of antiferromagnetic interaction indicates a weak covalent bond between Mn and bpy, which can be elucidated by the molecular orbital theory [6, 9].

Li[Mnbpy$_3$]·THF was obtained by reacting a solution of [Mnbpy$_3$] in tetrahydrofuran with lithium metal [7] or by reacting a solution of Mnbpy$_3$Br$_2$ in THF with Li$_2$bpy [2, 7], excluding air [7] or under N_2 [2]. It crystallized from the solvent as deep red [6] or brownish-violet, almost black crystals, which were washed and recrystallized from tetrahydrofuran [7]. The IR spectrum (recorded for Nujol mull in the 4000 to 100 cm^{-1} range) of the compound described as $Li[Mn bpy_3]\cdot nTHF$ shows characteristic ligand absorption bands at 1565, 1535, 944, ~415, 351, 250, and 218 cm^{-1}. The two bands at 235 and 184 cm^{-1} are assigned to the $Mn\text{-}N_{bpy}$ stretching vibrations. The bands at 1565 and 1535 cm^{-1}, which are shifted to lower frequencies than in $[Mn bpy_3]$ are assigned to $\nu(CC)$ and $\nu(CN)$ vibrations of bpy. The strong band at 944 cm^{-1}, which appears only in low-valency complexes of Mn, may indicate a negative charge on bpy, as found in $[Mn bpy_3]$ [2]. The effective magnetic moment is $\mu_{eff}=3.71$ μ_B at room temperature [7]. Susceptibility measurements at temperatures ranging from 320 to 1.6 K yield an effective magnetic moment $\mu_{eff}=3.6_3$ μ_B at 290 K [6]. The magnetic susceptibility does not obey the Curie-Weiss law over the temperature range investigated. A broad maximum of the susceptibility at about 6 K (due to a slight splitting of the degenerate ground level) can be explained with a radical-ion model in which three bpy radical ions ($\cdot bpy^-$) are coordinated as bidentate ligands to an Mn^{2+} ion, forming a distorted octahedron represented by the formula $[Mn^{2+}(\cdot bpy^-)_3]^-$ [6, 9].

The compound is readily soluble in tetrahydrofuran and dioxane, moderately soluble in ether and pyridine, and slightly soluble in benzene. When stored under N_2 at room temperature, the solid slowly loses THF (0.5 mol after 50 days). At 150°C in a vacuum, THF is lost quantitatively and some bipyridyl is also formed. The solid compound and its solutions are very air sensitive. It is decomposed by water and alcohol, forming a white flocculent precipitate which turns brown at the air. Careful oxidation of the compound covered by tetrahydrofuran with air provides the neutral complex $[Mn bpy_3]$ as a first reaction step. In pyridine three equivalents of iodine are required to form a yellow solution containing Mn^{2+} [7].

$Na_4[Mn bpy_3]\cdot 5C_4H_8O_2\cdot 3THF$. By treating $[Mn bpy_3]$ with excess sodium in tetrahydrofuran in the presence of sodium ketyl, a THF adduct of the complex $Na_4[Mn bpy_3]$ formed. It crystallized after removal of unreacted Na upon cooling to −78°C. By addition of dioxane at room temperature, rod- or cube-shaped black or blackish-violet crystals of the dioxane adduct were obtained. Susceptibility measurements at room temperature yielded an average magnetic moment $\mu_{eff}=5.99\pm0.02$ μ_B, which indicates a high-spin complex consisting of Mn^{2+} and bpy^{2-} ions. It is assumed that the Mn^{2+} orbitals are higher than the antibonding π orbitals of bpy, which take up the additional electrons introduced by Na. The compound loses all the solvent molecules upon heating in vacuo. At temperatures above 170°C a progressive decomposition with loss of bpy is observed. The compound is extremely air sensitive and loses tetrahydrofuran and bpy, brightening in color. It is very soluble in tetrahydrofuran but is decomposed instantly by deaerated proton-active solvents such as water and alcohols [8].

References:

[1] S. Herzog, M. Schmidt (Z. Chem. [Leipzig] **2** [1962] 24). – [2] Y. Saito, J. Takemoto, B. Hutchinson, K. Nakamoto (Inorg. Chem. **11** [1972] 2003/11). – [3] Y. Kaizu, T. Yazaki, Y. Torii, H. Kobayashi (Bull. Chem. Soc. Japan **43** [1970] 2068/71). – [4] E. Wulf, S. Herzog (Z. Anorg. Allgem. Chem. **387** [1972] 81/90, 82, 89). – [5] R. Taube, S. Herzog, (Z. Chem. [Leipzig] **2** [1962] 225/37, 233/4).

[6] M. Inoue, K. Hara, T. Horiba, M. Kubo (Bull. Chem. Soc. Japan **47** [1974] 2137/40). – [7] S. Herzog, M. Schmidt (Z. Chem. [Leipzig] **3** [1963] 392/3). – [8] S. Herzog, U. Grimm (Z. Chem. [Leipzig] **7** [1967] 432). – [9] K. Hara, M. Inoue, T. Horiba, M. Kubo (Chem. Letters **1974** 419/22).

16.1.31.2 $[Mn bpy_n]^{2+}$ Ions. Formation and Properties in Solution

The formation constants K_n (in L/mol) and β_n (in L^n/mol^n) with n = 1, 2, 3 obtained under various experimental conditions are listed in the following table:

method	t in °C	medium	ionic strength I (or pH)	log K_1	log K_2	log β_2	log K_3	Ref.
catalysis*)	25	H_2O	(7.5 to 8.5)	2.59	2.03	4.62	—	[1]
kinetics	25	H_2O	0.3 ($NaClO_4$)	2.59 ±0.08	—	—	—	[2]
distribution	30.3	H_2O/C_6H_{14}	1.0 (KNO_3)	2.54	1.85	4.39	1.51	[3]
distribution	25	$H_2O/CHCl_3$, H_2O/C_6H_{14}	0.1 (KCl)	2.62	2.00	4.62	~1.1	[5]
glass electrode	20	H_2O	0.1 ($NaNO_3$)	2.6	—	—	—	[4]
glass electrode	10	H_2O	0.1 (KNO_3)	2.72 ±0.02	1.99	4.71 ±0.08	(1.87)	[6]
glass electrode	25	H_2O	—	2.48 ±0.03	2.06	4.54 ±0.03	(1.25)	[6]
glass electrode	45	H_2O	—	2.35 ±0.03	1.82	4.17 ±0.10	(1.59)	[6]
glass electrode	25	H_2O/C_2H_5OH	0.1 (KNO_3)	2.42 ±0.03	1.82	4.24 ±0.05	(1.76)	[6]
kinetics	25	CH_3OH	0.20 ($NaClO_4$)	2.7	—	—	—	[8]
**)	—	—	—	2.55	1.9	4.45	1.45	[9]

*) Catalysis of the oxidation of Tiron (Na salt of 4,5-dihydroxybenzene-1,3-disulfonic acid) by H_2O_2 [1]. – **) Recommended values from [3 to 5].

For older data obtained by spectrophotometry, see [10 to 12]. Calculation from ESR data yields log β_3 = 6.03 in aqueous medium at room temperature [7]. The potentiometric values of [13] are rejected by [9].

The thermodynamic formation quantities ΔH, ΔG (in kcal/mol), and ΔS (in $cal \cdot mol^{-1} \cdot K^{-1}$) of the complexes $[Mn bpy_n]^{2+}$ from the reaction $[Mn(H_2O)_6]^{2+}$ + n bpy → $[Mn(H_2O)_{6-2n}bpy_n]^{2+}$ + 2n H_2O have been determined as follows:

method	t in °C	ionic strength I	n	−ΔH	−ΔG	ΔS	Ref.
calorimetry	30.3	1.0 (KNO_3)	1	5.73	3.53	−7.3	[3]
			2	6.11	6.10	0	
			3	6.23	8.19	6.5	
calorimetry	20	0.1 ($NaNO_3$)	1	3.5	3.5	0.0	[14]
calorimetry	25	1.0 ($NaClO_4$)	1	4.3	—	—	[13]*)
			2	8.6	—	—	
			3	12.9	—	—	

method	t in °C	ionic strength I	n	−ΔH	−ΔG	ΔS	Ref.
temperature	25	0.1 (KNO_3)	1	4.18 ± 0.6	3.38 ± 0.04	−2.7 ± 2.2	[6]
slope of β	—	—	2	6.32 ± 0.98	6.19 ± 0.03	−0.4 ± 3.4	

*) For ΔG and ΔS from incorrect formation constants, see [13].

The data given in [3, 8, 13, 14] were converted to values of ΔH and ΔG in kJ/mol and ΔS in $J \cdot mol^{-1} \cdot K^{-1}$ by [9].

The thermodynamic quantities of the mono complex formed in methanol at 25°C at ionic strength I = 0.2 mol/L ($NaClO_4$) calculated from kinetic data are ΔH = −2 kcal/mol and ΔS = 6 $cal \cdot mol^{-1} \cdot K^{-1}$ [8].

Acidified aqueous manganese salt solutions were reacted at 30.3°C in a calorimeter with solutions of bpy (large excess) which had been half neutralized with the acid corresponding to the anion of the manganese salt (or with HCl in the case of $MnSO_4$) to form the tris(bipyridyl) complexes, $[Mn(H_2O)_6]^{2+} + 3[bpyH]^+ \rightarrow [Mnbpy_3]^{2+} + 3H^+ + 6H_2O$. The following enthalpies of formation ΔH (in kcal/mol) depend little on the anion of the manganese salt [3]:

Mn salt	$Mn(NO_3)_2$	$MnCl_2$	$Mn(ClO_4)_2$	$MnSO_4$	$MnSO_4$*)
acid	HNO_3	HCl	$HClO_4$	HCl	HCl
−ΔH	4.60**)	4.69	4.74	5.82	4.96

*) In 50% aqueous ethanol. – **) 5.61 kcal/mol at I = 1.0 (KNO_3).

The chelate effect can be demonstrated by the reaction $Mnpy_{2n} + n\,bpy \rightarrow Mnbpy_n + 2n\,py$ where py = pyridine and n = 1, 2. By calorimetric measurement at 25°C the following values were determined [7]:

n	ΔH in kcal/mol	ΔG in kcal/mol	ΔS in $cal \cdot mol^{-1} \cdot K^{-1}$
1	+0.48	−0.82	+5.1
2	+0.96	−3.94	+17.2

Kinetic investigations by the temperature-jump relaxation method (A) [2] and the stopped flow method (B) [8, 15] yield the following reaction rate constants k_f in $L \cdot mol^{-1} \cdot s^{-1}$ and k_d in s^{-1} and parameters of activation ΔH^* in kcal/mol and ΔS^* in $cal \cdot mol^{-1} \cdot K^{-1}$ at 298.1 K for the formation (f) and dissociation (d) of the 1:1 complex at ionic strength I in L/mol ($NaClO_4$):

method	solvent	I	temp. range	log k_f	ΔH_f^*	ΔS_f^*	log k_d	ΔH_d	ΔS_d	Ref.
(A)	H_2O	0.3	283.1 to 311.1 K	5.45	9.2 ±1.0	−3 ±3	2.86	12.0 ±1.3	−5 ±4	[2]
(B)	CH_3OH	0.2	226.9 to 251.8 K	4.24	11.7 ±0.1	0.1 ±0.1	1.57a)	13.4 ±0.5	−6.4 ±1.9	[8]

a) The dissociation reaction in methanol was induced by Hg^{2+} [8].

For the exchange of H_2O molecules in the inner hydration sphere by bpy, a first-order exchange rate constant log k_{ex} = 7.36 (k_{ex} in s^{-1}) and the exchange activation enthalpy

$\Delta H^{*}_{ex} = 7.8$ kcal/mol per molecule of H_2O were reported. ΔH^{*}_{ex} is close to ΔH^{*}_{f}. Ring closure (chelation) possibly is the rate-determining step in complex formation [2] as was suggested earlier by [8]. The kinetic data recorded in aqueous [2] and methanolic [8] solutions suggest that complex formation with bpy as a bidentate, chelating ligand occurs first by addition of bpy to form an intermediate outer sphere complex followed by stepwise expulsion of two solvent molecules from the hexasolvated Mn^{2+} ion [2, 8].

Comparison of the stability of $[Mn\,bpy_n]^{2+}$ with that of other bipyridyl complexes gives the stability series Cu>Ni>Co>Zn>Cd>Ag>Mn [14], also see [5, 11].

The electronic spectrum of $[Mn\,bpy]^{2+}$ recorded in aqueous perchlorate solution at 27±1°C shows three absorption maxima, 306.0, 296.0, and 245.0 nm, with molar extinction coefficients $\varepsilon = 14500$, 16000, and 10200 $L \cdot mol^{-1} \cdot cm^{-1}$, respectively. The maxima show a considerable bathochromic shift with respect to the free ligand, and the first two indicate a splitting of the long-wave band. Thus it is assumed that the excited state of 2,2′-bipyridyl has a greater complexing tendency than its ground state [11].

Polarographic reduction of $[Mn\,bpy_3]^{2+}$ at −1.5 V in an ethanolic 1M LiCl solution at the dropping mercury cathode forms $[Mn\,bpy_3]^{+}$ [18]. Formation of $[Mn\,bpy_3]^{+}$, with $[Mn\,bpy_3]$ and $[Mn\,bpy_3]^{-}$ as further reduction steps, is also observed on polarographic reduction of $Mn\,bpy_3(ClO_4)_2$ in acetonitrile solutions, see p. 207.

The complexes exhibit catalytic properties in the decomposition of H_2O_2, which attains its maximum rate at an Mn:bpy mole ratio of 1:8. It is assumed that an Mn^{II} complex forms as an intermediate [16] and that $[Mn\,bpy_2]^{2+}$ is the most active species [17].

References:

[1] Ya. D. Tiginyanu (Zh. Fiz. Khim. **52** [1978] 2131; Russ. J. Phys. Chem. **52** [1978] 1234; Izv. Akad. Nauk Mold. SSSR Ser. Biol. Khim. Nauk **1977** No. 3, pp. 58/71, 65; C.A. **88** [1978] No. 12476). – [2] D. N. Hague, S. R. Martin (J. Chem. Soc. Dalton Trans. **1974** 254/8, 257). – [3] R. L. Davies, K. W. Dunning (J. Chem. Soc. **1965** 4168/85, 4174, 4182). – [4] G. Anderegg (Helv. Chim. Acta **46** [1963] 2397/410, 2409). – [5] H. Irving, D. H. Mellor (J. Chem. Soc. **1962** 5222/37, 5231).

[6] S. Cabani, M. Landucci (Ann. Chim. [Rome] **52** [1962] 1030/42, 1037/8). – [7] G. Atkinson, J. E. Bauman (Inorg. Chem. **2** [1963] 64/7, 66). – [8] D. J. Benton, P. Moore (J. Chem. Soc. Dalton Trans. **1973** 399/404, 402). – [9] W. A. E. McBryde (IUPAC Chem. Data Series No. 17 [1978] 1/78, 10, 27, 45, 72). – [10] P. G. Läuger, S. Fallab, H. Erlenmeyer (Helv. Chim. Acta **38** [1955] 92/6, 93).

[11] K. Sone, P. Krumholz, H. Stammreich (J. Am. Chem. Soc. **77** [1955] 777/80, 778). – [12] R. R. Miller, W. W. Brandt (J. Am. Chem. Soc. **77** [1955] 1384/5). – [13] G. Atkinson, J. E. Bauman (Inorg. Chem. **1** [1962] 900/4, 902). – [14] G. Anderegg (Helv. Chim. Acta **46** [1963] 2813/22, 2815/6). – [15] D. J. Benton, P. Moore (J. Chem. Soc. Chem. Commun. **1972** 717/8).

[16] A. Ya. Sychev, Ya. D. Tiginyanu (Zh. Fiz. Khim. **42** [1968] 1124/30; Russ. J. Phys. Chem. **42** [1968] 587/91). – [17] Ya. D. Tiginyanu, A. Ya. Sychev, V. M. Berdnikov (Zh. Fiz. Khim. **45** [1971] 1723/8; Russ. J. Phys. Chem. **45** [1971] 975/9). – [18] A. A. Vlček (Pure Appl. Chem. **10** [1965] 61/70, 62).

16.1.31.3 Isolated Mn bpy$_n$X$_2$ Complexes

Tris(bipyridyl) Compounds

Mn bpy$_3$X$_2$ (X = Cl, Br, I, ClO_4, BF_4). The chloride was prepared by dissolving $MnCO_3$ in hydrochloric acid and adding the stoichiometric amount of bpy. Preparation of Mn bpy$_3$Br$_2$ involved treating $MnCO_3 \cdot 4H_2O$ with a stoichiometric amount of bpy and then adding KBr until a precipitate formed [1]. It was also prepared by adding bpy (4.7 g) to a heated solution of $MnSO_4 \cdot 4H_2O$ (2.2 g) in H_2O (10 mL). Alcohol was then added drop by drop until all solids dissolved. The solution was filtered and cooled, and KBr (10 g in 50 mL water) was added. The yellow microcrystalline precipitate was recrystallized from hot water and air dried [2]. Mn bpy$_3$I$_2$ (yellow) was prepared in the same way, using KI instead of KBr [31]. To prepare Mn bpy$_3$(ClO$_4$)$_2$, solutions of $Mn(ClO_4)_2 \cdot 6H_2O$ in water and of bpy in methanol (slight excess) were mixed, and the precipitate of yellow crystals was washed with cold ethanol and dried in vacuo at 80°C for 24 h [33]. Hot aqueous solutions of manganese(II) sulfate and bpy were blended in a 1:3 mole ratio, and excess aqueous $KClO_4$ was added to precipitate the complex, which was washed with water and dried under 0.1 Torr at room temperature [3]. Solutions of $MnSO_4 \cdot H_2O$ (0.17 g) in water and bpy (0.62 g) in methanol were mixed and heated to dissolve the precipitate. After addition of an excess of solid $NaClO_4$ and cooling, a bright yellow crystalline precipitate formed. It was recrystallized from hot water [4] or aqueous ethanol and dried in vacuo at 80°C over P_2O_5 [34]. Mn bpy$_3$(BF$_4$)$_2$ can be prepared by adding a solution of bpy to a solution obtained by immersion of metallic manganese in a 1:1 v/v mixture of 48% aqueous HBF_4 and dimethyl sulfoxide [36].

For Mn bpy$_3$Br$_2$ a molar magnetic susceptibility $\chi_{mol} = 15155 \times 10^{-6}$ cm^3/mol (corrected for diamagnetism) has been measured at 20°C. The magnetic moment $\mu_{eff} = 5.98$ μ_B [2], $\mu_{eff} = 5.80 \pm 0.03$ μ_B [5] indicates a high-spin octahedral Mn^{II} (d^5) complex. The ESR spectrum of the solid shows a single broad line [5]; g = 2.01 and A = 83 G were measured on Mn bpy$_3$(ClO$_4$)$_2$ in acetonitrile at 23°C in the presence of 0.1M $(C_3H_7)_4NClO_4$ [32].

The far-IR spectra of Mn bpy$_3$X$_2$ (X = Br, I) recorded as Nujol mulls in the 400 to 100 cm^{-1} region show two very strong bands, at 224 and 191 cm^{-1} (bromide) or at 221 and 182 cm^{-1} (iodide), which are assigned to Mn-N$_{bpy}$ stretching vibrations, the second band being combined with ligand vibrations. For other (weak) ligand bands in this region and bands of Mn bpy$_3$I$_2$ in the 1600 cm^{-1} region, see [13]; for earlier IR data for Mn bpy$_3$Br$_2$, see [6]. The IR spectrum of Mn bpy$_3$(ClO$_4$)$_2$ was recorded in the 1800 to 650 cm^{-1} region by [4] and below 600 cm^{-1} by [3]. Three strong bands near 1600, 1450, and 760 cm^{-1} are observed by [4]. All could be ascribed to ligand vibrations. The four bands at 417, 409(sh), 356, and 239 cm^{-1} are assigned to (Mn-N) stretching vibrations; a crystal-field stabilization energy equal to zero is assumed [3].

On heating in vacuo at suitable temperatures, Mn bpy$_3$Cl$_2$ and Mn bpy$_3$Br$_2$ lose two molecules of bpy to give the 1:1 complexes [1]. Mn bpy$_3$Br$_2$ dissociates in two steps, the first occurring at about 45°C and the second at about 175°C. No consistent rate constants could be obtained because the dissociation isotherms are characterized by pronounced and irreproducible induction periods [9]. Mn bpy$_3$Br$_2$ is reduced by Li$_2$bpy in tetrahydrofuran (under N_2) to form [Mn bpy$_3$] or Li[Mn bpy$_3$] [31]. Catalytic activity of Mn bpy$_3$Br$_2$ has been studied in the decomposition of H_2O_2 [5] and in the oxidation of isopropylbenzene by O_2 at 70°C [11]. Mn bpy$_3$I$_2$ reacts with $LiAlH_4$ in tetrahydrofuran to give [Al bpy$_3$] [10].

Voltammetric investigations (direct current and in-phase alternating current polarography and cyclic voltammetry) on solutions of Mn bpy$_3$(ClO$_4$)$_2$ in acetonitrile containing 0.1M $(C_2H_5)_4NBF_4$ at 25°C revealed a three-step reduction wave and a one-step oxidation wave [34], also see [7]. In the presence of excess free bpy, d.c. polarography affords well defined reduction waves at $E_{1/2}=-1.314$, -1.496, and -1.722 V (versus Ag/AgCl, saturated aqueous NaCl). Each of these waves indicates a reversible one-electron reduction, which successively leads to the species $[Mn\,bpy_3]^+$, $[Mn\,bpy_3]$, and $[Mn\,bpy_3]^-$ [34], also see [7]. The reduction waves are not affected by excess bpy in acetonitrile or propylene carbonate solutions. However, they are greatly affected by free bpy when dimethylformamide and dimethyl sulfoxide are used as solvents. It is assumed that in the last two solvents (S) the solvates $[MnS_6]^{2+}$ form at low concentrations of bpy while at high concentrations mixed complexes such as $[Mn\,bpy_2S_2]^{2+}$ form. These are reduced in the same way as $[Mn\,bpy_3]^{2+}$, in three one-electron steps. In the absence of free bpy, the first two waves are ascribed to reduction of Mn^{II} to Mn^0 in a two-electron step, and then to Mn^- in a one-electron step [8].

Oxidation to manganese(III) in acetonitrile solutions containing 0.1 M $(C_2H_5)_4NBF_4$ [34] or $(C_3H_7)_4NClO_4$ [32] occurs at about +1.5 V (versus Ag/AgCl, 25°C) [34] or at −1.36 V (versus saturated calomel, 23°C) [32]. Trivalent Mn is produced in a (quasireversible) one-electron reaction, followed by a chemical reaction leading to a di-μ-oxo-dimanganese(III,IV) complex by partial disproportionation and subsequent dimerization. Controlled potential electrolysis at +1.60 V abstracts two electrons per Mn atom and yields dimeric $Mn^{(IV,IV)}$ species [32], see p. 221.

$Mn\,bpy_3(C_6H_2N_3O_7)_2$. The picrate was prepared by slow addition of 2,2′-bipyridyl in ethanol to an ethanolic solution of manganese(II) picrate in a 2:1 to 3:1 mole ratio. The mixture was refluxed for 0.5 h to complete the reaction, and the resulting yellow crystals were washed with ethanol and ether, then dried in vacuo. The compound melts at 212°C and has a magnetic moment $\mu_{eff}=5.93\ \mu_B$ at 303 K. The electronic reflectance spectrum (Nujol mulls) shows four absorption bands, at 21740, 27740, 35710, and 44440 cm^{-1}, which apparently are due to intra-ligand transitions. The IR spectrum (Nujol mulls) exhibits relevant absorption bands (ν in cm^{-1}) which are assigned as follows: 1595, 1520 bpy ring stretching; 1555 $\nu_{as}(NO_2)$ of picrate; 775, 758 ring deformation (or CH out of plane); and 225 $\nu(Mn\text{-}N_{bpy})$. The observed shifts of the ring stretching and deformation modes of bpy in the complex indicate coordination of bpy to Mn through the pyridine ring N atoms. The magnetic moment is that of a high-spin octahedral compound. The molar electrical conductance, $\Lambda=156.56\ \Omega^{-1}\cdot cm^2\cdot mol^{-1}$ for a solution in dimethylformamide, corresponds to that of a 1:2 electrolyte containing free picrate anions. The complex explodes on melting. It is only sparingly soluble in H_2O, $CHCl_3$, and CCl_4 but moderately soluble in dimethylformamide and dimethyl sulfoxide [39].

Bis(bipyridyl) Compounds

$Mn\,bpy_2(NO_3)_2$ precipitates upon mixing ethanolic solutions of $Mn(NO_3)_2$ or $Mn(NO_3)_2\cdot 6H_2O$ and bpy, mole ratio 1:2, or from a solution of $Mn(CO)_5NO_3$ and bpy (1:1.06) in nitromethane kept at −8°C for about two weeks after removal of $Mn(CO)_3bpyNO_3$, which crystallizes first. $Mn\,bpy_2(NO_3)_2$ is washed with alcohol and ether or recrystallized from hot nitromethane and dried in vacuo. The product from anhydrous $Mn(NO_3)_2$ is lime green, while in the other cases the crystals are yellow. The IR absorption spectra (recorded as Nujol or halogenocarbon mulls in the 2100 to 650 cm^{-1} region) are almost identical, differing only in the 700 to 800 cm^{-1} region. The differences are apparently due to a splitting of the ring hydrogen vibrations of bpy at ~760 cm^{-1} in the lime-green complex. Thus structures of symmetry D_{2h} and C_2 are tentatively proposed for the lime-green and yellow forms, respectively. $Mn\,bpy_2(NO_3)_2$ evidently contains monodentate nitrato groups. These probably occupy trans positions relative to bpy in the more stable yellow form [12].

Mn bpy$_2$Cl$_2$ separated as yellow crystals on evaporation of the combined solutions of $MnCl_2$ (2 mol) in H_2O and bpy (1 mol) in ethanol on a water bath [13]. Susceptibility measurements yield a magnetic moment $\mu_{eff} = 5.82 \pm 0.03\ \mu_B$, indicating a high-spin Mn^{II} complex. The ESR spectrum shows a single line with a line width of about 600 Øe, but on moistening the dry compound a hyperfine structure appears whose intensity increases with the amount of water added [5]. The electronic spectrum recorded for solutions in ethanol (0.005 to 0.1M) shows absorption maxima at 283 and 235 nm with molar extinction coefficients log ε_{max} of 4.60 and 4.50 (ε in $L \cdot mol^{-1} \cdot cm^{-1}$), respectively [14], in good agreement with earlier values measured in water by [13]. Mn bpy$_2$Cl$_2$ is slightly soluble in cold water, the solutions decomposing on dilution [13]. Catalytic activity of Mn bpy$_2$Cl$_2$ has been studied in the decomposition of H_2O_2 [5] and in the oxidation of isopropylbenzene by O_2 [11].

Mn bpy$_2$X$_2$ (X = NCO, CNO, NCS, NCSe, $CNCH_2COO$). The complex with X = NCO was prepared by reacting methanolic solutions of $Mn(NO_3)_2$ and bpy (mole ratio 1:2) with an aqueous methanolic solution of KOCN, mole ratio bpy:OCN^- = 1:1. Water was added dropwise until all KOCN dissolved. The crystals, which appeared after some hours, were washed with water until free from NO_3^-, then washed with alcohol and ether, and dried in vacuo [15]. For X = CNO an aqueous solution containing $MnSO_4$ and bpy in a 1:3 mole ratio was reacted at 50 to 60°C with an almost neutralized (pH 8 to 9) solution of NaCNO. The precipitate was cooled, washed with water, acetone, and ether, and dried in vacuo [16]. For X = NCS, hot ethanolic solutions of manganese(II) isothiocyanate and bpy were mixed, and the resulting solid was washed with ethanol and dried [37]. Single crystals of Mn bpy$_2$(NCS)$_2$ have been grown from nitromethane solution by slow evaporation [38]. For X = NCSe ethanolic solutions of $Mn(NCSe)_2$ and bpy were mixed in a 1:2 mole ratio, and the precipitate was washed with ethanol and dried over P_2O_5 [17]. For X = $CNCH_2COO$, the manganese(II) salt $Mn(O_2CCH_2CN)_2$ and bpy were reacted in 2,2-dimethoxypropane. After four days of standing at room temperature the precipitate was collected, washed with 2,2-dimethoxypropane, and dried in vacuo [18]. In the following table the magnetic moment μ_{eff} (in μ_B) and the molar electrical conductance Λ (in $\Omega^{-1} \cdot cm^2 \cdot mol^{-1}$) of the yellow complexes are listed for various anionic species X:

X	CNO	NCS	$CNCH_2COO$
μ_{eff}	5.79	5.99 (ambient)	5.99 at 24°C
Λ	18 at 20°C*)	—	23.0 at 25°C**)
Ref.	[16]	[37]	[18]

*) In dimethyl sulfoxide, dilution ν = 900 L/mol. – **) In acetonitrile, concentration c = 10^{-3} mol/L. The complex is essentially nonconducting.

Mn bpy$_2$(NCO)$_2$ crystallizes in an orthorhombic lattice of space group Pbcn-D_{2h}^{14} (No. 60) with lattice parameters a = 13.98(2), b = 9.85(2), and c = 15.73(3) Å; Z = 4. The density was measured to be 1.40 g/cm^3; the calculated density is 1.43 g/cm^3 [19], see also [20]. Structural data were collected from Weissenberg (rotation) photographs and refined to a reliability R = 0.081. A cis-octahedral structure of Mn bpy$_2$(NCO)$_2$ is found with the distances (in Å): $Mn \leftrightarrow N^{I}_{bpy}$ = 2.04(2), $Mn \leftrightarrow N^{II}_{bpy}$ = 2.06(2), and $Mn \leftrightarrow N_{NCO}$ = 2.12(2) within the octahedra. The angles N^{I}_{bpy}-Mn-N^{II}_{bpy}, N^{I}_{bpy}-Mn-N_{NCO}, and N^{II}_{bpy}-Mn-N_{NCO} are 93(2)°, 90(1)°, and 91(2)°, respectively. The NCO groups are almost linear (178(2)°) with Mn-N_{NCO} bonds making up angles close to unfolded MnNC (= 173(1)°), indicating an isocyanate structure. Distances within the NCO groups are (in Å) $N \leftrightarrow C$ = 1.20(2) and $C \leftrightarrow O$ = 1.22(2) [19]. Measured refractive indices: n_p = 1.620, n_m = 1.657, and n_g = 1.685; n_g-n_p = 0.065 [20]. The IR absorption spectrum (recorded as Nujol mulls in the 2400 to 400 cm^{-1} region) shows that the frequency assigned to the stretching vibration ν(CN) at 2212 cm^{-1} of the OCN group remains approximately the same as that found in the OCN^- ion of various inorganic cyanates, whereas ν(CO) at 1330 and 1342 cm^{-1} is increased by about 40 cm^{-1}. This also indicates an isocyanate ligand. The air-stable compound is insoluble in water, alcohol, acetone, ether, and chloroform. It is sparingly soluble in dimethylformamide, where the solubility increases on heating [15].

$Mnbpy_2(CNO)_2$ possibly may be formulated as $[Mnbpy_3]_2[Mn(CNO)_6]$ or $[Mnbpy_4]$-$[Mn(CNO)_4]$ [16]. The IR spectrum of $Mnbpy_2(CNO)_2$ (Nujol mull) shows strong absorption bands at 2082 and 1149 cm^{-1}, which are assigned to antisymmetric and symmetric stretching modes, respectively, of the CNO^- groups; the medium band at 494 cm^{-1} may indicate a CNO deformation mode. The compound does not explode but deflagrates vigorously. It is slightly soluble in water, fairly soluble in methanol and ethanol, and insoluble in ether and benzene [16].

According to single-crystal X-ray data $Mnbpy_2(NCS)_2$ crystallizes in an orthorhombic lattice, space group Pbca-D_{2h}^{15} (No. 61), with lattice parameters a = 16.858(9), b = 16.141(8), and c = 16.068(5) Å; Z = 8. The calculated density is 1.469 g/cm^3. For a list of atomic coordinates corresponding to a final anisotropic refinement (R = 0.045), see [38]. The atomic distances and some bonding angles around the Mn atom are shown in **Fig. 18.** The structure consists of a 6-coordinate Mn^{2+} ion, which is surrounded by two bidentate bipyridyl molecules coordinating through the ring N atoms and two monodentate terminal N-bonded thiocyanate groups. It is best described as pseudooctahedral with the isothiocyanate groups in cis position [38].

Fig. 18

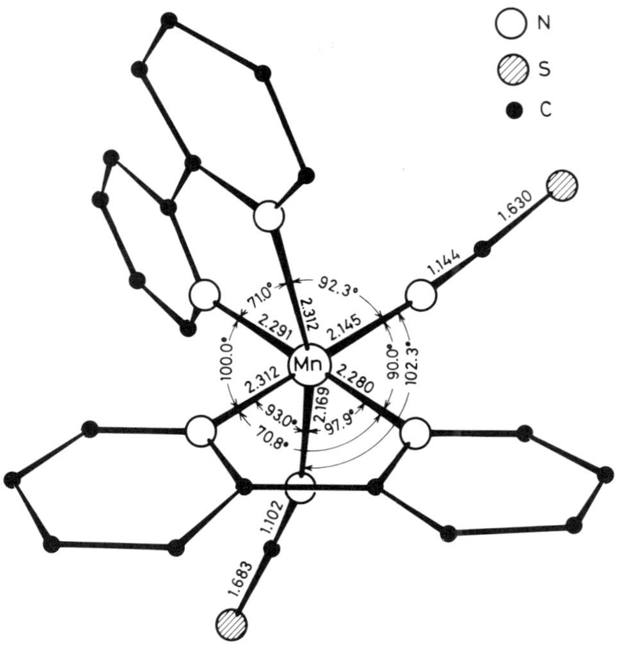

Structure of $Mnbpy_2(NCS)_2$ with bond lengths in Å and bond angles N-Mn-N.

The observed bands of the electronic spectrum (Nujol mull) are listed in the table below. These bands correspond to transitions from the $^6A_{1g}$ ground state to the excited states as follows:

ν in cm^{-1}	17540	22990	24420	29670
assignment, $^6A_{1g}\rightarrow$	$^4T_{1g}(^4G)$	$^4T_{2g}(^4G)$	^4E_g, $^4A_{1g}(^4G)$	$^4E_g(^4D)$

The Racah parameters $B = 730.8\ cm^{-1}$ and $C = 2923.3\ cm^{-1}$ were calculated using $^6A_{1g}\rightarrow{}^4E_g(^4D)$ and $^6A_{1g}\rightarrow{}^4E_g$, $^4A_1(^4G)$ transitions. The ligand field splitting parameter $Dq = 804\ cm^{-1}$ was calculated using the $^6A_{1g}\rightarrow{}^4E_g(^4G)$ transition. The nephelauxetic parameter is $h_x = 2.86$. All the parameters are consistent with an octahedral environment around the Mn^{2+} ion [37]. The IR spectrum of $Mn\,bpy_2(NCS)_2$ (recorded as Nujol mulls) shows absorption bands assigned to the vibration modes ν(CN) at 2050, ν(CS) at 792, δ(NCS) at 475, and ν(Mn-NCS) at 300 cm^{-1}. These are consistent with terminally N-bonded NCS groups and also confirm the octahedral environment of Mn. The fundamental frequencies of the uncoordinated bpy are altered little on coordination to Mn^{2+} [37].

X-ray diffraction diagrams of $Mn\,bpy_2(NCSe)_2$ indicate a trans configuration, where the linkage between Mn and the linear NCSe groups is made through the N atoms. The assumption of isoselenocyanato groups also agrees with the IR spectrum (liquid paraffin), which shows absorption bands at 2068 and ~2052 cm^{-1}, both assigned to ν(CN) stretching of the NCSe group, and bands at 410 and 425 cm^{-1}, both assigned to δ(NCSe) bending vibrations. Two bands, at 647 and 625 cm^{-1}, may be associated with a ν(CSe) stretching mode, which probably is overlapped by bpy vibrations [17].

The X-ray diagram of $Mn\,bpy_2(CNCH_2COO)_2$ indicates a low symmetry. In the IR spectrum (Nujol mulls) the position of ν(CN) at 2250 cm^{-1} and of δ(CCN) at 378 cm^{-1} of the cyanoacetate anion is evidence for the absence of cyanonitrogen-Mn bonds. The bands at 1635 cm^{-1} (sh) and 1625 cm^{-1} are assigned to ν_{as}(COO); bands at 1402 and 1385 cm^{-1} (sh), to ν_s(COO); and a band at 360 cm^{-1}, to COO rocking of the unidentate carboxylate groups. A band at ~290 cm^{-1} is assigned to ν(Mn-O) + ν(Mn-N) vibrations. For unassigned and ligand bands, see the paper. Conductivity (nonelectrolyte), magnetic and spectroscopic data suggest that $Mn\,bpy_2(CNCH_2COO)_2$ is a high-spin cis (less probably, trans) octahedral complex where the Mn atom is 6-coordinated, to four N atoms of two bpy molecules and to two oxygen atoms of the cyanoacetate anions. The complex decomposes on heating (4 K/min) with simultaneous loss of bpy and cyanoacetate groups, leaving MnO [18].

$[Mn\,bpy_2(H_2O)_2](C_6H_2N_3O_7)_2$. The picrate is prepared from equimolar amounts of manganese(II) picrate and 2,2′-bipyridyl in ethanol in the same way as $Mn\,bpy_3(C_6H_2N_3O_7)_2$ (p. 207). The yellow crystals melt at 198°C and are paramagnetic with $\mu_{eff} = 5.94\ \mu_B$, corresponding to a high-spin octahedral complex. The molar electrical conductance $\Lambda = 144.55\ \Omega^{-1}\cdot cm^2\cdot mol^{-1}$ measured in dimethylformamide is that of a 1:2 electrolyte containing free picrate anions. The solubility in water and organic solvents resembles that of $Mn\,bpy_3(C_6H_2N_3O_7)_2$ [39].

$Mn\,bpy_2Cr_2O_7$ was precipitated from an aqueous solution containing a manganese(II) salt and bpy (as base or as a salt with HCl) by adding dichromate solution dropwise and heating to about 80°C [21]. The compound is reddish brown. On heating, decomposition starts at 235°C. A peak in the DTA curve (maximum rate) is observed at 315°C; the end of reaction, at 320°C [22]. (A compound formulated as $Mn\,bpy_2(CrO_4)_2\cdot 4H_2O$ by [22] is probably a bipyridinium salt.)

Mono(bipyridyl) Compounds

$MnbpyX_2$ (X = Cl, Br). The pale yellow compounds were prepared from solutions of MnX_2 and bpy (stoichiometric amounts) in anhydrous acetone [23]. Ethanolic solutions of MnX_2 and bpy (mole ratio 1:1) were refluxed for 0.5 h, and the crystals which separated were washed with absolute ethanol and dried over $CaCl_2$ [24]. $MnbpyX_2$ resulted also from heating $Mnbpy_3X_2$ in vacuo until two bpy are lost [1]. The chloro compound can be obtained also by refluxing anhydrous $MnCl_2$ and bpy in dimethylformamide (DMF) for about ten minutes with stirring. After cooling to room temperature the precipitated solid was collected and washed with cold DMF and dry ether. As the compound at this stage contains one molecule of DMF it was heated under 0.01 Torr at 150°C to remove the solvent of crystallization [25]. It was also obtained from $[bpyH_2][MnCl_5]$ heated in vacuo for 3 h at 198°C [26], at 215°C in vacuo, or at 217°C in N_2 [33]. The following table lists the magnetic moment μ_{eff} (in μ_B). The temperature in K is given in parentheses, r.t. = room temperature. The Curie-Weiss constant $\Theta_p = -0.5$ K was obtained for the chloro compound [26]. Also given are the bands (in cm^{-1}) of the far-IR spectrum (Nujol mulls) assigned to the Mn-X and $Mn\text{-}N_{bpy}$ vibrations:

X	μ_{eff}	Ref.	ν_{as}(Mn-X)	ν_s(Mn-X)	$\nu(Mn\text{-}N_{bpy})$	ν(Mn-X) bridging	δ(Mn-X)	Ref.
Cl	5.87	[25]	259(sh)*)	—	245	176	159*)	[23, 27]
	5.82 (293)	[26]	259(sh)	248	239	—	147	[1]
	6.08 (r.t.)	[37]	—	250	—	—	—	[37]
Br	5.93 (296)	[23]	211*)	199*)	238	139	115**)	[23, 27]
	5.85 (305)	[24]	212	197	235	—	123	[1]

*) These bands have been assigned to ligand vibrations by [27]. – **) Unassigned in [27].

Bands assigned to ligand vibrations in the 4000 to 400 cm^{-1} region are reported in [1], the 400 to 40 cm^{-1} region is reported in [1, 23, 27], and the Raman spectrum (<300 cm^{-1}) is given in [27]. The assignment of the far-IR bands of $MnbpyX_2$ is only tentative as the spectra are very complicated and consist of a large number of bands [23, 27]. Thus a large part of the assignments (such as a chelate ring stretching mode at ~182 cm^{-1}) made by [1] could not be confirmed by [23, 27]. However, it is assumed that the complexes are polymeric octahedral [23, 24, 26], containing bridging (Mn-X-Mn) halogens [23, 27], also see [37]. Force constants in mdyn/Å, calculated using normal coordinate analysis and IR data, are K(Mn-N) = 0.37, K(N-C) = 5.35, p(chelate)resonance = 0.25, and p(py ring) = 0.32 [1]. These force constants suggest a normal chelate bonding in $MnbpyX_2$, without backbonding [1]. The magnetic moment of $MnbpyX_2$, which is temperature independent between 100 and 300 K for X = Cl [26], indicates a high-spin $Mn\text{-}d^5$ complex [24 to 26]. X-ray studies suggest isomorphism of $MnbpyCl_2$ with α-$CobpyCl_2$ [26].

In the following table the observed bands (in cm^{-1}) of the electronic spectrum of $MnbpyCl_2$ are listed together with the bands of two other 1:1 Mnbpy complexes, $MnbpySO_4$ (p. 213) and $Mnbpy(CH_3COO)_2$ (p. 213). The bands correspond to transitions from the $^6A_{1g}$ ground state to several excited states [37].

compound	assignment, $^6A_{1g}\rightarrow$				
	$^4T_{1g}(^4G)$	$^4T_{2g}(^4G)$	^4E_g, $^4A_{1g}(^4G)$	$^4E_g(^4D)$	$^4T_{1g}(^4P)$
$MnbpyCl_2$	19040	—	23360	26670	30770
$MnbpySO_4$	19230	23100	—	26670	30120
$Mnbpy(CH_3COO)_2$	18350	21980	—	27770	—

The ligand-field parameters B, C, Dq (in cm^{-1}), and h_x were calculated in [37].

compound	B	C	Dq	h_x
$MnbpyCl_2$	872.6	3490.5	960	3.42
$MnbpySO_4$	921.6	3686.4	1106	3.93
$Mnbpy(CH_3COO)_2$	764.6	3058.3	841	2.99

The electronic spectra and ligand field parameters of these complexes are consistent with an octahedral environment about the Mn^{2+} ions [37].

On heating in vacuo, $MnbpyCl_2$ loses 0.5 mol bpy at 365°C (in N_2 at 462°C) and 0.5 mol bpy at 437°C to leave $MnCl_2$ [26]. $MnbpyCl_2$ is hygroscopic and readily soluble in water and methanol where rapid disproportionation occurs, especially in water [25]. $MnbpyBr_2$ does not decompose or oxidize on heating even for hours at 120°C. It is a nonelectrolyte and poorly soluble in organic solvents [24].

$MnbpyCl_2\cdot DMF$ is formed on reacting $MnCl_2$ and bpy in DMF, see p. 211 [25].

$Mnbpy(N_3)_2$ and **$Mnbpy(NCO)_2\cdot 6H_2O$.** An aqueous solution of manganese(II) chloride or nitrate was treated with a slight excess of bpy dissolved in alcohol. The mixture was warmed and the stoichiometric amount of dilute aqueous NaN_3 or KOCN was added followed by refluxing for 30 min. The precipitate was collected, recrystallized from dimethyl sulfoxide, and dried in vacuo. The color, melting point (m.p.), effective magnetic moment μ_{eff} (in μ_B from susceptibility measurements), and IR absorption bands of the anions (KBr discs) are tabulated as follows:

compound	color	m.p. in °C	μ_{eff}	$\nu_{as}(N_3$ or NCO)	$\nu_s(N_3$ or NCO)
$Mnbpy(N_3)_2$	light yellow	281	5.30	2098, 2060(sh)	1315(sh)
$Mnbpy(NCO)_2\cdot 6H_2O$	light brown	280	5.88	2180 (very broad)	—

Both compounds are nonelectrolytes, having very low molar conductivity values. It is suggested from the magnetic moment and the IR spectrum that $Mnbpy(N_3)_2$ has an octahedral structure with azide bridges formed through a single N atom. $Mnbpy(NCO)_2\cdot 6H_2O$ is assumed to have a tetrahedral structure with terminal N-bonded cyanato groups. The very low solubility of both compounds results in weak electronic spectra with very weak ligand field bands, which could not be interpreted [28].

$(NH_4)_2[Mnbpy(NCS)_4]$. The complex was precipitated on slowly adding an aqueous-ethanolic solution of NH_4SCN to a hot aqueous-ethanolic solution containing Mn^{2+} and bpy in a 1:2 mole ratio. After cooling the precipitate was collected. The light yellow complex was first washed with alcohol, then with water, and dried at 50 to 60°C. A magnetic moment $\mu_{eff}=5.88\ \mu_B$ at room temperature results from susceptibility data, and indicates a high-spin octahedral complex. In the IR spectrum two types of Mn-N absorption bands are observed. These are due to the nitrogens of bpy and thiocyanate, respectively. Thus the thiocyanate groups are coordinated to Mn through the N atoms [40].

MnbpySO_4. A saturated aqueous solution of $MnSO_4$ was added to an ethanolic solution of 2,2'-bipyridyl, and the resulting precipitate was washed with ethanol and dried. The magnetic moment $\mu_{eff}=5.90$ μ_B results from susceptibility measurements at room temperature. The IR spectrum (recorded for Nujol mulls) shows absorption bands assigned to vibration modes of coordinated sulfate ions at 1150 to 1050 (ν_3), 975 (ν_1), 608 (ν_4), and 430 cm^{-1} (ν_2). The rock-salt region of the IR spectrum conclusively indicates bidentate chelation of 2,2'-bipyridyl. The sulfate ion, which retains its T_d symmetry, is regularly coordinated through all its oxygen atoms as a polydentate bridging ligand, thus leading to a probably polymeric octahedral structure. The electronic spectrum was recorded and interpreted together with that of MnbpyCl_2 on p. 212 [37].

Mnbpy$(CH_3COO)_2$ pecipitated on mixing hot ethanolic solutions of manganese(II) acetate and 2,2'-bipyridyl. At room temperature a magnetic moment $\mu_{eff}=5.77$ μ_B was observed. The IR spectrum (Nujol mull) shows absorption bands assigned to vibration modes of coordinated acetate at 1550 (ν_{as}OCO), 1450, 1410 (ν_sOCO), and 650 cm^{-1} (δOCO). The separation of the OCO symmetric and antisymmetric stretching modes (~140 cm^{-1}) indicates symmetrically coordinated bidentate acetate ions, and the splitting of the ν_s(OCO) mode may be due to a cis disposition. The IR and electronic (see p. 212) spectra suggest that Mnbpy$(CH_3COO)_2$ has a monomeric octahedral structure where the 6-coordinate Mn atom is surrounded by two nitrogens of bpy and four oxygens of two bidentate acetate ions in cis positions [37].

Mnbpy$(4\text{-}CH_3C_6H_4SO_2)_2 \cdot nH_2O$. The salt of p-toluenesulfinic acid with n = 1 was obtained by heating a mixture of $Mn(4\text{-}CH_3C_6H_4SO_2)_2 \cdot 2H_2O$ (4.74 mmol), bipyridyl (10.9 mmol), and tetrahydrofuran (70 to 100 mL) in a sealed tube at 60°C for 68 h. The anhydrous compound resulted on reacting the monohydrate and excess bpy in tetrahydrofuran at 120°C for 143 h. The anhydrous compound is yellow, the monohydrate is light yellow. Susceptibility measurements yield a magnetic moment $\mu_{eff}=6.15$ μ_B for the anhydrous compound. The IR spectrum (KBr discs) of the monohydrate shows very strong bands at 1015 and 975 cm^{-1}, assigned to antisymmetric and symmetric stretching vibrations, respectively, of the SO_2 group in the anion, and the two broad bands at 3405 and 1620 cm^{-1}, assigned to vibrations of lattice H_2O, which are absent in the anhydrous complex. It is assumed that the configuration of Mnbpy$(4\text{-}CH_3C_6H_4SO_2)_2$ is pseudooctahedral and the Mn atom is 6-coordinate, surrounded by the two N atoms of bpy and the four oxygen atoms of two bidentate p-toluenesulfinate anions as shown below:

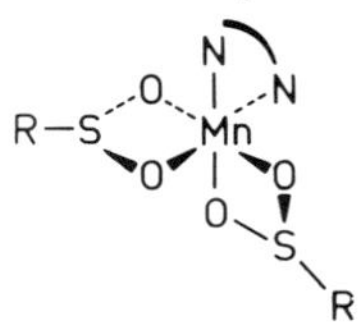

R = 4 − $CH_3C_6H_4$

This structure is confirmed by the electronic reflectance spectrum, which is that of a high-spin Mn(d^5) complex [30].

Mnbpy$(PH_2O_2)_2$. The phosphinate was prepared by mixing $Mn(PH_2O_2)_2 \cdot H_2O$ (20 mmol) in a minimum quantity of water with a concentrated alcohol solution of 2,2'-bipyridyl (20 mmol) and evaporation until crystals appeared. These were collected and dried in vacuo [29]. The pale yellow laths are elongated along c and belong to the monoclinic system as shown by Weissenberg photographs. The space group is C2/c-C_{2h}^6 (No. 15) with lattice constants a = 16.827(15), b = 10.745(20), c = 7.170(5) Å, and β = 91.46(3)°; Z = 4. The structure was resolved by the heavy atom method and refined by full-matrix-least-squares refinement (with anisotropic thermal parameters except for H) to R = 0.051. The final atomic parameters are listed in the original paper. The bond lengths in the neighborhood of Mn are shown in **Fig. 19.** Each 6-coordinate Mn atom is bonded to both N atoms of the 2,2′-bipyridyl ligand and to the four oxygen atoms of two $H_2PO_2^-$ groups in a distorted octahedral environment. The Mn atoms are linked each through a double Mn-O-P-O-Mn bridge to each of the Mn atoms related to it by the c glide, thus forming a chain polymer along c. The normal of the ligand mean plane (dihedral angle between the rings is 4.5°) is inclined 15° to the chain axis (c). A density of D = 1.75 g/cm³ was calculated from the X-ray data [35]. A magnetic moment μ_{eff} = 5.75 μ_B results from susceptibility measurements at room temperature. In the ESR spectrum a single band occurs at g = 2.048. Strong charge-transfer bands are observed in the visible spectrum. The IR spectrum shows strong absorption bands assigned to vibrations of the $PH_2O_2^-$ anion. Stretching vibrations in cm^{-1}: 2315 (PH_2), 1225, 1191, 1172 (PO_2, antisymmetric), and 1059 (PO_2, symmetric). Deformation vibrations in cm^{-1}: 1148, 1093, 810 (PH_2), and 451 (PO_2). The extensive splitting of the anion bands suggests differing functions for the phosphinate groups. An assumed bridging between two Mn atoms [29] was confirmed by the X-ray data [35].

Fig. 19

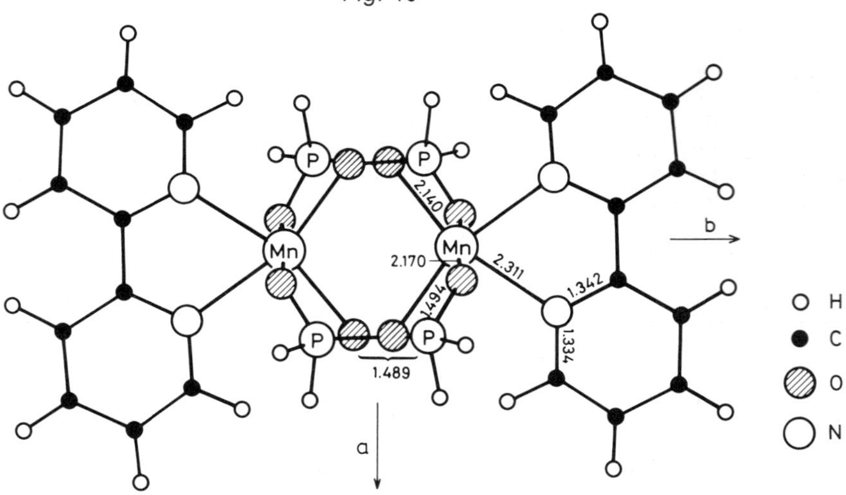

Axial (c) projection of one chain in Mnbpy$(PH_2O_2)_2$ with some atomic distances in Å.

References:

[1] J. S. Strukl, J. L. Walter (Spectrochim. Acta A **27** [1971] 223/38, 225/9, 237). – [2] F. H. Burstall, R. S. Nyholm (J. Chem. Soc. **1952** 3570/9, 3571, 3573, 3577). – [3] G. C. Percy, D. A. Thornton (J. Mol. Struct. **10** [1971] 39/48, 40/2). – [4] A. A. Schilt, R. C. Taylor (J. Inorg. Nucl. Chem. **9** [1959] 211/21, 216, 221). – [5] A. Ya. Sychev, Ya. D. Tiginyanu, Yu. S. Gromovoi (Zh. Fiz. Khim. **42** [1968] 2081/4; Russ. J. Phys. Chem. **42** [1968] 1097/9).

[6] R. J. H. Clark, C. S. Williams (Spectrochim. Acta A **23** [1967] 1055/9, 1057). – [7] Y. Sato, N. Tanaka (Bull. Chem. Soc. Japan **41** [1968] 2064/6). – [8] N. Tanaka (Electro-

chim. Acta **21** [1976] 701/10, 708). – [9] S. K. Dhar, F. Basolo (J. Inorg. Nucl. Chem. **25** [1963] 37/44, 39/40, 42). – [10] S. Herzog, H. Präkel (Z. Chem. [Leipzig] **5** [1965] 471).

[11] A. Ya. Sychev, I. M. Reibel, L. P. Podgornaya (Zh. Fiz. Khim. **43** [1969] 676/80; Russ. J. Phys. Chem. **43** [1969] 370/3). – [12] C. C. Addison, M. Kilner (J. Chem. Soc. A **1966** 1249/54, 1252). – [13] K. Yamasaki (Bull. Chem. Soc. Japan **13** [1938] 538/42). – [14] J. Császár, E. Horváth, L. Lehotai (Acta Univ. Szeged. Acta Phys. Chem. [2] **2** [1956] 49/55, 52; C.A. **1957** 16189), J. Császár, E. Horváth (Magy. Tud. Akad. Kem. Tud. Oszt. Közlemen. **14** [1960] 377/84, 378; C.A. **1961** 16138). – [15] A. M. Golub, G. V. Tsintsadze, A. M. Mamulashvili (Zh. Neorgan. Khim. **14** [1969] 3013/8; Russ. J. Inorg. Chem. **14** [1969] 1589/91).

[16] W. Beck, E. Schuierer (Chem. Ber. **95** [1962] 3048/55, 3051/4). – [17] P. Stancheva, V. V. Skopenko, G. V. Tsintsadze (Ukr. Khim. Zh. **35** [1969] 166/73; Soviet Progr. Chem. **35** No. 2 [1969] 39/44, 41). – [18] M. Cartwright, D. A. Edwards, J. M. Potts (Inorg. Chim. Acta **34** [1979] 211/20, 212, 218). – [19] G. V. Tsintsadze, T. I. Tsivtsivadze, F. V. Orbeladze (Zh. Strukt. Khim. **16** [1975] 320; J. Struct. Chem. [USSR] **16** [1975] 305). – [20] G. V. Tsintsadze, A. B. Tovbis, A. M. Mamulashvili, A. I. Kvitashvili (Tr. Gruz. Politekhn. Inst. **1969** No.6, pp. 13/5; C.A. **75** [1971] No. 11650).

[21] B. Lóránt (Thermochim. Acta **6** [1973] 205/14, 206). – [22] B. Lóránt (Thermochim. Acta **9** [1974] 67/75, 68). – [23] R. E. Wilde, T. K. K. Srinivasan, S. N. Ghosh (J. Inorg. Nucl. Chem. **35** [1973] 1017/21). – [24] A. Syamal (J. Indian Chem. Soc. **44** [1967] 989/90). – [25] J. A. Broomhead, F. P. Dwyer (Australian J. Chem. **14** [1961] 250/2).

[26] K. Akabori (J. Sci. Hiroshima Univ. A **39** [1975] 73/92, 76, 80, 86, 88; C.A. **83** [1975] No. 125417; Chem. Letters **1974** 1481/6). – [27] R. E. Wilde, T. K. K. Srinivasan (J. Inorg. Nucl. Chem. **36** [1974] 323/8, 327). – [28] R. C. Das, M. K. Mishra, B. K. Nanda, A. K. Sahu (J. Indian Chem. Soc. **54** [1977] 1027/9). – [29] J. Sala-Pala, R. Kergoat, J. E. Guerchais (Compt. Rend. C **274** [1972] 595/7). – [30] E. Lindner, G. Vitzthum (Chem. Ber. **102** [1969] 4062/9, 4063/5, 4068).

[31] Y. Saito, J. Takemoto, B. Hutchinson, K. Nakamoto (Inorg. Chem. **11** [1972] 2003/11). – [32] M. M. Morrison, D. T. Sawyer (Inorg. Chem. **17** [1978] 333/7). – [33] S. Sarkar, B. Sharma, R. Ch. Maurya, Sh. Ch. Chaurasia, Sh. K. Srivastava (Acta Ciencia Indica **4** [1978] 21/4, 23). – [34] M. C. Hughes, J. M. Rao, D. J. Macero (Inorg. Chim. Acta **35** [1979] L321/L324, L322). – [35] T. J. R. Weakley (Acta Cryst. B **34** [1978] 281/2).

[36] J. J. Habeeb, F. F. Said, D. G. Tuck (Inorg. Nucl. Chem. Letters **15** [1979] 113/6). – [37] I. S. Ahuja, R. Singh, C. L. Yadava (J. Mol. Struct. **74** [1981] 143/51, 145, 148). – [38] M. V. Veidis, B. Dockum, F. F. Charron, W. M. Reiff, T. F. Brennan (Inorg. Chim. Acta **53** [1981] L197/L199). – [39] B. D. Pandey, D. C. Rupainwar (Transition Metal Chem. [Weinheim] **6** [1981] 249/52). – [40] W. U. Malik, P. P. Bhargawa, M. M. Siddiqui (Acta Chim. Acad. Sci. Hung. **107** [1981] 155/9, 156).

16.1.31.4 Manganese(II) Complexes with 2,2'-Bipyridyl and Other Ligands

Complexes containing Mn^{2+}, bpy, and a second ligand L (anionic or neutral) in a 1:1:1 mole ratio may be formed by reaction of [MnL] complexes with bpy (formation constant K_a in L/mol), by reaction of $[Mn\,bpy]^{2+}$ with L (formation constant K_b in L/mol), or from Mn, L, and bpy (overall stability constant $\beta_{a,b}$ in $(L/mol)^2$). The equilibrium constant of the reaction $[Mn\,bpy]^{2+} + [MnL] \rightleftharpoons [Mn\,bpy\,L] + Mn^{2+}$ is given as ΔK, which can be calculated from $\log \Delta K = \log K_a - \log K_1$, where K_1 is the stability constant of the binary $[Mn\,bpy]^{2+}$ complex (see p. 203), or from $\log \Delta K = \log K_a + \log K_b - \log \beta_{a,b}$ [1]. The formation con-

second ligand	L	method	n	log K_a	log K_b	log $\beta_{a,b}$	Δ log K	Ref.
L_1 = triphosphate	$P_3O_{10}^{5-}$	kinetic	3	2.08 ± 0.08 (2.38)	—	—	—	[3]
L_2 = hydrogentriphosphate	$HP_3O_{10}^{4-}$	potentiometric	—	2.92	4.61	7.23 ± 0.03	0.30	[1]
L_3 = pyrocatecholate	$C_6H_4O_2^{2-}$	potentiometric	—	3.04	8.39	11.01 ± 0.03	0.42	[1]
L_4 = malonate	$C_3H_2O_4^{2-}$	potentiometric	—	2.64	2.74	5.36 ± 0.07	0.02	[1]
L_5 = 3-oxoglutarate	$C_5H_4O_5^{2-}$	potentiometric	—	—	1.18 ± 0.03	—	0.28	[2]
L_6 = ethylenediamine	$C_2H_8N_2$	potentiometric	—	≤1.6	≤1.7	≤4.3	≤−1.0	[1]
L_7 = diethylenetriamine	$C_4H_{13}N_3$	kinetic	3	2.68 ± 0.12 (2.98)	—	—	—	[3]
L_8 = triethylenetetraamine	$C_6H_{18}N_4$	kinetic	4	2.39 ± 0.08 (3.08)	—	—	—	[3]
L_9 = N,N′-bis(2-aminoethyl)-1,3-propanediamine	$C_7H_{20}N_4$	kinetic	4	2.35 ± 0.08 (2.83)	—	—	—	[3]
L_{10} = tris(2-aminoethyl)-amine	$C_6H_{18}N_4$	kinetic	4	2.45 ± 0.07 (2.93)	—	—	—	[3]
L_{11} = tropolone	$C_7H_6O_2$	potentiometric	—	—	6.14*)	—	—	[4]
L_{12} = 3-isopropyltropolone	$C_{10}H_{12}O_2$	potentiometric	—	—	6.16*)	—	—	[4]

*) The constant $K_d = [Mn\,bpy\,L]^{2+}/([Mn\,bpy_2]^{2+}[Mn\,L_2]^{2+})^{1/2}$ is 0.62 for the tropolone complex and 0.66 for the 3-isopropyltropolone complex. Because $K_d < 1$ (noncompatible ligands) both ternary complexes are less stable than the parent complexes [4].

stants (K_a, K_b, $\beta_{a,b}$) have been determined in aqueous solutions at 25°C [1 to 3] and in 50% aqueous dioxane at 30°C [4] by pH-potentiometric titration [1, 2, 4] for ionic strength I = 0.1 M ($NaClO_4$) [1, 4] or I = 0.6 M ($NaNO_3$) [2] and by kinetic measurements (temperature-jump relaxation technique) for I = 0.3 M ($NaClO_4$) [3]. In the table on p. 216, the relevant data for ternary complex formation are given for various ligands L; n is the number of coordination positions of Mn^{2+} assumed to be occupied by L. Values of log K_a given in parentheses are statistically corrected values obtained by multiplying K_a by 6/(6 − n) to allow for the number of H_2O available for substitution by bpy [3].

The second-order rate constants k_f in $L \cdot mol^{-1} \cdot s^{-1}$ of the reaction [MnL] + bpy → [MnLbpy] and the relevant activation parameters ΔH^* in kcal/mol and ΔS^* in $cal \cdot mol^{-1} \cdot K^{-1}$ (estimated errors in parentheses) are listed below.

ligand	$L_1 = P_3O_{10}^{5-}$	$L_7 = C_4H_{13}N_3$	$L_8 = C_6H_{18}N_4$	$L_9 = C_7H_{20}N_4$	$L_{10} = C_6H_{18}N_4$
log k_f 1)	5.28	6.03	5.81	5.75	5.80
ΔH^*	12.1 (1.8)	7.9 (1.1)	6.0 (1.1)	6.7 (1.0)	6.8 (1.6)
ΔS^* 1)	+6 (6)	−5 (4)	−12 (4)	−10 (4)	−9 (5)

1) Statistically corrected values.

The values are independent of pH generally in the pH range from 7.3 to 9.5 [3]. The results of kinetic investigations confirm the assumption that substitution of part of the inner hydration sphere of Mn^{2+} by the first ligand L generally has little influence on the kinetics of the reaction between Mn^{2+} and bpy [3].

References:

[1] H. Sigel, B. E. Fischer, B. Prijs (J. Am. Chem. Soc. **99** [1977] 4489/96, 4490). – [2] K. N. Leong, M. W. Lister (Can. J. Chem. **50** [1972] 3573/86, 3577, 3583). – [3] D. N. Hague, S. R. Martin (J. Chem. Soc. Dalton Trans. **1974** 254/8, 257). – [4] R. K. Kadian, J. L. Sharma, B. S. Garg, R. P. Singh (Indian J. Chem. A **19** [1980] 819/21).

16.1.31.5 Manganese(III) Complexes

Mnbpy(NO_3)$_3$. Large dark red crystals were formed during 24 h when solid 2,2′-bipyridyl was added at 5°C to the brown solution obtained after separation of solid $K_2[Mn(NO_3)_4]$ from a mixture of $KMnO_4$ and N_2O_4 in nitromethane. Equimolar amounts of Mn and bpy were used. The compound was also obtained by reacting $Mn(NO_3)_3$ with bpy in nitromethane under similar conditions, but with a much lower yield [1].

Weissenberg and precession photographs show that the compound belongs to the monoclinic space group $P2_1/c\text{-}C_{2h}^5$ (No. 14) with lattice constants a = 7.311(3), b = 18.022(7), c = 11.704(4) Å, and β = 108.73(3)° at 20°C; Z = 4. Structure determinations by three-dimensional Patterson and Fourier syntheses with refining up to R = 5.9% yield the atomic positions published in the paper (Mn at x = 0.0636(2), y = 0.1447(1), z = 0.2780(1)). The bond lengths Mn↔O, Mn↔N, N↔O, N↔C, and C↔C (in Å) are shown in **Fig. 20**, on p. 218. For bonding angles see the paper.

Mnbpy(NO_3)$_3$ is a monomeric molecular complex containing 7-coordinate Mn, and the configuration about Mn is that of an irregular pentagonal bipyramid. The Mn atom is coordinated in the equatorial plane to one nitrogen atom of bpy(N_5) and four oxygen atoms of two bidentate nitrate groups. The apical positions are occupied by the other nitrogen atom (N_4) of bpy and an oxygen atom of a monodentate nitrate group. The almost planar equatorial pentagon has a pseudo m symmetry along Mn-N_5 and is only slightly puckered. The two bidentate nitrate groups are quite dissimilar; one of them is bonded symmetrically to Mn (Mn↔O = 2.237(8) and 2.201(8) Å), while the other is bonded asymmetrically (Mn↔O = 2.112(8) and 2.400(8) Å). The Mn↔O bond length of the monodentate nitrate groups is considerably shorter (1.896(8) Å). The bidentate ligands ($2NO_3$, 1 bpy) are all planar, whereas the monodentate nitrate group is nonplanar [2]. The absence of nitrate ions and the presence of coordinated bpy are confirmed also by the IR spectrum [1]. The density D at 20°C measured by flotation is D = 1.82(2) g/cm^3; the density calculated from X-ray data is D = 1.81 g/cm^3 [2]. Magnetic measurements at 22°C yielded a molar susceptibility $\chi_{mol}^{corr} = 10530 \times 10^{-6}$ cm^3/mol (corrected for diamagnetism) and a magnetic moment $\mu_{eff} = 5.0 \pm 0.1\ \mu_B$, which indicates a high-spin manganese(III) complex [1]. The crystals are dichroic, appearing dark red to reflected light and dark green to transmitted light [2]. The room-temperature diffuse-reflectance electronic spectrum shows a band at 19700 cm^{-1}, assigned to a d-d transition. The IR spectrum (recorded for Nujol mulls in the 4000 to 350 cm^{-1} range) shows two strong absorption bands, at 1518 and 1265 cm^{-1}, which are very tentatively assigned to vibrations of bidentate nitrate groups. The two strong bands at 1500 and 1285 cm^{-1} are assigned to the monodentate nitrate group. Despite the nonequivalence of the three nitrate groups, only a single strong band at 923 cm^{-1} is observed for the lowest fundamental nitrate stretching region [1].

Fig. 20

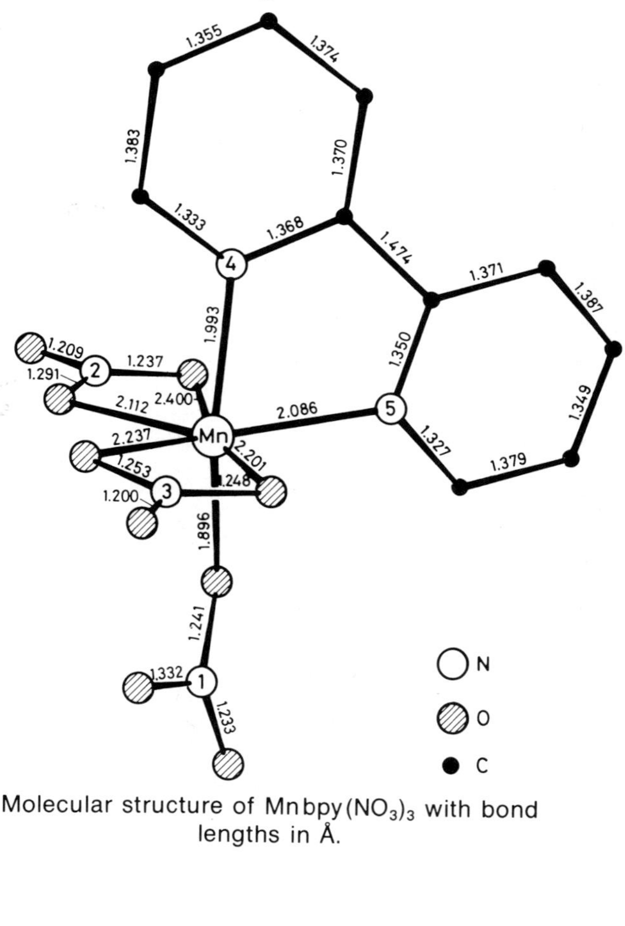

Molecular structure of Mn bpy(NO_3)$_3$ with bond lengths in Å.

Mn bpy$(NO_3)_3$ is decomposed by water to give a yellow-brown solution. It is soluble in acetonitrile, sparingly soluble in acetone (yellow solution), but insoluble in benzene, carbon tetrachloride, nitromethane, and ethyl acetate [1].

$[Mn bpy Cl_3]_n$ and $Mn bpy Cl_3 \cdot H_2O$. The anhydrous complex was prepared by controlled thermal decomposition of $[bpyH_2]MnCl_5$ at 100°C for 90 min [3] or at 130°C in vacuo for 2 h [4, p. 80]. The compound, formulated by [5] as $[Mn bpy Cl_3]_2$, was obtained when a suspension of MnO_2 (40 g) in CCl_4 was treated for two hours at −72°C with dry gaseous HCl, and the black residue was freed from excess acid and extracted with ether. To the purple solution so obtained an ethanolic solution of bpy (1.0 g) was added. The brick-red precipitate was collected, washed with ethanol and ether, and dried in vacuo [5]. The same compound was obtained when a solution of $MnCl_3$ (free from HCl) in ether was added dropwise at −5°C to a solution of 2,2'-bipyridyl in ether until the supernatant liquid was discolored. The precipitate, which crystallizes (dark red leaflets) on standing several hours at room temperature, was collected, washed with ether, and dried in flowing N_2 [12]. Mn bpy$Cl_3 \cdot H_2O$ was obtained by adding a filtered brown solution (15 mL) of $KMnO_4$ (10.0 g) in 10N HCl (100 mL) to a solution of bpy (1.0 g) in 2.5N HCl (20 mL) with stirring. A purple-brown precipitate appeared on scratching the walls of the container. The precipitate was washed with 2N HCl, small volumes of cold acetone and ether, and dried in vacuo [5]. The monoaquo complex was formed also on leaving $[bpyH_2]MnCl_5$ in moist air at room temperature [6].

The absorption band maximum in the electronic reflectance spectra is found at λ_{max} = 540 nm for the anhydrous complex and at λ_{max} = 550 nm for the monoaquo complex [5]. IR absorption bands in the range 500 to 200 cm^{-1} [3] and 700 to 200 cm^{-1} [8] observed for paraffin wax discs or Nujol mulls are given below:

complex	IR absorption bands								Ref.
$[Mn bpy Cl_3]_n$	421(m)	378(sh)	373(s)	361(s)	350(sh)	268(w)	240(mw)		[3]
Mn bpy$Cl_3 \cdot H_2O$	423(w)	384(s)	368(w)	357(s)		270(w)	263(w)	214(w)	[3]
Mn bpy$Cl_3 \cdot H_2O$		386(s)*)		356*)					[8]

*) Assigned to Mn-Cl stretching vibrations [8].

The Raman spectrum of $[Mn bpy Cl_3]_n$ shows two strong lines, one at 359 cm^{-1}, assigned to terminal Mn-Cl stretching, and one at 261 cm^{-1}, which may be due either to Mn-Cl-Mn bridging or Mn-N vibrations. The corresponding lines of the monoaquo complex are observed at 358 and 266 cm^{-1} [3]. Because the IR and Raman spectra of both $[Mn bpy Cl_3]_n$ and Mn bpy$Cl_3 \cdot H_2O$ are very similar, it is assumed by [3] that both compounds contain dimeric $[Mn bpy Cl_3]_2$ molecules, water being incorporated in the crystal lattice of Mn bpy$Cl_3 \cdot H_2O$ but not directly coordinated to Mn. This is confirmed by the loss of water on heating. However, two OH stretching bands, at 3450 and 3250 cm^{-1}, and the stability of Mn bpy$Cl_3 \cdot H_2O$ in vacuo over P_2O_5 favor the assumption of [5] that also the H_2O molecule is coordinated to Mn, thus completing an octahedral arrangement. The dimeric (or polymeric) nature of the anhydrous complex is also suggested by its magnetic moment (see p. 220), which is considerably lower than that of a high-spin Mn^{III} complex (μ_{eff} = 4.90 μ_B), and which is lower at lower temperatures (μ_{eff} = 3.48 μ_B at 118 K) [4, 5, 7]. On the other hand, the magnetic moment of the aquo complex is rather independent of temperature and corresponds to that of high-spin Mn^{III} [5, 7]. The following table lists the effective magnetic moments μ_{eff} (in μ_B) at the temperature (in K) given in parentheses and the Curie-Weiss constant Θ_p (in K) from susceptibility measurements in the temperature range beween 300 and 100 K [4, 7], where the Curie-Weiss law appeared to apply by both compounds [7]:

complex	μ_{eff}		Θ_p	
$[Mn\,bpy\,Cl_3]_n$	4.25 (292)		−154	−246
$Mn\,bpy\,Cl_3 \cdot H_2O$	4.96 (288)	4.98 ± 0.05	−25	
Ref.	[7]	[9]	[7]	[4]

The antiferromagnetic behavior of $[Mn\,bpy\,Cl_3]_2$ demonstrated by high values of Θ_p [4, 7] could be ascribed either to metal-metal interaction or to a superexchange mechanism through the bridging chlorine atoms. It is lost in the aquo complex, which may be a monomeric octahedral species [7], also see [5]. The ESR spectrum of $Mn\,bpy\,Cl_3 \cdot H_2O$ shows a single line with a width of about 600 Øe [9].

The anhydrous complex is air stable and decomposes on heating above 180°C [12]. According to DTA and TG studies it loses Cl_2 at 215°C in vacuo or at 217°C in N_2, leaving $Mn\,bpy\,Cl_2$ [4], also see [5]. It is readily soluble in water, giving dark brown solutions, from which all three Cl^- ions can be precipitated by $AgNO_3$. It is very sparingly soluble in methanol [12] and slightly soluble (with decomposition) in acetone, nitromethane, and nitrobenzene, giving green solutions [5]. The aquo complex is stable in a vacuum at room temperature over P_2O_5 [5] but loses its H_2O between 65 and 110°C when heated in air [3]. It can be converted to $[bpy\,H_2]MnCl_5$ by exposure to dry HCl gas [6]. The aquo complex also reacts with water [5]. A molar conductivity $\Lambda = 620\ \Omega^{-1} \cdot cm^2 \cdot mol^{-1}$ was measured for a 0.005 M solution at pH 2.6. This indicates that all Cl^- ions are released from the inner coordination sphere of the complex and are replaced by H_2O [9]. The catalytic activity of $Mn\,bpy\,Cl_3 \cdot H_2O$ in the decomposition of H_2O_2 [9] and the oxidation of isopropylbenzene by O_2 at 70°C [10] have been reported.

$K[Mn\,bpy(C_6H_4O_2)_2]$. The pyrocatecholato complex was prepared by mixing solutions of $K_3[Mn(C_6H_4O_2)_3] \cdot 3H_2O$ (1 g) in water (10 mL) and of bpy (0.28 g) in alcohol (2 mL). After heating on a water bath with continuous stirring a greenish-brown crystalline solid precipitated, which was washed with water and alcohol and air dried. It is assumed that complex formation in this case is preceeded by hydrolysis of $K_3[Mn(C_6H_4O_2)_3]$ to $K_3[Mn(C_6H_4O_2)_2(OH)_2]$, which then reacts with bpy to give $K[Mn\,bpy(C_6H_4O_2)_2]$. The complex is very slightly soluble in water and alcohol [11, pp. 370/2, 381].

References:

[1] D. W. Johnson, D. Sutton (Can. J. Chem. **50** [1972] 3326/31, 3327, 3330). – [2] F. W. B. Einstein, D. W. Johnson, D. Sutton (Can. J. Chem. **50** [1972] 3332/9, 3333, 3336). – [3] C. F. Bell, D. N. Waters (J. Inorg. Nucl. Chem. **39** [1977] 773/5). – [4] K. Akabori (J. Sci. Hiroshima Univ. A **39** [1975] 73/92, 76, 80, 85/90; Chem. Letters **1974** 1481/6; C. A. **83** [1975] No. 125417). – [5] H. A. Goodwin, R. N. Sylva (Australian J. Chem. **18** [1965] 1743/9, 1746/9).

[6] S. Sarkar, B. Sharma, R. C. Maurya, Sh. Ch. Chaurasia, Sh. K. Srivastava (Acta Ciencia Indica **4** [1978] 21/4). – [7] H. A. Goodwin, R. N. Sylva (Australian J. Chem. **20** [1967] 629/37, 631/4). – [8] S. N. Ghosh (J. Inorg. Nucl. Chem. **34** [1972] 1456/7). – [9] A. Ya. Sychev, Ya. D. Tiginyanu, Yu. S. Gromovoi (Zh. Fiz. Khim. **42** [1968] 2081/4; Russ. J. Phys. Chem. **42** [1968] 1097/9). – [10] A. Ya. Sychev, I. M. Reibel, L. P. Podgornaya (Zh. Fiz. Khim. **43** [1969] 676/80; Russ. J. Phys. Chem. **43** [1969] 370/3).

[11] P. Spacu, S. Popescu (Acad. Rep. Populare Romine Studii Cercetari Chim. **9** [1961] 367/95, 372, 381; C.A. **56** [1962] 15133/4). – [12] H. Funk, H. Kreis (Z. Anorg. Allgem. Chem. **349** [1967] 45/9, 48).

16.1.31.6 Manganese(III,IV) and Manganese(IV) Complexes

$[Mn_2O_2bpy_4]^{3+}$ and $[Mn_2O_2bpy_4]^{4+}$ Ions in Solution

Cyclic voltammetric studies and controlled potential electrolysis at +1.60 V on solutions of $Mnbpy_3(ClO_4)_2$ in acetonitrile containing 0.1M $(C_3H_7)_4NClO_4$ at 23°C (also see p. 207) reveal the presence of two di-μ-oxo complexes of composition $[bpy_2Mn^{III}O_2Mn^{IV}bpy_2]^{3+}$ and $[bpy_2Mn^{IV}O_2Mn^{IV}bpy_2]^{4+}$. It is assumed that the $Mn^{III,IV}$ complex is formed by the reversible reaction $2[Mnbpy_3]^{2+}+2H_2O \rightleftharpoons [Mn_2O_2bpy_4]^{3+}+4H^{+}+3e^{-}+2bpy$, which involves a possible disproportionation of an initially formed Mn^{III} species to Mn^{II} and Mn^{IV}. The existence of $[Mn_2O_2bpy_4]^{4+}$ is confirmed by controlled-potential electrolysis of $[Mnbpy_3]^{2+}$ at 1.60 V, which abstracts two electrons per Mn atom [1] and by cyclic voltammetry on acetonitrile solutions of previously isolated $Mn_2O_2bpy_4(ClO_4)_3$ [2 to 4]. Controlled-potential reduction of $[Mn_2O_2bpy_4]^{4+}$ at +0.80 V adds $0.5e^{-}$ per Mn atom to give $[Mn_2O_2bpy_4]^{3+}$. The corresponding half-wave potential is measured to be $E_{1/2}=+1.33$ V versus SCE [1].

The ESR spectrum of $[Mn_2O_2bpy_4]^{3+}$ is broadened due to superexchange with the oxo ligands and electron exchange between the two metal centers. The spectrum is described along with the properties of the perchlorate $Mn_2O_2bpy_4(ClO_4)_3 \cdot nH_2O$ on p. 223. The ESR spectrum of $[Mn_2O_2bpy_4]^{4+}$ confirms the equivalency of the two Mn atoms in the +4 oxidation state and the absence of significant electron pairing between them in solution (g = 2.03, A = 83 G) [1].

Isolated Compounds

$Mn_2O_2bpy_4(ClO_4)_3 \cdot nH_2O$ (n = 0, 2, 3). The dihydrate was first prepared by metathesis of $Mn_2O_2bpy_4(S_2O_8)_{1.5} \cdot 3H_2O$ (see p. 224) and excess $NaClO_4$ in water [6]. It was also prepared by adding an aqueous solution (60 mL) of $Mn(CH_3COO)_2 \cdot 4H_2O$ (17.5 mmol) to a solution of bpy (52.5 mmol) in acetone (30 mL). Then 80 mL of a 1M acetate buffer solution (pH 4.5) was added, and the pH adjusted to 4.5 by adding glacial acetic acid. After cooling to 0°C, a solution of $KMnO_4$ (7.5 mmol) in water (50 mL) was added dropwise with stirring, and stirring was continued for 15 min after addition of all $KMnO_4$. The green solution was filtered, and concentrated $NaClO_4$ solution was added to precipitate the complex as a green powder, which was washed with ethanol and diethyl ether. It can be recrystallized from acetonitrile or a 0.05M bipyridyl-bipyridinium nitrate buffer at pH 4.5 below 80°C. It can also be recrystallized at room temperature by slurrying it with Dowex-AGI-X8 in the nitrate form, thus dissolving it, and then by filtering off the resin and precipitating with $NaClO_4$ solution, which is added slowly until crystallization occurs. Combination of recrystallization from hot bipyridyl buffer and the resin technique at room temperature affords products of high purity [4]. The dihydrate can also be obtained by disproportionation of $MnbpyCl_3 \cdot H_2O$ in a bipyridyl buffer at pH 4.5 and precipitation with $NaClO_4$ [4]. Single crystals of the complex (assumed to be the trihydrate due to their loss in weight on heating at 130°C) were grown by diluting a saturated aqueous solution of bpy to 5 mL and stirring with 0.5 g of solid urea complex $Mn(CH_4N_2O)_6(ClO_4)_3$ for 15 min. The mixture then was filtered into a Petri dish, and 4 mL of ethanol (95%) was added. The greenish-black crystals appearing after 1 to 5 d of standing were collected and dried between filter paper [14].

The anhydrous complex resulted when the red methanolic solution (40 mL) containing bpy (15 mmol) and 5 mmol manganese(III) acetate dihydrate was precipitated with 60% $HClO_4$ (1.66 mL) in methanol or treated with HNO_3 (1.25 mL, d = 1.33 g/cm^3) and precipitated with $NaClO_4 \cdot H_2O$. After 3 h the dark green precipitate was washed with methanol and dried in vacuo [7].

The deep green single crystals of the trihydrate appearing as small rhomboids [14] or plates with well developed ab faces [8] are monoclinic, space group $P2_1/c$-C^5_{2h} (No. 14), with lattice constants a = 13.877(2), b = 13.936(2), c = 24.310(3) Å, and β = 103.56(1)°; Z = 4. The structure has been resolved by the heavy atom method and refined up to R = 0.080. The chelate portion for the cationic part $[Mn_2O_2bpy_4]^{3+}$ is represented in **Fig. 21** together with pertinent bond lengths (in Å). The two manganese atoms Mn(1) and Mn(2) are evidently different, and the smaller distances about Mn(2) indicate that this is an Mn^{4+} (d^3) ion. The lengthening of the Mn(2)↔N(6) and Mn(2)↔N(7) equatorial bonds relative to the Mn(2)↔N(5) and Mn(2)↔N(8) axial bonds indicates that N(6) and N(7) are trans to the μ-oxo bridge. The differences between the axial and equatorial bond lengths about Mn(1) are indicative of a high-spin Mn^{3+} (d^4) ion in a distorted environment [9]. The Mn(1)-Mn(2) axis lies almost parallel to the bc plane of the crystal and forms an angle of about 20° with the c axis [8]. The Mn(1)↔Mn(2) distance is 2.716 Å [9]. It is found to be about 0.039 Å shorter by the EXAFS (extended X-ray absorption fine structure) technique. Such shortening apparently confirms the di-μ-oxo configuration of the complex [13]. The density D = 1.62 g/cm³ was measured by flotation; D = 1.626 g/cm³ was calculated from X-ray data [9].

Fig. 21

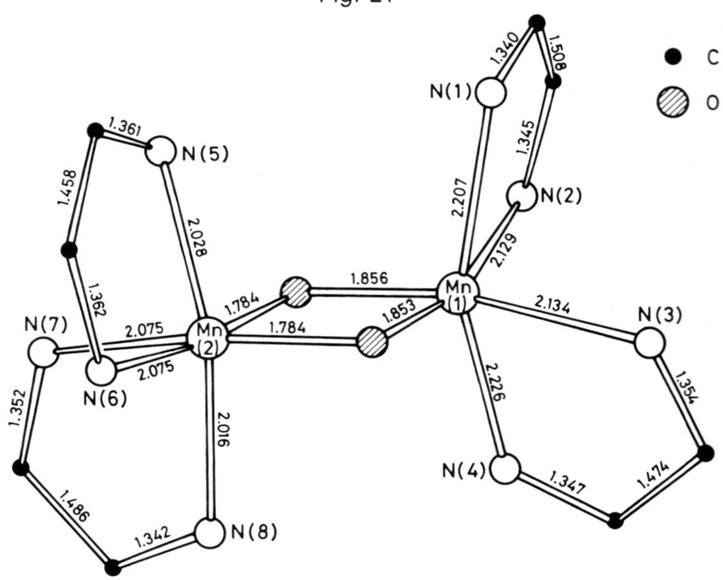

Structure of $[Mn_2O_2bpy_4]^{3+}$ in $Mn_2O_2bpy_4(ClO_4)_3 \cdot 3H_2O$ with bond lengths in Å.

The complex shows strong antiferromagnetic properties. Susceptibility measurements on the trihydrate (Gouy method) yield the magnetic moment $\mu_{eff} = 1.79 \pm 0.03\ \mu_B$ per Mn atom at room temperature, which decreases to $\mu_{eff} = 1.33\ \mu_B$ at 77 K, thus indicating an antiparallel alignment of the electrons on the Mn atoms induced by a superexchange coupling mechanism [9]. The dihydrate yields $\mu_{eff} = 1.66\ \mu_B$ per Mn atom at 298.8 K and 1.49 μ_B at 231.0 K [5]. For the anhydrous complex, $\mu_{eff} = 1.73\ \mu_B$ per Mn atom at 294.2 K was measured by the Faraday method [7]. The solution phase susceptibility of $[Mn_2O_2bpy_4]^{3+}$ in acetonitrile determined by NMR at room temperature and corrected for diamagnetism gives the effective magnetic moment $\mu_{eff} = 2.82 \pm 0.11\ \mu_B$ for the binuclear unit and $\mu_{eff}=1.81\pm0.07\ \mu_B$ (uncorrected) per Mn atom, which supports the conclusion that the di-μ-oxo-bridged species is stable in solution [2].

Low-temperature measurements (vibrating sample magnetometer), from 200 to 4 K, yielded $\mu_{eff}=2.04\ \mu_B$ at 196 K and 1.74 μ_B at 4 K for the dihydrate. The moment at 4 K is close to the $S=1/2$ spin-only moment of 1.73 μ_B, thus indicating high-spin Mn^{III} ions. The $Mn^{III,IV}$ complex can be described as an $S=(2, 3/2)$ system [10], in agreement with the optical [4] and structural data [9]. The temperature dependence of the magnetic data is in good agreement with the isotropic Heisenberg spin Hamiltonian $H=-2J\ S_1 \cdot S_2$ for the isotropic spin exchange energy $J=-150\pm7\ cm^{-1}$ [10], which differs significantly from the $J=-101\ cm^{-1}$ measured by [9]. The nonequivalency of the Mn nuclei was also confirmed by the ESR (X-band) spectra of solutions in acetonitrile between 298 and 12 K, which show a hyperfine structure with the parameters $A_1=167\pm3$ G and $A_2=79\pm3$ G (subscripts 1 and 2 refer to Mn^{III} and Mn^{IV}, respectively). The estimated average g value is 2.003. The hyperfine pattern with $|A_1|\approx2|A_2|$ and the small g anisotropy are consistent with high-spin Mn^{3+} (d^4) antiferromagnetically coupled to Mn^{4+} (d^3), producing an $S=1/2$ ground state [10]. ESR data (g=2.01, A=79 G) recorded on solutions of $[Mn_2O_2bpy_4]^{3+}$ in acetonitrile after controlled potential electrolysis of $[Mnbpy_3]^{2+}$ are given in [1]. The ESR spectrum of an $Mn_2O_2bpy_4(ClO_4)_3\cdot3H_2O$ single crystal shows a signal with hyperfine structure at g=2.00, the anisotropy of g being less than 0.01. In addition to this doublet ground state signal, there are also signals of the excited quartet state. An exchange integral $J/k=-207$ K was determined from the dependence of the signal intensities on temperature. The quartet-state spectrum exhibits a fine structure with estimated constants $|D|=0.39\ cm^{-1}$ and $|E|=0.12\ cm^{-1}$, which are interpreted in the terms of dipole-dipole interactions and crystal-field effects. The hyperfine structure of the doublet-state spectrum indicates an unequal distribution of odd electrons on the two manganese atoms of the complex [8].

The electronic diffuse reflectance spectrum of the dihydrate shows a broad asymmetric band between 10900 and 14700 cm^{-1} with two shoulders at 17900 and 22700 cm^{-1}, which could be assigned to a superposition of two absorptions corresponding to the electron transitions $^5B_{1g}\rightarrow{}^5A_{1g}$ and $^5B_{1g}\rightarrow{}^5B_{2g}$ of the Mn^{3+} ion in a tetragonally distorted octahedral ligand field [5]. The visible spectrum of $[Mn_2O_2bpy_4]^{3+}$ in 0.05 M aqueous bipyridyl buffer solution (pH=4.5) shows shoulders at 525 nm (≙19000 cm^{-1}) and 555 nm (≙18000 cm^{-1}) and a maximum at 684 nm (≙14500 cm^{-1}). A broad band with maximum at approximately 830 nm (≙12000 cm^{-1}) appears as a shoulder on the 684-nm band. The corresponding molar extinction coefficients are 530, 455, 561, and 476 $L\cdot mol^{-1}\cdot cm^{-1}$. The 830-nm band is considered to be due to a mixed valence or intervalence transfer absorption transition [4].

The IR spectrum of the dihydrate (recorded for KBr pellets or mineral oil mulls) in the ligand ring mode region shows absorption bands which are typical of bpy coordinated to metal ions. The band positions are given in a table [4]. Two bands in the 3500 and 1620 cm^{-1} region are assigned to vibrations of the water of crystallization (OH stretching and H-OH bending). Two very strong bands, at 1080 (broad) and 620 cm^{-1}, can be assigned to vibrations of an ionic perchlorate group with T_d symmetry. Three strong bands, at 686, 652, and 580 cm^{-1}, are assigned to vibrations of the double oxygen bridge between the two Mn atoms [5]. The band observed at 688 cm^{-1} by [4] is shifted to 676 cm^{-1} when the dihydrate is recrystallized at about 80°C from a bipyridyl buffer solution (pH=4.5) containing 53 at.% $H_2{}^{18}O$. The new band is shifted back to 688 cm^{-1} upon substitution by $H_2{}^{16}O$, thus establishing the 688 cm^{-1} band as one of the four expected bridging modes. Other IR bands are not sensitive to isotopic substitution [4].

Measurements on a 4.10×10^{-4} M aqueous solution of the dihydrate at 17°C yielded a molar electrical conductivity of $\Lambda = 283.8\ \Omega^{-1}\cdot cm^2\cdot mol^{-1}$, which lies between that of a 1:2 and a 1:3 electrolyte because of decomposition [5].

The solid dihydrate is stable at room temperature even when exposed to daylight and humid air for several days [5]. Irradiation with a Pyrex-filtered high-pressure Hg arc ($\lambda = 313$ nm) reduces it to an Mn^{II} complex via the unstable $Mn^{III,III}$ complex. No decomposition is observed for $\lambda > 320$ nm. Thus excitation of the metal-ligand charge-transfer bands is assumed to be effective in the photoreduction of the complex [3]. The compound is unstable in aqueous solution at temperatures above ~80°C but dissolves without decomposition in a 0.05 M aqueous bipyridyl buffer solution of pH = 4.5 to give a green solution, which upon acidification becomes red. The solution again becomes green (but less intensively) upon addition of a solution of CH_3COONa. The equilibrium $[Mn_2O_2bpy_4]^{3+} + H_2O + H^+ \rightleftharpoons [Mnbpy_2(OH)(OH_2)]^{2+} + 0.5[Mn_2O_2bpy_4]^{4+}$ is postulated from the pH dependence of the magnetic susceptibility and the extinction coefficient of the 830-nm band in aqueous bipyridyl(0.04 M)-$NaNO_3$(1 M) solution at 25°C. The acidified solution decomposes on prolonged storage at pH < 4, turning colorless as all Mn is reduced to the divalent state. The decomposition rate increases with increasing acidity. Decomposition producing hydrous manganese oxides is afforded by addition of strong bases such as NaOH unless they are added extremely slowly [4]. A 0.002 M solution of $[Mn_2O_2bpy_4](ClO_4)_3$ in a mixture of acetonitrile and 1 M H_2SO_4 (volume ratio 3:1) oxidizes anthracene (0.004 M) on irradiation with light of $\lambda > 360$ nm to give anthraquinone in a ~50 to 60% yield while the complex is reduced to a manganese(III, III) species, probably $[Mn_2O_2bpy_4]^{2+}$. No reduction of the manganese(III, IV) complex occurs with 1,10-phenanthroline or 9,10-dicyanoanthracene, but in the presence of the latter, phenylacetic acid is oxidized to give benzyl alcohol and benzaldehyde in considerable yields. It is assumed that the formation of benzyl alcohol implies a hydroxyl radical, generated by a one-electron oxidation of water. The complex decomposes on prolonged irradiation [12].

According to [1] the electrochemical reduction of $[Mn_2O_2bpy_4]^{3+}$ back to $[Mnbpy_3]^{2+}$ is possibly due to the availability of protons to form water from the oxo bridges and the availability of excess bpy to displace oxo ligands. The cathodic peak potential for the reduction of $[Mn_2O_2bpy_4]^{3+}$ to $[Mn_2O_2bpy_4]^{2+}$ was measured to be +0.38 V versus SCE [1]. However, $[Mn_2O_2bpy_4]^{2+}$ is immediately destroyed by a coupled chemical reaction and is not directly detectable, also see [2, 4].

$Mn_2O_2bpy_4(S_2O_8)_{1.5}\cdot3H_2O$ and $Mn_2O_2bpy_4(PF_6)_3\cdot2H_2O$. The persulfate trihydrate was prepared by adding $K_2S_2O_8$ (12 mmol) quickly to a stirred solution of $MnSO_4\cdot H_2O$ (7 mmol) and bpy (21 mmol) in water (60 mL) at 80°C. After 10 min, more $K_2S_2O_8$ (11.25 mmol) was added, and after a further 5 min the dark brown crystals were filtered off, washed with small portions of water, ethanol, and ether, and dried in vacuo over P_2O_5 [5]. The compound was also obtained when $MnSO_4\cdot4H_2O$ was dissolved in a warm water suspension of molten bpy in excess and oxidized with $K_2S_2O_8$. Large greenish-black crystals appeared on cooling. The complex was formed also by oxidation of $[Mnbpy_3]^{2+}$ with $(NH_4)_2S_2O_8$ in an acetate buffer solution [6]. The hexafluorophosphate dihydrate was obtained by reacting the red methanolic solution (40 mL) of manganese(III) acetate dihydrate (5 mmol) and bpy (15 mmol) with 1.25 mL HNO_3 ($d = 1.33$ g/cm^3) in 5 mL of methanol and adding a methanolic solution of $NaPF_6\cdot H_2O$ (2.74 g) for precipitation. After 3 h the green precipitate was washed with water and methanol and dried in a vacuum [7].

Magnetic susceptibility measurements yield the following magnetic moment μ_{eff} (in μ_B) per Mn atom; the temperatures in K are given in parentheses:

compound	μ_{eff}	μ_{eff}	Ref.
$Mn_2O_2bpy_4(S_2O_8)_{1.5}\cdot 3H_2O$	1.61 (293.86)	1.23 (78.28)	[5]
$Mn_2O_2bpy_4(PF_6)_3\cdot 2H_2O$	1.94 (292.5)	1.57 (78.0)	[7]

The diffuse reflectance electronic spectrum of the persulfate trihydrate shows a broad asymmetric band between 10500 and 14300 cm^{-1} with shoulders at 18000 and 23300 cm^{-1}. In the IR spectrum two strong broad bands of the $S_2O_8^{2-}$ ion are observed at 1270 and 660 cm^{-1} together with other bands at 1000 and 600 to 550 cm^{-1} [5]. The magnetic properties and spectroscopic bands as well as the molar electrical conductivity ($\Lambda = 292.0\ \Omega^{-1}\cdot cm^2\cdot mol^{-1}$ at 19°C) suggest a structure much like that of $Mn_2O_2bpy_4$-$(ClO_4)_3\cdot 2H_2O$ [5].

$Mn_2O_2bpy_4(S_2O_8)_{1.5}\cdot 3H_2O$ is stable at room temperature even when exposed to daylight and humid air for several days [5]. It loses its water of crystallization at 110°C [6].

MnbpyCl$_4$ was prepared by adding 15 mL of the brown solution, obtained from 10 g $KMnO_4$ in 100 mL of 10N HCl, to a solution of bpy (1.0 g) in 5N HCl (20 mL) while stirring the solution and scratching the walls of the vessel. The black crystalline precipitate was washed with water, then with small volumes of acetone and ether, and dried in vacuo. A pure product was obtained if crystals of this rather impure product were used as seeds in a second preparation [15]. The compound has a magnetic moment $\mu_{eff} = 3.82\ \mu_B$ at 291 K, which is almost temperature independent [11], corresponding to a high-spin Mn^{IV} complex [15]. The low Curie-Weiss value $\Theta_p = -8$ K suggests simple paramagnetism [11]. $MnbpyCl_4$ slowly loses Cl_2 on standing in air, its color changing to red brown. At 150 to 155°C $[MnbpyCl_3]_2$ is formed [15].

References:

[1] M. M. Morrison, D. T. Sawyer (Inorg. Chem. **17** [1978] 333/7). – [2] M. M. Morrison, D. T. Sawyer (J. Am. Chem. Soc. **99** [1977] 257/8). – [3] Y. Otsuji, K. Sawada, I. Morishita, Y. Taniguchi, K. Mizuno (Chem. Letters **1977** 983/6). – [4] S. R. Cooper, M. Calvin (J. Am. Chem. Soc. **99** [1977] 6623/30). – [5] R. Usón, V. Riera, M. Laguna (Transition Metal Chem. [Weinheim] **1** [1976] 21/5).

[6] R. S. Nyholm, A. Turco (Chem. Ind. [London] **1960** 74/5). – [7] R. Usón, V. Riera, M. A. Ciriano (Transition Metal Chem. [Weinheim] **1** [1976] 98/9). – [8] M. Inoue (Bull. Chem. Soc. Japan **51** [1978] 1400/3). – [9] P. M. Plaksin, R. C. Stoufer, M. Mathew, G. J. Palenik (J. Am. Chem. Soc. **94** [1972] 2121/2). – [10] S. R. Cooper, G. C. Dismukes, M. P. Klein, M. Calvin (J. Am. Chem. Soc. **100** [1978] 7248/52, 7250).

[11] H. A. Goodwin, R. N. Sylva (Australian J. Chem. **20** [1967] 629/37, 631/4). – [12] K. Mizuno, J. Ogawa, Y. Otsuji (Chem. Letters **1979** 945/6). – [13] J. A. Kirby, A. S. Robertson, J. P. Smith, A. C. Thompson, S. R. Cooper, M. P. Klein (J. Am. Chem. Soc. **103** [1981] 5529/37, 5531/3). – [14] P. M. Plaksin (Diss. Univ. Florida 1971, pp. 1/221, 14; Diss. Abstr. Intern. B **33** [1972] 619). – [15] H. A. Goodwin, R. N. Sylva (Australian J. Chem. **18** [1965] 1743/9, 1746/9).

16.1.32 With 4,4′-Dimethyl- and 4,4′-Diphenyl-2,2′-bipyridyl

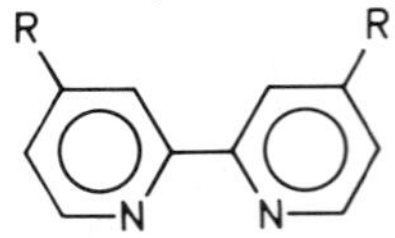

$R = CH_3$ $(=C_{12}H_{12}N_2)$
$R = C_6H_5$ $(=C_{22}H_{16}N_2)$

Complex Ions in Solution. Voltammetry of acetonitrile solutions of $Mn(C_{12}H_{12}N_2)_3(ClO_4)_2$ and $Mn(C_{22}H_{16}N_2)_3(ClO_4)_3$, described below, which contain 0.1M $(C_2H_5)_4NBF_4$ at 25°C revealed a three-step reduction wave and a one-step oxidation wave. An accurate potential determination could not be made for the oxidation [2]. In the presence of excess free ligand, d.c. polarography affords well-defined reduction waves corresponding to the species $[MnL_3]^+$, $[MnL_3]$, and $[MnL_3]^-$. The half-wave potentials occur at $E_{1/2} = -1.424$, -1.607, and -1.807 V for $Mn(C_{12}H_{12}N_2)_3(ClO_4)_2$ and at $E_{1/2} = -1.228$, -1.393, and -1.579 V for $Mn(C_{22}H_{16}N_2)_3(ClO_4)_2$ (versus Ag/AgCl, saturated NaCl). Each of these waves indicates a reversible one-electron reduction [2]. In acetonitrile solutions (23°C) of $MnL_3(ClO_4)_2$ containing 0.1M $(C_3H_7)_4NClO_4$ the oxidation waves occur at $E_{1/2} = +1.22$ V for $L = C_{12}H_{12}N_2$ and at +1.24 V for $L = C_{22}H_{16}N_2$ (versus SCE), producing Mn^{III} in a quasi-reversible one-electron reaction, which is followed by partial disproportionation and subsequent dimerization leading to the di-μ-oxo di-manganese(III, IV) complex species $[Mn_2O_2L_4]^{3+}$. These species can be oxidized reversibly to the corresponding $Mn^{IV,IV}$ species $[Mn_2O_2L_4]^{4+}$ at $E_{1/2} = +1.20$ V for $L = C_{12}H_{12}N_2$ and +1.21 V for $L = C_{22}H_{16}N_2$, or they can be reduced to an unstable $Mn^{III,III}$ species; the cathodic peak potentials are +0.25 V and +0.28 V, respectively, versus SCE.

ESR results (g, A) of the perchlorate complexes containing the cations $[MnL_3]^{2+}$, $[Mn_2O_2L_4]^{3+}$, and $[Mn_2O_2L_4]^{4+}$ in acetonitrile solution are listed in the following table:

ligand	$[MnL_3]^{2+}$		$[Mn_2O_4L_4]^{3+}$		$[Mn_2O_2L_4]^{4+}$	
	g	A(G)	g	A(G)	g	A(G)
$L = C_{12}H_{12}N_2$	2.01	85	2.01	78	2.01	77
$L = C_{22}H_{16}N_2$	2.01	unresolved	2.01	77	2.01	79

Thus the complexes behave like those of the parent ligand, 2,2′-bipyridyl, see p. 221, and the electrochemical and spectrographic results can be interpreted in the same way [1].

$Mn(C_{12}H_{12}N_2)_3(ClO_4)_2$ and **$Mn(C_{22}H_{16}N_2)_3(ClO_4)_2$** were prepared by adding an aqueous solution of $Mn(ClO_4)_2 \cdot 6H_2O$ to a slight excess of the ligand dissolved in ethanol [1] or by precipitation from a solution containing stoichiometric amounts of $MnSO_4$ and ligand in aqueous ethanol with $NaClO_4$, recrystallizing the precipitate from aqueous ethanol [2]. Crystals of the yellow complexes were washed with cold ethanol [1] and dried in a vacuum at 80°C for 24 h [1, 2].

References:

[1] M. M. Morrison, D. T. Sawyer (Inorg. Chem. **17** [1978] 333/7). – [2] M. C. Hughes, J. M. Rao, D. J. Macero (Inorg. Chim. Acta **35** [1979] L321/L324, L322).

16.1.33 With N-Oxides of 2,2′-Bipyridyl

16.1.33.1 With 2,2′-Bipyridyl N-Oxide

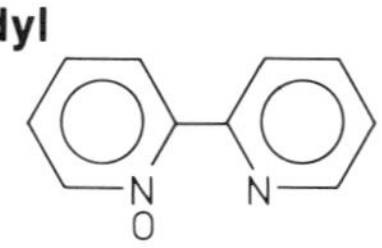

$(=C_{10}H_8N_2O)$

$Mn(C_{10}H_8N_2O)_3(ClO_4)_2$. Solutions of $Mn(ClO_4)_2 \cdot nH_2O$ and ligand, each in a mixture of ethanol and $HC(OC_2H_5)_3$ (volume ratio 7:3), were heated at 50 to 60°C for 30 min to remove water and mixed in a 1:3 mole ratio. After keeping the mixture at 50 to 60°C for about 5 to 10 min and slow cooling to room temperature, the precipitate was collected, washed with ethanol, and dried in vacuo over P_2O_5.

Susceptibility measurements at 300 K yielded a magnetic moment, $\mu_{eff}=5.73\ \mu_B$ (corrected for diamagnetism), characteristic for octahedral Mn^{II} compounds.

The electronic reflectance spectrum (Nujol mulls) shows strong bands at 235 (broad), 303, 366, 421 (broad), and 527 (shoulder) nm. The last two are considered as metal-to-ligand charge-transfer bands and cause the deep yellow color of the complex. The strongest absorption bands of a solution spectrum (6×10^{-4} M) in nitromethane are at 371, 432, 441, and 488 nm. The absorption bands in cm^{-1} of the IR spectrum (Nujol mulls) are tentatively assigned to stretching (ν) or deformation (δ) vibrations as follows:

ν(NO)	δ(NO)	ν(Mn-O)	ν(Mn-N)
1220(sh), 1213	842	355	265

Vibrations assigned to the perchlorate group at 1085 cm^{-1} (ν_3), 930 cm^{-1} (ν_1), and 621 cm^{-1} (ν_4) suggest the perchlorate to be ionic in the solid state. All other bands (see original paper) are assigned to ligand vibrations. The negative shifts of the ν(NO) and δ(NO) bands from the free-ligand values demonstrate that coordination of the ligand to Mn occurs through the oxygen. Coordination through the N atom of the pyridine fragment is suggested by the shifts and splittings of the pyridine ring modes (ν_{8a}, ν_{19a}, ν_{10b}). Conductivity measurements on 10^{-3} M solutions in nitromethane at 25°C yield a molar electrical conductivity $\Lambda=196\ \Omega^{-1}\cdot cm^2\cdot mol^{-1}$, corresponding to a 1:2 electrolyte. All measurements suggest that the ligand behaves as a bidentate O,N chelating agent, forming an octahedral arrangement around the 6-coordinate Mn atom. The ligand assumes a configuration close to that of the cis-form, with twisting of the two aromatic rings about the 2,2′ bond.

The complex gives a colorless solution in water, which most probably is due to dissociation into $[Mn(H_2O)_6]^{2+}$ ions and uncoordinated ligand. It is very sparingly soluble in ethanol and $CHCl_3$ but more soluble in nitromethane and dimethyl sulfoxide. The air-stable compound darkens at 240°C and decomposes at 250°C [1].

$Mn(C_{10}H_8N_2O)Cl_2\cdot C_2H_5OH$. Solutions of $MnCl_2\cdot nH_2O$ and ligand, each in a mixture of ethanol and $HC(OC_2H_5)_3$ (volume ratio 3:2), were heated to remove water and mixed. The crystalline precipitate was washed with ethanol and dried. Independent of the ligand-to-metal salt mole ratio, which may be changed from 3:1 to 1:2, the identical compound is obtained. The yellow-orange compound has a magnetic moment $\mu_{eff}=6.08\ \mu_B$ at 300 K (high spin). For d spacings determined from the X-ray powder pattern, see [2]. The electronic reflectance spectrum (Nujol mulls) shows strong bands at 203, 247, 278, 293 (broad), 325, and 435 (shoulder) nm. The solution spectrum (2.7×10^{-3} M in dimethyl sulfoxide) has a weak absorption band at 411 nm. The presence of strong Mn-to-ligand charge-transfer bands in the electronic spectrum is suggested by the intense color of the complex. The two aromatic rings of the ligand are twisted about the 2,2′-bond (as in the perchlorate complex).

The absorption bands in cm^{-1} of the IR spectrum are tentatively assigned to stretching (ν) or deformation (δ) vibrations as follows:

ν(NO)	δ(NO)	ν(Mn-O)	ν(Mn-Cl)	ν(Mn-N)
1241(sh), 1208	846(sh), 840	367, 360, 355	329, 324	271

The ν(OH) (ethanol) vibration (very broad band at 3550 to 3400 cm^{-1}) speaks in favor of the presence of lattice ethanol. All other bands (see the paper) are assigned to ligand vibrations. Coordination of the ligand through oxygen (of the N-oxide function) is demonstrated by the negative frequency shifts of the ν(NO) and δ(NO) bands, whereas coordination through the N atom of the pyridine fragment of the ligand is suggested by frequency shifts and splittings of the ring modes (ν_{8a}, ν_{19a}, ν_{10b}). Conductivity measurement for 10^{-3}M solutions in dimethylformamide at 25°C yields $\Lambda = 39\ \Omega^{-1} \cdot cm^2 \cdot mol^{-1}$, corresponding to a nonelectrolyte. This fact and the far-IR data – because ν(Mn-Cl) bands in similar positions occur also in tetrahedral manganese(II) chloride complexes – suggest a coordination number lower than six for Mn.

The complex loses its ethanol on being heated at 65°C and a reduced pressure of 60 Torr for 24 h as shown by the disappearance of the ν(OH) band of lattice ethanol. The air-stable compound is soluble in ethanol, $CHCl_3$, dimethylformamide, and dimethyl sulfoxide, but insoluble in nitromethane. It can be recrystallized from ethanol [2].

$Mn_2O_2(C_{10}H_8N_2O)_4(ClO_4)_3 \cdot 2H_2O$ and $Mn_2O_2(C_{10}H_8N_2O)_4(S_2O_8)_{1.5} \cdot 3H_2O$. To prepare the perchlorate of the type $[(C_{10}H_8N_2O)_2Mn^{III}O_2Mn^{IV}(C_{10}H_8N_2O)_2](ClO_4)_3$, an aqueous solution of $K_2S_2O_8$ (3.7 mmol) and $NaClO_4$ (41 mmol) was cooled to 0°C, and after removal of precipitated $KClO_4$, the filtrate was added dropwise to an aqueous solution containing $MnSO_4 \cdot H_2O$ (2 mmol) and ligand (6.6 mmol) at 74°C. After 10 min $NaClO_4$ (10 g) was added and after a further 10 min the precipitated crystals were washed with water, ethanol, and ether and dried in vacuo over P_2O_5. The perchlorate may be also obtained by reacting $[Mn_2O_2(C_{10}H_8N_2O)_4](S_2O_8)_{1.5} \cdot 3H_2O$ (0.09 mmol) with $NaClO_4$ (3 g) in H_2O (10 mL) at 20°C for 5 h and treating the resulting solid as above [3].

$Mn_2O_2(C_{10}H_8N_2O)_4(S_2O_8)_{1.5} \cdot 3H_2O$ was prepared by adding $K_2S_2O_8$ (4.3 mmol) to an aqueous solution of $MnSO_4 \cdot H_2O$ and ligand (mole ratio 1:3) at 45°C. After 2 min crystals appeared and more $K_2S_2O_8$ (1.1 g) was added. After a further 10 min the crystals were filtered off, and the filtrate was left at room temperature to precipitate purer crystals, which were treated in the same way as the perchlorate.

Both complex salts are green. In the following table are listed selected values of effective magnetic moment μ_{eff} in μ_B per Mn atom derived from susceptibility measurements between 65 and 300 K (temperatures in K are given in parentheses). Also given are the band frequencies in cm^{-1} of the electronic diffuse reflectance spectrum, of the stretching bands of the ligand, ν(NO), and of the main anion bands in the IR absorption spectrum of the complexes ($L = C_{10}H_8N_2O$).

complex	μ_{eff}	μ_{eff}	electronic bands	ν(NO)	anion bands
$[Mn_2O_2L_4](ClO_4)_3 \cdot 2H_2O$	1.88 (292.7)	1.13 (66.6)	12200(sh), 15300, 17900(sh)	1198	1080(b), 620
$[Mn_2O_2L_4](S_2O_8)_{1.5} \cdot 3H_2O$	1.82 (294.2)	1.17 (65.6)	12200(sh), 15500, 17900(sh)	1196	1270(b), 660

In the IR spectrum of the perchlorate complex, bands assigned to the $Mn\text{-}O_2\text{-}Mn$ oxygen bridge are observed at 692, 660(sh), 653, and 578 cm^{-1}. The corresponding bands of the persulfate complex are masked by bands characteristic of the anion. The shifts of the ν(NO) and δ(NO) bands and of the ring bands of the ligand (at 1600 and 760 cm^{-1}) from the free-ligand values indicate coordination to Mn through the oxygen atom of the amine oxide (N→O) group and through the nitrogen atom the pyridine ring fragment. The three $Mn\text{-}O_2\text{-}Mn$ bands of the perchlorate confirm double oxygen bridges with a symmetry other than D_{2h}. Such bridges are suggested also by the low magnetic moments of the complexes, which are even lower at lower temperature. The magnetic susceptibilities do not follow the Curie-Weiss law. The magnetic and spectral data can be interpreted as for the related complexes of 2,2′-bipyridyl.

Both complexes are stable in air at room temperature, but they decompose in solution [3].

References:

[1] A. N. Speca, N. M. Karayannis, L. L. Pytlewski, L. J. Winters, D. Kandasamy (Inorg. Chem. **12** [1973] 1221/6). – [2] A. N. Speca, N. M. Karayannis, L. L. Pytlewski (Inorg. Chim. Acta **17** [1976] 29/36). – [3] R. Usón, V. Riera, M. Laguna (Transition Metal Chem. [Weinheim] **1** [1976] 21/5).

16.1.33.2 With 2,2′-Bipyridyl N,N′-Dioxide

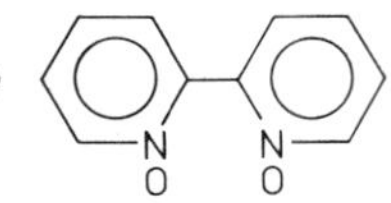

(= $C_{10}H_8N_2O_2$)

Manganese(II) Compounds

$Mn(C_{10}H_8N_2O_2)_3X_2 \cdot 2H_2O$ (X = ClO_4, Br) and **$Mn(C_{10}H_8N_2O_2)_3X_2 \cdot H_2O$** (X = ClO_4, I). The orange dihydrate of the perchlorate complex was prepared by mixing warm methanol solutions of $Mn(ClO_4)_2 \cdot nH_2O$ and ligand in stoichiometric amounts. It crystallized on cooling [1]. The monohydrate was obtained by mixing a hot aqueous solution of $MnClO_4 \cdot nH_2O$ with a boiling aqueous solution of the ligand in a 1:3 mole ratio. After the mixture was cooled to room temperature, the precipitate of orange needles was washed with small portions of cold water and recrystallized from boiling water in the dark – because it is very light sensitive, especially in solution [2]. The orange bromide dihydrate and iodide monohydrate were made in the same way as the perchlorate dihydrate, using $MnBr_2 \cdot nH_2O$ or $MnI_2 \cdot nH_2O$ instead of the perchlorate. All these crystals were washed with methanol and either ether or acetone, recrystallized from methanol, and dried in vacuo over P_2O_5 at room temperature [1]. Magnetic susceptibility measurements at 303 K yielded the magnetic moments $\mu_{eff} = 6.2\ \mu_B$ for $Mn(C_{10}H_8N_2O_2)_3Br_2 \cdot 2H_2O$ and $\mu_{eff} = 5.4\ \mu_B$ for $Mn(C_{10}H_8N_2O_2)_3I_2 \cdot H_2O$ [1]. The molar electrical conductivities Λ (in $\Omega^{-1} \cdot cm^2 \cdot mol^{-1}$) of the complexes were determined at 20°C in nitromethane solution for the perchlorate complex [3] and at 25°C in dimethylformamide solution for the bromo and iodo complexes. All these complexes are 1:2 electrolytes [1, 3].

complex	concentration ($M \cdot 10^3$)	Λ	Ref.
$Mn(C_{10}H_8N_2O_2)_3(ClO_4)_2 \cdot 2H_2O$	1.01	164	[3]
	0.145	200	
$Mn(C_{10}H_8N_2O_2)_3Br_2 \cdot 2H_2O$	1.00	133.7	[1]
	1.04	136.1	
$Mn(C_{10}H_8N_2O_2)_3I_2 \cdot H_2O$	0.97	144.5	[1]

The electronic spectra of $\sim 10^{-3}$ M solutions of the complex compounds in DMF show a band maximum at 410 nm with $\varepsilon = 750\ L \cdot mol^{-1} \cdot cm^{-1}$ [1]. An absorption maximum at 220 nm with shoulders at 241, 275, 301, and 428 nm (log ε = 4.9, 4.8, 4.5, 4.4, and 3.5, respectively) was observed by [4] for $[Mn(C_{10}H_8N_2O_2)_3]^{2+}$ in an acetonitrile solution containing 0.1M $(C_3H_7)_4NClO_4$. The band at 428 nm [4] or at 410 nm [1] can be assigned to a strong charge-transfer absorption [1, 4], which apparently is the cause of the orange color of the manganese(II) complexes [1]. The ESR X-band spectrum of $[Mn(C_{10}H_8N_2O_2)_3]^{2+}$ in acetonitrile yields the parameters g = 2.02 and A = 92 G. The value of the spectroscopic splitting parameter A indicates that the metal-ligand bonds have less covalent character than those of the related 2,2′-bipyridyl complex [4]. The measured NO stretching and bending frequencies, which agree quite well with those given by [5] for $Mn(C_{10}H_8N_2O_2)_3PtCl_4$ (see below), and the conductance and analytical data for the complexes show that each of the three 2,2′-bipyridyl N,N′-dioxide ligands is coordinated through both oxygen atoms to the same 6-coordinate Mn^{2+} ion, forming seven-membered rings [1]. The complexes are light sensitive [1]. The perchlorate hydrate complex decomposes very rapidly in light, turning black especially when finely powdered, but less rapidly in the dark. It is also very unstable in solution [2], but it seems to be stable in an oxygen-free atmosphere [3].

Electrochemical studies on solutions of $[Mn(C_{10}H_8N_2O_2)_3]^{2+}$ in acetonitrile containing 0.1M $(C_3H_7)_4NClO_4$ at 23°C reveal that the complex is oxidized reversibly in two one-electron steps at half-wave potentials $E_{1/2} = +0.87$ V and $E_{1/2} = +1.62$ V and is reduced irreversibly at a peak potential of −1.35 V. All potentials are measured versus SCE [4].

$Mn(C_{10}H_8N_2O_2)_3PtCl_4$ was prepared by dissolving the solid ligand in aqueous $MnCl_2 \cdot nH_2O$ at 45°C in a 3:1 mole ratio and, after cooling to room temperature, slowly adding the stoichiometric amount of aqueous K_2PtCl_4 in the dark. The precipitate was washed with small portions of cold water and finally with ethanol (95%) and dried in vacuo over P_2O_5 for 12 h in the dark. (The yield and purity may be increased by cooling the mother liquor for 1 h in a refrigerator.) The bright orange prisms begin to darken at 250°C but do not melt below 300°C. Susceptibility measurements at room temperature yield the magnetic moment $\mu_{eff} = 6.34$ μ_B [2]. In the IR spectrum recorded for the Nujol mull in the 4000 to 260 cm^{-1} region, the absorption at 1210 cm^{-1} is assigned to the NO stretching vibration, the doublet at 847 and 833 cm^{-1} to the NO bending vibration, and the absorption at 357 cm^{-1} to the Mn-O stretching vibration. The NO stretching vibration appears at a lower frequency in the complex than in the free solid ligand. This shift is attributed to a decrease of the NO double bond character as a result of the oxygen coordination to manganese. The presence of two NO bending modes, in both the complex and the free ligand, indicates that the ligand is in the gauche conformation [5]. The molar electrical conductivity $\Lambda = 266$ $\Omega^{-1} \cdot cm^2 \cdot mol^{-1}$ measured on a 1.022×10^{-3} M aqueous solution at 25°C corresponds to that of a 2:2 electrolyte. The complex is readily soluble in boiling water, moderately soluble in cold water, and practically insoluble in common organic solvents. The solid complex is stable in light for about 1 h, but solutions darken within a few minutes [2].

$Mn(C_{10}H_8N_2O_2)_2(NCS)_2$ and **$Mn(C_{10}H_8N_2O_2)Cl_2$.** The complexes were prepared by mixing hot ethanol solutions of manganese(II) thiocyanate or chloride and 2,2′-bipyridyl N,N′-dioxide (= L). The precipitates were washed with ethanol and dried. The magnetic moment μ_{eff} (in μ_B) at room temperature and the electronic and IR spectral wave numbers (in cm^{-1}, Nujol mulls) are given in the table on p. 231 together with the data of corresponding complexes formed with 4,4′-bipyridyl N,N′-dioxide ligand (= L′), which is described on p. 236. Dq is the ligand field splitting parameter in cm^{-1}. For other ligand field parameters, see [6]. The IR spectra show that the frequencies of the NO stretching (ν) and bending (δ) vibrations are lowered in going from the solid ligand 2,2′-bipyridyl N,N′-dioxide to the manganese complexes. This clearly indicates that both oxygen atoms of the ligand are bonded to a single Mn^{2+} ion with formation of a seven-membered chelate ring.

The magnetic moment and the IR and electronic spectra of the thiocyanate complex with ν(CN) at 2060, ν(CS) at 828, δ(NCS) at 480, and ν(Mn-NCS) at 290 cm^{-1} are indicative of a high-spin octahedral complex containing terminal N-bonded thiocyanate groups.

Table 1
Physical Properties of Complexes with 2,2′-Bipyridyl N,N′-Dioxide (= L) and 4,4′-Bipyridyl N,N′-Dioxide (= L′).

complex	μ_{eff}	electron transitions (in cm^{-1}) from $^6A_{1g}$ to					Dq	IR spectrum (in cm^{-1})		
		$^4T_{1g}(^4G)$	$^4T_{2g}(^4G)$	$^4E_g, ^4A_{1g}(^4G)$	$^4E_g(^4D)$	$^4T_{1g}(^4P)$		ν(NO)	δ(NO)	ν(Mn-X)
$MnL_2(NCS_2)_2$	5.89	18980	21500	24390	29680	32580	870	1225, 1205	840, 823	290
$MnL'(NCS)_2$	5.71	20000	—	23560	27630	31250	917	1205, 1175	830(sh), 825	270
$MnLCl_2$	5.97	18350	—	25970	—	—	1060	1220, 1208	845, 825	250
$MnL'Cl_2$	6.13	18870	21050	24390	29610	—	865	1230, 1183	850, 835	225
$Mn_2L'(SO_4)_2$	6.02	17790	20000	25300	27770	31250	993	1210, 1175	840, 825	—

Thus for $Mn(C_{10}H_8N_2O_2)_2(NCS)_2$ a monomeric structure with 6-coordinate Mn atoms surrounded by four oxygens of two 2,2′-bipyridyl N,N′-dioxide molecules and two N atoms of the terminally bonded thiocyanate groups is reasonable. $Mn(C_{10}H_8N_2O_2)Cl_2$ probably has a polymeric octahedral structure, with bidentate chelating ligand molecules and bridging Cl atoms. The compounds are quite stable [6].

Manganese(III) and Manganese(IV) Compounds

$[Mn(C_{10}H_8N_2O_2)_3]^{3+}$ and $[Mn(C_{10}H_8N_2O_2)_3]^{4+}$ Ions. These species were formed by electrochemical oxidation of $Mn(C_{10}H_8N_2O_2)_3(ClO_4)_2$ in acetonitrile solution at +0.87 and +1.62 V (versus SCE) [4]. Solutions of the complex ions in acetonitrile containing 0.1M $(C_3H_7)_4NClO_4$ yield electronic absorption spectra showing one maximum with several shoulders (wavelengths in nm). The extinction coefficients, as log ε, are in parentheses.

complex	maximum	shoulders				
$[Mn(C_{10}H_8N_2O_2)_3]^{3+}$	214(5.0)	260(4.7)	280(4.6)	362(4.0)	525(2.9)	605(2.6)
$[Mn(C_{10}H_8N_2O_2)_3]^{4+}$	213(4.9)		275(4.6)	375(4.0)	540(3.4)	

The 362-nm and 375-nm bands are probably due to charge-transfer processes of the Mn^{III} and Mn^{IV} complexes, respectively. The ESR X-band spectrum of $[Mn(C_{10}H_8N_2O_2)_3]^{4+}$ gives g = 2.00 and a splitting parameter A = 96 G. The electrochemistry of these complexes indicates that monomeric Mn^{III} and Mn^{IV} species can be stabilized by the ligand 2,2′-bipyridyl N,N′-dioxide. This is in contrast to the dimerization of Mn^{III} and Mn^{IV} when coordinated to ligands of bipyridyl and phenanthroline types, see pp. 221/5, 264. The reactivity of $[Mn(C_{10}H_8N_2O_2)_3]^{4+}$ towards water is slow and probably involves reaction with OH^- ions [4].

$Mn(C_{10}H_8N_2O_2)_3(ClO_4)_3 \cdot 3H_2O$ and $Mn(C_{10}H_8N_2O_2)_3(S_2O_8)_{1.5} \cdot 4H_2O$. The perchlorate was prepared by adding an aqueous solution of $Na_2S_2O_8$ (obtained from 3 g $K_2S_2O_8$ (0.011 mol) and 15 g $NaClO_4$ at 0°C) in small portions to a solution of $MnSO_4 \cdot 4H_2O$ (0.006 mol) and ligand (0.020 mol) in water at 90°C. After 10 min 30 g solid $NaClO_4$ was added and the temperature was kept at 80°C for 5 to 10 min, then lowered to 30°C. The dark red crystals were washed with water, ethanol, and ether, and stored in a desiccator. To prepare the persulfate, $K_2S_2O_8$ (0.012 mol) was added to an aqueous solution of $MnSO_4 \cdot 4H_2O$ (0.007 mol) and ligand (0.021 mol) at 75°C. After 10 min other $K_2S_2O_8$ (3.0 g) was added, and the solution was heated further for a few minutes. After cooling, the dark red crystals were washed carefully with water (4 × 4 mL), ethanol, and ether. Susceptibility measurements at 15°C yielded the magnetic moments μ_{eff} = 4.97 μ_B for the perchlorate hydrate and 5.00 μ_B for the persulfate hydrate, indicating high-spin manganese(III) complexes (four unpaired d electrons). Conductivity measurements on solutions of the perchlorate trihydrate in nitromethane at 20°C give the molar electrical conductivities Λ = 288 for 1.24×10^{-4} M and 227 $\Omega^{-1} \cdot cm^2 \cdot mol^{-1}$ for 8.42×10^{-4} M solutions, thus indicating 1:3 electrolytes. Both compounds are indefinitely stable in the solid state but are easily hydrolyzed by water, especially in dilute solutions. The anhydrous compounds could not be prepared, some of the H_2O molecules are tenaciously retained although their actual coordination to Mn is not likely. The perchlorate trihydrate loses part of its H_2O at 110°C (weight loss 2.1% in 10 h). Also H_2O is still present after heating the persulfate compound in vacuo at 90°C for 12 h over P_2O_5, as was shown by IR spectra [3].

References:

[1] S. K. Madan, W. E. Bull (J. Inorg. Nucl. Chem. **26** [1964] 2211/7). – [2] P. G. Simpson, A. Vinciguerra, J. V. Quagliano (Inorg. Chem. **2** [1963] 282/6). – [3] R. S. Nyholm, A. Turco

(J. Chem. Soc. **1962** 1121/2; Ric. Sci. Rend. A [2] **1** No. 1 [1961] 97/8; C.A. **57** [1962] 6870). – [4] M. M. Morrison, D. T. Sawyer (Inorg. Chem. **17** [1978] 338/9). – [5] A. Vinciguerra, P. G. Simpson, Y. Kakiuti, J. V. Quagliano (Inorg. Chem. **2** [1963] 286/9).

[6] I. S. Ahuja, R. Singh, C. L. Yadava (J. Mol. Struct. **74** [1981] 143/51, 145/6).

16.1.34 With 4,4'-Bipyridyl 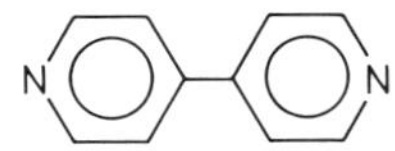 (= $C_{10}H_8N_2$)

A mixed-ligand complex with 4,4'-bipyridyl and acetylacetone is described in "Mangan" D 1, 1979, p. 80.

Anhydrous Compounds

The complexes $\mathbf{Mn(C_{10}H_8N_2)_2(NCS)_2}$ and $\mathbf{Mn(C_{10}H_8N_2)X_2}$ (X = Cl, Br, CH_3COO) can be prepared by mixing hot ethanol solutions of the manganese salt (thiocyanate, chloride, acetate) and 4,4'-bipyridyl. The sulfate $\mathbf{Mn_2(C_{10}H_8N_2)(SO_4)_2}$ was obtained by adding a saturated aqueous solution of $MnSO_4$ to the ethanol solution of the ligand [1]. The halides were also produced from a solution of $MnX_2 \cdot nH_2O$ in absolute ethanol dehydrated with triethyl orthoformate by adding an ethanol solution of the ligand. The colorless precipitates were washed with absolute ethanol and diethyl ether and dried in a vacuum over $CaCl_2$ [2]. The magnetic moment μ_{eff} in μ_B from susceptibility measurements at room temperature [1] and at 20°C [2], absorption maxima (in cm^{-1}) of the electronic (diffuse reflectance), and far-IR spectra (recorded for Nujol mulls in the 600 to 200 cm^{-1} range) [1, 2] are listed in the table on p. 235 together with the tentative assignment of electron transitions from the $^6A_{1g}$ ground state, the ligand-field splitting parameter Dq in cm^{-1} [1], and modes of Mn-X and Mn-N vibrations [1, 2].

In the isothiocyanato complex the IR bands assigned to fundamental vibration modes of coordinated thiocyanate occur at 2070 (ν-CN), 840 (ν-CS), 470 (δ-NCS), and 285 cm^{-1} (νMn-NCS), thus indicating terminally N-bonded NCS groups. In view of the magnetic moment and electronic spectrum, a polymeric complex with 6-coordinate Mn in an octahedral environment of N atoms may be assumed. Four N atoms originate from two bridging bidentate 4,4'-bipyridyl molecules, and two originate from the NCS groups. The physical properties of the halide complexes indicate that manganese(II) is surrounded octahedrally by a polymeric, halogen-bridged structure [1, 2] and two bridging 4,4'-bipyridyl molecules, much like the pyridine complexes [2].

The acetato complex shows IR absorption bands due to vibrations of coordinated acetate at 1580 (ν_{as}-OCO), 1430 (ν_s-OCO), and 665 cm^{-1} (δ-OCO). The separation of the symmetric and antisymmetric OCO stretching modes ($\Delta = 150\ cm^{-1}$) suggests the exclusive presence of symmetrically bonded bidentate acetate groups, which together with bidentate 4,4'-bipyridyl bridging between two Mn atoms, form a polymeric octahedral complex where the acetate groups are trans [1]. The 2:1 sulfato complex shows IR bands due to coordinated sulfate at 1120 to 1050 (ν_3), 1000 (ν_1), 615 (ν_4), and 450 cm^{-1}(ν_2). Other IR bands observed in the NaCl region are indicative of bidentate bridging 4,4'-bipyridyl. Thus, a polymeric octahedral structure of the sulfato complex, with extensive bridging through the sulfate oxygens, all coordinated to Mn, was tentatively proposed. The SO_4 groups essentially retain their T_d symmetry [1]. The halide complexes are insoluble in the common polar and nonpolar solvents [2].

Hydrates

An aqueous solution of $Mn(SCN)_2$ and excess ligand in ethanol was used to prepare the thiocyanato complex, $Mn(C_{10}H_8N_2)_2(NCS)_2 \cdot 2H_2O$. It was purified and dried like the sulfato complex. The nitrato and iodo complexes, $[Mn(C_{10}H_8N_2)(H_2O)_2X_2]$ ($X = NO_3$, I), were prepared by reacting solutions of manganese(II) nitrate or iodide in absolute ethanol with a solution of 4,4'-bipyridyl in a minimum amount of the same solvent at 60°C. A mole ratio lower than 1:2 (i.e., excess ligand) was used. The precipitates obtained on cooling were collected, washed with ethanol and ether, and dried in vacuo. $[Mn(C_{10}H_8N_2)(H_2O)_4]SO_4$ was obtained by reacting a warm aqueous solution of $MnSO_4$ with a hot ethanolic solution of excess ligand. The precipitate was washed with water, ethanol, and ether, and dried in a vacuum [3].

The color and IR bands in cm^{-1}, assigned to stretching vibrations ν(Mn-N) and ν(Mn-X) and to vibrations of the anions NCS^-, ONO_2^-, and SO_4^{2-} are listed below. For other IR bands see [3].

compound $L = C_{10}H_8N_2$	color	(Mn-N)	(Mn-X)	anion bands
$MnL_2(NCS)_2 \cdot 2H_2O$	pale yellow	203	245	2087 (ν_{CN}), 820 (ν_{CS}), 475 (δ_{NCS})
$[MnL(H_2O)_2(NO_3)_2]$	flesh	210 (234)	—	1460 (ν_4), 1310 (ν_1), 1050 (ν_2), 820 (ν_3)*)
$[MnL(H_2O)_2I_2]$	light brown	210	235	—
$[MnL(H_2O)_4]SO_4$	pale yellow	212 (255)	—	1120 (ν_3), 632 (ν_4)

*) Obscured by ligand bands.

It is concluded from the IR spectra that H_2O is coordinated to Mn in the nitrato, iodo, and sulfato complexes but not in the thiocyanato complex. In the nitrato complex the NO_3^- group is coordinated as a monodentate ligand. The NCS^- groups in the thiocyanato complex are coordinated to Mn through their N atoms. The iodo complex may have an octahedral configuration. In the IR spectrum of $[Mn(C_{10}H_8N_2)(H_2O)_4]SO_4$, the two single bands ($\nu_3$ (broad), ν_4) indicate the presence of uncoordinated SO_4^{2-} ions. Thus it is assumed that the complex contains the polymeric cation $[Mn(C_{10}H_8N_2)(H_2O)_4]_n^{2n+}$, with 4,4'-bipyridyl as a bridging bidentate ligand. As all these complexes are insoluble in polar and nonpolar solvents (except the nitrato complex, which dissolves slightly in acetonitrile to give nonconducting solutions), it is assumed that the other complexes are also polymeric, containing 4,4'-bipyridyl as a bridging ligand [3].

References:

[1] I. S. Ahuja, R. Singh, C. L. Yadava (J. Mol. Structure **74** [1981] 143/51, 145/6). – [2] A. Anagnostopoulos (J. Inorg. Nucl. Chem. **35** [1973] 3366/9). – [3] A. Anagnostopoulos (Inorg. Nucl. Chem. Letters **12** [1976] 225/34).

Table 2
Physical Properties of Complexes with 4,4'-Bipyridyl.

complex ($L = C_{10}H_8N_2$)	μ_{eff}	electron transitions (in cm^{-1}) from $^6A_{1g}$ to $^4T_{1g}(^4G)$	$^4T_{2g}(^4G)$	^4E_g, $^4A_{1g}(^4G)$	$^4E_g(^4D)$	$^4T_{1g}(^4P)$	Dq[3]	IR spectrum (in cm^{-1}) ν(Mn-X)	ν(Mn-N)	Ref.
$MnL_2(NCS)_2$	6.14	19610	22990	25000	27390	30190	1023	285	—	[1]
$MnLCl_2$	5.77	—	—	—	29100[1]	36500[1]	873	235	208[2]	[1,2]
$MnLBr_2$	5.77	18600[1]	—	—	—	35700[1]	—	(230)	208	[2]
$MnL(CH_3COO)_2$	5.90	—	—	—	26670	—	1053	—	—	[1]
$Mn_2L(SO_4)_2$	5.75	—	—	25640	—	—	1003	—	—	[1]

[1] Band maxima given by [2] were not assigned. – [2] Only given by [2]. – [3] For other ligand field parameters see [1]. Strong unassigned bands in the far-IR spectra are found at 491, 390, and 278 cm^{-1} for $Mn(C_{10}H_8N_2)Cl_2$ and at 398 and 270 cm^{-1} for $Mn(C_{10}H_8N_2)Br_2$ [2].

16.1.35 With 4,4′-Bipyridyl N,N′-Dioxide 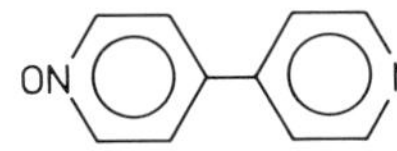 (= $C_{10}H_8N_2O_2$)

$Mn(C_{10}H_8N_2O_2)X_2$ (X = Cl, NCS) and **$Mn_2(C_{10}H_8N_2O_2)(SO_4)_2$** were prepared from manganese(II) chloride, thiocyanate, or sulfate and 4,4′-bipyridyl N,N′-dioxide in the same way as the corresponding complexes of 4,4′-bipyridyl (p. 233). Their magnetic moments and electronic and IR spectral wave numbers (NO and Mn-X bands) have been recorded, together with those of the complexes of 2,2′-bipyridyl N,N′-dioxide (p. 231). The steric position of the two oxygen atoms in the ligand precludes chelation, but they can coordinate to different manganese ions, forming a bridge and leading to polymer chain structures. The chloro complex apparently has a polymeric octahedral structure involving bridging Cl atoms in addition to the bridging 4,4′-bipyridyl N,N′-dioxide ligands. The magnetic moment and electronic and IR spectra of the isothiocyanate are indicative of a high-spin polymeric structure with bridging NCS groups (ν(CN) at 2090, ν(CS) at 800, δ(NCS) at 475, ν(Mn-NCS) at 270 cm^{-1}) and bidentate bridging 4,4′-bipyridyl N,N′-dioxide molecules. The Mn atom is surrounded by an octahedron formed by two N and two S atoms of bridging NCS groups and two oxygen atoms of the bridging organic ligand. In the binuclear sulfato complex, a polymeric octahedral structure is assumed: 4,4′-bipyridyl N,N′-dioxide also acts as a bidentate bridging ligand. The sulfate groups are coordinated to Mn through all their oxygen atoms, essentially retaining their T_d symmetry, as suggested by the IR spectrum, which shows bands assigned to SO_4 vibrations at 1150 to 1060 (ν_3), 1010 (ν_1), 610 (ν_4), and 480 cm^{-1} (ν_2), I. S. Ahuja, R. Singh, C. L. Yadava (J. Mol. Struct. **74** [1981] 143/51).

16.1.36 With 2,2′,2″-Terpyridyl and Related Compounds

16.1.36.1 With 2,2′,2″-Terpyridyl 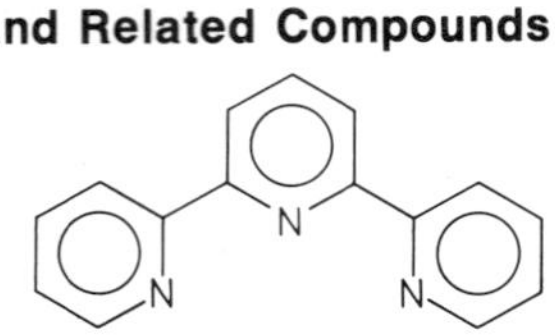 (= 2,2′:6′,2″-Terpyridine = $C_{15}H_{11}N_3$)

Complexes in Solution

Complexes exhibiting the lower oxidation states of manganese, $[Mn(C_{15}H_{11}N_3)_2]^+$, $[Mn(C_{15}H_{11}N_3)_2]$, $[Mn(C_{15}H_{11}N_3)_2]^-$, and $[MnC_{15}H_{11}N_3]^-$, are reported to form by electrochemical reduction of $Mn(C_{15}H_{11}N_3)_2(ClO_4)_2$ in acetonitrile solution. $[Mn(C_{15}H_{11}N_3)_2]^{3+}$ is formed by the electrochemical oxidation of $Mn(C_{15}H_{11}N_3)_2(ClO_4)_2$ in acetonitrile solution, see p. 238.

The stability constant (K in L/mol) of $[MnC_{15}H_{11}N_3]^{2+}$ at 298.5 K was determined by kinetic measurements in anhydrous methanolic solution of $Mn(ClO_4)_2$ at ionic strength I = 0.2 M ($NaClO_4$) and in aqueous solution. Values of log K and approximate values of the formation enthalpy ΔH in kcal/mol and the formation entropy in cal·mol^{-1}·K^{-1} are given below:

solvent	log K	ΔH	ΔS	Ref.
methanol	5.0	−1	+19	[1, 4]
water	4.4	−5.7	0*)	[2, 3]

*) Calculated by [1].

Rapid dissociation of $[MnC_{15}H_{11}N_3]^{2+}$ in aqueous solution at 298 K and pH ~ 7 was observed by [6]. The second-order rate constant of formation k_f in L·mol^{-1}·s^{-1}, the rate constant of dissociation k_d in s^{-1}, and the corresponding activation parameters ΔH^* in kcal/mol and ΔS^* in $cal \cdot mol^{-1} \cdot K^{-1}$ at 298.15 K are listed in the following table:

solvent	log k_f	log k_d 2)	ΔH_f^*	ΔH_d^*	ΔS_f^*	ΔS_d^*	Ref.
water	5.0	0.6	6.6	12.3	(−15.7) 1)	(−16.1) 1)	[2]
methanol	3.72	−1.33	12.8±1.0	13.7±0.6	+1.3±3.9	−18±2	[1, 4]

1) Values calculated by [1]. – 2) Dissociation of $[MnC_{15}H_{11}N_3]^{2+}$ is induced by cation exchange with Co^{2+} in water [2] or Hg^{2+} in methanol [1].

The pseudo-first-order decomposition rate constants are found to be largely independent of the concentration of the added metal ion [1, 2] and to increase considerably with temperature [1].

The solvent exchange rate constants log k_{ex} = 7.49 and 5.57 (k_{ex} in s^{-1}) at 298.1 K for water and methanol, respectively, are reported in [5]. Mixed complexes of terpyridyl and dimethyl sulfoxide formed in DMSO solution are described along with the pure DMSO complexes.

References:

[1] D. J. Benton, P. Moore (J. Chem. Soc. Dalton Trans. **1973** 399/404, 402). – [2] R. H. Holyer, C. D. Hubbard, S. F. A. Kettle, R. G. Wilkins (Inorg. Chem. **5** [1966] 622/5). – [3] C. D. Hubbard, P. Moore, R. G. Wilkins (Proc. 8th Intern. Conf. Coord. Chem., Vienna 1964, pp. 286/9; C.A. **67** [1966] No. 26152). – [4] D. J. Benton, P. Moore (J. Chem. Soc. Chem. Commun. **1972** 717/8). – [5] D. M. W. Buck, P. Moore (J. Chem. Soc. Dalton Trans. **1976** 638/42).

[6] R. Hogg, R. G. Wilkins (J. Chem. Soc. **1962** 341/50, 345, 349).

Isolated Compounds

$Mn(C_{15}H_{11}N_3)_2(ClO_4)_2$. Yellow crystals of the complex were formed on adding an aqueous solution of $Mn(ClO_4)_2 \cdot 6H_2O$ to a slight excess of ligand in ethanol [1] (also see [2]) or from a mixture of $MnCl_2 \cdot 4H_2O$ (0.5 mmol) dissolved in water and ligand (1.2 mmol) in hot absolute ethanol by precipitation with $NaClO_4$ [3]. The crystals were washed with cold ethanol [1] or with aqueous ethanol (50 vol%) and purified with absolute ethanol [3]. They were dried in vacuo at 80°C [1, 3] for 24 h [1] or 6 h over P_2O_5 [3]. Susceptibility measurements at room temperature (corrected for diamagnetism) yield a magnetic moment of $\mu_{eff} = 6.08\ \mu_B$, indicating a high-spin d^5 octahedral configuration. The electronic spectrum of a 6.68×10^{-6} M solution in acetonitrile shows four absorption maxima, at 331, 302, 280, and 232 nm, with molar extinction coefficients of 16800, 23100, 37100, and 46500 $L \cdot mol^{-1} \cdot cm^{-1}$, respectively. These maxima are accompanied by shoulders at 314, 275, 267, 249, and 239 nm [3]. The ESR spectrum of an acetonitrile solution containing 0.1M $(C_3H_7)_4NClO_4$ shows an unresolved band at g = 2.03 [1].

Voltammetric studies (d.c. and a.c. polarography, cyclic voltammetry) on solutions of $Mn(C_{15}H_{11}N_3)_2(ClO_4)_2$ in acetonitrile 0.1M in various supporting electrolytes reveal two one-electron diffusion-controlled reduction waves in the lower potential range from 0 to about −1.8 V versus saturated calomel electrode (SCE), which lead to the species $[Mn(C_{15}H_{11}N_3)_2]^+$ and $[Mn(C_{15}H_{11}N_3)_2]$ [2 to 4].

supporting electrolyte	reference electrode	$E_{1/2}$ first step	$E_{1/2}$ second step	Ref.
$(C_3H_7)_4NClO_4$	SCE	−1.132	−1.495	[2]
$(C_2H_5)_4NClO_4$	SCE	−1.155	−1.506	[4]
$(C_2H_5)_4NBF_4$	Ag/AgCl, satd. NaCl	−1.115	−1.455	[3]

Two further reduction waves are observed at $E_{1/2} \sim -1.85$ and −1.962 V versus SCE. They are supposed to correspond to the formation of the anionic species $[MnC_{15}H_{11}N_3]^-$ and $[Mn(C_{15}H_{11}N_3)_2]^-$. The 1:1 complex $[MnC_{15}H_{11}N_3]^-$ results from dissociation of the 1:2 complexes in the 0 or −1 oxidation state [4]. Two reduction waves for $E_{1/2} > -2$ V versus SCE correspond to the reduction of free terpyridyl [2, 4]. The oxidation wave leading to $[Mn(C_{15}H_{11}N_3)_2]^{3+}$ occurs at $E_{1/2} = +1.278$ V versus SCE and is obtained by d.c. polarography at the rotating Pt electrode (a.c. polarography yields $E_{1/2} = +1.298$ V) [3] in good agreement with $E_{1/2} = +1.28$ V versus SCE given by [1]. The oxidation to the Mn^{III} species is assumed to be quasireversible [3]. Electrolysis at +1.60 V yields a nonstoichiometric value for the number of electrons (>2) transferred per manganese, a value which increases on extending the electrolysis time. Thus the electrolytic oxidation of $Mn(C_{15}H_{11}N_3)_2(ClO_4)_2$ does not occur by a single rearrangement (dimerization as in the complex with 2,2′-bipyridyl, see pp. 207, 221), although follow-up chemical reactions are also assumed. A simple dimerization is probably prevented on stereochemical grounds [1].

$Mn(C_{15}H_{11}N_3)_2Br_2 \cdot nH_2O$ (n = 0, 1). $MnCl_2 \cdot 4H_2O$ (1 mmol), 2,2′,2″-terpyridyl (2 mmol) and water (10 mL) were boiled and shaken until the ligand has dissolved. To the cooled solution a saturated solution of KBr was added. Yellow crystals of the bromide complex precipitated. They were recrystallized from hot water and air dried. At about 56°C (in vacuo) the water of hydration was lost to give the anhydrous complex, $Mn(C_{15}H_{11}N_3)_2Br_2$. Susceptibility measurements at 20°C yield the magnetic moment $\mu_{eff} = 5.9\ \mu_B$ for the anhydrous complex, indicating high-spin. The anhydrous compound loses one molecule of terpyridyl when heated in vacuo at 80°C for 1.5 h to give pale yellow $Mn(C_{15}H_{11}N_3)Br_2$ [5].

$Mn(C_{15}H_{11}N_3)_2I_2$ was obtained by reacting an aqueous solution of $Mn(C_{15}H_{11}N_3)Cl_2$ with excess KI [6].

$Mn(C_{15}H_{11}N_3)X_2$ (X = Cl, Br, I). To prepare the chloro complex, equimolar amounts of $MnCl_2$ and ligand in aqueous solution were refluxed for one hour. The solution then was evaporated to induce crystallization. The bromo complex was obtained by reacting an aqueous solution of $Mn(C_{15}H_{11}N_3)Cl_2$ with excess KBr [6] or by heating $Mn(C_{15}H_{11}N_3)_2Br_2 \cdot H_2O$ in vacuo at 80°C for 1.5 h [5]. The iodo complex resulted upon heating $Mn(C_{15}H_{11}N_3)_2I_2$ in vacuo at 180°C for several hours [6].

Single crystals of the chloro complex were grown by slow dropwise addition of an ethanolic solution of the ligand to a boiling ethanolic solution containing a slight excess of $MnCl_2$. The yellow monoclinic crystals are elongated along [001]. According to Weissenberg and precession photographs the space group is $P2_1/c\text{-}C_{2h}^5$ (No. 14) with lattice constants a = 10.97, b = 8.25, c = 16.35 Å, and β = 94.0°; Z = 4. The unit cell volume V is 1476 Å^3. It is assumed that Mn is bonded to the three N atoms of the ligand and to two Cl atoms to form a distorted trigonal bipyramid [7]. The five-coordinate manganese in $Mn(C_{15}H_{11}N_3)X_2$ for X = Cl, Br, and I was established by comparison with the patterns of analogous Zn compounds previously shown to be 5-coordinate. Susceptibility measurements for $Mn(C_{15}H_{11}N_3)X_2$ at 298 K (X = Cl, Br) and 300 K (X = I) yield the magnetic moments μ_{eff} = 5.96, 5.98, and 5.94 μ_B for X = Cl, Br, and I, respectively, indicating high-spin. The compounds solvolyze and disproportionate quickly when in solution [6].

$Mn(C_{15}H_{11}N_3)Cl_3$. A solution of 2,2′,2″-terpyridyl (0.40 g) in 2.5 M aqueous HCl (20 mL) was added to a solution obtained by addition of 10 M HCl (15 mL) to $KMnO_4$ (1.5 g). Crystallization was induced by scratching. The crystals were collected 30 min after the initiation of precipitation, washed with 2 M HCl, acetone, and ether, and dried in vacuo over P_2O_5. Susceptibility measurements in the temperature range from 83 to 350 K (corrected for diamagnetism) show that the compound appears to obey the Curie-Weiss law with $\Theta_p = -10$ K. The magnetic moment μ_{eff} is 4.74 μ_B at 83 K and 4.91 μ_B at 310 K. The room temperature moment is approximately the spin-only value. The compound probably has an octahedral configuration with 6-coordinate Mn atoms. It is insoluble in or decomposed by all common solvents [8].

References:

[1] M. M. Morrison, D. T. Sawyer (Inorg. Chem. **17** [1978] 333/7). – [2] R. P. Bonomo, S. Musumeci, E. Rizzarelli, S. Sammartano (Gazz. Chim. Ital. **104** [1974] 1067/75, 1073). – [3] J. M. Rao, M. C. Hughes, D. J. Macero (Inorg. Chim. Acta **18** [1976] 127/31). – [4] M. Aihara, H. Kishita, S. Misumi (Bull. Chem. Soc. Japan **48** [1975] 680/3). – [5] R. Hogg, R. G. Wilkins (J. Chem. Soc. **1962** 341/50, 341/2, 345).

[6] J. S. Judge, W. M. Reiff, G. M. Intille, P. Ballway, W. A. Baker (J. Inorg. Nucl. Chem. **29** [1967] 1711/6). – [7] C. M. Harris, T. N. Lockyer, N. C. Stephenson (Australian J. Chem. **19** [1966] 1741/3). – [8] H. A. Goodwin, R. N. Sylva (Australian J. Chem. **20** [1967] 629/37, 631/2, 637).

16.1.36.2 With Phenyl Derivatives of Terpyridyl

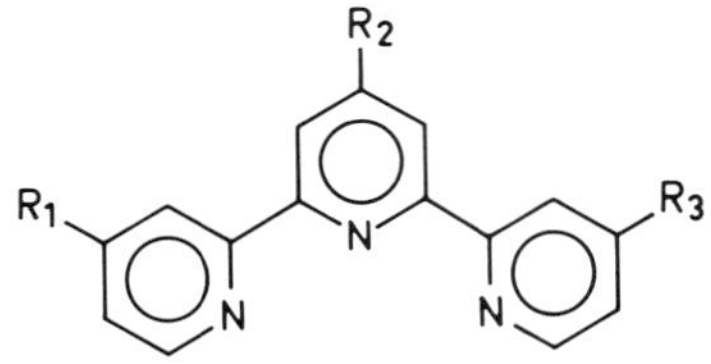

ligand I with $R_1, R_3 = H, R_2 = C_6H_5$ (= $C_{21}H_{15}N_3$)
ligand II with $R_1, R_2, R_3 = C_6H_5$ (= $C_{33}H_{23}N_3$)

$Mn(C_{21}H_{15}N_3)_2(ClO_4)_2$. The preparation of the dark yellow crystalline complex is analogous to that described for $Mn(C_{15}H_{11}N_3)_2(ClO_4)_2$ on p. 237. The room-temperature magnetic moment of 6.02 μ_B corresponds to a high-spin d^5 octahedral configuration. The electronic spectrum of a 1.15×10^{-5} M solution of this Mn^{II} complex in acetonitrile shows absorption maxima at 283 and 251 nm with ε = 66100 and 52300 $L \cdot mol^{-1} \cdot cm^{-1}$, which are accompanied by shoulders at 334, 321, 278, 271, and 247 nm. Both d.c. and a.c. polarography of acetonitrile solutions containing 0.1 M $(C_2H_5)_4NBF_4$ afford two well-defined reduction waves, $E_{1/2} = -1.071$ and -1.344 V for d.c. and -1.069 and -1.349 V for a.c., and one oxidation wave at $E_{1/2} = +1.223$ V for d.c. and $+1.230$ V for a.c. (versus Ag/AgCl, saturated NaCl solution). Each indicates a quasireversible, diffusion-controlled one-electron reaction, leading to the species $[Mn(C_{21}H_{15}N_3)_2]^+$, $[Mn(C_{21}H_{15}N_3)_2]$, and $[Mn(C_{21}H_{15}N_3)_2]^{3+}$, respectively, just as for the related complexes with terpyridyl [1].

$Mn(C_{33}H_{23}N_3)_2(ClO_4)_2$. Solutions of $MnCl_2 \cdot 4H_2O$ (0.07 mmol) in aqueous ethanol (50 vol%) and ligand II (0.15 mmol) in a hot mixture of ethanol and chloroform (50%) were blended, and $NaClO_4$ was added to precipitate the yellow complex, which was washed with aqueous ethanol (50%) and dried in vacuo at 80°C for 24 h. The room-temperature magnetic moment of 5.87 μ_B corresponds to a high-spin octahedral configuration. The electronic spectrum of a 6.94×10^{-6} M solution in acetonitrile shows the four absorption maxima at 343, 331, 284, and 265 nm, with $\varepsilon = 46500$, 46000, 98700, and 99400 $L \cdot mol^{-1} \cdot cm^{-1}$, respectively. A shoulder is observed at 239 nm ($\varepsilon = 87500$ $L \cdot mol^{-1} \cdot cm^{-1}$). Polarographic measurements (d.c.), as for $Mn(C_{15}H_{11}N_3)_2(ClO_4)_2$, see pp. 237/8, revealed two well-defined reduction waves, at $E_{1/2} = -1.043$ V and -1.348 V, and a poorly defined oxidation wave at +1.19 V (versus Ag/AgCl in saturated NaCl solution). Each reduction wave corresponds to a reversible diffusion-controlled one-electron reaction, leading to the species $[Mn(C_{33}H_{23}N_3)_2]^+$ and $[Mn(C_{33}H_{23}N_3)_2]$. The ill-defined oxidation wave may be due to some electrode poisoning, possibly caused by the oxidation product $[Mn(C_{33}H_{23}N_3)_2]^{3+}$ [2].

References:

[1] J. M. Rao, M. C. Hughes, D. J. Macero (Inorg. Chim. Acta **18** [1976] 127/31). – [2] M. C. Hughes, D. J. Macero, J. M. Rao (Inorg. Chim. Acta **49** [1981] 241/5).

16.1.36.3 With 2,2′,2″-Terpyridyl N,N′,N″-Trioxide (= $C_{15}H_{11}N_3O_3$)

$[Mn(C_{15}H_{11}N_3O_3)_2]^{n+}$ (n = 2, 3, 4). **Complexes in Solution**

By dissolving $Mn(C_{15}H_{11}N_3O_3)_2(ClO_4)_2 \cdot 2H_2O$ in acetonitrile, the stable cationic species $[Mn(C_{15}H_{11}N_3O_3)_2]^{2+}$ is formed. Electrochemical studies in acetonitrile solution containing 0.1M $(C_3H_7)_4NClO_4$ at 23°C reveal that this complex ion is oxidized in two reversible one-electron steps, at half-wave potentials $E_{1/2} = +1.06$ and $+1.77$ V versus SCE, to the stable Mn^{III} and Mn^{IV} complex species. Absorption maxima (in nm) with shoulders in the electronic spectra of the solutions are listed in the following table. The molar extinction coefficients, as log ε, are given in parentheses.

complex cation	maximum	shoulders
$Mn(C_{15}H_{11}N_3O_3)_2^{2+}$	220 (4.8)	289 (4.5), 338 (3.74), 432 (3.4)
$Mn(C_{15}H_{11}N_3O_3)_2^{3+}$	215 (4.9)	286 (4.6), 322 (4.1), 375 (3.9), 550 (3.1)
$Mn(C_{15}H_{11}N_3O_3)_2^{4+}$	210 (4.9)	285 (4.6), 331 (4.1), 385 (4.0), 560 (3.5)

The bands at 432, 375, and 385 nm are probably due to charge-transfer absorptions of the Mn^{II}, Mn^{III}, and Mn^{IV} complex species, respectively. The ESR X-band spectrum of $[Mn(C_{15}H_{11}N_3O_3)_2]^{2+}$ gives the parameters g = 1.98 and A = 92 G, while that of $[Mn(C_{15}H_{11}N_3O_3)_2]^{4+}$ gives g = 2.01 and A = 93 G. The spectroscopic splitting parameter A for $[Mn(C_{15}H_{11}N_3O_3)_2]^{2+}$ indicates that its manganese-ligand bond has a more limited covalent character than the 2,2′,2″-terpyridyl complex [1].

For $[Mn(C_{15}H_{11}N_3O_3)_2]^{2+}$, an irreversible reduction occurs at a peak potential of −1.28 V, and controlled-potential coulometry at +2.00 V yields irreproducible results involving more than one electron. $[Mn(C_{15}H_{11}N_3O_3)_2]^{2+}$ was not observed to be light sensitive in acetonitrile solution. The $[Mn(C_{15}H_{11}N_3O_3)_2]^{4+}$ complex ion slowly reacts with water [1].

Isolated Compounds

$Mn(C_{15}H_{11}N_3O_3)_2(ClO_4)_2 \cdot 2H_2O$ was prepared by [1] in the manner described by [2] for the related compound **$Mn(C_{15}H_{11}N_3O_3)Cl_2 \cdot 3H_2O$**, which was obtained by adding a slight excess of $MnCl_2 \cdot 4H_2O$ in alcohol to a heated slurry of the ligand. The mixture was refluxed for several hours and filtered while hot. The precipitated solid was washed with anhydrous alcohol and dried in vacuo [2]. Both compounds are bright orange and light sensitive [1, 2]. $Mn(C_{15}H_{11}N_3O_3)Cl_2 \cdot 3H_2O$ decomposes to a greenish-black substance [2].

$Mn(C_{15}H_{11}N_3O_3)_2(ClO_4)_3$ was prepared from $MnSO_4 \cdot 4H_2O$, $Na_2S_2O_8$, and ligand in essentially the same way as described by [3] for the preparation of $Mn(C_{10}H_8N_2O_2)_3(ClO_4)_3$ (see p. 232). Its high magnetic moment, $\mu_{eff} \sim 6\ \mu_B$, is most likely indicative of significant amounts of Mn^{2+} impurity. The electronic spectrum of the black compound exhibits band maxima at 6250, 7270, and 17950 cm^{-1} [2].

References:

[1] M. M. Morrison, D. T. Sawyer (Inorg. Chem. **17** [1978] 338/9). – [2] W. M. Reiff, W. A. Baker (Inorg. Chem. **9** [1970] 570/6). – [3] R. S. Nyholm, A. Turco (J. Chem. Soc. **1962** 1121/2).

16.1.37 With 1,10-Phenanthroline

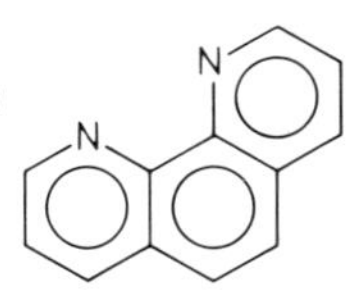

$(=C_{12}H_8N_2 = phen)$

Mixed complexes with 1,10-phenanthroline and diketones are described in "Mangan" D 1, 1979, pp. 70, 82, 114, 118, 120, and 124.

16.1.37.1 $[Mnphen_n]^{2+}$ Ions. Formation and Properties in Solution

The formation constants K_n (in L/mol) and β_n (in L^n/mol^n) for n = 1, 2, 3 obtained under various experimental conditions are listed in the table:

method	t in °C	medium	I(salt) pH	log K_1	log K_2	log β_2	log K_3	Ref.
catalysis[a)]	25	H_2O	0.05 (borate buffer) pH 7.5 to 8.5	4.01	3.83	7.84	—	[1]
catalysis[b)]	20	H_2O	0.1 pH 8.08	3.95	3.71	7.66	—	[2]
catalysis[c)]	25	H_2O	pH 5.20	4.00	3.50	7.50	3.00	[3]
pot.[d)]	25	H_2O	0.05 (K_2SO_4)	4.18 ± 0.01	2.91	7.09 ± 0.04	3.41	[4]
pot.[d)]	25	H_2O/ C_2H_5OH (50%)	0.1 (KNO_3)	4.06 ± 0.03	4.16	8.22 ± 0.04	—	[4]
distribution	25	H_2O/ $CHCl_3$	0.1 (KCl)	4.10	3.10	7.20	3.20	[5]
pot. (redox)	25	H_2O	0.1 (K_2SO_4)	3.5	3.25	6.75	3.0	[6]
pot. (glass)	20	H_2O	0.1 ($NaNO_3$)	4.13	3.48	7.61	2.7	[7]
pot.[d)]	25	H_2O	0.1	3.88	3.16	7.04	3.07	[8]
kinetics	11	H_2O	~0 pH 6.0 to 6.8	3.6	—	—	—	[13]
kinetics	25	CH_3OH	0.2 ($NaClO_4$)	3.8	—	—	—	[10]

method	t in °C	medium	I(salt) pH	log K_1	log K_2	log β_2	log K_3	Ref.
kinetics	18.6	$(CH_3)_2SO$ (DMSO)	0.2 ($NaClO_4$)	2.63	—	—	—	[11]
recommended values from [4 to 8]	25	H_2O	~0.1	4.1	3.2	7.3	3.1	[12]

a) Catalysis of the oxidation of Tiron by H_2O_2 [1]. – b) Catalysis of the oxidation of indigo carmine by H_2O_2 [2]. – c) Catalysis of the oxidation of iodide by periodate [3]. – d) Ag-phenanthroline electrode [4, 8].

Less reliable values of K_n (n = 1, 2, 3) [14, 15] and β_3 [16] have also been reported.

The thermodynamic formation data ΔH, ΔG (in kcal/mol), and ΔS (in cal·mol^{-1}·K^{-1}) of $[Mn\,phen_n]^{2+}$ complexes for the reaction $[Mn(L)_6]^{2+} + n\,phen \rightarrow [Mn(L)_{6-2n}phen_n]^{2+} + 2n\,L$ with L = H_2O or CH_3OH, (I) from calorimetric measurements on aqueous solutions [17] or (II) from the kinetics in methanol solutions [10], are:

method	t in °C	medium	I	n	−ΔH	−ΔG	ΔS	Ref.
I	20	H_2O	0.1 ($NaNO_3$)	1	3.5	5.5	6.8	[17]
I	20	H_2O	0.1 ($NaNO_3$)	2	7.0	10.2	10.9	[17]
I	20	H_2O	0.1 ($NaNO_3$)	3	9.0	13.8	10.4	[17]
II	25	methanol	0.2 ($NaClO_4$)	1	3.0	(5.4)	8	[10]

The data given by [10, 17] are converted to values of ΔH and ΔG in kJ/mol and ΔS in J·mol^{-1}·K^{-1} by [12, p. 69].

Stopped-flow kinetics also yielded reaction rate constants k_f in L·mol^{-1}·s^{-1} and k_d in s^{-1} and activation parameters ΔH* in kcal/mol and ΔS* in cal·mol^{-1}·K^{-1} for the formation (f) and dissociation (d) of the 1:1 complex at ionic strength I ($NaClO_4$) and temperature T in K:

solvent	I	T	log k_f	ΔH_f^*	ΔS_f^*	log k_d	ΔH_d^*	ΔS_d^*	Ref.
H_2O	*)	284.1	5.1	—	—	1.5	9.8**)	−16.6**)	[13]
CH_3OH	0.2 ($NaClO_4$)	298.1	4.95	10.6 ±0.4	−0.4 ±1.6	1.15	13.5 ±0.3	−8.2 ±0.9	[10, 18]

*) pH 6.0 to 6.8. – **) Calculated by [10].

Formation reactions were followed under pseudo-first-order conditions with Mn^{2+} in excess [10, 13]. The dissociation kinetics of $[Mn\,phen]^{2+}$ is induced by cation exchange with Cu^{2+} in water [13] or H^+ in methanol [10]. The solvent exchange rate constants k_{ex} in s^{-1} in water, methanol, and DMSO at 298.1 K are the same as those with 2,2′,2″-terpyridyl [11]; see p. 236. The mixed complex $[Mn\,phen(DMSO)_4]^{2+}$ is described along with the DMSO complexes. ^{17}O NMR of solutions containing Mn^{2+} and phenanthroline in water enriched to about 7% in ^{17}O yielded the water exchange rate constants k_{ex} in s^{-1} of the complexes $[Mn\,phen(H_2O)_4]^{2+}$ and $[Mn\,phen_2(H_2O)_2]^{2+}$ (pH 4 to 5) at 0°C. The activation parameters ΔH_{ex}^* in kcal/mol and ΔS_{ex}^* in cal·mol^{-1}·K^{-1} per molecule of H_2O were also reported, see p. 243 [19].

The electronic spectrum of $[Mn\,phen]^{2+}$ recorded in aqueous perchlorate solution at 27 ± 1°C shows three absorption maxima, 290, 268, and 226 nm, with the molar extinction coefficients ε_{max} = 11500, 36000, and 37000 L·mol^{-1}·cm^{-1}, respectively. The maxima are almost unshifted from their free-ligand values [9].

complex	10^{-6} k_{ex}	$\Delta H^{\ast}_{ex}$	$\Delta S^{\ast}_{ex}$
$[Mnphen(H_2O)_4]^{2+}$	13 ± 2	9 ± 2	6 ± 7
$[Mnphen_2(H_2O)_2]^{2+}$	31 ± 3	9 ± 2	8 ± 9
$[Mn(H_2O)_6]^{2+}$	5.9	8.8	5

The kinetic data can be interpreted like those of the complexes of bpy (p. 204) [10, 13] or 2,2′,2″-terpyridyl (p. 236) [10, 11].

ESR spectra of the manganese(II)-phenanthroline system were recorded at room temperature for phen : Mn^{2+} mole ratios from 1:1 to 6:1. Up to a mole ratio of 1.5:1, the spectrum is that of $[Mn(H_2O)_6]^{2+}$, but with a markedly reduced intensity. For a ratio of 2:1, the spectrum is a superposition of the $[Mn(H_2O)_6]^{2+}$ spectrum and a broad line spectrum. When there are 4 to 6 ligand molecules per Mn^{2+} ion, the spectra consist of six broad overlapping bands, assumed to be due to $[Mnphen_2(H_2O)_2]^{2+}$ and $[Mnphen_3]^{2+}$, whereas $[Mnphen(H_2O)_4]^{2+}$ seems to have such a short electronic relaxation time that it does not contribute to the ESR spectra [19].

The complexes, especially $[Mnphen_2]^{2+}$, exhibit catalytic activity in decomposing H_2O_2 to $H_2O + O_2$, but it is only about a third of that of $[Mnbpy_2]^{2+}$ [20] and can be explained by the formation of an unstable intermediary peroxo complex [21 to 23]. $[Mnphen]^{2+}$ is much less active, and $[Mnphen_3]^{2+}$ is almost inactive [15]. The catalytic activity of the complexes $[Mnphen_n]^{2+}$ (n = 1, 2) have been studied for the oxidation of Tiron by H_2O_2 [1], of indigo carmine by H_2O_2 [2], and of iodide by periodate [3]. For the catalytic activity of $[Mnphen]^{2+}$ at 25°C in the decarboxylation of 2,2-dimethyl-3-oxobutanedioic acid (= dimethyloxaloacetic acid) to give 3-methyl-2-oxobutanoic acid, see [24].

References:

[1] Ya. D. Tiginyanu (Izv. Mold. SSR Ser. Biol. Khim. Nauk **1977** No. 3, pp. 58/71, 65; C.A. **88** [1978] No. 12476). – [2] Ya. D. Tiginyanu, V. P. Lozovan (Izv. Vysshikh Uchebn. Zavedenii Khim. Khim. Tekhnol. **21** No. 8 [1978] 1150/4; C.A. **89** [1978] No. 186775). – [3] V. I. Oprya (Issled. Obl. Farm. Khim. Sb. Nauchn. Statei **1975** 88/9; C.A. **84** [1976] No. 141184). – [4] Y. Bokra, G. Berthon (J. Chim. Phys. **69** [1972] 414/20, 417, 419). – [5] W. A. E. McBryde, D. A. Brisbin, H. Irving (J. Chem. Soc. **1962** 5245/53, 5252).

[6] R. J. P. Williams (personal communication, cited in H. Irving, D. H. Mellor, J. Chem. Soc. **1962** 5232). – [7] G. Anderegg (Helv. Chim. Acta **46** [1963] 2397/410, 2409). – [8] J. M. Dale, C. V. Banks (Inorg. Chem. **2** [1963] 591/4, 503). – [9] K. Sone, P. Krumholz, H. Stammreich (J. Am. Chem. Soc. **77** [1955] 777/80). – [10] D. J. Benton, P. Moore (J. Chem. Soc. Dalton Trans. **1973** 399/404, 402).

[11] D. M. W. Buck, P. Moore (J. Chem. Soc. Dalton Trans. **1976** 638/42, 642). – [12] W. A. E. McBryde (IUPAC Chem. Data Ser. No. 17 [1978] 1/78, 9, 23, 40, 69). – [13] R. H. Holyer, C. D. Hubbard, S. F. A. Kettle, R. G. Wilkins (Inorg. Chem. **4** [1965] 929/35, 932/4). – [14] H. Irving, D. H. Mellor (J. Chem. Soc. **1962** 5222/37, 5232). – [15] R. P. Pantaler, L. D. Alfimova, M. E. Globus, A. M. Bulgakova (Zh. Neorgan. Khim. **22** [1977] 1559/63; Russ. J. Inorg. Chem. **22** [1977] 847/9).

[16] R. R. Miller, W. W. Brandt (J. Am. Chem. Soc. **77** [1955] 1384/5). – [17] G. Anderegg (Helv. Chim. Acta **46** [1963] 2813/22, 2815). – [18] D. J. Benton, P. Moore (J. Chem. Soc. Chem. Commun. **1972** 717/8). – [19] M. Grant, H. W. Dodgen, J. P. Hunt (Inorg. Chem. **10** [1971] 71/3). – [20] Ya. D. Tiginyanu, A. Ya. Sychev, V. M. Berdnikov (Zh. Fiz. Khim. **45** [1971] 1723/8; Russ. J. Phys. Chem. **45** [1971] 975/9).

[21] A. Ya. Sychev, S. S. Budnikov, S. M. Bardina (Zh. Fiz. Khim. **44** [1970] 2420/1; Russ. J. Phys. Chem. **44** [1970] 1374/5). – [22] A. Ya. Sychev, Ya. D. Tiginyanu (Katal. Reakts. Zhidk. Faze Tr. 2nd Vses. Konf., Alma-Ata 1966 [1967], pp. 496, 500; C.A. **69** [1968] No. 46342). – [23] Ya. D. Tiginyanu, A. Ya. Sychev (Sb. Nauchn. Rab. Aspir. Kishinev Gos. Univ. **1968** 67/70; C.A. **74** [1971] No. 46231). – [24] K. G. Claus, J. V. Rund (Inorg. Chem. **8** [1969] 59/63, 60).

16.1.37.2 Isolated Manganese(II) Complexes

Tetrakis (1,10-phenanthroline) Compounds

$Mnphen_4X_2 \cdot nH_2O$ ($X = ClO_4$, I, BF_4, $B(C_6H_5)_4$; n = 0, 2, 4). The anhydrous perchlorate complex was obtained by heating freshly prepared tetrahydrate at 100 to 105°C [1]. $Mnphen_4I_2$ appeared immediately on mixing solutions of $MnSO_4 \cdot 4H_2O$ (5 mmol), 1,10-phenanthroline monohydrate (20 mmol), and KI (20 mmol) in 20% methanol [11]. It was also obtained on heating the tetrahydrate at 110°C for 24 h [1]. The perchlorate dihydrate, which is formed on exposing the tetrahydrate in air [1], was prepared also from an aqueous solution of manganese(II) perchlorate and a methanol solution of 1,10-phenanthroline monohydrate. The mixture must be heated [3]. $Mnphen_4(ClO_4)_2 \cdot nH_2O$ (n = 4 [1] or 2 [2]), $Mnphen_4I_2 \cdot 4H_2O$, and $Mnphen_4(BF_4)_2 \cdot 2H_2O$ were obtained by the method of [2] on treating a hot aqueous solution of $MnSO_4 \cdot H_2O$ and 1,10-phenanthroline monohydrate with sodium or ammonium perchlorate, iodide, or tetrafluoroborate [1]. According to [1] $Mnphen_4(ClO_4)_2 \cdot 4H_2O$ forms on mixing any manganese(II) salt with 1,10-phenanthroline in mole ratios from 1:1 to 1:5, provided that both excess perchlorate ion and water are present. $Mnphen_4[B(C_6H_5)_4]_2 \cdot 2H_2O$ was obtained by mixing concentrated ethanol solutions of manganese(II) tetraphenylborate and ligand in a 1:5 mole ratio [1]. Recrystallization of the precipitated complexes from water affords colorless powders [1] or needles ($X = ClO_4$) [2, 3].

The effective magnetic moments μ_{eff} were found [1]:

complex	$Mnphen_4(ClO_4)_2 \cdot 4H_2O$	$Mnphen_4I_2 \cdot 4H_2O$	$Mnphen_4(BF_4)_2 \cdot 2H_2O$	$Mnphen_4[B(C_6H_5)_4]_2 \cdot 2H_2O$
μ_{eff} in μ_B	6.0	6.0	5.9	6.0

The IR spectrum of the perchlorate dihydrate (Nujol mulls in the 2000 to 600 cm^{-1} region) shows characteristic bands at 1607, 1519, 1432, 1144, 1093, 838, and 726 cm^{-1} [2]. The strong band at 1093 cm^{-1} is assigned to ν_3 vibrations of the uncoordinated ClO_4^- ion [1, 2]; the other bands of the rather complex spectrum are at least partly due to ligand vibrations. Thus the bands at 1519 and 1432 cm^{-1}, which occur at higher frequencies ($\Delta\nu = +14$ and $+9$ cm^{-1}) than in the free ligand, suggest coordination of the ring N atoms [2] to probably 8-coordinate manganese [1, 2].

According to thermogravimetric analysis, freshly prepared $Mnphen_4(ClO_4)_2 \cdot 4H_2O$ readily loses two water molecules when allowed to stand in air. The dihydrate loses its H_2O between 75 and 119°C [1] or at 80°C in vacuo over P_2O_5 [3]. One molecule of phenanthroline is released at about 180°C, leaving yellow $Mnphen_3(ClO_4)_2$ [1, 2] (reported to be $Mnphen_2(ClO_4)_2$ by [3]). $Mnphen_4I_2 \cdot 4H_2O$ loses all H_2O at 110°C within 24 h [1]. $Mnphen_4I_2 \cdot 3H_2O$, for which no preparation was given, is said to form the dihydrate between 50 and 70°C. The anhydrous complex forms between 125 and 155°C. It apparently decomposes above 190°C since it behaves then like a mixture of $Mnphen_2I_2$ and 1,10-phenanthroline [12]. The perchlorate dihydrate is said to be fairly soluble in ethanol, glacial acetic acid, nitrobenzene, and pyridine, forming yellow solutions [3]. However, the

References on p. 254

yellow color of the solutions in water and the other solvents indicates the presence of $[Mnphen_3]^{2+}$ ions, and recrystallization of the perchlorate dihydrate from n-butanol affords $Mnphen_3(ClO_4)_2 \cdot H_2O$ [1]. Colorless precipitates from aqueous solutions are obtained with Na_2CO_3 and alkalies. A red color is produced with $FeSO_4$ [3].

$Mnphen_4(BrC_{10}H_{14}O\text{-}SO_3)_2 \cdot 6H_2O$. The 3-bromo-8-camphorsulfonate was synthesized by mixing hot aqueous solutions of hydrated manganese(II) 3-bromo-8-camphorsulfonate and phen·H_2O in a 1:3 to 1:4.5 mole ratio and filtering while hot [13, p. 285]. On cooling, colorless [1] or light yellow needle-shaped silky crystals appeared. These were washed with cold water and dried in air; they melt above 90°C [13]. The compound has a magnetic moment $\mu_{eff}=6.0\ \mu_B$ [1]. It loses three molecules of H_2O over boiling benzene and is rather sparingly soluble in cold water [13].

Tris(1,10-phenanthroline) Compounds

$Mnphen_3(ClO_4)_2 \cdot nH_2O$ (n = 0, 1) and **$Mnphen_3X_2$** (X = BF_4, $B(C_6H_5)_4$). The anhydrous perchlorate was obtained by heat treatment of $Mnphen_4(ClO_4)_2 \cdot 2H_2O$ at 180°C for 25 min [1], also see [2]. It was prepared also by adding excess $KClO_4$ solution to a hot aqueous solution containing $MnSO_4$ and ligand in a 1:3 mole ratio. The precipitate was washed with water and dried at 0.1 Torr between 20 and 25°C [4]. The tris(phenanthroline) perchlorate apparently formed when an aqueous solution containing a Mn^{II} salt, excess ligand, and $NaClO_4(+CH_3COONH_4)$ was extracted with nitrobenzene at pH 7.0 [5]. The monohydrate was obtained by recrystallizing $Mnphen_4(ClO_4)_2 \cdot nH_2O$ (n = 2, 4) from acetone, n-butanol, nitromethane [1], or ethanol [6] or by reacting manganese(II) perchlorate and 1,10-phenanthroline in a 1:3 mole ratio as ethanol or acetone solutions [1]. $Mnphen_3(BF_4)_2$ can be prepared by adding a solution of 1,10-phenanthroline to a solution obtained by immersion of manganese metal in a 1:1 v/v mixture of 48% aqueous HBF_4 and dimethyl sulfoxide [40]. $Mnphen_3[B(C_6H_5)_4]_2$ may be prepared by reacting $MnSO_4 \cdot H_2O$, ligand, and alkali tetraphenylborate [1].

All the yellow compounds exhibit the magnetic moment $\mu_{eff}=5.9\ \mu_B$ (not measured for $Mnphen_3(BF_4)_2$) [1]. The electronic spectrum of a complex formulated as $Mnphen_2$-$(ClO_4)_2$ [7, 8] but most probably $Mnphen_3(ClO_4)_2$ shows four absorption maxima, at 1650, 510, 267, and 228 nm with molar extinction coefficients log ε = 2.19, 1.22, 4.68, and 4.79 (ε in $L \cdot mol^{-1} \cdot cm^{-1}$). The solvent was 98% ethanol [7, 8]. A band at 590 nm (log ε = 1.53) may be due to the T_d symmetry of the ligand field [8]. The IR spectrum of $Mnphen_3(ClO_4)_2$, which shows characteristic absorption bands at 1520, 1431, 1090, 848, and 724 cm^{-1}, closely resembles that of $Mnphen_4(ClO_4)_2 \cdot 2H_2O$ (p. 244) [2] and may be interpreted in the same way [1, 2]. The far-IR spectrum at <600 cm^{-1} exhibits four bands assigned to Mn-N stretching vibrations, at 513, 424, 280, and 234 cm^{-1}. These are not affected by crystal field stabilization as the calculated stabilization energy is equal to zero [4].

The X-ray photoelectron spectrum of $Mnphen_3(ClO_4)_2$ recorded in vacuo (p = 10^{-7} Torr) at room temperature, using AlK_α radiation, yielded a binding energy of E = 399.6 eV for the N(1s) electrons, a shift of +0.7 eV relative to free 1,10-phenanthroline. It is assumed that all N atoms of the ligand have equivalent charge densities in this chelate [9].

The compound is a 1:2 electrolyte in nitromethane as is shown by the conductivity (Λ between 160 and 180 $\Omega^{-1} \cdot cm^2 \cdot mol^{-1}$) [1]. Its ESR spectrum in acetonitrile containing 0.1M $(C_3H_7)_4NClO_4$ consists of a single line at g = 2.01 with A = 86 G [6].

$Mnphen_3(ClO_4)_2$ decomposes on heating above 190°C with indeterminate loss in weight as shown by thermogravimetry. It explodes near 300°C [1]. The catalytic activity of

$Mnphen_3(ClO_4)_2$ in the photo-induced cyclodimerization of electron-rich aromatic olefins such as phenyl vinyl ether in acetonitrile was studied by [41]. A boiling solution of $Mnphen_3(ClO_4)_2$ in acetonitrile reacted with the lithium salt of 7,7,8,8-tetracyano-1,4-quinodimethane (=TCNQ) in a 1:2 mole ratio yields a precipitate of composition $Mnphen_3(C_{12}H_4N_4)_2$. If the compound is reacted with TCNQ in acetonitrile and LiTCNQ in ethanol in a 1:2:2 mole ratio, a compound of composition $Mn^{2+}phen_3(C_{12}H_4N_4^-)_2(C_{12}H_4N_4^\circ)_2$ results [10].

Cyclic voltammetric studies on solutions of $Mnphen_3(ClO_4)_2$ in acetonitrile containing 0.1M $(C_3H_7)_4NClO_4$ at 23°C revealed an oxidation wave at $E_{1/2}=1.28$ V (versus saturated calomel), which produces tervalent Mn in a quasi-reversible one-electron reaction followed by a chemical reaction leading to the di-μ-oxo-dimanganese(III, IV) complex $[Mn_2O_2phen_4]^{3+}$ by partial disproportionation and subsequent dimerization. Controlled potential electrolysis at +1.60 V produces the dimeric $Mn^{IV,IV}$ species $[Mn_2O_2phen_4]^{4+}$ by abstraction of two electrons per Mn atom [6].

$Mnphen_3(C_{17}H_{33}COO)_2$. The oleate was obtained on adding the stoichiometric amount of 1,10-phenanthroline to an aqueous solution of manganese(II) acetate followed by addition of aqueous $C_{17}H_{33}COOK$ [35].

$Mnphen_3(C_6H_2(NO_2)_3O)_2$. The picrate was prepared from ethanol solutions of manganese(II) picrate and 1,10-phenanthroline, the latter added slowly to the former (mole ratios 1:2 up to 1:3). The mixture was refluxed for 0.5 h to complete the reaction, and the resulting crystals were washed with ethanol and ether, then dried in vacuo. The orange crystals melt and explode at 258°C. The magnetic moment $\mu_{eff}=5.99\ \mu_B$ at 303 K indicates a high-spin octahedral complex. The electronic reflectance spectrum (Nujol mulls) shows four absorption bands, 21730, 27930, 33330, and 45040 cm^{-1}, which apparently are due to intra-ligand transitions within the anion or phen. The IR spectrum (Nujol mulls) exhibits absorption bands assigned to phenanthroline ring stretching at 1630 and 1605 cm^{-1}, CH out-of-plane or ring deformations at 775 and 720 cm^{-1}, $\nu_{as}(NO_2)$ vibration of the picrate ion at 1555(sh) cm^{-1}, and the $\nu(Mn\text{-}N_{phen})$ vibration at 219 cm^{-1}, which indicate coordination of Mn to phen through the N atoms of the pyridine rings as in $Mnbpy_3(C_6H_2(NO_2)_3O)_2$ (p. 207). The molar conductivity of the complex in dimethylformamide is 157.12 $\Omega^{-1}\cdot cm^2\cdot mol^{-1}$and is that of a 1:2 electrolyte. The compound is sparingly soluble in water, $CHCl_3$, and CCl_4 but moderately soluble in dimethylformamide and dimethyl sulfoxide [43].

$Mnphen_3(C_{10}H_{15}O\text{-}SO_3)_2$. The 10-camphorsulfonate apparently forms in solution on adding phenanthroline to an aqueous solution of $Mn(OSO_2\cdot C_{10}H_{15}O)_2$ as suggested by the lower rotation of light after addition of the ligand [36].

$Mnphen_3(C_2H_5OCS_2)_2$. The ethyl xanthogenate was prepared by reacting aqueous $MnCl_2$, $C_2H_5OCS_2K$, and phenanthroline in stoichiometric amounts by adding the xanthogenate solution to the solution of the other components. The yellow precipitate was washed with warm water and dried in vacuo. Its magnetic moment $\mu_{eff}=6.00\ \mu_B$ at room temperature is that of a normal high-spin Mn^{II} complex. The X-ray powder diffraction pattern is identical with those of the related complexes of Fe, Co, Ni, and Zn, which are probably isostructural with the manganese complex. The compound decomposes on heating [39].

Bis(1,10-phenanthroline) Compounds

$Mnphen_2X_2$ ($X=NO_2$, F, Cl, Br, I). To prepare the nitrito complex, warm aqueous solutions containing $MnSO_4\cdot 4H_2O$ and 1,10-phenanthroline monohydrate in a 1:2 mole ratio and excess KNO_2 were mixed, and the resulting precipitate was washed with water and dried in air [1]. The fluoro complex was obtained by heating $Mnphen_2F_2 \cdot 6H_2O \cdot ^1/_2DMF$ (p. 248) in vacuo at 120°C over P_2O_5 [11, 12]. $Mnphen_2Cl_2$ and $Mnphen_2Br_2$ were prepared by reacting manganese(II) chloride or bromide and 1,10-phenanthroline in a 1:2 mole ratio [1, 11, 15] in either acetone or ethanol [1, 15] or in 20% methanol or DMF [11]. The precipitates were recrystallized from 20% methanol and dried in vacuo at 120°C [11]. The chloro complex crystallized upon evaporation of an aqueous solution containing $MnCl_2 \cdot 4H_2O$ and 1,10-phenanthroline in a 1:3.5 mole ratio on a water bath [13] or upon salting out from an aqueous solution containing the $[Mnphen_3]^{2+}$ ion [14]. The iodo complex was synthesized by mixing warm ethanol solutions of MnI_2 and ligand in a 1:2 mole ratio [1], by reacting manganese(II) iodide and ligand in acetone [15], or by reacting manganese(II) acetate, ligand, and excess lithium iodide in warm ethanol [1]. It was obtained also from $Mnphen_4I_2$, which upon extraction with absolute ethanol in a Soxhlet apparatus gave a solution from which $Mnphen_2I_2$ crystallized. The crystals were dried at 120°C in a vacuum over P_2O_5 [11].

The nitrito and halogeno complexes are yellow [1, 11, 13]. The iodo complex is greenish yellow [11]. The complexes show the following magnetic moments μ_{eff} in μ_B at 24°C and the characteristic far-IR absorptions (wax discs) and Raman (R) bands in the 400 to 40 cm^{-1} region according to [11].

X	μ_{eff} [1)]	ν(Mn-X)	ν(Mn-N)	ν(Mn-N) or phen*[2)] (R)	δ(Mn-X_2)
F	5.93	382, 356.5	202.6, 161[2)]	—	—
Cl	5.95[3)]	255, 224	213, 176[2)]	178, 167	110.5
Br	5.95	210, 186(sh)	214(sh), 182[2)]	176, 159	91.5
I	6.10	194.5, 172(sh)	212, 178[2)]	—	85

[1)] μ_{eff} = 6.0 μ_B for X = NO_2, Cl, Br, and I [1]. – [2)] May also be assigned to phen* where phen* refers to vibrations of the ligand activated by complex formation. – [3)] μ_{eff} = 5.64 μ_B (at 291 K) calculated from earlier susceptibility data in [16].

ESR spectra were recorded at room temperature on powdered polycrystalline $Mnphen_2X_2$ samples (X = Cl, Br, I) and interpreted for values of the zero-field splitting parameters D and E of the spin Hamiltonian. The ratio λ = E/D is closer to $^1/_3$ (rhombic symmetry) than zero (axial symmetry) for all compounds [15], also see [17]. The X-band spectrum of $Mnphen_2Cl_2$ shows a very strong band at 1950 G and several weaker bands to higher field, though none above 6000 G, which suggests a low D value. The Q-band spectrum shows three main bands. A very strong band at g_{eff} = 2 confirms the low D value postulated. The main feature of the X-band spectra of $Mnphen_2X_2$ (X = Br, I) is a very strong band at g_{eff} = 4.3. For X = I also see [17]. The Q-band spectrum of $Mnphen_2Br_2$ shows a multitude of bands between 0 and 12650 G; that of $Mnphen_2I_2$ mainly shows a strong band near 6400 G (g_{eff} = 4.3) and near 12000 G (g_{eff} = 2). Comparison of the observed spectra with calculated transitions yields D in cm^{-1} and λ [15].

X	Cl	Br	I
D	0.14	0.395	0.975
λ	0.267	0.267	0.3

The X-ray diffraction (d spacings) [11] and magnetic, far-IR [11], and ESR properties [15] show that the halogeno complexes $Mnphen_2X_2$ are monomeric [11] and contain high-spin 6-coordinate Mn^{II} [1, 11] in a cis-octahedral configuration [11]; for X=Cl, Br also see [15]. This conclusion is especially confirmed by two ν(Mn-X) IR bands [11]. A cis-octahedral configuration was assumed for X=I by [18], whereas a distorted trans arrangement is suggested by the D value [15]. The bromo and iodo complexes appear to be isostructural, but such a conclusion cannot be made for the fluoro and chloro complexes even though their X-ray diffraction patterns are quite similar [11]. The iodo complex apparently is isomorphous with the related complexes of Fe, Co, and Ni [18].

According to thermal analyses (DTA, TG, DTG) the compounds $Mnphen_2X_2$ lose phenanthroline as follows: the fluoro complex loses one phen molecule between 200 and 305°C, leaving an apparently equimolar mixture of MnF_2 and $Mnphen_2F_2$. The residual ligand is lost between 305 and 340°C. The chloro complex loses one molecule between 300 and 420°C to give $MnphenCl_2$; the bromo complex loses one molecule between 370 and 460°C to give $MnphenBr_2$. The iodo complex loses two molecules between 320 and 430°C to give Mn_2O_3 [12]. Cold saturated aqueous solutions of $Mnphen_2Cl_2$ are light yellow. On adding Na_2CO_3 or bases, white precipitates of manganese(II) carbonate or hydroxide form. MnS is precipitated by ammonium sulfide, and $MnO_2 \cdot aq$ is precipitated by $NH_3+H_2O_2$. $FeSO_4$ turns the solution blood red [13]. The complex is destroyed by strong acids [19].

$Mnphen_2F_2 \cdot 6H_2O \cdot 0.5DMF$ was synthesized by adding anhydrous MnF_2 and 1,10-phenanthroline hydrate in a 2:5 mole ratio to 50% aqueous DMF. After 1 h of stirring the solution was filtered and allowed to evaporate to a small volume, whereupon large yellow hexagonal crystals formed. These were washed rapidly with a small volume of the ice-cold solvent [11]. Heating to 70 to 95°C yields the dihydrate, which loses water between 100 and 130°C, leaving the anhydrous complex [12].

$Mnphen_2(OH)ClO_4 \cdot 0.5H_2O$ was precipitated from an aqueous solution containing $[Mnphen_3]^{2+}$, malic acid, and excess base upon addition of perchlorate. The dried yellow complex has a magnetic moment $\mu_{eff}=3.9\ \mu_B$, which suggests a possibly binuclear structure. It is insoluble in most polar and nonpolar solvents (except formamide and acetone, in which it partly dissolves) [19].

$Mnphen_2XClO_4 \cdot H_2O$ (X=Cl, Br, I). The compounds were formulated as $[Mnphen_2X\text{-}H_2O]ClO_4$. The chloro complex was prepared by reacting warm ethanol solutions of $MnCl_2$ and $NaClO_4$ with phenanthroline in a 1:2 mole ratio. The bromo complex was obtained by adding excess NaBr to a warm aqueous solution containing manganese(II) perchlorate and 1,10-phenanthroline monohydrate in a 1:2 mole ratio. The iodo complex was made by adding a solution of manganese(II) perchlorate and excess lithium iodide in acetone to a warm solution of $Mnphen_4(ClO_4)_2$ in acetone. The complexes also formed on reacting MnX_2 with $NaClO_4$ and ligand in ethanol. The yellow solids were washed with ethanol (X=Cl, Br) or acetone (X=I) and dried in air. Magnetic moments of $\mu_{eff}=5.9$, 5.9, and 6.0 μ_B were found for X=Cl, Br, and I, respectively [1].

$Mnphen_2SO_4 \cdot nH_2O$ (n=8, 0). The hydrated complex was obtained by evaporating an aqueous solution containing $MnSO_4 \cdot 4H_2O$ and phenanthroline hydrate in a 1:3.5 mole ratio on a water bath. Coarse yellow crystal clusters form. These are dried in air on porous clay. Drying over boiling benzene and xylene affords the anhydrous compound. The yellow aqueous solutions behave like those of $Mnphen_2Cl_2$ [13].

$Mnphen_2X_2$ (X=CN, NCO, NCS, NCSe). The cyano complex was prepared by heating $Mnphen_2(CN)_2 \cdot 3H_2O$ (p. 250) to 100°C [11]. The isocyanato, isothiocyanato, and isoselenocyanato complexes were prepared by adding solutions of excess KOCN, KSCN, or

KSeCN (4 mol per mole Mn) in 20% aqueous methanol to the solution of $MnSO_4 \cdot 4H_2O$ and ligand in 20% methanol. The pale yellow precipitates were washed with acetone, dried at 60°C, recrystallized from dimethylformamide, and dried in vacuo at 120°C [11]. According to [23] the pale yellow isocyanato complex was washed with water, alcohol, and ether, and dried in vacuo. The isothiocyanato complex was also prepared by reaction of an aqueous solution of phen·HCl·H_2O in excess with an almost boiling, slightly acidified aqueous solution of manganese(II) chloride or sulfate and NH_4SCN. After about 10 min the sandy-colored precipitate was filtered off while warm; washed with a warm solution of NH_4SCN and ligand, then with water, a mixture of ethanol and ether (1:3), and finally ether; and dried in vacuo [24]. The isoselenocyanato complex was synthesized by reacting an ethanol solution of $Mn(NCSe)_2$ with an ethanol solution of the ligand in a 1:2 mole ratio. The cream-colored precipitate was washed with alcohol and dried over P_2O_5 [25]. In the following table are listed the magnetic moment μ_{eff} in μ_B (from susceptibility measurements at 24°C) and characteristic absorption bands in cm^{-1} of the IR spectra recorded on Nujol [11] or liquid paraffin mulls [23] for frequencies >400 cm^{-1} and on wax discs for the range 400 to 40 cm^{-1}. Vibrations of phen itself and of phen* (activated by coordination) are given in [11].

X	μ_{eff}	ν(CN)	ν(CO), (CS), (CSe)	δ(NCO), (NCS), (NCSe)	ν(Mn-X)	ν(Mn-N)	Ref.
CN	5.95[1)]	2109, 2079	—	—	293(broad)	191, 168[2)]	[11]
NCO	5.96	2188, 2176	1335, 1318	630, 623	282	210, 170[2)]	[11]
NCO	—	2198, 2180	1342, 1325	630, 623	—	—	[23]
NCS	5.98	2062, 2049	805	483	235(sh)	199.5, 162[2)]	[11]
NCSe	5.98	2062, 2055	620 to 625 (broad)	433.5	218.5	200.5, 162[2)]	[11]

[1)] μ_{eff} of $Mnphen_2(CN)_2 \cdot 3H_2O$. – [2)] May also be assigned to phen*.

The magnetic moments are typical of high-spin octahedral complexes [11]. The X-ray data (d spacings given in the paper) and IR data suggest that the isothiocyanato, isoselenocyanato, and probably also the isocyanato complexes are isostructural [11] and have the same isocyanato structure as the related cyanato and thiocyanato complexes of Co, Ni, and Zn [23]. $Mnphen_2(NCS)_2$ appears to be isomorphous with the related complexes of Co, Ni, Fe, and Cu but not with that of Zn [18]. The Raman and IR spectra (for Raman frequencies see the paper) show the Mn atom in $Mnphen_2(CN)_2$ to be coordinated to the C atoms of the CN groups [11], whereas in $Mnphen_2(NCO)_2$ [11, 23], $Mnphen_2(NCS)_2$ [11], and $Mnphen_2(NCSe)_2$ [11, 25] it is coordinated to the pseudohalogen N atom [11, 23, 25]. A cis-octahedral configuration is consistent with the splitting of the ν(CN) vibrations [11]. However, this was not observed for $Mnphen_2(NCSe)_2$ by [25]. Both the ESR-Q and X band spectra of powdered $Mnphen_2(NCS)_2$ samples at room temperature show a very strong band at $g_{eff} = 2$, with shoulders toward higher and lower fields, corresponding to a distortion parameter $D = 0.05 \pm 0.003$ cm^{-1}, while λ is assumed to be close to $^1/_3$ (for D and λ see p. 247) [15].

The isocyanato [23], isothiocyanato [24], and isoselenocyanato [25] complexes are air stable [23 to 25]. Thermolysis of $Mnphen_2X_2$ (X = CN, NCO, NCS, NCSe) was studied by TG, DTG, and DTA in air. The order of stabilities is observed to be $Mnphen_2(CN)_2 < Mnphen_2(NCO)_2 < Mnphen_2(NCSe)_2 < Mnphen_2(NCS)_2$. Formation of mono complexes $MnphenX_2$ was not observed. $Mnphen_2(CN)_2$ and $Mnphen_2(NCO)_2$ decompose at >170 and >300°C, respectively, leaving Mn_2O_3. The weight loss of $Mnphen_2(NCS)_2$ in a first decomposition state >300°C corresponds to that calculated for one phen molecule and the conversion of the coordinated thiocyanate to $(CN)_2$, S, and S^{2-}. Solid products such as MnphenS seem to be present. Oxidation of the remaining phenanthroline produces a second intermediate, stable between 600 and 750°C, which may be $2MnSO_4 \cdot Mn_2O_3$. Finally, the intermediate is converted to Mn_2O_3. $Mnphen_2(NCSe)_2$ seems to decompose >300°C in a way similar to the isothiocyanato complex. $MnSeO_2$ is assumed to be formed in a first decomposition stage, finally it is converted to Mn_2O_3. The much smaller weight losses observed by thermography in N_2 correspond to volatilization of some 1,10-phenanthroline and conversion of the pseudohalides to volatile products. Undefined polymeric substances remain [42].

$Mnphen_2(NCO)_2$ and $Mnphen_2(NCSe)_2$ are insoluble in water, alcohol, and acetone [23, 25]. $Mnphen_2(NCO)_2$ is insoluble in ether and chloroform and sparingly soluble in dimethylformamide, where the solubility is increased on heating [23]. $Mnphen_2(NCS)_2$ is nonhygroscopic and may be heated to constant weight at 110°C within 30 min [24].

$Mnphen_2(CN)_2 \cdot 3H_2O$ was prepared by mixing solutions of $MnSO_4 \cdot 4H_2O$ and 1,10-phenanthroline in 20% aqueous methanol in a 1:2 mole ratio. The mixture was heated to about 80°C under N_2 and an aqueous methanol solution of KCN (3 mol per mole Mn) was added gradually while a temperature of 80°C was maintained. The precipitated orange-yellow complex was dehydrated when heated at 100°C. Magnetic properties (μ_{eff} given in the table on p. 249) measured at −60 and −120°C show little variation with temperature [11].

$Mnphen_2(CNO)_2 \cdot 0.33H_2O$. The light yellow fulminato complex was obtained from $MnSO_4$, 1,10-phenanthroline hydrate, and NaCNO in the same way as $Mnbpy_2(CNO)_2$, p. 208. Magnetic measurements at 290, 195, and 77 K yielded an average value of $\mu_{eff} = 5.79\ \mu_B$; the paramagnetic Curie temperature is $\Theta_p = -6$ K. In the IR spectrum, bands of the symmetric and antisymmetric stretching vibrations for the CNO group are observed at 1140 cm^{-1} and 2076 cm^{-1}. Bands of the deformation vibration are at 497 cm^{-1} and 492 cm^{-1}. A molar conductivity of $\Lambda = 28\ \Omega^{-1} \cdot cm^2 \cdot mol^{-1}$ was measured at 20°C for dimethyl sulfoxide solution. The complex is sparingly soluble in water, moderately soluble in methanol and ethanol, and insoluble in ether and benzene. It does not explode but does burn vigorously [22].

$Mnphen_2Cr_2O_7$ was precipitated from an aqueous solution containing an Mn^{2+} salt and 1,10-phenanthroline (or its hydrochloride) by adding $Cr_2O_7^{2-}$ solution dropwise and heating to about 80°C [20]. The compound is reddish brown and on heating starts to decompose at 255°C. A peak in the DTA curve is observed at 350°C, the end of the reaction in the first stage is 345°C, and complete decomposition occurs at 430°C [21]. (A compound formulated as $Mnphen_2(CrO_4)_2$ by [21] is probably a phenanthrolinium salt.)

$Mnphen_2(CH_3COO)_2$ (preparation not given) is soluble in water, methanol, propanediol, and acetone, in each solvent dissociating into ions. Its virostatic effect against the influenza virus was studied [34].

$[Mnphen_2(H_2O)_2](C_6H_2(NO_2)_3O)_2$. The picrate was obtained in low yield from ethanol solutions of manganese(II) picrate and 1,10-phenanthroline in the same way as the tris complex, see p. 246, but with Mn^{2+}:phen in a 1:1 mole ratio. The light orange crystals melt and explode at 234°C. The magnetic moment $\mu_{eff} = 6.02\ \mu_B$ at 303 K is indicative of a high-spin octahedral complex. The molar conductivity of a solution in dimethylformamide, $\Lambda = 153.45\ \Omega^{-1} \cdot cm^2 \cdot mol^{-1}$, is that of a 1:2 electrolyte. The compound is sparingly soluble in water, $CHCl_3$, and CCl_4 but moderately soluble in dimethylformamide and dimethyl sulfoxide [43].

Mnphen$_2$(C$_{10}$H$_7$-SO$_3$)$_2$·2H$_2$O. The 2-naphthalenesulfonate crystallizes from aqueous solutions of manganese(II) 2-naphthalenesulfonate and 1,10-phenanthroline on cooling. The yellow crystals are washed with water and dried in air. The compound loses phenanthroline on heating above 80°C in a vacuum. Its solubility in water is low [13].

A complex of composition **Mnphen$_2$(C$_{19}$H$_8$Br$_4$O$_5$S)** was formed in aqueous solutions of manganese(II) salt and bromophenol blue (= $C_{19}H_{10}Br_4O_5S$) with a large excess of 1,10-phenanthroline. The visible spectrum of such solutions has an absorption maximum at 610 nm with an extinction coefficient of 1.0×10^5 L·mol^{-1}·cm^{-1} at a phen-to-Mn^{2+} mole ratio of 3×10^4. The complex can be extracted by $CHCl_3$ [37, 38], with optimum efficiency at pH 7 to 8 [38].

Mono(1,10-phenanthroline) Compounds

MnphenX$_2$ (X = N_3, Cl, Br, I). The azido complex was synthesized by adding a stoichiometric amount of dilute aqueous NaN_3 to a warmed mixture of manganese(II) chloride or nitrate in water and ligand (slight excess) in ethanol. After 30 min of refluxing (ensuring complete reaction) the precipitate was collected, recrystallized from dimethyl sulfoxide, and dried in vacuo [29]. The chloro and bromo complexes were prepared by heating Mnphen$_2$X$_2$ (X = Cl, Br) in air to constant weight, for the chloro complex at 310 or for the bromo complex at 345°C or in vacuo at 250°C over P_2O_5 [12]. In addition, they can be synthesized by mixing solutions of MnX_2 and phenanthroline (stoichiometric amounts) in anhydrous acetone [26]. The chloro complex was obtained also by heating MnphenCl$_3$ at 252°C in N_2 or at 245°C in vacuo [30] or by adding 1,10-phenanthroline to excess $MnCl_2$ dissolved in dimethylformamide and refluxing the mixture. The complex precipitated with one molecule of DMF, which could be removed only by heating at 100 to 150°C in a vacuum [27]. According to [12], unsolvated MnphenCl$_2$ was isolated by reacting $MnCl_2$ and phen in a 1:0.75 mole ratio in dry dimethylformamide. The bromo complex can be made by refluxing equimolar amounts of $MnBr_2$ and ligand in ethanol solution for $^1/_2$ h. The resulting precipitate was washed with absolute alcohol and dried over $CaCl_2$ [28]. To prepare the iodo complex, solutions of anhydrous MnI_2 and ligand, mole ratio 1:0.4, were combined in dry acetone and refluxed. The precipitated crystals were washed with dry acetone and dried in vacuo at 45°C [12].

The following table lists the magnetic moments μ_{eff} in μ_B at temperatures T (in K) and principal absorption bands (frequencies in cm^{-1}) in the far-IR spectrum of the yellow [26 to 30] complexes (yellowish green for MnphenI$_2$ [12]) with assignments:

X	μ_{eff}	T	Ref.	ν(Mn-X)	bridging X[1)]	δ(Mn-X)	ν(Mn-N$_{phen}$)	Ref.
N_3	5.85	—	[29]	—	—	—	—	—
Cl	5.88	299	[30]	242	181(sh)	160	282[2)]	[12,31]
	5.83	—	[27]	243[3)]	—	158	283	[26]
Br	5.83	296	[26]	209	—	126, 118	281[2)]	[12]
	5.86	305	[28]	210[3)]	141	120	282	[26]
I	5.84	296	[12]	201	—	112, 95	282[2)]	[12]

[1)] Assigned by [31]. – [2)] Left unassigned by [12]. – [3)] The bands at 244 cm^{-1} (MnphenCl$_2$) and 211 cm^{-1} (MnphenBr$_2$) are assigned to ν(Mn-N$_{phen}$) by [31].

The magnetic moment of MnphenCl_2 is almost independent on temperature (μ_{eff} = 5.80 μ_B at 105.5 K); the Curie-Weiss constant is Θ_p = −8 K [30]. The Raman spectra of MnphenCl_2 and MnphenBr_2 are reported in [12, 31]. The IR spectrum of Mnphen$(N_3)_2$ shows two absorption bands, at 2050 and 1280 cm^{-1}, which are assigned to azide antisymmetric and symmetric stretching vibrations, respectively [29]. The IR, Raman, and ESR-X band spectra (single line at g_{eff} = 2) suggest that the halogeno complexes MnphenX_2 (X = Cl, Br, I) have polymeric halogen-bridged octahedral structures [12], also see [26, 31]. Such structures were confirmed by the X-ray powder patterns of MnphenCl_2 [26, 30] and MnphenBr_2 [26], which show that these complexes are isostructural with the ones of Fe, Co, Ni, and Cu [26] and that MnphenCl_2 and α-CophenCl_2 are isomorphous [30]. The azido complex has a terminal and single N-bridged octahedral structure (IR and magnetic properties) and is probably polymeric (m.p. 280°C and almost insoluble). Conductivity measurements show that it is a nonelectrolyte in methanol/dimethylformamide solution [29]. MnphenBr_2 is also nonconducting [28].

Thermal analysis (DTA, TG, DTG) shows that the thermal degradation of MnphenX_2 in air occurs in two steps, leading to solids of approximate composition Mnphen$_{0.75}X_2$ (X = Cl, Br) or MnphenO (X = I) in a first step and Mn_2O_3 in a second step in the following temperature ranges (in °C):

degradation steps	X = Cl	X = Br	X = I
first	430/520	475/520	230/400
second	530/690	530/700	410/620

The products "Mnphen$_{0.75}X_2$" are black and show an IR spectrum with two broad regions of appreciable absorption, from 1700 to 950 cm^{-1} and from 650 to 640 cm^{-1}, and several weak maxima. The IR spectrum is different from that of coordinated 1,10-phenanthroline and may indicate the presence of polymeric pyrolysis products of the ligand. Thermally MnphenI_2 is much less stable than Mnphen$_2I_2$ (p. 248), MnphenBr_2, or MnphenCl_2 [12]. Thermal degradation of MnphenCl_2 in vacuo at 460°C yields $MnCl_2$ [30].

MnphenCl_2 is hygroscopic and readily soluble in water and methanol, although it disproportionates, especially in water [27]. MnphenBr_2 is air stable for hours even at 120°C. It is high melting and only slightly soluble in organic solvents [28]. MnphenI_2 gradually turns brown in air due to slow formation of free I_2 [12].

Mnphen(NCO)$_2$·3H$_2$O was prepared in the same way as Mnphen$(N_3)_2$ (p. 251, using KOCN instead of NaN_3). The pale yellow compound melts at 275°C and has a magnetic moment μ_{eff} = 5.40 μ_B. The IR spectrum (KBr discs, 4000 to 650 cm^{-1} region) shows two strong absorption bands, 2210 and 2160 cm^{-1}. These are assigned to antisymmetric vibrations of the NCO groups, which are assumed to be terminally N-bonded to Mn, thus producing a tetrahedral configuration. The compound is a nonelectrolyte as its molar conductivity in dimethyl sulfoxide/methanol mixtures is very low [29].

Mnphen(NCS)$_2$. Solutions of $Mn(NCS)_2$ (0.006 mol) and 1,10-phenanthroline (0.002 mol, added reactant) in dry acetone were combined and refluxed for an additional 10 min. The yellow solid was collected, washed with acetone, and heated at 60°C to remove solvent. The IR spectrum (Nujol mulls, 2500 to 400 cm^{-1} region) shows two absorption bands, at 2112 and 2092 cm^{-1}, assigned to ν(CN) vibrations. They suggest the occurrence of bridging NCS groups in the probably polymeric complex as the frequency of the ν(CN) bands is increased from that in Mnphen$_2$(NCS)$_2$. The compound decomposes in air at about 300°C to give a polymer of approximate stoichiometry Mn_2phenS_3, as shown by thermal analysis (TG, DTG, DTA). Further heating forms a product of composition $2MnSO_4 \cdot Mn_2O_3$ at about 500°C; Mn_2O_3 is the final product above ~800°C. Thermal degradation under N_2 occurs in one step, between 270 and 440°C, leaving a nonvolatile residue containing Mn, C, N, and S [42].

$(NH_4)_2[Mnphen(NCS)_4]$ was precipitated on adding a solution of freshly prepared NH_4SCN in aqueous 25% ethanol to a hot solution of $MnSO_4$ and phen (1:2 mole ratio) in aqueous 25% ethanol. After cooling and 2 to 3 h, the precipitate was collected, washed with alcohol, then with water, and dried at 50 to 60°C. The light yellow compound has a magnetic moment $\mu_{eff} = 5.89\ \mu_B$ at room temperature, indicating a high-spin octahedral complex. The IR spectrum shows absorption bands that are assigned to vibrations of phen and thiocyanate groups coordinated to Mn through the N atoms [44].

$Mnphen(PH_2O_2)_2$. A saturated aqueous solution of $Mn(PH_2O_2)_2 \cdot H_2O$ was mixed with a concentrated alcohol solution of 1,10-phenanthroline in an equimolar ratio and evaporated until crystals appeared. These were collected and dried in vacuo over P_2O_5 [32]. The pale yellow laths are elongated along c. They are monoclinic, as shown by Weissenberg photographs. The space group is $P2_1/c-C^5_{2h}$ (No. 14), with lattice constants a = 10.791(9), b = 11.173(8), c = 13.15(2) Å, and β = 118.65(3)°; Z = 4. The structure was solved by the heavy atom method and refined by full-matrix least-squares refinement (with anisotropic thermal parameters, except for H) to R = 0.063. The final atomic parameters are listed in the paper. The interatomic distances in the neighborhood of Mn are shown in **Fig. 22**; for bonding angles see [33]. Each 6-coordinate Mn atom is bonded to both N atoms of the organic ligand and four oxygen atoms of four $PH_2O_2^-$ groups in a distorted octahedral environment. The phosphinate groups form Mn-O-P-O-Mn bridges to four other Mn atoms so that a puckered network parallel to the bc plane results. The plane of the

Fig. 22

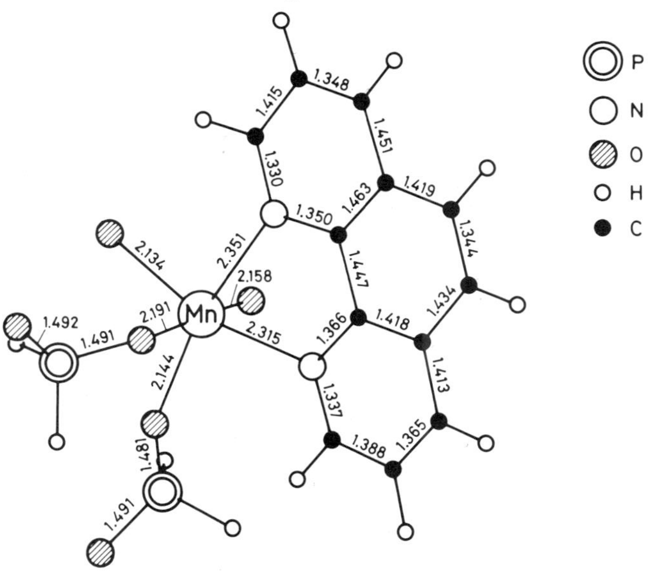

Coordination about one Mn atom in $Mnphen(PH_2O_2)_2$ with atomic distances in Å.

References on p. 254

phenanthroline molecule is tilted 7.2° relative to the plane of the Mn-N bonds, apparently to minimize contacts between the molecules in adjacent layers. A density of 1.74 g/cm^3 was calculated from the X-ray results [33]. Susceptibility measurements at room temperature yielded a magnetic moment of $\mu_{eff} = 5.81\,\mu_B$. The ESR spectrum of the solid at room temperature exhibits a single band at g = 2.024. Strong charge-transfer bands appear in the visible spectrum. The IR spectrum shows strong absorption bands assigned to vibrations of the $PH_2O_2^-$ anion. Stretching vibrations in cm^{-1}: 2325, 2305 (PH_2), 1220, 1183 (PO_2, antisymmetric), and 1070 and 1058 (PO_2, symmetric). Deformation vibrations in cm^{-1}: 1157, 1087, 810 (PH_2), and 482 (PO_2). The extensive splitting of the anion bands suggests differing functions for the phosphinate groups, as in Mn bpy$(PH_2O_2)_2$. The low molar conductivity $\Lambda = 0.5\ \Omega^{-1} \cdot cm^2 \cdot mol^{-1}$ of a 0.001M solution of Mn phen$(PH_2O_2)_2$ in nitrobenzene confirms the coordination of the $PH_2O_2^-$ groups to Mn [32].

References:

[1] B. Chiswell, E. J. O'Reilly (Inorg. Chim. Acta **7** [1973] 707/12, 710). – [2] A. A. Schilt, R. C. Taylor (J. Inorg. Nucl. Chem. **9** [1959] 211/21, 215, 220). – [3] P. Pfeiffer, B. Werdelmann (Z. Anorg. Allgem. Chem. **261** [1950] 197/209, 204). – [4] G. C. Percy, D. A. Thornton (J. Mol. Struct. **10** [1971] 39/48, 40/2). – [5] A. A. Schilt, R. L. Abraham, J. E. Martin (Anal. Chem. **45** [1973] 1808/11).

[6] M. M. Morrison, D. T. Sawyer (Inorg. Chem. **17** [1978] 333/7, 334). – [7] J. Császár, E. Horváth, L. Lehotai (Acta Univ. Szeged. Acta Phys. Chem. [2] **2** [1956] 49/55, 52; C.A. **1957** 16189; Magy. Tud. Akad. Kem. Tud. Oszt. Kozlemen. **14** [1960] 377/84, 379; C.A. **1961** 16138). – [8] Á. Kiss, J. Császár (Acta Chim. Acad. Sci. Hung. **38** [1963] 405/19, 410, 415). – [9] T. Yoshida, S. Sawada (Bull. Chem. Soc. Japan **48** [1975] 3379/80). – [10] D. S. Acker, D. C. Blomström, E. I. du Pont de Nemours & Co. (U.S. 3162641 [1962/64] 1/54, 35; C.A. **63** [1965] 550).

[11] R. E. Morcom, C. F. Bell (J. Inorg. Nucl. Chem. **35** [1973] 1865/74, 1866/9). – [12] C. F. Bell, R. E. Morcom (J. Inorg. Nucl. Chem. **36** [1974] 3689/94, 3690, 3693). – [13] B. Pfeiffer, F. Tappermann (Z. Anorg. Allgem. Chem. **215** [1933] 273/87, 276, 284). – [14] R. R. Miller, W. W. Brandt (J. Am. Chem. Soc. **77** [1955] 1384/5). – [15] R. D. Dowsing, J. F. Gibson, M. Goodgame, P. J. Hayward (J. Chem. Soc. A **1970** 1133/8).

[16] L. Cambi, A. Cagnasso (Atti Reale Accad. Lincei [6] **19** [1934] 458/61, 460; C.A. **1934** 6650/1). – [17] R. D. Dowsing, J. F. Gibson, D. M. L. Goodgame, M. Goodgame, P. J. Hayward (Nature **219** [1968] 1037/8). – [18] E. D. McKenzie (Coord. Chem. Rev. **6** [1971] 187/216, 204/5). – [19] N. Ahmad, S. Kirschner (Pakistan J. Sci. Res. **16** [1973] 159/63; C.A. **80** [1974] No. 152250). – [20] B. Lóránt (Thermochim. Acta **6** [1973] 205/14, 206).

[21] B. Lóránt (Thermochim. Acta **9** [1974] 67/75, 68). – [22] W. Beck, E. Schuierer (Chem. Ber. **95** [1962] 3048/55, 3051/4). – [23] A. M. Golub, G. V. Tsintsadze, A. M. Mamulashvili (Zh. Neorgan. Khim. **14** [1969] 3013/8; Russ. J. Inorg. Chem. **14** [1969] 1589/91). – [24] T. I. Pirtea (Z. Anal. Chem. **184** [1961] 252/5). – [25] P. Stancheva, V. V. Skopenko, G. V. Tsintsadze (Ukr. Khim. Zh. **35** [1969] 166/73; Soviet Progr. Chem. **35** No. 2 [1969] 39/44, 41).

[26] R. E. Wilde, T. K. K. Srinivasan, S. N. Ghosh (J. Inorg. Nucl. Chem. **35** [1973] 1017/21). – [27] J. A. Broomhead, F. P. Dwyer (Australian J. Chem. **14** [1961] 250/2). – [28] A. Syamal (J. Indian Chem. Soc. **44** [1967] 989/90). – [29] R. C. Das, M. K. Mishra, B. K. Nanda, A. K. Sahu (J. Indian Chem. Soc. **54** [1977] 1027/9). – [30] K. Akabori (J. Sci. Hiroshima Univ. A **39** [1975] 73/92, 76, 85/6, 88; Chem. Letters **1974** 1481/6).

[31] R. E. Wilde, T. K. K. Srinivasan (J. Inorg. Nucl. Chem. **36** [1974] 323/8, 327). – [32] J. Sala-Pala, R. Kergoat, J. E. Guerchais (Compt. Rend. C **274** [1972] 595/7). – [33] T. J. R. Weakley (Acta Cryst. B **34** [1978] 3756/8). – [34] A. Shulman, D. O. White (Chem. Biol. Interactions **6** [1973] 407/13, 409; C.A. **79** [1973] No. 74160). – [35] D. G. Nicholson (Ind. Eng. Chem. **34** [1942] 1175/9, 1177).

[36] P. Pfeiffer, Y. Nakatsuka (Ber. Deut. Chem. Ges. **66** [1933] 415/8). – [37] M. M. Tananaiko, L. I. Gorenshtein (Ukr. Khim. Zh. **43** [1977] 1319/21; Soviet Progr. Chem. **43** No. 12 [1977] 89/91). – [38] M. M. Tananaiko, L. I. Gorenshtein (Ukr. Khim. Zh. **41** [1975] 996/7; Soviet Progr. Chem. **41** No. 9 [1975] 94/5). – [39] D. G. Holah, C. N. Murphy (Inorg. Nucl. Chem. Letters **8** [1972] 1069/72). – [40] J. J. Habeeb, F. F. Said, D. G. Tuck (Inorg. Nucl. Chem. Letters **15** [1979] 113/6, 115).

[41] K. Mizuno, J. Ogawa, M. Kamura, Y. Otsuji (Chem. Letters **1979** 731/2). – [42] C. F. Bell, R. E. Morcom (J. Therm. Anal. **17** [1979] 7/18). – [43] B. D. Pandey, D. C. Rupainwar (Transition Met. Chem. **6** [1981] 249/52). – [44] W. U. Malik, P. P. Bhargawa, M. M. Siddiqui (Acta Chim. Acad. Sci. Hung. **107** [1981] 155/9, 156).

16.1.37.3 Manganese(III) Complexes

Mnphen$(NO_3)_3$. The brick-red crystalline compound was prepared from $Mn(NO_3)_3$ and 1,10-phenanthroline in the same way as Mnbpy$(NO_3)_3$ (p. 217). It is not isomorphous with the bipyridyl complex although the X-ray powder patterns are quite similar at low angles. Magnetic measurements at 22°C yield an effective moment of $\mu_{eff} = 5.0 \pm 0.1\ \mu_B$. The IR spectrum and the solubility resemble that of Mnbpy$(NO_3)_3$, p. 218 [1].

Mnphen$F_3 \cdot H_2O$. A filtered solution of Mn_2O_3 (1 g) in aqueous HF (20%, 10 mL) was diluted with water (10 mL) and added to an aqueous solution of phenanthroline monohydrate (1 g, 30 mL H_2O) and HF (40%, 2 or 3 drops) with stirring until crystallization occurred. After 30 min the dull orange-brown crystals were collected, washed with a small volume of water, acetone, and ether, and dried in vacuo over P_2O_5. Susceptibility measurements at temperatures ranging from 318 to 83 K yield the magnetic moments μ_{eff} = 4.89, 4.86, and 4.68 μ_B at 318, 278 to 182, and 83 K, respectively. The Curie-Weiss law is apparently obeyed; $\Theta_p = -13$ K. The compound is a simple paramagnet. The room-temperature moment approximates the spin-only value. The IR spectrum shows an absorption band at 1520 cm^{-1}, indicating coordination of phenanthroline. Dehydration causes extensive decomposition [2].

[MnphenCl_3]$_n$ and **Mnphen$Cl_3 \cdot H_2O$.** The anhydrous compound, for which a dimeric or more highly polymerized structure is proposed [2, 5], was prepared by controlled thermal decomposition of [H_2phen]$MnCl_5$ at 100°C for 90 min [3] or at 160°C in vacuo for 2 h [4]. It may be synthesized in exactly the same way as described for [MnbpyCl_3]$_n$ (p. 219) using phenanthroline monohydrate or phenanthroline in ether [5, 6]. Red-brown leaflets are obtained [6].

The monohydrate was obtained by adding 100 mL of the brown solution used in the preparation of Mnbpy$Cl_3 \cdot H_2O$ (p. 219) to a solution of phenanthroline hydrate (3.0 g) in 30 mL of aqueous 10N HCl. The red-brown crystals were collected, washed with 5N HCl, acetone, and ether, and dried in vacuo [5]. The monohydrate forms also on leaving [H_2phen]$MnCl_5$ in moist air at room temperature [7].

The absorption band maximum in the electronic reflectance spectra in $BaSO_4$ is found at 540 nm for [MnphenCl_3]$_n$ and at 550 nm for Mnphen$Cl_3 \cdot H_2O$, the same as for the corresponding complexes with bipyridyl [5]. IR absorption bands in the range 500 to 180 cm^{-1} observed on paraffin discs or Nujol mulls by [3] are

for $[MnphenCl_3]_n$ 436(m), 368(s), 348(sh), 230(w), and 210(mw)

for $MnphenCl_3 \cdot H_2O$ 436(ms), 370(s,br)*⁾, 356(sh), 256(w), 232(sh), 212(m), and 190(s)

*⁾ The strong broad band at 371 cm^{-1} is assigned to the Mn-Cl stretching vibration by [8].

The band observed at 1520 cm^{-1} in the IR spectrum of $MnphenCl_3 \cdot H_2O$ indicates coordinated phenanthroline [5]. The Raman spectrum of $[MnphenCl_3]_n$ shows two strong bands, 363 and 228 cm^{-1}, which are assigned to terminal Mn-Cl stretching and to either Mn-Cl-Mn bridging or Mn-N vibrations as in $[MnbpyCl_3]_n$. Since the far-IR and Raman spectra of $[MnphenCl_3]_n$ and $MnphenCl_3 \cdot H_2O$ are very similar, it is assumed that both compounds contain dimeric $[MnphenCl_3]_2$ units. It is also assumed that the water is incorporated in the crystal lattice but not directly coordinated to Mn, as the water is lost on heating [3]. However, two OH stretching bands at 3450 and 3280 cm^{-1} and the stability of $MnphenCl_3 \cdot H_2O$ in vacuo over P_2O_5 favor the assumption of [5] that the H_2O is coordinated to Mn, thus completing an octahedral arrangement. The dimeric (or polymeric) nature of $[MnphenCl_3]_n$ is also suggested by its magnetic moment (see below). It decreases at lower temperatures ($\mu_{eff}=3.31$ μ_B at 119 K [2]) and is considerably less than that of a high-spin Mn^{III} complex ($\mu_{eff}=4.90$ μ_B) [2, 4, 5], whereas that of the hydrated complex is rather independent on temperature and almost truely corresponds to that of high-spin Mn^{III} complexes [2, 5]. The following table lists the effective magnetic moment μ_{eff} in μ_B at the temperature in K given in parentheses and the Curie-Weiss constant Θ_p in K from the susceptibility in the range between 300 and 100 K [2, 4]:

complex	μ_{eff}		Θ_p	
$[MnphenCl_3]_n$	4.16(294)		−183	−248
$MnphenCl_3 \cdot H_2O$	4.75(293)	4.98 ± 0.05	−7	
Ref.	[2]	[9]	[2]	[4]

Both compounds appeared to obey the Curie-Weiss law [2].

The antiferromagnetic behavior of $[MnphenCl_3]_n$ demonstrated by the high values of Θ_p [2, 4] may be interpreted like for $[MnbpyCl_3]_n$ on p. 220 [2]. The ESR spectrum of $MnphenCl_3 \cdot H_2O$ resembles that of $MnbpyCl_3 \cdot H_2O$ (p. 220) [9].

$[MnphenCl_3]_n$ is air stable and decomposes on heating above 180°C [6]. According to DTA and TG studies it loses Cl_2 in vacuo at 245°C, or at 252°C under N_2, leaving $MnphenCl_2$ [4]. Its solubility is the same as that of $[MnbpyCl_3]_n$ [5, 6]. The aquo complex is stable in vacuo at room temperature over P_2O_5 [5], but it loses its H_2O between 85 and 130°C when heated in air at a rate of 10°C/min [3]. Loss of water is accompanied by decomposition [5]. It can be reconverted to $[H_2phen]MnCl_5$ by exposure to dry HCl gas [7]. The aquo complex is either almost insoluble in common solvents or dissolves with decomposition (solvolysis). In water, Cl^- ions dissociate rapidly [5]. A molar conductivity $\Lambda=645$ $\Omega^{-1} \cdot cm^2 \cdot mol^{-1}$ measured in 0.005M solutions at pH 2.6 indicates that all Cl^- ions are released from the inner coordination sphere of the complex and are replaced by H_2O [9]. $MnphenCl_3 \cdot H_2O$ dissolved in a slightly acidified aqueous solution containing phenanthroline monohydrate gives a dark green-brown solution. On addition of $HClO_4$, a deep red-brown crystalline solid of stoichiometry $[MnOphen_2](ClO_4)_2 \cdot 0.5H_2O$ can be isolated. This product is best formulated as $Mn_2O_2phen_4(ClO_4)_4 \cdot H_2O$ (p. 259) and contains manganese(IV), apparently due to disproportionation of $MnphenCl_3 \cdot H_2O$ [2]. The green intermediate, a compound containing the $[phen_2Mn^{III}O_2Mn^{IV}phen_2]^{3+}$ cation, was isolated by adding $MnphenCl_3 \cdot H_2O$ to phenanthroline buffer at pH 4.5 [11]; see p. 258. An aqueous solution of $MnphenCl_3 \cdot H_2O$ exhibits about the same catalytic acitvity in the decomposition of H_2O_2 at pH=6.5 as the phenanthroline complexes of divalent manganese [9].

$K[Mnphen(C_6H_4O_2)_2]$. The pyrocatecholato complex was prepared from an aqueous solution of $K_3[Mn(C_6H_4O_2)_3] \cdot 3H_2O$ (1.3 g) and an alcoholic solution of phenanthroline (0.28 g) in the same way as $K[Mnbpy(C_6H_4O_2)_2]$ (p. 220). The greenish brown crystals are very sparingly soluble in water and alcohol [10].

References:

[1] D. W. Johnson, D. Sutton (Can. J. Chem. **50** [1972] 3326/31, 3327, 3330). – [2] H. A. Goodwin, R. N. Sylva (Australian J. Chem. **20** [1967] 629/37, 630/2, 637). – [3] C. F. Bell, D. N. Waters (J. Inorg. Nucl. Chem. **39** [1977] 773/5). – [4] K. Akabori (J. Sci. Hiroshima Univ. A **39** [1975] 73/92, 76, 80, 85/90; Chem. Letters **1974** 1481/6; C.A. **83** [1975] No. 125417). – [5] H. A. Goodwin, R. N. Sylva (Australian J. Chem. **18** [1965] 1743/9, 1746, 1749).

[6] H. Funk, H. Kreis (Z. Anorg. Allgem. Chem. **349** [1966] 45/9, 48). – [7] S. Sarkar, B. Sharma, R. C. Maurya, Sh. Ch. Chaurasia, Sh. K. Srivastava (Acta Ciencia Indica **4** No. 1 [1978] 21/4). – [8] S. N. Ghosh (J. Inorg. Nucl. Chem. **34** [1972] 1456/7). – [9] A. Ya. Sychev, Ya. D. Tiginyanu, Yu. S. Gromovoi (Zh. Fiz. Khim. **42** [1968] 2081/4; Russ. J. Phys. Chem. **42** [1968] 1097/9). – [10] P. Spacu, S. Popescu (Acad. Rep. Populare Romine Studii Cercetari Chim. **9** [1961] 367/95, 372, 381).

[11] S. R. Cooper, M. Calvin (J. Am. Chem. Soc. **99** [1977] 6623/30).

16.1.37.4 Manganese(III,IV) and Manganese(IV) Complexes

$[Mn_2O_2phen_4]^{3+}$ and $[Mn_2O_2phen_4]^{4+}$ Ions in Solution

Cyclic voltammetry and controlled potential electrolysis (at +1.60 V) on solutions of $Mnphen_3(ClO_4)_2$ in acetonitrile containing 0.1M $(C_3H_7)_4NClO_4$ at 23°C (also see p. 246) revealed the presence of two dimeric di-μ-oxo species, $[phen_2Mn^{III}O_2Mn^{IV}phen_2]^{3+}$ and $[phen_2Mn^{IV}O_2Mn^{IV}phen_2]^{4+}$, which are formed by the reversible oxidation-reduction reactions discussed for manganese complexes of 2,2′-bipyridyl on p. 221 [1]. Controlled potential reduction of red $[Mn_2O_2phen_4]^{4+}$ at +0.80 V adds 0.5 e^- per manganese to give green $[Mn_2O_2phen_4]^{3+}$. The half-wave potential is at $E_{1/2}$ = +1.34 V versus SCE [1, 2]. Similar results were obtained by [3, 4] with $(C_2H_5)_4NClO_4$ supporting electrolyte. The quasi-reversible one-electron reduction of $[Mn_2O_2phen_4]^{3+}$ is assumed to give unstable $[Mn_2O_2phen_4]^{2+}$ [1 to 4], which exists long enough to be reoxidized to $[Mn_2O_2phen_4]^{3+}$ [4].

$[Mn_2O_2phen_4]^{4+}$ in acetonitrile is reduced to $[Mn_2O_2phen_4]^{3+}$ by one equivalent of OH^- or Cl^- ions. Further addition of Cl^- or OH^- ions causes the decomposition of the di-μ-oxo bridges, as shown in the cyclic voltammogram and by the increased magnetic moment of the solutions. Reaction of $[Mn_2O_2phen_4]^{4+}$ with catechols forms quinones and an Mn^{II} species. The color changes from red to yellow. These reactions do not appear to generate O_2 from water [1].

The ESR spectra of the electrogenerated species $[Mn_2O_2phen_4]^{3+}$ and $[Mn_2O_2phen_4]^{4+}$ in acetonitrile and other physical properties of these ions in solution are described along with the properties of the perchlorates $Mn_2O_2phen_4(ClO_4)_3 \cdot nH_2O$, see below, and $Mn_2O_2phen_4(ClO_4)_4 \cdot H_2O$, p. 259.

Isolated Compounds

$Mn_2O_2phen_4(ClO_4)_3 \cdot nH_2O$ (n = 4, 2, 1). From a methanol solution containing manganese(III) acetate dihydrate and 1,10-phenanthroline, 60% $HClO_4$ in methanol precipitates the tetrahydrate, $Mn_2O_2phen_4(ClO_4)_3 \cdot 4H_2O$, while treatment with HNO_3 (d = 1.33 g/cm^3) in methanol followed by $NaClO_4 \cdot H_2O$ yields the dihydrate [6]. The dihydrate can also be obtained by oxidizing an aqueous manganese(II) sulfate solution and ligand with $NaIO_4$ and adding aqueous $NaClO_4$ or by reaction of $Mn_2O_2phen_4(S_2O_8)_{1.5} \cdot 3H_2O$, see p. 259, with excess $NaClO_4$ in water at 20°C. The resulting precipitates are washed with water, ethanol, and ether [7] or methanol [6], and dried in vacuo [6, 7]. According to [2], the monohydrate results by metathesis of $Mn_2O_2phen_4(S_2O_8)_{1.5} \cdot 3H_2O$ with $NaClO_4$.

An extended X-ray absorption fine structure study of the dihydrate yields the atomic distance $Mn^{III} \leftrightarrow Mn^{IV} = 2.699$ Å by comparison with the bipyridyl complex $[Mn_2^{III,IV}O_2bpy_4](ClO_4)_3 \cdot 2H_2O$ (p. 222) [8]. Susceptibility measurements by the Faraday method (corrected for diamagnetism) yield a magnetic moment $\mu_{eff} = 2.12$ μ_B per Mn atom at 291.9 K [6], a value which is considerably lower than that obtained for the dihydrate prepared by oxidation of manganese(II) sulfate ($\mu_{eff} = 2.79$ μ_B at 298.6 K) [7]. The higher value may be due to some contamination with Mn^{2+} ions [6]. The magnetic susceptibilities measured between 298 and 67 K do not obey the Curie-Weiss law. The magnetic moment, which decreases somewhat at lower temperatures, is much lower than that expected for a high-spin complex. It is assumed that there are magnetic exchange interactions between the Mn^{3+} and Mn^{4+} ions transmitted by the oxygen bridges [7]. The solution-phase susceptibility of $Mn_2O_2phen_4(ClO_4)_3 \cdot H_2O$ in acetonitrile determined by NMR at room temperature and corrected for diamagnetism gives the magnetic moment $\mu_{eff} = 2.51 \pm 0.08$ μ_B for the binuclear unit and 1.56 μ_B (uncorrected) per Mn atom [2]. ESR data (g = 2.01 and A = 78 G) recorded on an electrogenerated solution of $[Mn_2O_2phen_4]^{3+}$ in acetonitrile containing 0.1M $(C_3H_7)_4NClO_4$ are given in [1].

The electronic diffuse reflectance spectrum of the dihydrate shows a broad asymmetric band between 11100 and 14800 cm^{-1}, with two shoulders at 18500 and 22000 cm^{-1}, which were assigned to a superposition of two absorptions corresponding to the same electron transitions as in $[Mn_2O_2bpy_4](ClO_4)_3 \cdot 2H_2O$, see p. 223. The strong to medium absorption bands in the IR spectrum at 690, 655, and 582 cm^{-1} are assigned to vibrations of the double oxygen bridge between the Mn atoms. Two bands in the 3500 and 620 cm^{-1} regions are assigned to vibrations of the water of crystallization, and two other bands, at 1080 and 620 cm^{-1}, are assigned to vibrations of free ClO_4^- ions.

Measurements on a 5.43×10^{-4} M aqueous solution of the dihydrate at 18°C yielded a molar conductivity of $\Lambda = 278.1$ $\Omega^{-1} \cdot cm^2 \cdot mol^{-1}$, which lies between those of 1:2 and 1:3 electrolytes.

The dihydrate is stable at room temperature, even on exposure to daylight and moist air for several days, but it slowly decomposes in solution [7].

$Mn_2O_2phen_4(ClO_4)_3 \cdot (CH_3)_2CO$ was prepared by oxidation of the solution resulting from aqueous $Mn(CH_3COO)_2 \cdot 4H_2O$ and an acetone solution of the ligand: The pH was adjusted to 4.5, the solution oxidized with aqueous $KMnO_4$, and aqueous $NaClO_4$ added. The complex was also prepared by adding $MnphenCl_3 \cdot H_2O$ to a phenanthroline buffer of pH 4.5. The complex was recrystallized in the following manner: The raw product was dissolved in 0.01M phenanthroline buffer (pH 4.5) at 70°C. The solution was filtered, cooled to room temperature, and made 20% by volume in acetone. Black crystals deposited overnight and were washed with ice-cold acetone [4].

The magnetic susceptibility was measured and corrected for diamagnetism. It is characteristic of an S = (2, $^3/_2$) pair with strong antiferromagnetic coupling; J, the isotropic spin exchange energy, is -134 ± 5 cm^{-1} [5]. The magnetic moment decreases as the temperature is lowered:

μ_{eff} in μ_B	2.66	2.14	1.90	1.80	1.75
T in K	298	215	135	55	11.1
Ref.	[4]	[5]	[5]	[5]	[5]

The similar magnetic behavior of the 1,10-phenanthroline and 2,2′-bipyridyl complexes indicate that the phenanthroline complex also has a di-μ-oxo structure [5].

The ESR spectrum of an acetonitrile solution is essentially identical between 12 and 298 K with that of the bipyridyl complex (p. 223) [5]. The electronic spectrum of a solution (in 0.01M phenanthroline buffer at pH 4.5) is also essentially identical with that of the bipyridyl complex, obeying Beer's law from 10^{-5} to 10^{-1} M [4].

The IR absorption spectrum of $Mn_2O_2phen_4(ClO_4)_3 \cdot (CH_3)_2CO$ has an intense band at 1697 cm^{-1} due to the acetone of crystallization. This band is lost on heating the complex at 110°C for 3 h. The bands in the ligand-ring region are typical of phenanthroline coordinated to metal ions. A band at 686 cm^{-1} is assigned to a stretching mode of the di-μ-oxo bridge [4].

The aqueous solution chemistry of the phenanthroline complex resembles that of $Mn_2O_2bpy_4(ClO_4)_3$ [4].

$Mn_2O_2phen_4(S_2O_8)_{1.5} \cdot 3H_2O$ was prepared by adding $K_2S_2O_8$ to a solution of $MnSO_4 \cdot H_2O$ (3.5 mmol) and 1,10-phenanthroline monohydrate in water at 74°C. The instantaneously formed yellow precipitate (the Mn^{II} complex of the ligand) slows the oxidation but is partly converted into a dark brown precipitate by adding two 0.5 g portions of $K_2S_2O_8$ at intervals of 10 min. After a further 10 min the mixture was filtered off with suction, and the yellow solid was washed with small portions of water, ethanol, and ether, and dried in vacuo over P_2O_5. The magnetic susceptibilities corrected for diamagnetism yield the magnetic moments μ_{eff} (in μ_B) per Mn atom: 2.77 at 299.78 K, 2.70 at 77.56 K. The electronic diffuse reflectance spectrum shows a broad asymmetric band between 11000 and 14800 cm^{-1}, with two shoulders at 20000 and 22000 cm^{-1}. In the IR spectrum two strong, broad absorption bands of the $S_2O_8^{2-}$ ion are at 1270 and 660 cm^{-1}. Other bands are at 1000 and 600 to 550 cm^{-1}. These magnetic and spectroscopic results as well as the molar conductivity ($\Lambda = 316.9$ $\Omega^{-1} \cdot cm^2 \cdot mol^{-1}$ at 18°C) suggest a structure much like that of $Mn_2O_2bpy_4(ClO_4)_3 \cdot 2H_2O$ [7].

$Mn_2O_2phen_4(PF_6)_3 \cdot 2H_2O$ was obtained in the same way as the perchlorate, p. 258, except that $NaPF_6 \cdot H_2O$ in methanol is used instead of $NaClO_4 \cdot H_2O$ [6]. The magnetic susceptibilities corrected for diamagnetism yield the magnetic moments μ_{eff} (in μ_B) per Mn atom: 2.05 at 293.7 K, 1.56 at 77.9 K [6].

$Mn_2O_2phen_4(ClO_4)_4 \cdot H_2O$ was prepared by adding $MnphenCl_3 \cdot H_2O$ (1 g) to a slightly acidified aqueous solution (80 mL) of 1,10-phenanthroline monohydrate (1 g). The dark green-brown solution resulting after all solid has dissolved was filtered and added slowly to a cooled, stirred mixture of $HClO_4$ (70%, 26 mL) and H_2O (40 mL). The deep red-brown crystalline precipitate was collected after 15 min, washed with ethanol and ether, and dried in vacuo over $CaCl_2$. On the basis of the chemical analysis, the compound was originally formulated as $MnOphen_2(ClO_4)_2 \cdot 0.5H_2O$ [9].

The X-ray absorption fine structure technique and comparison with $Mn_2O_2bpy_4$-$(ClO_4)_3 \cdot 2H_2O$ yields an atomic distance for $Mn^{IV} \leftrightarrow Mn^{IV}$ of 2.717 Å [8]. Susceptibility measurements yield the magnetic moments $\mu_{eff} = 1.86$, 1.66, and 1.36 μ_B at 293, 177, and 94 K, respectively. The Curie-Weiss law is obeyed; $\Theta_p = -205$ K [9]. The observed low magnetic moment suggests a binuclear bridged structure, which shows a strong antiferromagnetic spin coupling of the manganese ions as indicated by the high Θ_p value [9], also see [2, 5]. An ESR spectrum in acetonitrile (containing 0.1M $(C_3H_7)_4NClO_4$) of electrogenerated $[Mn_2O_2phen_4]^{4+}$ showed g = 2.01 [1], whereas no ESR signal was found from 298 to 12 K for $Mn_2O_2phen_4(ClO_4)_4 \cdot H_2O$ dissolved in acetonitrile [5].

The absorption spectrum of the complex in aqueous phenanthroline solution had no bands in the visible region [4]. In the IR absorption spectrum of $[Mn_2O_2phen_4]$-$(ClO_4)_4 \cdot H_2O$, 48 bands between 3400 and 225 cm^{-1} were observed. The bands in the ligand-mode region were typical of 1,10-phenanthroline coordinated to metal ions [4], also see [9]. The band at 692 cm^{-1} was assigned to a stretching mode of the di-μ-oxo bridge [4]. Preliminary IR data suggest that the Mn-O bonds in this binuclear di-μ-oxo bridged Mn^{IV}-O_2-Mn^{IV} species are stronger than in the Mn^{IV}-O_2-Mn^{III} species [2].

References:

[1] M. M. Morrison, D. T. Sawyer (Inorg. Chem. **17** [1978] 333/7). – [2] M. M. Morrison, D. T. Sawyer (J. Am. Chem. Soc. **99** [1977] 257/8). – [3] Y. Otsuji, K. Sawada, I. Morishita, Y. Taniguchi, K. Mizuno (Chem. Letters **1977** 983/6). – [4] S. R. Cooper, M. Calvin (J. Am. Chem. Soc. **99** [1977] 6623/30, 6624/7). – [5] S. R. Cooper, G. C. Dismukes, M. P. Klein, M. Calvin (J. Am. Chem. Soc. **100** [1978] 7248/52, 7251).

[6] R. Usón, V. Riera, M. A. Ciriano (Transition Metal Chem. [Weinheim] **1** [1976] 98/9). – [7] R. Usón, V. Riera, M. Laguna (Transition Metal Chem. [Weinheim] **1** [1976] 21/5, 23). – [8] J. A. Kirby, A. S. Robertson, J. P. Smith, A. C. Thompson, S. R. Cooper, M. P. Klein (J. Am. Chem. Soc. **103** [1981] 5529/37, 5531/3). – [9] H. A. Goodwin, R. N. Sylva (Australian J. Chem. **20** [1967] 629/37).

16.1.38 With Derivatives of 1,10-Phenanthroline

Mixed ligand complexes with acetylacetone and 5-nitro-1,10-phenanthroline or 2,9-dimethyl-1,10-phenanthroline are described in "Mangan" D 1, 1979, p. 83.

16.1.38.1 Complexes in Solution

2-Methyl- and 5-methyl-1,10-phenanthroline form complexes with Mn^{2+} and ligand in 1:1, 1:2, and 1:3 mole ratio [1, 2], whereas 2,9-dimethyl-1,10-phenanthroline apparently forms only the 1:1 complex [1]. The formation constants log K_n (in L/mol) and log β_n (in L^n/mol^n) were determined at 25°C from the distribution of the ligands between an organic solvent such as n-hexane, chloroform, or carbon tetrachloride and an aqueous phase containing known concentrations of Mn^{2+} at an ionic strength I = 0.1M (KCl):

ligand	formula	log K_1	log K_2	log β_2	log K_3	log β_3	Ref.
2-methyl derivative	$C_{13}H_{10}N_2$	3.0	2.5	5.5	2.4	7.9	[1,3]
5-methyl derivative	$C_{13}H_{10}N_2$	4.28	3.33	7.61	3.60	11.21	[2,3]
2,9-dimethyl derivative	$C_{14}H_{12}N_2$	≪3*)	—	—	—	—	[1,3]

*) No precise value is available because of the low degree of complex formation even at high ligand concentrations [1].

For comparison of the Mn complex stabilities with those of other metals see the papers [1, 2].

The 1:1 complexes with 2,9-dimethyl-1,10-phenanthroline and other 2-substituted phenanthroline derivatives (R = Cl, Br, OCH_3, OC_6H_5, CN) exhibit catalytic activity in the decarboxylation of dimethyloxaloacetic acid to give 2-oxo-3-methylbutanoic acid and CO_2. However, this activity is less than that of the unsubstituted phenanthroline (1:1) complex, decreasing in the order R = H > OCH_3 > Cl > CN = OC_6H_5 > Br > 2,9-$(CH_3)_2$ [4].

References:

[1] H. Irving, D. H. Mellor (J. Chem. Soc. **1962** 5237/45, 5240). – [2] W. A. E. McBryde, D. A. Brisbin, H. Irving (J. Chem. Soc. **1962** 5245/53, 5251). – [3] W. A. E. McBryde (IJPAC Chem. Data Ser. No. 17 [1978] 1/78, 31/2). – [4] K. G. Claus, J. V. Rund (Inorg. Chem. **8** [1969] 59/63, 60).

16.1.38.2 Isolated Compounds

With 5-Nitro-1,10-Phenanthroline (= $C_{12}H_7N_3O_2$)

$Mn(C_{12}H_7N_3O_2)Cl_2$, which was prepared from $MnCl_2$ and ligand by the same way as $Mn(C_{14}H_{12}N_2)Cl_2$ (see below), is a yellow crystalline solid with μ_{eff} = 5.75 μ_B at 305 K. Its other physical and chemical properties resemble those of Mn phen Br_2 (p. 251), A. Syamal (J. Indian Chem. Soc. **44** [1967] 989/90).

With 2,9-Dimethyl-1,10-Phenanthroline (= $C_{14}H_{12}N_2$)

$Mn(C_{14}H_{12}N_2)_2X_2$ (X = Cl, Br, ClO_4). Concentrated ethanol solutions of MnX_2 and the ligand were mixed in a 1:5 mole ratio by adding the ligand to MnX_2. The precipitate is pale yellow (X = Cl) or yellow (X = Br, ClO_4). Susceptibility measurements yield the magnetic moments μ_{eff} = 6.0 μ_B for X = Cl and 5.9 μ_B for X = ClO_4. The nonexistence of tris and tetrakis ligand complexes is presumably due to steric hindrances of the methyl groups. The chloro and bromo complexes are virtually nonelectrolytes in nitromethane [1].

$Mn(C_{14}H_{12}N_2)Cl_2$ was prepared by refluxing $MnCl_2$ and 2,9-dimethyl-1,10-phenanthroline in dimethylformamide for 20 min. After the solution had stood in a sealed flask at 14°C for 48 h, large light brown crystals formed. These were washed with cold CH_2Br_2 and anhydrous ether. There is one molecule of DMF of crystallization, which is lost on heating in vacuo at 150°C for 2 h [2]. In a second method, the components (1 mol each) were refluxed in ethanol for 30 min, and the resulting crystalline solid was washed with absolute ethanol and dried over $CaCl_2$ [3]. The unsolvated fawn [2] or pale yellow compound has a susceptibility at 305 K corresponding to a magnetic moment μ_{eff} = 5.99 μ_B, indicating five unpaired electrons [3]. It is nonconducting in nitrobenzene, and its X-ray powder pattern differs from that of similar metal complexes [2]. The other physical and chemical properties resemble those of Mn phen Br_2 (p. 251), suggesting a polymeric octahedral structure [3].

References:

[1] B. Chiswell, E. J. O'Reilly (Inorg. Chim. Acta **7** [1973] 707/12, 711). – [2] H. S. Preston, C. H. L. Kennard, R. A. Plowman (J. Inorg. Nucl. Chem. **30** [1968] 1463/7, 1466). – [3] A. Syamal (J. Indian Chem. Soc. **44** [1967] 989/90).

With 3,5,6,8- and 3,4,7,8-Tetramethyl-1,10-phenanthroline (= $C_{16}H_{16}N_2$)

The 3,5,6,8-tetramethyl-1,10-phenanthroline complexes $Mn(C_{16}H_{16}N_2)_3Cl_2 \cdot nH_2O$ and $Mn(C_{16}H_{16}N_2)_3SO_4 \cdot nH_2O$ were obtained by adding an aqueous solution of $MnCl_2 \cdot 4H_2O$ (1.95 g) or $MnSO_4 \cdot 7H_2O$ (2.74 g) to a solution of the ligand monohydrate in aqueous ethanol (25%) while heating on a water bath to bring about complete dissolution. After 15 min the ethanol was boiled off, and the previously filtered orange solutions were evaporated to dryness. The yellow crystalline powders were recrystallized from a little hot water and dried in air or in an oven. The degree of hydration depends on the extent of drying. One of the ligands may dissociate in water. The compounds exhibit bactericidal and fungicidal properties. They form clathrates with aromatic hydrocarbons such as xylenes.

The complex with 3,4,7,8-tetramethyl-1,10-phenanthroline, $Mn(C_{16}H_{16}N_2)_2(CH_3COO)_2 \cdot nH_2O$, was obtained by adding an aqueous solution of $Mn(CH_3COO)_2 \cdot 4H_2O$ (367 g in 500 mL water) over 3 h to a solution of the ligand (710 g) in acetone (2.5 L) at 50°C with stirring. The solution was filtered and added to acetone (7.5 L). On cooling to 5°C pale cream crystals form. These were collected and dried at 50°C. The water content depended on the extent of drying.

Reference:

Australian National University (Brit. 956241 [1961/64] 1/14, 4, 6/8; Belg. 617989 [1962/62] 1/34, 14; C.A. **59** [1963] 11531/2).

With 4,7-Diphenyl-1,10-phenanthroline (= $C_{24}H_{16}N_2$)

$Mn(C_{24}H_{16}N_2)_3(ClO_4)_2$. The complex was formed on adding an aqueous solution of $Mn(ClO_4)_2 \cdot 6H_2O$ to a slight excess of ligand in ethanol. The yellow crystals were washed with cold ethanol and dried in vacuo at 80°C for 24 h. Cyclic voltammetry on solutions of $Mn(C_{24}H_{16}N_2)_3(ClO_4)_2$ in acetonitrile containing 0.1M $(C_3H_7)_4NClO_4$ at 23°C revealed an oxidation wave at $E_{1/2} = 1.22$ V versus SCE. This oxidation produces trivalent manganese in a quasi-reversible one-electron reaction, followed by a reaction leading to a di-μ-oxo-dimanganese(III, IV) complex. This species can reversibly be oxidized to give a dimeric $Mn^{IV,IV}$ species at $E_{1/2} = +1.26$ V or can be reduced to an unstable dimeric $Mn^{III,III}$ species at a cathodic peak potential $E_{p,c} = +0.35$ V. The binuclear $Mn^{IV,IV}$ species is also formed by controlled potential electrolysis of $[Mn(C_{24}H_{16}N_2)_3]^{2+}$ at +1.60 V, which abstracts two electrons per magnanese atom. The g-factor of the unresolved ESR spectrum of $Mn(C_{24}H_{16}N_2)_3(ClO_4)_2$ in acetonitrile solution containing 0.1M $(C_3H_7)_4NClO_4$ is 2.01; g = 2.01 and A = 78 G are given for the electrogenerated binuclear $Mn^{IV,IV}$ species $[Mn_2O_2(C_{24}H_{16}N_2)_4]^{4+}$ in acetonitrile solution, M. M. Morrison, D. T. Sawyer (Inorg. Chem. **17** [1978] 333/7).

With 2,9-Dihydrazino-1,10-phenanthroline and Its 1,1′-Dimethyl Derivative

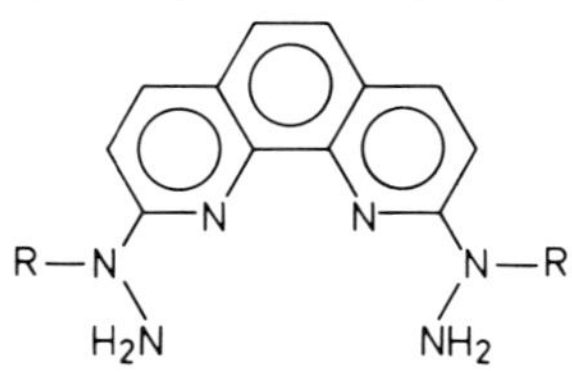

R = H (= $C_{12}H_{12}N_6$)
R = CH_3 (= $C_{14}H_{16}N_6$)

$Mn(C_{12}H_{12}N_6)Cl_2$ and **$Mn(C_{14}H_{16}N_6)X_2$** (X = Cl, NCS) were prepared by dissolving the ligands in hot water or methanol and adding a solution of $MnCl_2 \cdot 4H_2O$. The isothiocyanato complex $Mn(C_{14}H_{16}N_6)(NCS)_2$ was obtained by adding thiocyanate ions after cooling the solution. The resulting precipitates were washed and vacuum dried.

The IR spectrum of all complexes shows two absorption bands in the NH stretching region between 3300 and 3100 cm^{-1}, one NH bending mode at about 1650 cm^{-1}, and a strong band at about 850 cm^{-1}, the CH deformation of the phenanthroline hydrogens. The IR spectra indicate that the coordination of the quadridentate ligands to manganese(II) is trans-equatorial. The sole resonance at 2080 cm^{-1} recorded for $Mn(C_{14}H_{16}N_6)(NCS)_2$ is indicative of N-bonded isothiocyanato groups. The molar conductivity Λ of 0.001 M solutions of the complexes in dimethyl sulfoxide measured at 22°C is that of 1:2 electrolytes, the values (in $\Omega^{-1}\cdot cm^2\cdot mol^{-1}$) being 50 for $Mn(C_{12}H_{12}N_6)Cl_2$, 59 for $Mn(C_{14}H_{16}N_6)Cl_2$, and 67 for $Mn(C_{14}H_{16}N_6)(NCS)_2$, indicating complete dissociation.

Reference:

J. Lewis, T. D. O'Donoghue (J. Chem. Soc. Dalton Trans. **1980** 736/42).

16.1.39 With 1,10-Phenanthroline 1-Oxide

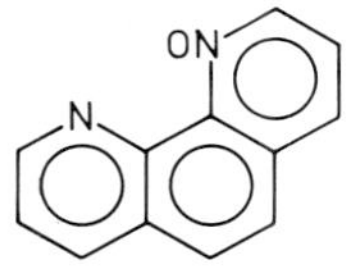

(= $C_{12}H_8N_2O$)

$Mn(C_{12}H_8N_2O)_3(ClO_4)_2$. Solutions of $Mn(ClO_4)_2\cdot 6H_2O$ and ligand (mole ratio 1:3.2), each in ethanol/$HC(OC_2H_5)_3$ (60:40 v/v), were heated to about 50 to 60°C and stirred for about 20 to 30 min. Mixing the warm solutions produced a light brick-red precipitate, which was washed with ethanol and dried in vacuo over P_2O_5 [1].

The susceptibility at 300 K corrected for diamagnetism gives a magnetic moment of $\mu_{eff} = 5.76\ \mu_B$, characteristic for 6-coordinate Mn^{II} [1].

The IR spectra of hexachlorobutadiene (4000 to 1300 cm^{-1}) or Nujol mulls (1350 to 200 cm^{-1}) show the following absorption bands (frequencies in cm^{-1}): ν(NO) 1251, 1234; δ(NO) 821, 816; ν(Mn-O) 368; and ν(Mn-N) 330, 310. Absorptions at 1090 and 624 cm^{-1} are assigned to ν_3 and ν_4 of the ClO_4^- ion. For ligand bands see the paper. The ν(NO) ligand bands, occurring also as a doublet in the free ligand, are shifted toward lower frequencies (~15 cm^{-1}) in the complex, thus suggesting coordination of the bidentate ligand to 6-coordinate Mn through the oxygen of the amine oxide (NO) group. Coordination through the nitrogen of the pyridine fragment of the ligand is suggested by frequency shifts of several bands associated with vibrations of the aromatic amine ring. The presence of ionic perchlorate in the solid is confirmed by the value of the molar conductivity, $\Lambda = 148\ \Omega^{-1}\cdot cm^2\cdot mol^{-1}$ for a 0.001 M solution in dimethyl sulfoxide at 25°C, which is that of a 1:2 electrolyte. The X-ray powder pattern (d spacings in the paper) does not suggest isomorphism with other related metal complexes, such as those of Fe, Co, and Zn.

The electronic reflectance spectrum of Nujol mulls shows a very strong broad band at 282 nm and shoulders at 218, 247, 312, 322, 360, 398, and 516 nm. The spectrum of a 10^{-3} M solution in dimethyl sulfoxide shows very strong absorption below 400 nm, while maxima occur at 476 and 487 nm with ε = 375 and 380 $L\cdot mol^{-1}\cdot cm^{-1}$, respectively. The color of the Mn^{II} complex results from the presence of charge-transfer bands in the 470 to 520 nm region. Other bands are possibly d-d transitions and ligand bands [1].

$Mn(C_{12}H_8N_2O)_3(ClO_4)_2$ is insoluble in water and chloroform, sparingly soluble in nitromethane, nitrobenzene, and dimethyl sulfoxide, but readily soluble in dimethylformamide [1].

$Mn(C_{12}H_8N_2O)_2Cl_2$ and **$Mn(C_{12}H_8N_2O)Cl_2 \cdot H_2O$** were prepared from $MnCl_2 \cdot 4H_2O$ and 1,10-phenanthroline 1-oxide, in a mole ratio 1:3.2 for the former and 2:1 for the latter, proceeding in the same way as indicated for $Mn(C_{12}H_8N_2O)_3(ClO_4)_2$, except that $HC(OC_2H_5)_3$ was mixed with the washing liquid.

Below are listed the color, the magnetic moment μ_{eff} (in μ_B) at 300 K, and frequencies of the most important IR absorption bands (in cm^{-1}) along with the tentative assignment:

compound	color	μ_{eff}	ν(NO)	δ(NO)	ν(Mn-O)	ν(Mn-N)	ν(Mn-Cl)
$Mn(C_{12}H_8N_2O)_2Cl_2$	red	5.60	1247, 1230	819(sh), 812	378	328(sh), 321	245(sh), 240
$Mn(C_{12}H_8N_2O)Cl_2 \cdot H_2O$	light orange	5.75	1258, 1231	821, 811, 802	385	339, 330(sh)	260, 255, 244

The IR spectra of both complexes resemble that of $Mn(C_{12}H_8N_2O)_3(ClO_4)_2$ and indicate that 1,10-phenanthroline 1-oxide is coordinated as a bidentate ligand to Mn through the oxygen of the amine oxide (NO) group and through the nitrogen of the other pyridine ring, as shown by the frequency shifts of the ν(NO) and ring vibrations. The position of the ν(Mn-Cl) bands suggest the presence of 6-coordinate Mn^{II} in $Mn(C_{12}H_8N_2O)_2Cl_2$ (octahedral symmetry) and of 5-coordinate Mn^{II} in $Mn(C_{12}H_8N_2O)Cl_2 \cdot H_2O$, which apparently has a binuclear chlorine-bridged structure (three strong ν(Mn-Cl) bands) and may be formulated as $[Cl(C_{12}H_8N_2O)MnCl_2Mn(C_{12}H_8N_2O)Cl] \cdot 2H_2O$ [2].

The X-ray powder pattern of $Mn(C_{12}H_8N_2O)_2Cl_2$ (d spacings in the paper) resembles that of $Co(C_{12}H_8N_2O)_2Cl_2 \cdot H_2O$ and $Ni(C_{12}H_8N_2O)_2Cl_2 \cdot 2H_2O$ closely enough such as to classify them as being almost isostructural. The electronic spectra (Nujol mulls) were recorded. Strong charge-transfer bands (possibly the very broad band at 512 nm of $Mn(C_{12}H_8N_2O)_2Cl_2$ and the band at 468 nm with shoulder at 498 nm of $Mn(C_{12}H_8N_2O)Cl_2 \cdot H_2O$), which are presumably metal-to-ligand in origin, mask the d-d transition bands. Bands in the 200 to 360 nm region are probably ligand bands ($\pi \rightarrow \pi^*$, $n \rightarrow \pi^*$) [2].

Both compounds are air stable, insoluble or sparingly soluble in water, ethanol, nitromethane, chloroform, dimethyl sulfoxide, dimethylformamide, and most other organic solvents [2].

$Mn_2O_2(C_{12}H_8N_2O)_4(ClO_4)_3 \cdot 2H_2O$ and **$Mn_2O_2(C_{12}H_8N_2O)_4(S_2O_8)_{1.5} \cdot 2H_2O$.** The perchlorate was prepared by adding $Mn_2O_2(C_{12}H_8N_2O)_4(S_2O_8)_{1.5} \cdot 2H_2O$ (0.4 g) to a solution of $NaClO_4$ (5.5 g) in water (35 mL) with stirring at 20°C for 5 h. The resulting solid was washed with water, ethanol, and ether, and dried in vacuo over P_2O_5.

$Mn_2O_2(C_{12}H_8N_2O)_4(S_2O_8)_{1.5} \cdot 2H_2O$ was synthesized by adding $K_2S_2O_8$ (2.10 g) to a solution of $MnSO_4 \cdot H_2O$ (0.78 g) and ligand (2.72 g) in water (150 mL) at 75°C, whereupon a yellowish-red precipitate appears. It gradually turned green. After 10 min, more $K_2S_2O_8$ (2.10 g) was added to the deep green solution, which then was allowed to stand another 10 min at 75°C. The solid was collected and treated with water at 75°C.

On the next page selected values of the magnetic moments per Mn atom μ_{eff} (in μ_B) at temperatures T (in K) derived from susceptibility measurements between 300 and 65 K, band frequencies (in cm^{-1}) of the electronic diffuse reflectance spectrum, and frequencies of the NO stretching vibrations of the ligand and main anion bands in the IR absorption spectra of both di-μ-oxo complexes are listed.

Both complexes are stable at room temperature, even when exposed to moist air and daylight, but they decompose in solution [3].

complex ($L = C_{12}H_8N_2O$)	μ_{eff} (T)		electronic bands		IR bands $\nu(NO)$*)	anion bands
$Mn_2O_2L_4(ClO_4)_3 \cdot 2H_2O$	2.10 (292)	1.33 (66)	14600	18200(sh)	1251, 1232	1080, 620
$Mn_2O_2L_4(S_2O_8)_{1.5} \cdot 2H_2O$	2.12 (295)	1.44 (67)	14300 to 16700	17900(sh) 21200(sh)	1236	1270, 660

*) Designated with $\gamma(NO)$ by [3].

In the IR spectrum of the perchlorate dihydrate the bands at 680, 648, and 583 cm^{-1} were assigned to vibrations of the $Mn\text{-}O_2\text{-}Mn$ bridge. The shifts of the NO vibrations ($\Delta\nu(NO) \approx -18$ cm^{-1}) and the shifts of the ring vibrations of the ligand (1600 and 850 cm^{-1}) indicate coordination to Mn through the oxygen of the amine oxide (NO) group and through the nitrogen atom of the pyridine fragment. The three vibrations of the double oxygen bridge between the two Mn atoms as well as the magnetic results are interpreted like those of the complexes with 2,2′-bipyridyl N-oxide (p. 228) [3].

References:

[1] A. N. Speca, L. L. Pytlewski, N. M. Karayannis (J. Inorg. Nucl. Chem. **36** [1974] 1227/34). – [2] A. N. Speca, L. L. Pytlewski, N. M. Karayannis (Z. Anorg. Allgem. Chem. **422** [1976] 182/92). – [3] R. Usón, V. Riera, M. Laguna (Transition Metal Chem. [Weinheim] **1** [1976] 21/5).

16.1.40 With Naphthyridines

16.1.40.1 With 1,5-Naphthyridine-4,8-diol

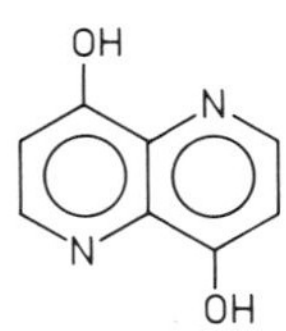

($= C_8H_6N_2O_2$)

$Mn(C_8H_4N_2O_2) \cdot 1.5H_2O$. A warm aqueous solution of the ligand obtained by adding an equimolar amount of Na_2CO_3 was mixed with $MnSO_4 \cdot H_2O$ dissolved in water. The brown precipitate was collected after several hours, washed with water, and dried in vacuo over P_2O_5 at room temperature. The IR absorption spectrum of the complex (strongest bands at $\nu = 3300$, 1550, 1502, 1415, and 1213 cm^{-1}) suggests chelation of the ligand: The band of the free ligand at 3050 cm^{-1}, which is due at least partly to NH stretching vibrations, disappears, and the very strong broad $\nu(CO)$ band at 1545 cm^{-1} is replaced by a sharp, very strong peak at 1502 cm^{-1}, corresponding apparently to an aromatic ring vibration. According to thermogravimetric analysis in N_2/O_2 atmosphere (heating rate 5 K/min) the compound loses its water between 30 and 210°C and decomposes with loss of ligand between 255 and 300°C to leave Mn_3O_4. The slightly hygroscopic complex is apparently polymeric as it has a high melting point and is insoluble in common solvents. The high electrical resistivity of the powder ($>10^{12}$ $\Omega \cdot cm^{-1}$) apparently reflects the low crystallinity, S. B. Brown, M. J. S. Dewar (J. Inorg. Nucl. Chem. **42** [1980] 140/3).

16.1.40.2 With 1,8-Naphthyridine and Its 2,7-Dimethyl Derivative

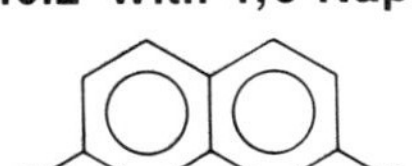

R = H ($= C_8H_6N_2$)
R = CH_3 ($= C_{10}H_{10}N_2$)

$Mn(C_8H_6N_2)_4(ClO_4)_2$. To a methanol solution of $Mn(ClO_4)_2 \cdot 6H_2O$ which had been dehydrated by refluxing with 2,2-dimethoxypropane for 30 min, a methanol solution of

1,8-naphthyridine was added in a 1:4 mole ratio. The clear colorless solution was evaporated after 4 d under reduced pressure to about one-fourth of its volume (10 mL), then a mixture of ethanol and methanol (3:1, 25 mL) was added. The solution was concentrated again to a fourth of its volume under reduced pressure, then cooled in a refrigerator for 3 d. The resulting light yellow crystals were collected and dried in vacuo for one week. The air-stable compound appears to be isomorphous with the related complexes of Fe, Co, Ni, Cu, and Pd as the X-ray powder patterns are similar [1], also see [2].

The IR spectrum recorded between 4000 and 200 cm^{-1} [1] and 400 and 100 cm^{-1} [3] consists of ligand modes which have been shifted to higher frequencies on complexation. Absorption bands at 1580, 1135, and 775 cm^{-1} can be assigned to naphthyridine skeletal stretching or bending (1135 cm^{-1}) vibrations; those at 1072, 415 [1], 205, and 193 cm^{-1} [3] arise from ligand vibrations. The two strong bands at 1100 and 616 cm^{-1} are assigned to vibrations of the ClO_4^- ion (ν_3, ν_4) [1], and the one strong band at 215 cm^{-1} [1], or 223 cm^{-1} [3], is assigned to ν(Mn-N) vibrations [1, 3]. As there are no IR bands attributable to noncoordinated ligand and the $\nu(ClO_4)$ bands are those of ionic perchlorate (T_d), it is assumed that Mn is eight-coordinate, bonded to both N atoms of each of the four bidentate 1,8-naphthyridine ligands which form four-membered chelate rings in a distorted dodecahedral geometry [1, 3]. The molar conductivity $\Lambda = 178\ \Omega^{-1} \cdot cm^2 \cdot mol^{-1}$ of a 7.1×10^{-4} M solution in nitromethane measured at 25°C is that of a 1:2 electrolyte [1]. The electronic spectrum of a solution in acetonitrile shows four absorption maxima, 32700, 33200, 33900, and 39200 cm^{-1}, which are assigned as ligand $\pi^* \leftarrow \pi$ bands [1, p. 36].

$Mn(C_8H_6N_2)_2(NO_3)_2$. A solution of 1,8-naphthyridine (2 mmol) in absolute ethanol (10 mL) was added to a solution of $Mn(NO_3)_2 \cdot 6H_2O$ (0.5 mmol) in the same solvent (15 mL), which had been dehydrated by stirring (3 h) with 2,2-dimethoxypropane (5 mL), and the mixture was stirred for 2 d. The resulting colorless precipitate was washed with anhydrous diethyl ether (60 mL) and dried in vacuo at 110°C over P_2O_5 for 2 to 5 d. The IR spectrum (recorded for Nujol mulls in the 4000 to 200 cm^{-1} region) shows strong absorption bands at 1581, 1136, 1031, and 777 cm^{-1}, which can be assigned to naphthyridine skeletal stretching or bending (at 1136 cm^{-1}) modes shifted upon coordination to higher frequencies. The two strong bands at 620 and 420 cm^{-1} are assigned to ligand deformation vibrations. A broad band at 233 cm^{-1} with shoulder at 210 cm^{-1} apparently is due to ν(Mn-N), and a shoulder at 264 cm^{-1} is due to ν(Mn-O) vibrations. The six bands at 1459 (broad), 1031, 820, 1294, 744 (740), and 700 cm^{-1} are ascribed to vibrations of the nitrate groups (ν_1 to ν_6), which are coordinated as bidentate ligands of C_{2v} symmetry. Thus the Mn atom is 8-coordinate, bonded to four nitrogens of two 1,8-naphthyridine and to four oxygens of two bidentate nitrate groups. The low molar conductivity of $\Lambda = 25\ \Omega^{-1} \cdot cm^2 \cdot mol^{-1}$ for a 5.7×10^{-4} M solution in nitromethane at 25°C corresponds to a nonelectrolyte, thus confirming the coordination of both nitrate groups to Mn [4].

$Mn(C_{10}H_{10}N_2)_3(ClO_4)_2$. A solution of 2,7-dimethyl-1,8-naphthyridine in absolute methanol was added to a solution of $Mn(ClO_4)_2 \cdot 6H_2O$ in the same solvent (which had been dehydrated by refluxing with 2,2-dimethoxypropane for several hours) in a 3:1 mole ratio. After 3 to 5 d of stirring, the solution was concentrated under reduced pressure to about a tenth of its volume, and cooled. The resulting product was recrystallized from absolute ethanol, washed with $CHCl_3$, and dried in vacuo at 110°C over P_2O_5 for several days [5, 6]. The far-IR spectrum which was recorded for Nujol mulls in the 400 to 100 cm^{-1} region, shows two medium bands, at 202 and 152 cm^{-1}, assigned to Mn-N stretching, and three other bands due to ligand vibrations, which on coordination are shifted toward higher frequencies. The Mn-N stretching frequencies observed in $[Mn(C_{10}H_{10}N_2)_3](ClO_4)_2$ are lower than those of the complex with 2,2′-bipyridyl (see p. 206), apparently a result of strain in the four-membered chelate ring of the naphthyridine complex [5].

References:

[1] R. L. Bodner, D. G. Hendricker (Inorg. Chem. **12** [1973] 33/7). – [2] R. L. Bodner, D. G. Hendricker (Inorg. Nucl. Chem. Letters **6** [1970] 421/3). – [3] B. Hutchinson, A. Sunderland, M. Neal, S. Olbricht (Spectrochim. Acta A **29** [1973] 2001/4, 2002). – [4] D. G. Hendricker, R. J. Foster (Inorg. Chem. **12** [1973] 349/52). – [5] B. Hutchinson, A. Sunderland (Inorg. Chem. **11** [1972] 1948/50).

[6] D. G. Hendricker, R. L. Bodner (Inorg. Chem. **9** [1970] 273/7, 274).

16.1.41 With Codeine

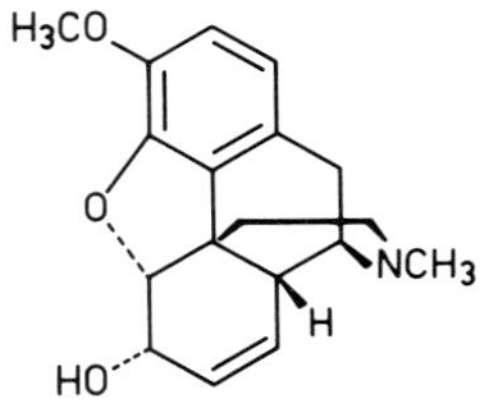

$(=C_{18}H_{21}NO_3)$

$\mathbf{Mn(C_{18}H_{21}NO_3)_2Cl_2 \cdot H_2O}$ was obtained on mixing methanol solutions of manganese(II) chloride and codeine. The pink product decomposes at 254 to 255°C. The manganese atom apparently is coordinated only to the oxygen of the secondary OH group in the ligand as suggested by the shift of the IR absorption band assigned to OH stretching vibrations. No such band is found in the 3600 to 3500 cm^{-1} region, probably because of merging with an intramolecular hydrogen-bond frequency which arises on coordination to Mn. The tertiary NCH_3 group and furan ring oxygen evidently are not coordinated. The compound is readily soluble in water and mineral acids but is insoluble in common organic solvents, O. Farooq, A. Malik (Acta Chim. Acad. Sci. Hung. **85** [1975] 389/94, 390).

16.1.42 With Berberine

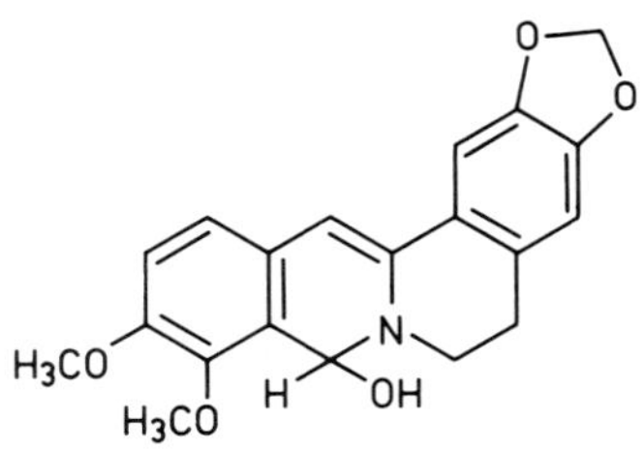

$(=C_{20}H_{19}NO_5)$

Manganese(II) chloride and berberine form a 1:1 complex in a mixture of n-butanol and acetic acid, volume ratio 100:7, that has been saturated with water [1]. On heating the product to 90°C the anhydrous complex was obtained [2]. It forms coarse yellow crystals melting at 252°C. For X-ray d spacings see the paper [1]. The refractive indices obtained by the immersion method are $n_\alpha = 1.591$, $n_\beta = 1.681$, and $n_\gamma > 1.780$, the crystals being doubly refracting, $n_\gamma - n_\alpha > 0.189$. Optically the crystals are biaxial positive [1]. At 250 to 260°C endothermic decomposition and loss of CH_3Cl occur. An exothermic effect is observed between 290 and 300°C. This could be due to a change of aggregation. Above 300°C complete decomposition and burning are observed [2]. The compound is only slightly soluble in water [1].

References:

[1] S. B. Davidyants, Yu. D. Sadykov, D. D. Ikrami, V. A. Degtyarev, K. T. Poroshin (Dokl. Akad. Nauk Tadzh.SSR **9** No. 11 [1966] 20/3; C.A. **66** [1967] No. 85914). – [2] D. D. Ikrami, S. B. Davidyants, Yu. D. Sadykov, K. T. Poroshin (Dokl. Akad. Nauk Tadzh.SSR **9** No. 12 [1966] 24/7; C.A. **66** [1967] No. 98519).

16.1.43 With Lactams

Complexes with 4-butanelactam (= 2-pyrrolidinone) and its derivatives are described in Chapters 16.1.2 and 16.1.3 on pp. 77 to 80.

16.1.43.1 With 6-Hexanelactam

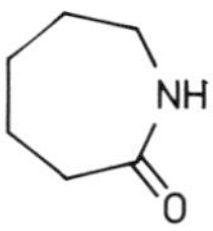

(= ε-Caprolactam = $C_6H_{11}NO$)

$Mn(C_6H_{11}NO)_6(ClO_4)_2$. Anhydrous $Mn(ClO_4)_2$ obtained by dehydrating the hexahydrate with 2,2′-dimethoxypropane was reacted with ε-caprolactam solution in a minimum of anhydrous methanol in a 1:6 mole ratio. An equal volume of ether was added to precipitate the complex. The product was washed with ether and dried in vacuo over P_4O_{10} at room temperature. The colorless crystalline compound melts at 174°C. Susceptibility measurements at 295 K yield the magnetic moment μ_{eff} = 6.25 μ_B, corresponding to five unpaired electrons. The IR spectrum (recorded for Nujol mulls) shows an absorption band assigned to CO stretching at 1627 cm^{-1}. The shift of -33 cm^{-1} due to complexation indicates coordination of the ligand to Mn through the oxygen of the carbonyl group. Isomorphism of $Mn(C_6H_{11}NO)_6(ClO_4)_2$ with the related complexes of Fe, Co, Ni, Cu, Zn, Cd, Mg, and Ca is suggested by very similar X-ray powder patterns. Solid $Mn(C_6H_{11}NO)_6$-$(ClO_4)_2$ is remarkably stable to air and light, but in aqueous solution it is much less stable and tends to decompose quickly, precipitating hydrolysis products. The molar conductivity $\Lambda = 182.4\ \Omega^{-1}\cdot cm^2\cdot mol^{-1}$ of a 0.74 M solution in nitromethane at 25°C indicates a 1:2 electrolyte [1].

$Mn(C_6H_{11}NO)_6(BF_4)_2$. A solution of $Mn(BF_4)_2\cdot 6H_2O$ (0.005 mol) was dehydrated by heating it in $HC(OC_2H_5)_3$ (15 mL) at 50°C for 15 min. After filtration, a solution of ε-caprolactam (0.03 mol) in a minimum amount of ethanol was added. The mixture was evaporated in vacuo until crystals formed. These were sucked off, washed repeatedly with diethyl ether, and dried in vacuo.

The pink complex compound, which melts at 172°C (168°C DTA), is crystalline (X-ray d values in the paper) and apparently isomorphous with the corresponding complexes of Fe, Co, Ni, Cu, and Zn. The IR spectrum recorded in the 2000 to 400 cm^{-1} region shows an absorption band at 1620 cm^{-1} assigned to ν(CO) vibrations which are shifted toward lower frequencies (-38 cm^{-1}) upon complexation, thus indicating coordination of the ligand through oxygen as in $Mn(C_6H_{11}NO)_6(ClO_4)_2$. The bands at 1030 and 530 to 520 cm^{-1} are assigned to vibrations (ν_4, ν_3) of the noncoordinated BF_4^- ion (symmetry T_d). The molar conductivity $\Lambda = 173\ \Omega^{-1}\cdot cm^2\cdot mol^{-1}$ of a 0.001M solution in nitromethane indicates a 1:2 electrolyte and confirms the structure [2].

Thermal decomposition of the solid compound starts at 193°C and ends at 570°C: $Mn(C_6H_{11}NO)_6(BF_4)_2 \rightarrow MnF_2 + 2BF_3 + 6C_6H_{11}NO$ is suggested by the analysis of the decomposition products [2].

$Mn(C_6H_{11}NO)_4(NCS)_2$. A solution of manganese(II) thiocyanate obtained by stirring a solution of manganese(II) chloride in triethoxyacetic acid and KSCN (mole ratio 1:2) at 50 to 60°C for 30 min and filtering was reacted with an anhydrous ethanol solution of ε-caprolactam in a 1:4 mole ratio at 50 to 60°C for 30 min. Addition of excess ether precipitated colorless crystals overnight. These were collected, washed with anhydrous ethanol and ether, and dried in vacuo [3].

The magnetic susceptibility at 25°C gives a magnetic moment $\mu_{eff} = 5.8\ \mu_B$, indicative of 6-coordinate Mn^{II} and a high-spin octahedral configuration. In the IR spectrum (Nujol mull), a shift relative to the free ligand of the ν(CO) vibration to lower frequencies ($\sim 25\ cm^{-1}$) indicates coordination of the ligand to Mn through the oxygen of the carbonyl group; ν(CN) of the complexed ligand is shifted to higher frequency. The positive shift of the thiocyanate group from the 2060 cm^{-1} position in free NCS^- shows that the complex contains terminally N-bonded thiocyanate groups. The absorption band frequencies (in cm^{-1}) observed in the electronic spectra (mull or 0.01M solution in $CHCl_3$ plus lactam) of $Mn(C_6H_{11}NO)_4(NCS)_2$ and their assignment are given along with the corresponding frequencies of $Mn(C_8H_{15}NO)_4(NCS)_2$ (p. 270) and $Mn(C_{12}H_{23}NO)_4(NCS)_2$ (p. 271) from [3]:

		electron transitions from $^6A_{1g}$			
		$\rightarrow {}^4T_{2g}(D)$	$\rightarrow {}^4E_g$ $\rightarrow A_{1g}(G)$	$\rightarrow {}^4T_{2g}(G)$	$\rightarrow {}^4T_{1g}(G)$
$Mn(C_6H_{11}NO)_4(NCS)_2$	solid	28600	24900	—	18500
	solution	28800 (600)	24500 (500)	23000 (300)	—
$Mn(C_8H_{15}NO)_4(NCS)_2$	solid	28400	25000	—	18200
	solution	28600 (500)	25000 (500)	23200 (300)	—
$Mn(C_{12}H_{23}NO)_4(NCS)_2$	solid	27900	24600	—	18100
	solution	28400 (600)	24800 (500)	23000 (300)	—

The conductivity of a 10^{-3}M solution in acetone is that of a 1:1 electrolyte. The complex is stable to air and light but is hydrolyzed in aqueous solution [3].

$Mn(C_6H_{11}NO)Cl_2$ was prepared by reacting $MnCl_2 \cdot 4H_2O$ and ε-caprolactam in a mixture of methanol and ethyl acetate. It forms pale pink hygroscopic crystals [4].

References:

[1] S. K. Madan, H. H. Denk (J. Inorg. Nucl. Chem. **27** [1965] 1049/58, 1052/4). – [2] F. Kůtek, B. Janalík, B. Dušek (Sb. Vys. Sk. Chem. Technol. Praze Anorg. Chem. Technol. B **23** [1978] 51/62; C.A. **93** [1980] No. 60137). – [3] C. D. Rao, B. K. Mohapatra (J. Inorg. Nucl. Chem. **39** [1977] 689/91). – [4] K. Stürzer (Z. Naturforsch. **17b** [1962] 197).

16.1.43.2 With 8-Octanelactam and 12-Dodecanelactam

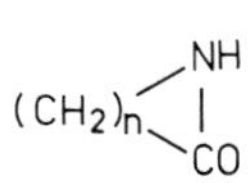

n = 7 (= $C_8H_{15}NO$)
n = 11 (= $C_{12}H_{23}NO$)

$Mn(C_8H_{15}NO)_6X_2$ (X = ClO_4, Br, I, BF_4). The perchlorato, bromo, and tetrafluoroborato complexes were prepared by heating solutions of $Mn(ClO_4)_2 \cdot 6H_2O$, $MnBr_2 \cdot 4H_2O$, or $Mn(BF_4)_2 \cdot 6H_2O$ (0.005 mol) in a mixture of triethyl orthoformate (15 mL) and ethanol

(5 mL) at 50°C for 20 min to assure dehydration and then adding ethanol solutions of 8-octanelactam (0.02 mol for the bromo compound and 0.03 mol for the others). The crystals which appear on cooling were washed with diethyl ether and dried in vacuo. The iodo complex was obtained from an anhydrous ethanol solution of $MnCl_2 \cdot 4H_2O$ (0.005 mol) and KI (0.01 mol) dehydrated with triethyl orthoformate. The mixture was concentrated to precipitate KCl, which was filtered off. The ethanol solution of the ligand (0.03 mol) was added. The crystals resulting on evaporation were washed and dried as above [1].

The compounds are colorless, except the iodide, which is yellow, and crystalline, as shown by their X-ray powder diagrams (d spacings in the paper). The following table lists the melting points (m.p.) and starts of decomposition (dec.), both in °C from DTA and TG, the shifts ($\Delta\nu$) of the ν(CO) absorption bands (cm^{-1}) in the IR spectrum (KBr discs, 4000 to 200 cm^{-1} region) from ν(CO) of the free ligand, the magnetic moments μ_{eff} (in μ_B) from the Faraday method at 25°C, and the molar conductivities Λ of 0.001M solutions in dimethylformamide (DMF), nitromethane (CH_3NO_2), and nitrobenzene ($C_6H_5NO_2$).

compound	m.p.	dec.	$\Delta\nu$	μ_{eff}	Λ in $\Omega^{-1} \cdot cm^2 \cdot mol^{-1}$		
					(DMF)	(CH_3NO_2)	($C_6H_5NO_2$)
$Mn(C_8H_{15}NO)_6(ClO_4)_2$	146	190	−28	5.76	140	134	38.6
$Mn(C_8H_{15}NO)_6Br_2$	105	170	−40	5.71	121.0	insol.	insol.
$Mn(C_8H_{15}NO)_6I_2$	85	170	−30	5.67	146.0	42	12
$Mn(C_8H_{15}NO)_6(BF_4)_2$	149	180	−29	5.78	136	149	46

The shift of the ν(CO) band in the IR spectra of the complexes indicates coordination of the 8-octanelactam to Mn through the oxygen of its carbonyl group. The values of μ_{eff} are indicative for 6-coordinate Mn complexes with high-spin octahedral configuration. The compounds are 1:2 electrolytes in DMF and, except $Mn(C_8H_{15}NO)_6I_2$ and $Mn(C_8H_{15}NO)_6Br_2$, in nitromethane and nitrobenzene. $Mn(C_8H_{15}NO)_6I_2$ behaves as a nonelectrolyte in nitrobenzene [1].

The perchlorate explodes at 295°C, and the tetrafluoroborate decomposes, releasing BF_3 and 8-octanelactam. The final product is MnF_2 or, in the other cases, Mn_3O_4. All complexes are air stable and soluble in water, ethanol, and DMF. The perchlorate and tetrafluoroborate also dissolve in nitromethane and nitrobenzene [1].

$Mn(C_8H_{15}NO)_4(NCS)_2$ was prepared in the same way as $Mn(C_6H_{11}NO)_4(NCS)_2$ (p. 268), but using 8-octanelactam instead of ε-caprolactam [2]. It was also prepared by a method similar to that given for $Mn(C_8H_{15}NO)_6I_2$ (see above): $MnCl_2 \cdot 4H_2O$ was converted into the isothiocyanate in anhydrous ethanol solution, and 8-octanelactam was added in a 1:2 or 1:4 mole ratio [1].

The colorless crystalline [1, 2] compound melts at 111°C [1]. The magnetic susceptibility at 25°C yields an effective magnetic moment of 5.6 μ_B [2] or 5.80 μ_B [1]. The IR spectrum agrees with that of $Mn(C_6H_{11}NO)_4(NCS)_2$ (p. 268) [2]. $\Delta\nu$(CO) is -20 cm^{-1} [1]. For the electronic absorption spectrum see p. 268. Thermal decomposition starts at 180°C [1]. The complex is stable to air and light but is hydrolyzed in aqueous solution [2]. It is also soluble in ethanol, dimethylformamide, nitromethane, and nitrobenzene [1]. The molar conductivity Λ in $\Omega^{-1} \cdot cm^2 \cdot mol^{-1}$ of 0.001M solutions at 25°C is $\Lambda = 11.0$ in acetone [2] and 85.7 in dimethylformamide [1], corresponding to a 1:1 electrolyte [1, 2]. The complex is almost nonconducting in nitromethane ($\Lambda = 36$) and nitrobenzene ($\Lambda = 9$) [1].

$Mn(C_8H_{15}NO)_nCl_2$ ($n=1$, 2). The compounds are prepared in the same way as $Mn(C_8H_{15}NO)_6X_2$ (p. 269) using $MnCl_2 \cdot 4H_2O$ and 8-octanelactam in a 1:1 or 1:2 mole ratio. Both compounds are colorless and crystalline as shown by their X-ray powder diagrams. The melting points (m.p.) and starts of decomposition (dec.) in °C, the magnetic moments μ_{eff} in μ_B at 25°C, and the shift of the ν(CO) absorption band in the IR spectrum from that of the free ligand are given below:

compound	m.p.	dec.	μ_{eff}	$\Delta\nu$(CO) in cm^{-1}
$Mn(C_8H_{15}NO)Cl_2$	250	255	5.14	−30
$Mn(C_8H_{15}NO)_2Cl_2$	80	200	5.44	−30

Both compounds are almost nonconducting in DMF. They are insoluble in nitromethane and nitrobenzene. The rather high melting point, very low conductivity, and the magnetic moment of $Mn(C_8H_{15}NO)Cl_2$, which is notably lower than those of other manganese(II)-lactam complexes, point to a binuclear tetrahedral structure with two bridging ligand molecules, which may be formulated as $[Mn(C_8H_{15}NO)Cl_2]_2$. $Mn(C_8H_{15}NO)_2Cl_2$ also appears to be tetrahedral to judge from its low magnetic moment. The other properties of the chloro complexes resemble those of $Mn(C_8H_{15}NO)_6X_2$ [1].

$Mn(C_{12}H_{23}NO)_4(NCS)_2$. The colorless crystalline complex is prepared in the same way as $Mn(C_6H_{11}NO)_4(NCS)_2$ (p. 268) using 12-dodecanelactam instead of ε-caprolactam. Magnetic susceptibility measurements at 25°C yield $\mu_{eff}=5.7\ \mu_B$. The IR spectrum agrees with that of $Mn(C_6H_{11}NO)_4(NCS)_2$. For the electronic absorption spectra, see p. 268. The compound is stable to air and light but is hydrolyzed in aqueous solution. The molar conductivity Λ of a 1×10^{-3} M acetone solution is $\Lambda = 8.0\ \Omega^{-1} \cdot cm^2 \cdot mol^{-1}$ and corresponds to that of a 1:1 electrolyte [2].

References:

[1] F. Kůtek, B. Dušek (Sb. Vys. Sk. Chem. Technol. Prace Anorg. Chem. Technol. B **23** [1978] 37/49; C.A. **93** [1980] No. 60136).– [2] C. D. Rao, B. K. Mohapatra (J. Inorg. Nucl. Chem. **39** [1977] 689/91).

16.2 Complexes with N-Heterocycles Containing Two N Atoms in the Ring

16.2.1 With 1H-Pyrazole 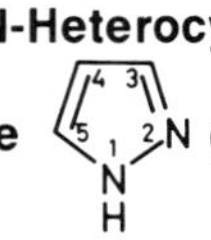 (=$C_3H_4N_2$) and Its Methyl Derivative (=$C_4H_6N_2$)

16.2.1.1 Complexes in Solution

The stability constants of the 1H-pyrazole complexes $Mn(C_3H_4N_2)^{2+}$ and $Mn(C_3H_4N_2)_2^{2+}$ have been determined polarographically in aqueous solution at 25°C and concentrations of 0.1 mol/L $NaNO_3$: $K_1=1.8$ L/mol, $\beta_2=2.2\ L^2/mol^2$. The manganese(II) complexes are much less stable than the complexes of Fe, Co, Ni, Cu, and Zn [1].

16.2.1.2 Isolated Compounds

In free methyl-1H-pyrazole, the methyl group may be located at either the 3 or the 5 position of the heterocyclic ring:

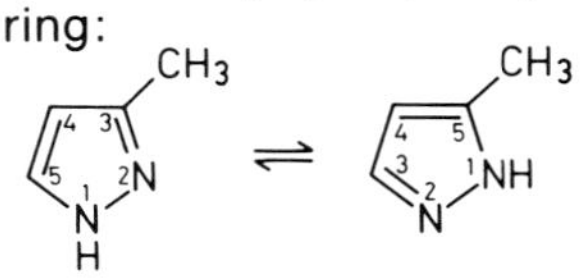

Steric considerations, comparison with complexes of other substituted pyrazoles, and vibrational spectra indicate that the coordinated ligand is present as 5-methyl-1H-pyrazole rather than 3-methyl-1H-pyrazole [12, 13]. X-ray powder patterns give evidence for isomorphism between several of the Mn compounds and those of other first-row transition metals [4, 11, 12].

MnL_6X_2 and $MnL_6X_2 \cdot L$ Complexes

The pyrazole compounds **$Mn(C_3H_4N_2)_6X_2$** ($X = ClO_4$, BF_4) [2] and the methylpyrazole adduct **$Mn(C_4H_6N_2)_6(ClO_4)_2 \cdot C_4H_6N_2$** [12] were prepared by mixing ethanolic solutions of the Mn^{II} salts with ethanolic solutions of the pyrazoles. The crystals appeared on standing or on addition of ether. Thus a 1:1 solution of pyrazole (0.07 mol) in ethanol was added to a solution of the salt hydrate (0.01 mol) in 10 mL of ethanol and 3 mL $HC(OC_2H_5)_3$, the last for dehydration. The crystals were filtered, washed three times with dry ether, and dried in vacuo [2]. X-ray powder patterns show the compounds $Mn(C_3H_4N_2)_6X_2$ to be isomorphous [2]. The properties of the compounds are shown in the following table; the IR data (in Nujol) are given in cm^{-1}, the magnetic moment μ_{eff} in μ_B:

compound	color	m.p. in °C	ν(Mn-N)	μ_{eff} at 293 K	Ref.
$Mn(C_3H_4N_2)_6(ClO_4)_2$	cream	254 to 255	221	5.82	[2]
$Mn(C_3H_4N_2)_6(BF_4)_2$	colorless	260 (dec.)	221	5.81	[2]
$Mn(C_4H_6N_2)_6(ClO_4)_2 \cdot C_4H_6N_2$	—	135 to 136	209 [14]	5.92	[12, 14]

The magnetic moments are those of high-spin complexes with O_h symmetry [2, 12]. For $Mn(C_4H_6N_2)_6(ClO_4)_2 \cdot C_4H_6N_2$, results of ESR measurements ($g = 2.00 \pm 0.01$, $A = 86$ G) as well as ligand-field spectra of the isomorphous Fe, Co, and Ni compounds further confirm the presence of fairly regular octahedral MnL_6^{2+} groups [12]. The IR spectrum (Nujol) shows the ring stretching vibration R 6 at 947 cm^{-1} (free ligand, 937 cm^{-1}). Further ligand bands (only a few are given) are similar to those reported on p. 273 for the halogeno complexes $Mn(C_4H_6N_2)_4X_2$. Unsplit ν_3 and ν_4 bands and forbidden ν_1 and ν_2 bands are evidence for uncoordinated anions [12]. The presence of uncoordinated ligand in $Mn(C_4H_6N_2)_6(ClO_4)_2 \cdot C_4H_6N_2$ is concluded from the visible spectra of isomorphous Co and Fe compounds [12] and is confirmed by weak bands in the IR spectrum at positions of uncomplexed methylpyrazole. It therefore should be formulated as $MnL_6(ClO_4)_2 \cdot L$ [14]. Coordination to Mn evidently occurs via the double bonded N atom (=N-) [2, 12]. The Mn-N bonds are somewhat stronger than in $Mn\text{-}NH_3$ complexes and are highly covalent, as shown by susceptibility and ESR measurements [2].

MnL_4X_2 Complexes

MnL_4X_2 ($X = NO_3$, Cl, Br, I). The pyrazole complexes ($L = C_3H_4N_2$) [3] and the methylpyrazole ($L = C_4H_6N_2$) complexes [12] were prepared as described above for $Mn(C_3H_4N_2)_6X_2$ ($X = ClO_4$, BF_4). The deuterated methylpyrazole compound with X = Br is prepared by repeated exchange from D_2O and precipitation with $HC(OC_2H_5)_3$ and $(C_2H_5)_2O$ [14]. $Mn(C_3H_4N_2)_4(NO_3)_2$ melts at 180°C [4]. The melting points of $Mn(C_4H_6N_2)_4X_2$ are 132 to 133°C for $X = NO_3$ [4, 12], 169 to 173°C for X = Cl, 180 to 182°C for X = Br, and 171 to 173°C for X = I [12]. For $Mn(C_4H_6N_2)_4Br_2$, which crystallizes from alcoholic solution as colorless needles, the measured and calculated densities were found to be $D_m = 1.600(5)$ and $D_c = 1.593$ g/cm^3, respectively [13].

ESR spectra of the solid compounds at Q- and X-band frequencies can be interpreted on the basis of a slightly distorted D_{4h} symmetry with D ranging from 0.05 to 1.5 cm^{-1} and λ (equal to E/D, D = axial splitting parameter, E = rhombic splitting parameter) ranging from 0.01 to 0.1 cm^{-1}. The D and λ values increase in the order $NO_3 < Cl < Br < I$, indicating hydrogen bonding between the anion and the pyrazole ligand of increasing strength in this order [3]. The results of more recent ESR, heat-capacity, paramagnetic-susceptibility, and magnetization measurements (on p. 274) are consistent with D = $+0.98 \pm 0.02$ and E < 0.01 cm^{-1} for $Mn(C_3H_4N_2)_4I_2$ [5]. X-ray powder patterns show the methylpyrazole compounds $Mn(C_4H_6N_2)_4X_2$ to belong to different X-ray types for the various anions [12]. The IR and Raman spectra of $Mn(C_4H_6N_2)_4X_2$ and the deuterated bromo compound (Br(D)) were obtained on samples dispersed in Nujol (IR) or on solids (Raman) in the range 4000 to 150 cm^{-1}. In the following table assignments are given for the bands of the complexes and the free ligand (selected values). The complex bands were assigned on the basis of C_{2h} symmetry for the whole molecule [14]:

assignment	Cl	Br(H)	Br(D)	Br (Raman)*)	I	ligand (liquid)
N-H, D stretching	3270(vs) 3260(vs)	3300(vs) 3290(sh)	2475(vs) —	— —	3110(vs) 3295(vs)	3200(br) —
ring stretching R_1	1568(s)	1568(s)	1560(s)	1568(m)	1560(s)	1582(s)
ring stretching R_3	1335(w)	1335(m)	1340(m)	1336(s)	1335(m)	1360(s)
ring stretching R_4	1293(vs)	1294(s)	1288(m)	1286(s)	1286(s)	1307(m), 1276(m)
ring bending R_6	946(s)	944(s)	943(s)	948(w)	944(s)	937(s)
N-H, D deformation	716(s)	700(vs)	700(m)	695(m)	686(vs)	—
N-H, D deformation	—	586	487(vs)	584(m)	573(m)	575(vbr)
$Mn-N_{ligand}$ and Mn-X stretching	200(vs, br) —	202(s) 180(m)	200(s) 180(m)	190(w) 165(m)	212(s), 204(s) 185(m)	 —

*) Single crystals.

Some of the bands are considerably shifted, and split, upon complexation, the shift of the ν(NH) band largely depending on the anion. This suggests hydrogen bonding between the ligand and the anion, bonding strongest for X = Cl and decreasing in the sequence X = Cl > Br > NO_3 > I [14]. In the electronic spectra of $Mn(C_4H_6N_2)_4X_2$, taken between λ = 250 and 750 nm, no significant tetragonal splitting is observed, even though it is expected theoretically [15]. Figures showing the spectra of the Mn complex, as well as those of the Fe, Co, Ni, and Cu compounds, which are partially isomorphous with the Mn compounds [4, 12], are presented in [15].

For the 5-methylpyrazole complex $Mn(C_4H_6N_2)_4Br_2$, a detailed X-ray study was carried out [13]. The crystal system is triclinic, the space group $P\bar{1}-C_i^1$ (No. 2). The lattice parameters are a = 8.802(6), b = 9.695(5), c = 7.613(8) Å, α = 105.12(4)°, β = 114.98(4)°, and γ = 92.90(3)°; Z = 1. Bond lengths (in Å) are given in **Fig. 23**, p. 274. Bond angles: 88.8° and 91.5° for Br-Mn-N, 90.9° for N-Mn-N, 122.9° and 123.6° for Mn-N-N, and 133.3° and 133.0° for Mn-N-C. The coordination polyhedron around the Mn atom is slightly distorted from D_{4h} symmetry, first by Br-Mn-N angles deviating from 90°, second by the unequal Mn-N bond lengths. The compound consists of $Mn(C_4H_6N_2)_4Br_2$ units of crystal symmetry $C_i(\bar{1})$, the units being separated by normal van der Waals distances. Some structural features may be due to hydrogen bonding [13].

Fig. 23

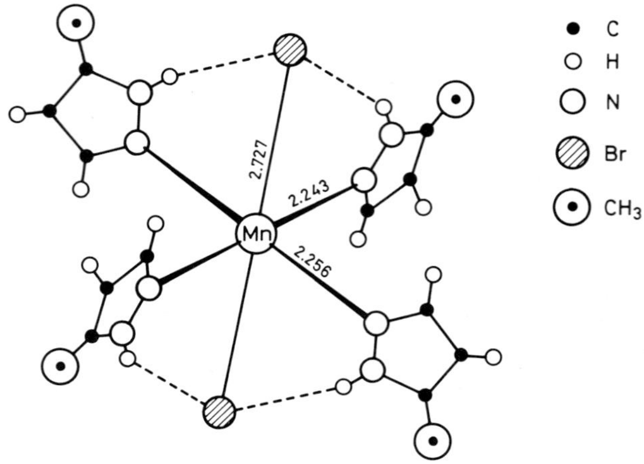

Molecular structure of the methylpyrazole complex $Mn(C_4H_6N_2)_4Br_2$.

For $Mn(C_3H_4N_2)_4I_2$, heat capacity (1.3 to 80 K), paramagnetic susceptibility (2 to 80 K), and isothermal magnetization (2 to 20 K) were measured. For graphs see the paper. The C_p versus T curve suggests a Schottky anomaly: there is a minimum at 4.2 K and a broad maximum (4.2 $J \cdot mol^{-1} \cdot K^{-1}$) at 1.65 K [5]. $Mn(C_3H_4N_2)_4I_2$ darkens on standing in air, possibly due to oxidation [5].

$Mn(C_3H_4N_2)_4SO_4 \cdot H_2O$ and **$Mn(C_4H_6N_2)_4SO_4$.** Pyrazole in hot aqueous solution or neat 5-methylpyrazole were added to a hot solution of the hydrated manganese sulfate. The colorless crystals appeared on cooling to room temperature. $\mu_{eff} = 5.92\ \mu_B$ is found for both compounds. The complexes decomposed explosively above 200°C [6].

$MnL_4(NCS)_2$. The pyrazole ($L = C_3H_4N_2$) and the 5-methylpyrazole ($L = C_4H_6N_2$) compounds were prepared similarly to $Mn(C_3H_4N_2)_6X_2$ ($X = ClO_4$, BF_4, p. 272), except that ethanol or acetone could be used as solvents for $Mn(NCS)_2$ and that neat pyrazoles were used [4]. $Mn(C_3H_4N_2)_4(NCS)_2$ melts at 176°C [4]. Depending on the solvent (no details given), $Mn(C_4H_6N_2)_4(NCS)_2$ can be isolated in two dimorphic forms as shown by differing IR spectra and X-ray powder diagrams. One form, melting at 128°C, is isomorphous with the Co and Ni compounds, the other, melting at 129°C, is isomorphous with the Fe and a second form of the dimorphic Ni complex. The IR spectra of the compounds are characterized by a single ν(CN) band of the NCS^- anion near 2060 cm^{-1}, indicative for trans-NCS groups coordinated via nitrogen. The ν(CS) band could not be observed due to overlap with ligand bands. The δ(NCS) bands appeared as rather weak bands near 475 cm^{-1} [4].

MnL_2X_2 Complexes

$Mn(C_3H_4N_2)_2X_2$ ($X = Cl$, Br). The slightly pink-colored pyrazole compounds were prepared as described on p. 272 for $Mn(C_3H_4N_2)_6X_2$ ($X = ClO_4$, BF_4) [7 to 9]. They are stable in air [8]. X-ray powder patterns show the compounds to be isomorphous with one another, with the Co and Ni compounds, and with the pyridine compounds $Mn py_2X_2$, p. 92 [11]. $Mn(C_3H_4N_2)_2Cl_2$ melts above 300°C [11]. The enthalpy of formation resulting from solution enthalpy measurements is $\Delta H = -15.19 \pm 0.19$ kcal/mol [11]. The measured and calculated density is $D_m = 1.832(3)$ and $D_c = 1.838$ g/cm³ [10].

For $Mn(C_3H_4N_2)_2Cl_2$ an X-ray study was carried out. The complex crystallizes in the monoclinic system, space group C2/c-C^6_{2h} (No. 15) with a=18.26(2), b=3.761(2), c=13.839(2) Å, and β=94.78(8)°; Z=4 [10]. Intramolecular distances are shown in **Fig. 24.** The bond angles around the Mn atom are 91.9(3)° for Cl-Mn-N, 89.2(3)° for Cl-Mn-N′, 87.1(1)° for Cl-Mn-Cl′, and 178.5(3)° for N-Mn-N′. Further details are given in the paper. The coordination symmetry of Mn is C_2. The compound consists of Cl-bridged metal chains of the type 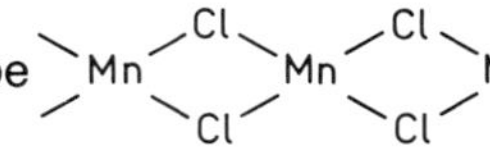 with pyrazole completing the distorted octahedral coordination. The chains extend in the direction [010], and neighboring chains are displaced by ½,0,0, −½,0,0, 0,0,½, and 0,0,−½. There is evidence for hydrogen bonding between N-H and Cl [10]. In the IR spectrum of $Mn(C_3H_4N_2)_2Cl_2$ (polyethylene tablets, recorded between 400 and 100 cm^{-1}), overlapping strong bands at 247, 237, and 225 cm^{-1} are assigned to the ν(Mn-N) vibrations [11].

Fig. 24

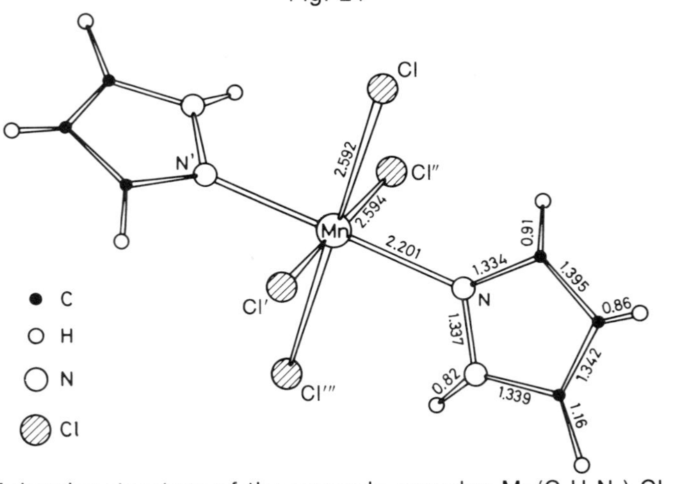

Molecular structure of the pyrazole complex $Mn(C_3H_4N_2)_2Cl_2$.

Magnetic susceptibility data in the 2 to 80 K temperature range [7, 9] and heat capacity data in the 1.3 to 80 K temperature range [8] indicate a magnetic behavior at low temperatures similar to that described for the pyridine compounds $Mnpy_2X_2$ on p. 93. The paramagnetic Curie temperatures Θ_p (in K) [7], the temperatures for the transitions to short-range ordering T_{tr} (in K), the Néel temperatures T_N (in K), and the corresponding magnetic energies ΔE (in J/mol) and entropies ΔS (in $J \cdot mol^{-1} \cdot K^{-1}$) were reported as follows [8]:

compound	Θ_p	T_{tr}	T_N	ΔE (T_{tr})	ΔE (T_N)	ΔS (T_{tr})	ΔS (T_N)
$Mn(C_3H_4N_2)_2Cl_2$	−12(1)	3.1±0.1	<1.3	98	<0.8	14.9±0.1	<0.8
$Mn(C_3H_4N_2)_2Br_2$	−13(1)	3.35±0.05	1.930±0.005	102	2.1	14.9±0.1	1.3

The compound $Mn(C_3H_4N_2)_2Cl_2$ does not exhibit a transition to a long-range ordered state down to 1.3 K, which suggests that the inter-chain interactions are very weak [8]. The following values for the intra-chain exchange parameter J/k (in K) were evaluated from the magnetic data [7, 8] and from ESR line-widths at 110 and 300 K [7] with the isotropic antiferromagnetic Heisenberg linear-chain model [7, 8]. In addition, a moderately large single-ion anisotropy was assumed [8]. The ratio between the inter- and intrachain coupling |J'/J| was also derived [7]:

compound	J/k(χ)	J/k(C_p)	J/k(ESR) at 110 K	J/k(ESR) at 300 K	|J'/J|
$Mn(C_3H_4N_2)_2Cl_2$	~−0.9	−0.78	−0.6(1)	−0.52(8)	$<0.6\times10^{-2}$
$Mn(C_3H_4N_2)_2Br_2$	~−1.0	−0.82	~−1.0	−0.75(9)	1.2×10^{-2}

The values from ESR show that J/k decreases with temperature between 100 and 300 K for both compounds [7].

Other Compounds

$[Mn_4F_4(C_4H_6N_2)_{12}](BF_4)_4$ was prepared by adding 5-methylpyrazole (30 mmol) to an ethanolic solution of $Mn(BF_4)_2\cdot6H_2O$ (4 mmol) containing triethyl orthoformate for dehydration. Upon addition of diethyl ether a colorless solid precipitates. It was washed with diethyl ether and dried in vacuo. The compound melts above 250°C. Its X-ray powder diagram (strongest spacings at d = 14.2, 11.0, 10.0, and 3.68 Å) suggests isomorphism with the analogous complexes of Co, Ni, and Cd. Susceptibility measurements at temperatures between 2 and 90 K sugggest a maximum susceptibility <2 K. The effective magnetic moments are μ_{eff} = 5.81 μ_B at 50.5 K, 3.82 μ_B at 10.4 K, and 2.39 μ_B at 2.3 K. The decrease of μ_{eff} at low temperature and the ESR spectrum showing an isotropic signal centered at g = 2.00 are in accord with a tetrameric structure where each Mn atom is bound to three (magnetically equivalent) triply bridging fluoride ions as concluded from the X-ray data. The IR spectrum (recorded for Nujol mulls in the 4000 to 180 cm^{-1} region) shows two absorption bands, at 316 and 237 cm^{-1}, assigned to ν(Mn-F) or ν($Mn\text{-}N_{ligand}$) vibrations (stretching), respectively, and a ligand band at 352 cm^{-1}. The BF_4^- ion is not coordinated, as shown by the normal unsplit, antisymmetric B-F stretching (ν_3) and bending (ν_4) vibrations [16].

References:

[1] D. R. Crow, J. V. Westwood (J. Inorg. Nucl. Chem. **30** [1968] 179/87, 185). – [2] J. Reedijk (Rec. Trav. Chim. **88** [1969] 1451/70, 1454, 1460, 1468). – [3] R. D. Dowsing, B. Nieuwenhuijse, J. Reedijk (Inorg. Chim. Acta **5** [1971] 301/4). – [4] J. Reedijk, J. A. Smit (Rec. Trav. Chim. **90** [1971] 1135/40, 1137). – [5] J. Reedijk, F. W. Klaaijsen, H. T. Witteveen (J. Chem. Soc. Faraday Trans. II **69** [1973] 1537/41).

[6] A. Anagnostopoulos, D. Nicholls, K. R. Seddon (J. Inorg. Nucl. Chem. **36** [1974] 2235/8). – [7] H. T. Witteveen, B. Nieuwenhuijse, J. Reedijk (J. Inorg. Nucl. Chem. **36** [1974] 1535/41). – [8] F. W. Klaaijsen, H. W. J. Blöte, Z. Dokoupil (Physica B+C **81** [1976] 1/14, 4/7, 12). – [9] H. T. Witteveen, J. Reedijk (Solid State Commun. **12** [1973] 557/60). – [10] S. Gorter, A. D. van Ingen Schenau, G. C. Verschoor (Acta Cryst. B **30** [1974] 1867/71).

[11] M. S. Barvinok, L. G. Lukina (Zh. Neorgan. Khim. **22** [1977] 2167/70; Russ. J. Inorg. Chem. **22** [1977] 1173/5). – [12] J. Reedijk (Rec. Trav. Chim. **89** [1970] 605/18, 608, 613). – [13] J. Reedijk, B. A. Stork-Blaisse, G. C. Verschoor (Inorg. Chem. **10** [1971] 2594/9). – [14] J. Reedijk (Rec. Trav. Chim. **90** [1971] 117/36, 124, 131/3). – [15] J. Reedijk (Rec. Trav. Chim. **89** [1970] 993/1016, 1011).

[16] R. W. M. Ten Hoedt, J. Reedijk (Inorg. Chim. Acta **51** [1981] 23/7).

16.2.2 With 3,5-Dimethyl-1H-pyrazole

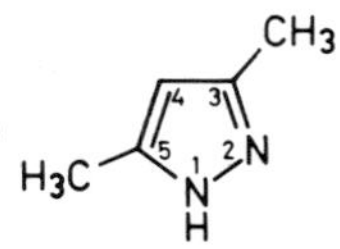

($=C_5H_8N_2$)

On account of steric hindrance, the maximum coordination number of manganese in the 3,5-dimethylpyrazole complex is four, and not six as for the complexes with pyrazole and methylpyrazole, see p. 272 [1].

$Mn(C_5H_8N_2)_4(NCS)_2$ was prepared by addition of excess 3,5-dimethylpyrazole dissolved in hot C_2H_5OH to a solution of $Mn(NCS)_2$ in a C_2H_5OH/H_2O mixture. The complex melts at 142.5°C. The IR spectrum is characterized by a single ν(CN) band of the NCS^- anion at 2085 cm^{-1}, indicative for trans-NCS groups coordinated via nitrogen. The ν(CS) band was observed at 828 cm^{-1}. The δ(NCS) band appeared at 485 cm^{-1} [2].

$Mn(C_5H_8N_2)_4(NCS)_2 \cdot 2C_5H_8N_2$. The 3,5-dimethylpyrazole adduct was prepared as described on p. 272 for $Mn(C_3H_4N_2)_6X_2$ ($X=ClO_4$, BF_4). It melts at 148 to 150°C. The IR spectrum (in Nujol; only far-IR data reported in detail) shows a single CN stretching band at 2085 cm^{-1}. Two peaks, at 1025 and 1010 cm^{-1}, of a quartet between 1050 and 1000 cm^{-1} appear to be due to uncoordinated 3,5-dimethylpyrazole. A strong band at 231 cm^{-1} is assigned to the Mn-N stretching vibration. The ESR spectrum points to a very small zero-field splitting ($D=0.03$ cm^{-1}) and nearly axial symmetry. ESR and IR spectrum are in agreement with the proposed structure containing trans-NCS anions [1].

$Mn(C_5H_8N_2)_3(NO_3)_2$ and **$Mn(C_5H_8N_2)_3Br_2$.** The compounds were prepared as described on p. 272 for $Mn(C_3H_4N_2)_6X_2$ ($X=ClO_4$, BF_4). The colorless nitrate melts at 142 to 144°C; the buff colored bromide, at 154 to 155°C. The IR spectrum of the nitrate indicates the presence of coordinated NO_3^- ions. This is concluded from the split antisymmetric NO stretching vibration (not given), the split combination band at about 1750 cm^{-1}, and the appearance of the forbidden band at about 1040 to 1020 cm^{-1}. However, a conclusion concerning uni- or bidentate coordination of NO_3^- is not possible. A very strong and broad band in the far-IR is assigned to the Mn-O stretching vibration. For the bromide, strong bands at 221 and 196 cm^{-1} are ascribed to the Mn-N and Mn-Br stretching vibrations, respectively. Tetrahedral coordination of the manganese atom is ruled out by the IR spectra for both compounds. Splittings of the ligand vibrations (not given) indicate a very low symmetry for $MnL_3(NO_3)_2$. The ESR spectrum is almost identical with that of the 1,2-dimethylimidazole complex ($D=0.07$ cm^{-1}, p. 301) [1].

$Mn(C_5H_8N_2)_2X_2$ (X=F, Cl, Br, NCS). The complex compounds with X=Cl, Br, NCS were prepared as described on p. 272 for $Mn(C_3H_4N_2)_6X_2$ ($X=ClO_4$, BF_4) [1]. The fluoro compound was prepared by reacting 3,5-dimethylpyrazole with a solution of $Mn(BF_4)_2$. In the following table the magnetic exchange parameter J is derived from susceptibility [3] or ESR line width [1]. Frequencies ν are given in cm^{-1}.

X	color	m.p. in °C	ν(Mn-N)	ν(Mn-X)	J in cm^{-1}	Ref.
F	colorless	—	260, 250	375, 355	∼−2.5	[3]
Cl	colorless	320 to 322	230, 206	176, 162	∼0.4	[1]
Br	buff	180 to 182	227	259	—	[1]
NCS	colorless	198 to 200	264	184	∼0.2	[1]

IR and ESR Q- and X-band spectra [1, 3] suggest a linear chain structure for the compounds with X=F [3], Cl, and NCS [1], as described on p. 275 for the pyrazole compound $Mn(C_3H_4N_2)_2Cl_2$, with an antiferromagnetic interaction of the Mn atoms via the bridging X atoms. For $Mn(C_5H_8N_2)_2F_2$ the very strong antiferromagnetic interaction is deduced also by the low-temperature susceptibility, which has a maximum at about 15 K. For this compound the NH bands, occurring as a multiplet around 3100 cm^{-1}, indicate a strong hydrogen bond between ligand and F^- ions [3]. X-ray powder diagrams show that the bromo compound, $Mn(C_5H_8N_2)_2Br_2$, is not isomorphous with the chloro compound. A distorted tetrahedral structure similar to that of the isomorphic Ni, Zn, and Co bromo compounds is assumed. The ESR spectrum indeed indicates a geometry very much distorted from axial [1].

References:

[1] M. A. Guichelaar, J. A. M. van Hest, J. Reedijk (Delft Progr. Rept. **2** No. 1 [1976] 51/63, 54, 57/8, 61). – [2] A. Anagnostopoulos (J. Inorg. Nucl. Chem. **38** [1976] 435/7). – [3] M. A. Guichelaar, J. A. M. van Hest, J. Reedijk (Inorg. Nucl. Chem. Letters **10** [1974] 999/1004, 1002).

16.2.3 With 1,1-Methylenebis(3,5-dimethyl-1H-pyrazole)

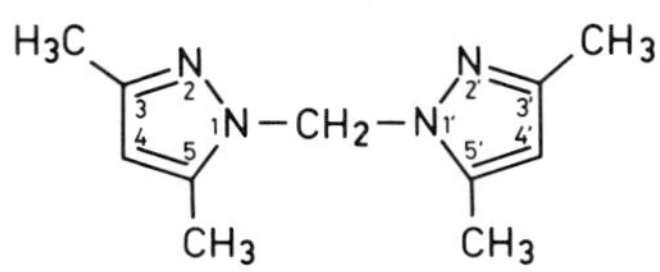

($=C_{11}H_{16}N_4$)

$Mn(C_{11}H_{16}N_4)_2(ClO_4)_2$. The colorless compound was prepared as described on p. 272 for $Mn(C_3H_4N_2)_6X_2$ compounds ($X=ClO_4$, BF_4). It melts at 262°C. It is isomorphous with the Cd compound. The IR spectrum (in Nujol, from 4000 to 20 cm^{-1}) exhibits a complex ClO_4 pattern, with split ν_3 bands at 1155, 1100, and 1035 cm^{-1}, ν_4 bands at 627, 619, and 615 cm^{-1}, and a single ν_1 band at 935 cm^{-1}, all indicative of coordinated bidentate ClO_4^- ions. Seven bands observed in the far-IR spectrum, at 362m, 352m,br, 342m, 300m, 268w, 220m,br, and 182 cm^{-1}(s), were not assigned. The molar electrical conductivity measured in nitromethane is $\Lambda=142\ \Omega^{-1}\cdot cm^2\cdot mol^{-1}$, suggesting exchange of coordinated ClO_4^- with nitromethane [1].

$[Mn_2F_2(C_{11}H_{16}N_4)_4](BF_4)_2$ was prepared by adding an ethanolic solution of $Mn(BF_4)_2 \cdot 6H_2O$ (3 mmol) containing excess triethyl orthoformate for dehydration to an ethanolic solution of the ligand (6 mmol). The precipitate of fine crystals was recrystallized from ethanol, washed with diethyl ether, and dried in vacuo. The colorless compound appears to be isomorphous with the analogous complexes of Fe and Cd according to X-ray powder diagrams. The IR spectrum (recorded for Nujol mulls in the 4000 to 400 cm^{-1} region and on polyethylene discs in the 500 to 20 cm^{-1} (far-IR) region) confirms such isomorphism. In the far-IR two very strong absorption bands, at 347 and 218 cm^{-1}, are tentatively assigned to ν(Mn-F) and $\nu(Mn\text{-}N_{ligand})$ vibrations, respectively, whereas the other bands in that region apparently are due to ligand vibrations. The positions of the ν(Mn-F) and ν(Mn-N) vibrations are indicative of octahedral difluoro-bridged chain structures. The molar electrical conductivity $\Lambda=172\ \Omega^{-1}\cdot cm^2\cdot mol^{-1}$ of a 0.001M solution in nitromethane is that of a 1:2 electrolyte consisting of $[(C_{11}H_{16}N_4)_2MnF_2Mn(C_{11}H_{16}N_4)_2]^{2+}$ cations and BF_4^- anions [2].

References:

[1] J. Reedijk, J. Verbiest (Transition Metal. Chem. [Weinheim] **4** [1979] 239/43). – [2] J. Verbiest, J. A. C. Van Coijen, J. Reedijk (J. Inorg. Nucl. Chem. **42** [1980] 971/5).

16.2.4 With 4-Nitro-1H-pyrazole

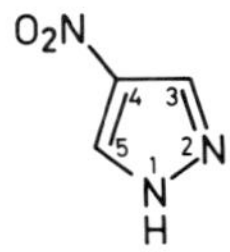
(=$C_3H_3N_3O_2$)

$Mn(C_3H_2N_3O_2)_2 \cdot NH_3 \cdot H_2O$. The ammineaquo compound was precipitated on addition of $MnCl_2 \cdot 4H_2O$ to 4-nitropyrazole, both dissolved in aqueous ammonia (25%), in an N_2 atmosphere. The precipitate was washed with water, ethanol, and ether and dried in a vacuum at room temperature. The IR spectrum of the yellow compound was recorded for Nujol mulls and CsI pellets in the range between 4000 and 180 cm^{-1}. The IR absorption ranges were given for a series of investigated transition metal complexes, together with the corresponding band positions of the free neutral ligand and the assignments. The following distinct IR absorptions were given for the manganese complex and assigned as follows:

$MnL_2 \cdot NH_3 \cdot H_2O$	550	470*)	270, 245
assignment	NH_3 rocking	CN bending (ν_{17})	metal-to-ligand vibrations

*) In the free ligand at 438 cm^{-1}.

Assignments of IR bands were made with the aid of deuterated nitropyrazole. For both the neutral ligand and the coordinated anion in the solid complex, a C_{2v} symmetry can be concluded from the IR and Raman spectra (the latter given only for the Zn complex and the free ligand). For the anion this is in agreement with a bidentate coordination via two nitrogen atoms. Coordination of the NH_3 molecule is concluded from the NH_3 rocking frequencies, which are very sensitive to the nature of the metal in the series of investigated compounds. Coordination of the water molecule is also assumed on account of the similarity of the vibrational spectra with those of similar metal complexes and the ligand-field spectra for these complexes (not given for the manganese complex), which indicate octahedral coordination.

The compound appeared to be sensitive to oxygen. It does not decompose on heating up to 250°C.

Reference:

G. Nieuwpoort, J. G. Vos, W. L. Groeneveld (Inorg. Chim. Acta **29** [1978] 117/23).

16.2.5 With Tris(3,5-dimethyl-1H-pyrazol-1-ylethyl)amine

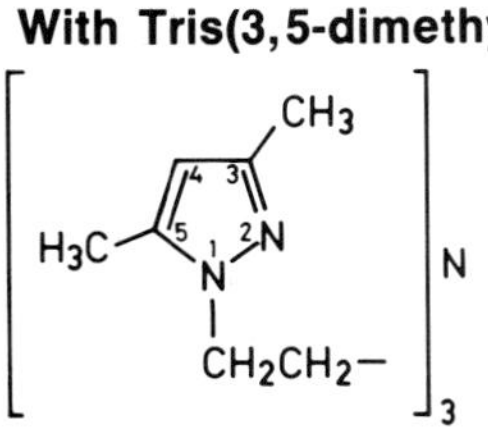
(=$C_{21}H_{33}N_7$)

$[MnBr(C_{21}H_{33}N_7)]B(C_6H_5)_4 \cdot 0.5\,CH_2Cl_2$. The compound was obtained by adding a solution of the ligand and $NaB(C_6H_5)_4$ in dichloromethane to a warm solution of manganese(II) bromide while excluding air. Susceptibility measurements at 298 K yielded the magnetic moment $\mu_{eff} = 5.86\ \mu_B$, indicating a high-spin $Mn^{II}(d^5)$ complex. The electronic reflectance spectrum shows only three very weak bands, at 17400, 22400, and above 25000 cm^{-1}, assigned to spin-forbidden electronic transitions from the ground sextet to the first excited quartet states. The spectrum is comparable to those of other high-spin 5-coordinate Mn^{II} complexes, F. Mani (Inorg. Nucl. Chem. Letters **17** [1981] 45/9, 46).

16.2.6 With 1-Acetyl-1H-pyrazole

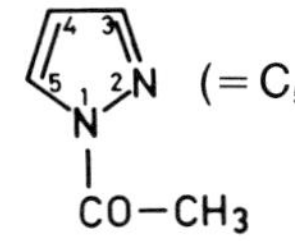

$(=C_5H_6N_2O)$

$[Mn(C_5H_6N_2O)_3](BF_4)_2$. To prepare the complex, $Mn(BF_4)_2 \cdot 6H_2O$ (0.01 mol) was first dehydrated with excess triethyl orthoformate (0.08 mol), and most of the solvent was evaporated under reduced pressure. The almost dry residue was dissolved in nitromethane and reacted with 1-acetylpyrazole (0.03 mol). The mixture was stirred for several hours, and the resulting solution was concentrated by evaporating nitromethane under reduced pressure at about 50°C. Upon cooling, colorless crystals separated. These were collected and washed with n-pentane, then evacuated for $^1/_2$ h to remove any solvent. All operations were done under anhydrous conditions. The complex melts at 202°C. Its IR spectrum recorded for Nujol mulls in the 4000 to 180 cm^{-1} region shows a broad absorption band at 3405 cm^{-1}, assigned to ν(OH) vibrations of the enolic form of 1-acetylpyrazole. A band at 1477 cm^{-1} is assigned to ν(C=C) vibrations of the enolized acetyl group. Five other bands, at 1529, 1293, 914, 688, and 593 cm^{-1}, are assigned to ring vibrations (2 ν, 3 δ), while two bands at 1050 and 524 cm^{-1} are associated with ν(BF) and δ(BF) vibrations, respectively, of the tetrahedral BF_4^- ion. One band at 220 cm^{-1} is ascribed to ν(Mn-ligand) vibrations. It was concluded from the IR and proton NMR spectra that Mn is coordinated to the enolic oxygen and one ring nitrogen (N2) of the bidentate enolic ligands to produce an octahedral complex cation. The tetrahedral BF_4^- ions are not coordinated.

$Mn(C_5H_6N_2O)Cl_2$ was prepared by reacting anhydrous $MnCl_2$ (0.01 mol) and 1-acetylpyrazole (excess) in nitromethane for 24 h. The resulting colorless solid is collected, washed with n-pentane, and dried by evacuation. It melts above 300°C. The IR spectrum (Nujol mull) shows an absorption band at 1704 cm^{-1}, assigned to ν(CO) vibrations of the acetyl group of the ligand (present in the keto form). Bands at 1534, 1372, 1292, 1074, 947, 663, and 601 cm^{-1} are assigned to ring modes (3 ν, 3 δ, breathing at 1074 cm^{-1}), and one band at 252 cm^{-1} is ascribed to ν(Mn-Cl) vibrations. It is concluded from the IR spectrum that Mn is coordinated to the oxygen of the acetyl group, as shown by the lowering of the ν(CO) band frequency ($\Delta\nu = -38$ cm^{-1}), to one nitrogen (N2) of the pyrazole ring (shift of ring modes), and to the Cl atoms, thus forming a six-coordinate, probably polymeric complex.

Reference:

W. L. Driessen, P. L. A. Everstijn (Inorg. Chim. Acta **41** [1980] 179/83, 180).

16.2.7 With Pyrazolinones

16.2.7.1 With Antipyrine (=1,5-Dimethyl-2-phenyl-1,2-dihydro-3H-pyrazol-3-one = Phenazone = $C_{11}H_{12}N_2O$)

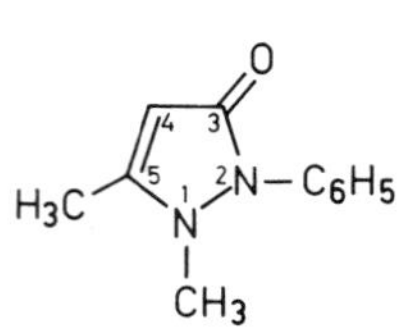

and Its 4-Amino Derivative $(=C_{11}H_{13}N_3O)$

16.2.7.1.1 Complexes in Solution

The stepwise formation of manganese(II) complexes in aqueous solution was determined potentiometrically at 25°C at ionic strength $I = 0.5$ (KNO_3):

ligand	log β_1	log β_2	log β_2
antipyrine	0.57	0.89	1.02
4-aminoantipyrine	1.07	1.83	—

Decreasing stability in the ligand order 4-aminoantipyrine > pyramidon (see p. 284) > antipyrine was found for the complexes $MnL_n(H_2O)^{2+}_{6-n}$. Manganese is 6-coordinate.

Reference:

B. Lenarcik, M. Wiśniewski, M. Gabryzewski (Polish J. Chem. **54** [1980] 1869/74, 1872).

16.2.7.1.2 Isolated Compounds

$Mn(C_{11}H_{12}N_2O)_6X_2$ (X = I, ClO_4, BF_4). The iodide was prepared by dissolving $MnCl_2 \cdot 4H_2O$, KI, and antipyrine in water or acetonitrile in a 1:2:6 mole ratio and evaporating the solution slowly [1, 2]. In a previous preparation, hydrated $Mn(NO_3)_2$ and crystalline NH_4I were used, and some small crystals of $Na_2S_2O_4$ added (to reduce I_2) [3]. The precipitate was recrystallized from a minimum amount of water containing a trace of $Na_2S_2O_4$ [3] or from acetonitrile [1]. The perchlorate [1, 4, 7] and the tetrafluoroborate [8] may be prepared similarly in aqueous solution with the use of any manganese(II) salt [4] and ammonium perchlorate or tetrafluoroborate. Samples of the manganese(II) perchlorate complex doped into isomorphous metal perchlorate complexes were prepared with direct use of the hydrated metal perchlorates [5, 6]. The crystals of $Mn(C_{11}H_{12}N_2O)_6X_2$ grown from aqueous solutions are pale pink [1, 4, 8] with well developed faces [1, 3, 6]. The perchlorate is nonhygroscopic [7], forming hexagonal [6], elongated (up to 3 cm) needles [4] or plates [1, 6]. It is isomorphous with the complexes $M(C_{11}H_{12}N_2O)_6(ClO_4)_2$ (M = Mg, Ca, Zn [10], Pb [11], and Co [5]), which belong to the hexagonal space group $P\bar{3}$-C^1_{3i} (No. 147) [10, 11]. The densities D, molar volumes V_m, and melting points (m.p.) of the compounds $Mn(C_{11}H_{12}N_2O)_6X_2$ are:

complex	D^{25}_4 in g/cm^3	V_m in cm^3/mol	m.p. in °C	Ref.
$Mn(C_{11}H_{12}N_2O)_6I_2$	1.42 [9]	—	171	[3]
$Mn(C_{11}H_{12}N_2O)_6(ClO_4)_2$	1.362	1016	197 (sharp)	[4]
$Mn(C_{11}H_{12}N_2O)_6(BF_4)_2$	1.359	1041	180 to 181	[8]

The IR spectrum of $Mn(C_{11}H_{12}N_2O)_6(ClO_4)_2$ exhibits an absorption band at 1631 cm^{-1}, which is assigned to a CO stretching mode shifted to lower frequencies by 35 cm^{-1} relative to free antipyrine, thus indicating a CO-Mn bond. The ν_3 and ν_4 bands of the ClO_4^- ion are unsplit (symmetry T_d) and occur at about 1100 to 1090 cm^{-1} (ν_3) and 630 to 620 cm^{-1} (ν_4), respectively. The electronic spectrum in acetonitrile shows two intense bands at about 35000 and 42000 cm^{-1}, both due to the coordinated ligand. Free antipyrine absorbs at 35970 and 41680 cm^{-1} in acetonitrile. A magnetic moment of $\mu = 5.54\ \mu_B$ at 300 K is obtained from susceptibility measurements, indicating a high-spin octahedral manganese(II) complex [7]. Susceptibility measurements on $Mn(C_{11}H_{12}N_2O)_6I_2$ by the Faraday technique yield the specific susceptibility $\chi = 8.68 \times 10^{-6}$ cm^3/g at room temperature [9]. Measurements on spherical and rod-shaped samples of the perchlorate at 295, 77, and 4.58 K yielded the magnetic anisotropy $\chi_{\parallel} - \chi_{\perp}$ (in 10^{-12} cm^3/g, not corrected for diamagnetic anisotropy) [5]:

sample shape	$\chi_{\parallel}-\chi_{\perp}$ at 295 K	$\chi_{\parallel}-\chi_{\perp}$ at 77 K	$\chi_{\parallel}-\chi_{\perp}$ at 4.58 K
sphere	2250	3380	157000
rod	2390	3220	227000

The observed shape anisotropy effect is due to field inhomogenities and nonuniform magnetization of the crystal. Therefore, serious errors may result with nonspherical crystals. The zero-field splitting parameter D (~ -0.015 cm^{-1} at T<20 K) is about three times larger than that obtained by ESR measurements of Mn^{2+} doped into a series of isomorphous compounds. The interpretation of the magnetic anisotropy above 20 K is complicated by the large diamagnetic contributions from the ligands. Presumably, significant differences exist in the orientation of the ligands in the isomorphous Mn and Mg complexes [5].

The ESR spectrum at X-band frequencies of powdered $Mn(C_{11}H_{12}N_2O)_6X_2$ (X = I, ClO_4) is broad and complicated, extending over 700 G [1]. For X = I a Lorentzian line shape (line width $\Delta H = 699.66$ Oe) was observed, and the splitting factor g = 2.0027 was derived [9]. However, single crystals show a partially resolved spectrum in a general direction and a 6-line spectrum along two directions inclined at an angle of 54° on either side of the crystallographic c axis. Crystals isomorphously diluted by Mg^{2+} (1:100) show spectra that are simple 6-line equispaced, yielding $g = 2.012 \pm 0.004$, the hyperfine coupling constant $A = (89.12 \pm 4.0) \times 10^{-4}$ cm^{-1}, and the axial field splitting parameter $D = (45.56 \pm 4.0) \times 10^{-4}$ cm^{-1}. If the magnetic field is parallel to the c axis, a 10-line spectrum results. If the field is in the (ab) plane, a 15-line isotropic spectrum results. Both spectra are of varying intensities. Their nature and the X-ray results indicate an axially distorted octahedral crystal field at the Mn^{2+} site with the distortion axis along c, as required by the trigonal crystal symmetry. A further dilution (1:7000) yields a well resolved isotropic 6-line spectrum with $g = 2.005 \pm 0.002$ and $A = (89.28 \pm 2.0) \times 10^{-4}$ cm^{-1} for a powder, while single crystals of dilution (1:1000) exhibit a 24-line spectrum along the (crystallographic) c axis and an 18-line spectrum in the (ab) plane. The parameters g, A, D, and (a − F) recorded from ESR spectra of diluted single crystals aiong the c axis are [1]:

dilution	g	10^4A, cm^{-1}	10^4D*), cm^{-1}	10^4(a − F)*) cm^{-1}
1:1000	2.003 ± 0.002	−88.20 ± 2.0	−42.97 ± 2.0	5.57 ± 2.0
1:7000	2.003 ± 0.002	−88.41 ± 2.0	−43.95 ± 2.0	6.9 ± 2.0

*) D and F are zero-field splitting parameters due to the axial field; constant a is the zero-field splitting parameter due to the cubic field.

The parameters A, D, and (a − F) suggest strong hyperfine interaction and a small axial crystal field superimposed over a weak cubic crystal field [1]. The spin-Hamiltonian ESR parameters of Mn^{2+} ions doped into isomorphous hexakis(antipyrene) perchlorate complexes of other divalent metals such as Mg, Ca, Co, Cd, and Pb reveal a marked decrease of D with increasing size of the host ions, while g, A, and B are nearly unaffected, and the effect on (a − F) is small and random. The hyperfine coupling constant A is unaffected by changes in the lattice and indicates some covalency of the Mn-O bond. Solutions of $Mn(C_{11}H_{12}N_2O)_6(ClO_4)_2$ in CH_2Cl_2 exhibit an isotropic hyperfine coupling constant in agreement with the single crystal data [6]. The ESR X-band spectrum of $Mn(C_{11}H_{12}N_2O)_6I_2$ dissolved in pyridine consists of six resolved hyperfine components in the range from 3099.2 to 3599.1 Gauss, the components having different line widths in good agreement with the computed values [2].

The conductivity of aqueous solutions of the perchlorate (molar conductivity Λ = 254 $\Omega^{-1}\cdot cm^2\cdot mol^{-1}$) shows that the complex is a 1:2 electrolyte. The cryoscopic molecular weight indicates that its antipyrine ligands are replaced by H_2O. Differential thermal analysis of mixtures with Al_2O_3 (1:10, heating rate 10°C/min) yield two exothermic peaks, at 337 and 475°C, the first one due to the decomposition of the complex (explosive if undiluted) and the second due to the oxidation of carbon formed on decomposition [7]. All complexes $Mn(C_{11}H_{12}N_2O)_6X_2$ are soluble in water [3, 4, 7, 8], the solubility in 100 mL of solution at 20°C being 3.63 g for X = ClO_4 [4] and 5.4 g for X = BF_4 [8]. The iodide is soluble also in $CHCl_3$ and slightly soluble in alcohol and acetone [3]; the perchlorate is soluble in acetonitrile [7].

$Mn(C_{11}H_{12}N_2O)_2X_2$ (X = Cl, Br, I, NCS). The chloro compound was prepared by reacting a boiling solution of antipyrine (25 g) in amyl alcohol (60 mL) with 15 g of dehydrated $MnCl_2$. After one day a light pinkish plastic mass deposited and yellowish rosette-like aggregations of fine needles of the desired compound were formed. The needles can be separated from the soft amorphous phase mechanically. The soft phase may be recrystallized repeatedly. The compound was dried by pressing on filter paper. It melts at 133°C and is slightly soluble in alcohol and acetone. The bromo complex was obtained on cooling a hot solution of dehydrated $MnBr_2$ (25 g) and antipyrine (35 g) in 60 mL of absolute alcohol. If no crystallization occurred, the alcohol was partially evaporated, quickly as the solution soon turns brown in air. The compound was recrystallized twice from alcohol, in which it is notably soluble. The slightly yellowish crystals start decomposing above 110°C, turning brown, and melt at 140°C. They are soluble in alcohol and acetone. The iodo compound was prepared much like the hexakis complex, $Mn(C_{11}H_{12}N_2O)_6I_2$ (p. 281), but without excess antipyrine. It is obtained as a colorless powder, which on exposition to air quickly turns yellow. It is also sensitive to moisture. The isothiocyanato complex was obtained by dissolving antipyrine in a concentrated aqueous solution of $Mn(NO_3)_2$ and adding a solution of NH_4SCN. The precipitate which became crystalline after several hours was recrystallized from water in the absence of air. The slightly yellowish pink compound becomes soft and begins to decompose on heating above 115°C. It is soluble in alcohol and acetone, the solutions turning brown in air [3].

$Mn(C_{11}H_{12}N_2O)_2(C_7H_5O_3)_2$. The salicylato compound was precipitated on reacting an aqueous solution of manganese(II) salicylate with antipyrine in a 1:1 to 1:3 mole ratio. The X-ray powder diagrams of several preparations with various mole ratios are identical. The patterns differ from those of the starting substances. The precipitate seems to be composed of two different phases having directions of light extinction of 34° to 38° and 14° [12].

$MnX_2\cdot 2CdX_2\cdot mC_{11}H_{12}N_2O\cdot nH_2O$ (X = Cl, m = 6, n = 0; X = Br, I, m = 8, n = 2; X = NCS, m = 7, n = 1). The chloro compound is prepared by adding $CdCl_2$ to an aqueous solution saturated with $MnCl_2$ and antipyrine, leaving a slight excess of $MnCl_2$ (relative to $CdCl_2$). Upon standing overnight, fine colorless crystals of $MnCl_2\cdot 2CdCl_2\cdot 6C_{11}H_{12}N_2O$ appear, which melt at 176°C and are only slightly soluble in cold alcohol, less soluble in acetone, and insoluble in chloroform.

The bromo compound was obtained by adding KBr and antipyrine to an aqueous solution of $MnCl_2$ (50% by weight) and heating to boiling. Then a solution of $Cd(NO_3)_2$ was added in a quantity leaving a large excess of Mn^{2+} relative to Cd^{2+}. If the solution is very concentrated two liquid layers form, which are stirred vigorously to yield an emulsion. Upon cooling, fragile colorless crystals of $MnBr_2\cdot 2CdBr_2\cdot 8C_{11}H_{12}N_2O\cdot 2H_2O$ appear. These are dried over $CaCl_2$. The compound is unstable. It melts at 132°C and decomposes at higher temperatures, turning viscous and yellow, then brown. It dissolves in alcohol, somewhat at room temperature and readily on heating. It is only slightly soluble in acetone and chloroform. The iodo compound is obtained by using KI instead of KBr. If a second liquid layer appears, it is removed by adding water. The colorless powder of composition $MnI_2 \cdot 2CdI_2 \cdot 8C_{11}H_{12}N_2O \cdot 2H_2O$ soon turns yellow. On heating to 100°C, it becomes brown and melts at 158°C. It is somewhat soluble in alcohol and acetone, more soluble upon warming, and is insoluble in chloroform.

The isothiocyanato complex is prepared like the bromo complex (above), replacing KBr by KSCN. A soft viscous precipitate forms. It hardens after several days and may be powdered. The compound of composition $Mn(NCS)_2 \cdot 2Cd(NCS)_2 \cdot 7C_{11}H_{12}N_2O \cdot H_2O$ is slightly pinkish. It melts at 121°C, turning brown. Its solubility resembles that of the bromo complex [13].

$Mn(C_{11}H_{13}N_3O)_2Br_2 \cdot 4H_2O$. The compound with 4-aminoantipyrine was prepared by reacting a solution of $MnBr_2$ in acetone with an excess of powdered ligand. The precipitate formed after dissolution of the base. It was filtered off and washed with acetone. The colorless crystalline compound loses its H_2O at about 105°C, turning bright yellow as a consequence of partial decomposition of the organic ligand. It is soluble in alcohol and chloroform [3].

References:

[1] R. Srinivasan, C. K. Subramanian (Indian J. Pure and Appl. Phys. **9** [1971] 21/3; Proc. Nucl. Phys. Solid State Phys. Symp., Calcutta 1965, Pt. A, pp. 22/9; C.A. **65** [1966] 11578). – [2] B. N. Misra, S. D. Sharma, S. K. Gupta (Nuovo Cimento Soc. Ital. Fis. B [11] **19** [1974] 129/34, 133; C.A. **80** [1974] No. 102017). – [3] P. Souchay (Bull. Soc. Chim. France [5] **7** [1940] 835/75, 864, 867). – [4] E. Wilke-Dörfurt, O. Schliephake (Z. Anorg. Allgem. Chem. **183** [1929] 301/10, 305). – [5] D. J. Mackey (Australian J. Chem. **30** [1977] 1461/5, 1463).

[6] G. M. Woltermann, J. R. Wasson (J. Phys. Chem. **77** [1973] 945/9, 947). – [7] A. Ravi, J. Gopalakrishnan, C. C. Patel (Indian J. Chem. **5** [1967] 356/9). – [8] E. Wilke-Dörfurt, H. G. Mureck (Z. Anorg. Allgem. Chem. **184** [1929] 121/34, 125). – [9] B. N. Misra, S. K. Gupta (J. Magn. Resonance **12** [1973] 126/33, 127, 132). – [10] M. Vijayan, M. A. Viswamitra (Z. Krist. **122** [1965] 153/5).

[11] M. Vijayan, M. A. Viswamitra (Acta Cryst. **21** [1966] 522/32, B **24** [1968] 1067/76). – [12] W. Hückel, R. Rosmus (Arch. Pharm. **293** [1960] 159/69, 165). – [13] P. Souchay (Bull. Soc. Chim. France [5] **7** [1940] 875/94, 892/3).

16.2.7.2 With Pyramidon

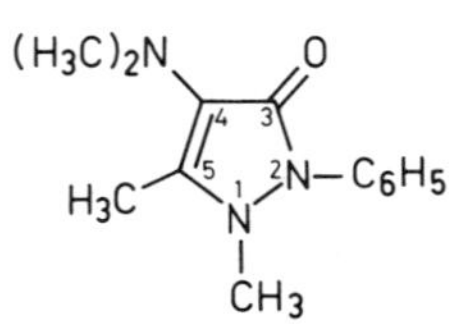

$(=C_{13}H_{17}N_3O)$

Complex Formation in Solution. Potentiometric measurements on aqueous solutions of manganese(II) nitrate and pyramidon at 25°C gave $\log K_1 = 0.74$ as the stability constant of the 1:1 complex at an ionic strength $I = 0.5$ (KNO_3) [6].

$Mn(C_{13}H_{17}N_3O)_2X_2$ (X = Cl, Br). The chloro compound was prepared in the same way as the antipyrine complex $Mn(C_{11}H_{12}N_2O)_2Cl_2$ (p. 283). The colorless crystalline product was washed with amyl alcohol. It melts at 166°C, is soluble in alcohol and chloroform, and slightly soluble in acetone. Ion migration suggests the configuration $[Mn(C_{13}H_{17}N_3O)_2]^{2+}(Cl^-)_2$, the cation mobility being 21.4 $cm^2 \cdot \Omega^{-1}$ [1].

The bromo compound was obtained easily by mixing concentrated solutions of $MnBr_2$ and pyramidon in acetone. The light pinkish solid is more stable than the antipyrine complex. On heating, the complex becomes brownish at 187°C and melts at 192°C. It is slightly soluble in alcohol and acetone [1].

$Mn(C_{13}H_{17}N_3O)_2I_2 \cdot 2H_2O$ is prepared from the components in the same way as the antipyrine complex $Mn(C_{11}H_{12}N_2O)_2I_2$ (p. 284). It is rather unstable, turning yellow easily in light and air. On heating, it melts above 115°C, turning brown [1].

$Mn(C_{13}H_{17}N_3O)_2(NCS)_2$ was prepared by dissolving pyramidon in a concentrated aqueous solution of $Mn(NO_3)_2$, adding several drops of acetic acid, and a concentrated solution of NH_4SCN. The precipitate, which sometimes appeared only upon standing overnight, was washed with very little water, and dried in an oven [1] at about 100°C [2]. The compound crystallizes as fine colorless [1] or slightly greenish [2] needles, which melt at 199°C [1] or at 176°C, forming a clear melt [2]. The IR spectrum of the compound embedded in KBr discs or dissolved in a mixture of acetonitrile (7%) and CCl_4 has been recorded in the range 4000 to 400 cm^{-1}. The most notable absorption bands (wave numbers in cm^{-1}) are assigned to the following vibrations:

in KBr discs	2063	—	1632, 1590, 1510(sh), 1494	1318
in solution	—	1710/1683	1600	—
assignment	ν(NC)	ν(CO)	stretching vibrations of the aromatic ring system	ν(CN)

The position of the ν(NCS) band at 2063 cm^{-1} suggests a coordination of the NCS^- ion to Mn via the N atom, while pyramidon is coordinated to the manganese atom as a bidentate ligand via its exocyclic N atom and the carbonyl oxygen. The absence of the CO absorption band for the solid indicates coordination of the pyramidon molecule via the CO group. The appearance of a strong CO band in solution points to the rupture of the metal-oxygen bond under the action of the solvent. Values of the CN stretching frequencies did not allow definite conclusions about the interaction of the noncyclic nitrogen atom with the metal. But on the basis of earlier investigations of pyramidonium metal compounds it can be concluded that the exocyclic nitrogen atom occupies a coordination position, with formation of a five-membered chelate ring [3].

The complex is slightly soluble in acetone and chloroform [1]. The solubilities (g/100 mL) in other solvents are 0.14 in water, 0.11 in ethanol, and 0.038 in ether. The molar electrical conductivities of the compound in aqueous solution at 15°C and dilutions of 1:1000 and 1:2000 are $\Lambda = 218$ and 241 $\Omega^{-1} \cdot cm^2 \cdot mol^{-1}$, respectively [2]. The saturated solutions are stable towards HNO_3 and H_2SO_4. Upon addition of 0.1N $AgNO_3$, a slight precipitate of AgSCN is formed, which increases on adding HNO_3. The transient violet color appearing in neutral solutions is not observed after acidification. Upon addition of a dilute solution of $FeCl_3$, the solutions turn cherry red, which indicates free SCN^- ions [2]. Potentiometric investigations of solutions in acetone or acetonitrile revealed that the complex behaves as an ampholyte, being divalent both as base and acid [4].

$Mn_2(C_{13}H_{17}N_3O)_3Cl_4 \cdot H_2O$. To a hot solution of pyramidon (0.01 mol) in acetone (10 mL) 0.005 mol of $MnCl_2 \cdot 4H_2O$ was added, followed by ether to reduce the solubility of the complex. The precipitate was treated with ether until it becomes powdery, then filtered off, washed with ether, and dried at room temperature.

The IR spectrum was recorded in the 3600 to 400 cm^{-1} range for samples pelleted with KBr. The most notable absorption bands (wave numbers in cm^{-1}) and their assignments are given in the paper. The observed shifts of $\nu(C\text{-}N\text{-}CH_3)$ (+30 cm^{-1}) and $\nu(CO)$ (−40 cm^{-1}) suggest coordination of the ligand through the oxygen atom of the CO group and the tertiary N atom of the side chain, thus forming a five-membered ring [5].

References:

[1] P. Souchay (Bull. Soc. Chim. France [5] **7** [1940] 835/75, 865/6). – [2] V. I. Kumov (Zh. Obshch. Khim. **19** [1949] 1236/49, 1240; C.A. **1955** 3393; J. Gen. Chem. [USSR] **19** [1949] 1231/42). – [3] B. E. Zaitsev, V. K. Akimov, A. I. Busev, I. A. Emel'yanova (Zh. Obshch. Khim. **40** [1970] 2453/5; J. Gen. Chem. [USSR] **40** [1970] 2439/41). – [4] V. K. Akimov, A. I. Busev, I. A. Emel'yanova (Zh. Analit. Khim. **25** [1970] 40/4; J. Anal. Chem. [USSR] **25** [1970] 30/2). – [5] Kh. Kh. Khakimov, N. T. Alimkhodzhaeva, O. F. Khodzhaev (Koord. Khim. **3** [1977] 1214/7; Soviet J. Coord. Chem. **3** [1977] 943/5).

[6] B. Lenarcik, M. Wiśniewski, M. Gabyszewski (Polish J. Chem. **54** [1980] 1869/74, 1872).

16.2.7.3 With 4-Acyl Derivatives of 5-Methyl-2-phenyl-2,4-dihydro-3H-pyrazol-3-one

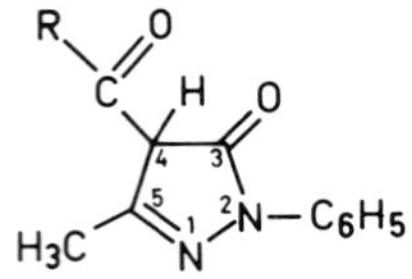

$R = C_nH_{2n+1}$ (n = 1 to 5)
$R = C_6H_5$ ($HL = C_{17}H_{14}N_2O_2$)

Complexes in Solution

For the chelates of manganese(II) with $R = C_nH_{2n+1}$ (n = 1 to 5) ligands in the two-phase water-chloroform system, apparent mean complex stability constants K_{av} were determined by a special titration procedure. K_{av} is used to characterize the extractive behavior of the specified metal ion. K_{av} is related to the oxidation number of the metal ion; N, the accumulative stability constant in the aqueous phase; β_N, and the distribution coefficient, λ, by the equation $\log K_{av} = 1/N \log \beta_N(1+\lambda)$. It is assumed that N is also the maximum coordination number of the metal ion for the ligand. The values of log K_{av} at 22°C and ionic strength I = 0.1M (ClO_4^-) depend on the chain length n of the acyl group:

R	CH_3	C_2H_5	C_3H_7	C_4H_9	C_5H_{11}
formula of ligand	$C_{12}H_{12}N_2O_2$	$C_{13}H_{14}N_2O_2$	$C_{14}H_{16}N_2O_2$	$C_{15}H_{18}N_2O_2$	$C_{16}H_{20}N_2O_2$
log K_{av} (±0.15) ...	3.85	4.40	4.50	5.15	5.4

So-called extraction constants (not defined) were calculated from the data of [1] by [2]. The chelate of manganese(II) formed with 4-benzoyl-5-methyl-2-phenyl-3-pyrazolone (= $C_{17}H_{14}N_2O_2$) was extracted quantitatively by a 0.01M solution of the ligand in isopentyl alcohol from aqueous solutions containing 1.5×10^{-5} mol/L Mn^{2+} with I = 1.0 M (ClO_4^-) at pH ≧ 4.6. Similar pH values were found for cyclohexanone and ethyl acetate as solvents, for phthalate-buffered solutions, and for solutions with additional foreign salts by use of a ^{54}Mn radiotracer [3]. Maximum extraction (95 to 100%) of Mn^{2+} (10^{-4} mol/L, I = 1M ($NaNO_3$)) by 0.01M ligand solutions in $CHCl_3$ occurred at pH 9 in isopentyl alcohol and isopentyl acetate [5], at pH 6.5 to 9 in methyl isobutyl ketone [5, 6], and at pH 8 to 9 in nitrobenzene [5]. The pH values fo 50% extraction of Mn^{2+} by ligand solutions in isopentyl alcohol, ethyl acetate, cyclohexanone, chloroform, and benzene are reported in [3 to 5]. The kinetics were studied by [3]. The extraction constants were not stated for the Mn complex because of its rather low stability in solution [5].

Isolated Compounds

$Mn(C_{17}H_{13}N_2O_2)_2(H_2O)_2$ was prepared by adding a concentrated ethanolic solution of the ligand ($R=C_6H_5$, slight excess) dropwise to an ethanolic solution of manganese(II) nitrate. After adding a solution of sodium acetate the mixture was refluxed for 1 h on a water bath and concentrated to half its original volume to precipitate the complex, which was collected, washed with several small portions of absolute ethanol in vacuo, and dried at 50°C. A magnetic moment of $\mu_{eff}=5.90\ \mu_B$ results from susceptibility measurements. The electronic reflectance spectrum of the solid shows two very weak absorption bands at 14300 and 16950 cm^{-1}. The absorption spectrum of the ethanolic solution is similar. The low intensity apparently is due to the doubly spin-forbidden nature. The IR spectrum shows a very broad absorption band in the 3400 to 3000 cm^{-1} region, which confirms the presence of water molecules in the complex. The ν(CO) band of the ligand appears at 1600 to 1550 cm^{-1} in the complex (shift toward lower frequency). The ν(CN) ring band is broad with structure, thus suggesting coupling of the bending mode of coordinated H_2O with ν(CN). The molar electrical conductivity in ethanol is that of a nonelectrolyte. The physical data thus show that $Mn(C_{17}H_{13}N_2O_2)_2(H_2O)_2$ has a high-spin octahedral structure involving bidentate chelating ligands and coordinated water molecules [7].

References:

[1] B. S. Jensen (Acta Chem. Scand. **13** [1959] 1890/6, 1895). – [2] O. Navratil (Radiokhimiya **19** [1977] 626/41; Soviet Radiochem. **19** [1977] 516/30, 519, 522). – [3] Yu. A. Zolotov, V. G. Lambrev (Radiokhimiya **8** [1966] 627/32; Soviet Radiochem. **8** [1966] 572/6, 574). – [4] Yu. A. Zolotov, V. G. Lambrev, M. K. Chmutova, N. T. Sizonenko (Dokl. Akad. Nauk SSSR **165** [1965] 117/20; Dokl. Chem. Proc. Acad. Sci. USSR **160/165** [1965] 1075/8). – [5] Z. Skorko-Trybuła, C. Różycki, E. Kosiarska (Chem. Anal. [Warsaw] **22** [1977] 311/8, 314; C.A. **87** [1977] No. 207295):

[6] E. Ivanova, S. Mareva, N. Jordanov (Z. Anal. Chem. **288** [1977] 62). – [7] A. K. Rana, J. R. Shah (Indian J. Chem. A **20** [1981] 615/7).

16.2.7.4 With Picrolonic Acid

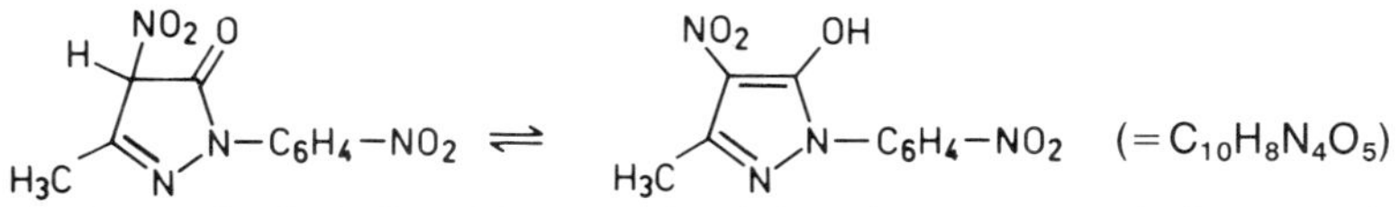

Complexes in Solution. The conductivity of aqueous solutions containing various ratios of $MnSO_4\cdot4H_2O$ and picrolonic acid indicates the presence of a 1:2 complex [1].

$Mn(C_{10}H_7N_4O_5)_2\cdot2.5H_2O$. The compound was obtained as a yellowish precipitate upon mixing aqueous solutions of $MnCl_2\cdot4H_2O$ (3 g of Mn^{2+} per liter) and picrolonic acid (0.01 mol/L, twofold excess). The mixture was heated to boiling, and the precipitate was filtered off while hot and washed with water and ethanol, each 0.0005 M in picrolonic acid, then with ether, and dried at 100°C in a vacuum [2]. The crystals obtained from neutral or weak acid solutions are yellow pleochroïtic needles with pale orange, pink, and green lusters [3]. Thermogravimetric analyses (heating rate 3°C/min) show two degradation steps, the first at 130 to 150°C due to the loss of all H_2O (2.5 mol) and the second at

210 to 230°C due to an explosive decomposition, which at 260 to 280°C leaves a residue consisting probably of manganic oxides and carbon [4]. The compound is only slightly soluble in water or alcohol. It may be used for gravimetric analysis of manganese [2].

References:

[1] D. P. Joshi, D. V. Jain (J. Indian Chem. Soc. **41** [1964] 711/4, 712, **40** [1963] 242/3). – [2] D. Becherescu (Bul. Stiint. Tehnic. Inst. Politehn. Timisoara [2] **1** [1956] 281/4; C.A. **1961** 7160). – [3] F. R. Sheyanova, K. N. Sidorova (Uch. Zap. Gor'k. Gos. Univ. im. N. I. Lobachevskogo No. 32 [1958] 97/108, 99, 103). – [4] D. Becherescu (Acad. Rep. Populare Romine Baza Cercetari Stiint. Timisoara Studii Cercetari Stiinte Chim. **6** No. 1/2 [1959] 115/22, 119/20; C.A. **1960** 24068/9).

16.2.7.5 With Other Pyrazolinones

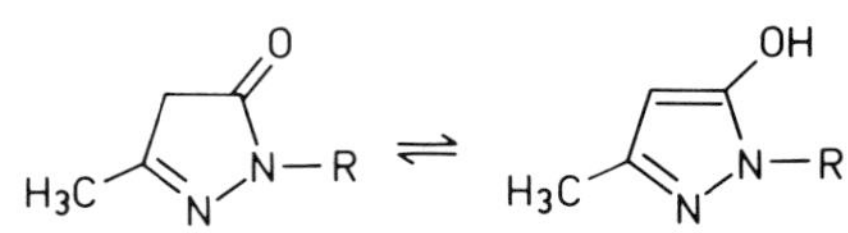

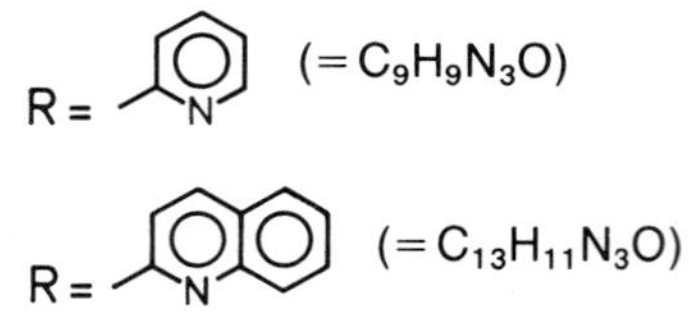

$Mn(C_9H_8N_3O)_2$ and **$Mn(C_{13}H_{10}N_3O)_2$** were obtained by reacting aqueous solutions of manganese(II) acetate with an aqueous solution of ligand I or an ethanolic solution of ligand II in a 1:2 mole ratio. A slight excess of ligand was used. The precipitate was washed with hot water and ether. The complex with ligand I is pale violet, while the complex with ligand II is pale brown. It is assumed that in these complexes Mn is coordinated to the ring N atom of the heterocyclic substituent and to the deprotonated oxygen atom of the enolic ligand tautomers, E. Suenaga (Yakugaku Zasshi **79** [1959] 205/7; C.A. **1959** 13136).

16.2.7.6 With Di(4-antipyrinyl)methane

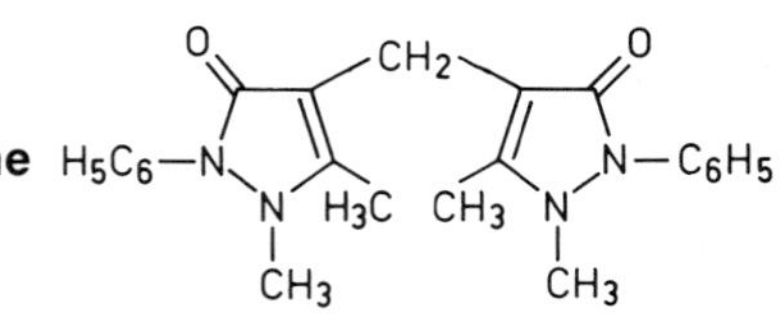

($=C_{23}H_{24}N_4O_2$)

$Mn(C_{23}H_{24}N_4O_2)_2(NCS)_2$. The compound was first prepared by [1] from the components. The IR spectrum of its solution in dichloroethane exhibits an absorption band at 2059 cm^{-1} and the solid in KBr discs has a band at 2058 cm^{-1}. These bands are assigned to an antisymmetric NCS stretching mode. As the intensity of the band is about twice that of the free NCS^- ion, it is assumed that the NCS groups are located in the inner coordination sphere of the complex and are bonded to Mn via their N atoms [2, 3].

References:

[1] V. P. Zhivopistsev (Diss. Perm Univ. 1965, cited in V. K. Akimov, A. I. Busev, Zh. Analit. Khim. **26** [1971] 134/42; J. Anal. Chem. [USSR] **26** [1971] 116/26, 118). – [2] B. E. Zaitsev, V. K. Akimov, W. Santos, A. K. Molodkin (Zh. Neorgan. Khim. **24** [1979] 2469/72; Russ. J. Inorg. Chem. **24** [1979] 1369/71). – [3] B. E. Zaitsev, W. Santos, A. D. Arynov, N. P. Vasil'eva, V. K. Akimov, A. K. Molodkin (Zh. Neorgan. Khim. **24** [1979] 414/8; Russ. J. Inorg. Chem. **24** [1979] 228/30).

16.2.8 With 1H-Indazol-5-amine

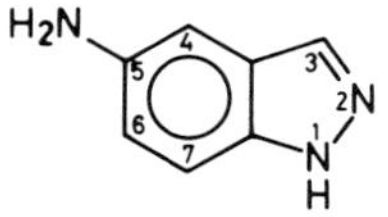

$(= C_7H_7N_3)$

$Mn(C_7H_7N_3)Cl_2$. The compound was prepared by mixing ethanolic solutions of $MnCl_2$ and the ligand. The pink precipitate was digested on a hot water bath for $^1/_2$ h, then filtered off, washed thoroughly with water, ethanol, and ether, and dried in vacuo [1]. The air-stable compound melts at 310°C with decomposition; it is insoluble in nearly all the usual organic solvents and thus probably polymeric [1, 2]. The IR spectrum, which was recorded for KBr discs in the 4000 to 700 cm^{-1} range, shows six strong absorption bands (wave numbers in cm^{-1}), which are assigned as follows:

band positions	3196	1580	1519	1435	1329	1095
assignment	ν(NH)	ν(CN) + ν(ring)	ν(ring)	ν(ring)	ν(ring)	ν(NN)

There are no marked band shifts relative to the free ligand, excepting ν(NH), which is shifted by 71 cm^{-1} towards higher frequencies. The ligand appears to be monodentate, coordinating only through the tertiary N atom. As the magnetic moment, $\mu_{eff} = 5.00\ \mu_B$ from susceptibility measurements at 24°C, is somewhat low for a compound with five unpaired electrons, a polymeric bridging structure is assumed [1]. The low molar conductivity ($\Lambda \sim 10^{-6}\ \Omega^{-1} \cdot cm^2 \cdot mol^{-1}$) of solutions in dimethylformamide indicates a nonelectrolyte [1, 2].

References:

[1] S. A. A. Zaidi, A. S. Farooqi (J. Inorg. Nucl. Chem. **35** [1973] 4320/3). – [2] S. A. A. Zaidi, A. S. Farooqi (Proc. Chem. Symp., Aligarh, India, 1972, Vol. 2, pp. 55/62, 56, 59/60; C.A. **82** [1975] No. 30497).

16.2.9 With N-Oxides of Pyrazole Derivatives

16.2.9.1 With 3,5-Diphenyl-1H-pyrazol-1-ol 2-Oxide

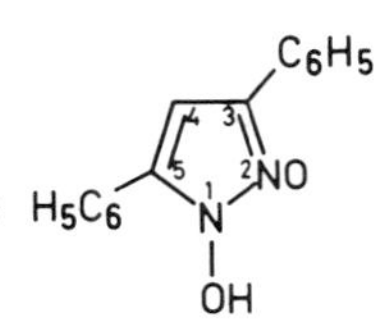

$(= C_{15}H_{12}N_2O_2)$

$Mn(C_{15}H_{11}N_2O_2)_2$ was prepared by slow addition of the ligand to $Mn(ClO_4)_2 \cdot 6H_2O$ (mole ratio 1:2, both in aqueous solution). The mixture was heated at 80°C for 1 h to remove ClO_4^- from the precipitated cream-colored solid, which then was washed with methanol and anhydrous diethyl ether to apparent dryness and stored in vacuo over $CaSO_4$. The IR spectrum (recorded for KBr discs and Nujol mulls) shows three characteristic absorption bands 1261, 818, and 370 cm^{-1} (br). These are assigned to ν(NO) and δ(NO) vibrations of the ligand and ν(Mn-O) vibrations, respectively. The observed shift of the NO bands (ν, δ) indicates coordination of the bidentate pyrazole derivative to Mn through the oxygen of the N-oxide group. Thermal analysis (DSC, TGA, DTG) shows a weak exothermic effect at 227°C, which suggests a polymorphic transition, and a much stronger effect at 266°C, with a loss of 44% by weight, indicating partial decomposition of the complex. The total weight loss is 78% at 600°C. The compound is a nonelectrolyte in dimethylformamide and is effectively insoluble in the other solvents commonly used for conductivity studies, D. X. West, A. A. Johnson (J. Inorg. Nucl. Chem. **41** [1979] 1101/4).

16.2.9.2 With 5-Methyl-3-phenyl-1H-pyrazol-1-ol 2-Oxide

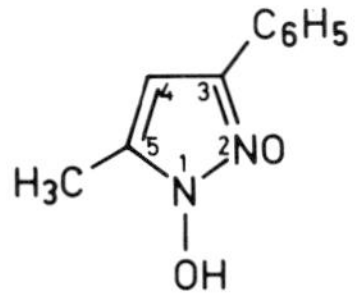

$(=C_{10}H_{10}N_2O_2=HL)$

$Mn(C_{10}H_9N_2O_2)_2$. An aqueous solution of $Mn(ClO_4)_2$ was added slowly to an aqueous solution of the ligand and heated to 80°C for about 1 h. The precipitate was filtered off, washed with boiling water, methanol, and anhydrous ether, and dried in a vacuum over $CaSO_4$. The IR spectrum of the tan-colored compound was obtained on KBr discs and Nujol mulls. The characteristic absorption bands of the manganese complex MnL_2 and the potassium salt KL were assigned as follows:

MnL_2	1270(s)	1045(m), 1029(m)	858(s)	379(m)
KL	1261(s)	1032(m), 1020(m)	850(s)	—
assignment	ν(NO)	ν(NN)	δ(NO)	ν(Mn-ON)

The shift of the ν(NO) band in the complex indicates coordination of the ligand via the NO groups and the formation of a five-membered chelate ring with two nearly equivalent NO-to-Mn bonds. The electronic spectrum recorded for Nujol mulls exhibits a strong maximum at 22880 cm^{-1}. The complex decomposes starting at about 330 K (=57°C) in an endothermic reaction.

Reference:

J. F. Hansen, S. P. McCormick, D. X. West (J. Inorg. Nucl. Chem. **39** [1977] 1231/3).

16.2.9.3 With 3,4,5-Trimethyl-1-ol pyrazol 2-Oxide

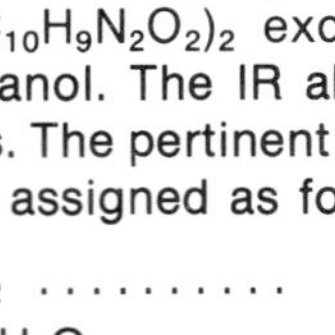

$(=C_6H_{10}N_2O_2=HL)$

$Mn(C_6H_9N_2O_2)_2$. The tan colored compound was prepared from stoichiometric amounts of $Mn(ClO_4)_2$ and the ligand in aqueous solution. It was isolated like $Mn(C_{10}H_9N_2O_2)_2$ except that the precipitate was washed with $HC(OC_2H_5)_3$ instead of methanol. The IR absorption and elctronic spectra (in cm^{-1}) were recorded for Nujol mulls. The pertinent absorption bands of the complex MnL_2 and the sodium salt $NaL \cdot H_2O$ were assigned as follows:

MnL_2	1326(s)	866(s)	358(s)	40160(s)	22990(m)
$NaL \cdot H_2O$	1317(s)	869(s)	—	39840(s)	—
assignment	ν(NO)	δ(NO)	ν(Mn-O)	intra-ligand	charge transfer

The ligand is considered to be bidentate and coordinated like the previous ligand.

The compound decomposes at 279°C and is a nonelectrolyte in dimethylformamide as shown by conductivity measurements.

Reference:

D. X. West, M. A. Vanek (J. Inorg. Nucl. Chem. **40** [1978] 1027/9).

16.2.10 With 1H-Imidazole 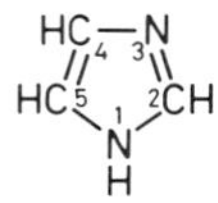 (= $C_3H_4N_2$)

Mixed-ligand complexes with 1H-imidazole and acetylacetone or 1-phenyl-1,3-butanedione are described in "Mangan" D 1, 1979, pp. 83 and 119.

16.2.10.1 Complexes in Solution

Potentiometric measurements on aqueous solutions of $Mn(NO_3)_2$ and imidazole at 25°C and ionic strength I = 0.16 ($NaNO_3$) suggest the formation of two weak complexes, $[Mn(C_3H_4N_2)]^{2+}$ and $[Mn(C_3H_4N_2)_2]^{2+}$, having the stability constants log K_1 = 1.65 and log K_2 = 1.25 (K in L/mol) [1].

The ^{13}C nuclear relaxation rate in the NMR spectrum of aqueous solutions containing imidazole, NaCl, and Mn^{2+} at 31°C (pH between 2.6 and 11.0) indicate that the ^{13}C transverse relaxation rates of imidazole in the presence of Mn^{2+} (pH > 5) are dominated completely by the scalar mechanism. Therefore it is impossible to obtain the geometry of Mn-imidazole complexes from ^{13}C line width measurements [2]. Measurements of the spin-lattice and transverse relaxation times, T_1 and T_2, in the ^1H and ^{13}C NMR spectra on solutions of imidazole in D_2O in the presence of Mn^{2+} show that the dipolar term contributes less than 50% to the broadening of the proton lines. Thus selective-broadening experiments do not allow determination of the binding site of small ligands such as imidazole. The ratios of the scalar coupling constants are quite constant for H2, H4, and H5 in the complexes of Mn [3].

References:

[1] R. B. Martin, J. T. Edsall (J. Am. Chem. Soc. **80** [1958] 5035/5). – [2] R. E. Wasylishen, M. R. Graham (Can. J. Chem. **54** [1976] 617/23, 621). – [3] W. G. Espersen, R. B. Martin (J. Phys. Chem. **80** [1976] 161/4, 163).

16.2.10.2 Isolated Compounds

$Mn(C_3H_4N_2)_6(NO_3)_2$ and **$Mn(C_3H_4N_2)_6(ClO_4)_2$** were prepared by reacting stoichiometric quantities of the Mn^{II} salts and imidazole in ethanol solution followed by precipitation with ether. The precipitates were filtered off under suction, washed with alcohol and ether, and dried in vacuo [1]. The perchlorate can also be prepared from the components in aqueous solution [2].

The melting points (m.p.), the molar conductivities Λ (in $\Omega^{-1}\cdot cm^2\cdot mol^{-1}$) of acetone solutions, the effective magnetic moments μ_{eff} in μ_B (from susceptibility measurements), and assigned IR absorption bands (in cm^{-1}, (Nujol and hexachlorobutadiene mulls, 5000 to 650 cm^{-1} [1], 4000 to 150 cm^{-1} [2]) are as follows:

complex	m.p. in °C	Λ	μ_{eff}	ν(NH)	ν(NH)	δ(NH)	Ref.
$Mn(C_3H_4N_2)_6(NO_3)_2$	9.5	11.3	5.70	—	3275(m)	1426(w)	[1]
$Mn(C_3H_4N_2)_6(ClO_4)_2$	22.8	267.0	5.67	3450(br)	3200(s)	1422(s)	[1]
$Mn(C_3H_4N_2)_6(ClO_4)_2$	—	—	—	3437	3350	1223	[2]

In respect to the free-ligand values, ν(NH) shifts to higher frequencies, whereas δ(NH) shifts to slightly lower frequencies [1]. The bands of the perchlorate at 1535, 1486, 1421, 1325, and 1156 cm^{-1} are assigned to ν(ring) vibrations; bands at 936 and 845 cm^{-1} are assigned to δ(ring) vibrations; and bands at 656, 607, and 595 cm^{-1} are assigned to γ(ring) vibrations. Bands due to $\nu(ClO_4)$ vibrations occur at 1130, 1111, 1084 (ν_3), 621 (ν_4), and 464 cm^{-1} (ν_2). The band occurring at 211 cm^{-1} can be assigned to ν(Mn-N), and those at 278 and 168 cm^{-1} belong to δ(Mn-N-C) vibrations [2]. More main IR bands are given by [1]. For the spectrum of the deuterated complex, see [2].

$Mn(C_3H_4N_2)_6(NO_3)_2$ is brownish white, crystalline, and soluble in common organic solvents; it is a nonelectrolyte in acetone. Its magnetic moment indicates five unpaired electrons and is compatible with a 6-coordinate Mn^{II} complex. The coordination of the NO_3 groups to Mn is indicated by the low conductance in acetone. Either a 6-coordinate structure with two imidazole molecules loosely held in the crystal lattice or an 8-coordinate structure is possible as the IR spectrum does not show bands of free imidazole. $Mn(C_3H_4N_2)_6(ClO_4)_2$ is an 1:2 electrolyte, and its IR spectrum shows a broad band of ionic ClO_4^- in the 1137 to 1045 cm^{-1} region which, like its magnetic moment, indicates a 6-coordinate Mn complex [1]. The perchlorate apparently is isostructural with the analogous complexes of Fe, Co, and Ni [2].

The ESR spectrum of Mn^{2+} doped in $Cd(C_3H_4N_2)_6(NO_3)_2$, which was recorded at 300 and 77 K, shows a marked dependence of anisotropy on the temperature. At 300 K the spectrum is dominated by a cubic field with parameters g = 2.0005, A = −89 G, and a = 2.67 G. At 77 K it shows axial character with A = −89 G, $B_2^0(D) = -14.4$ G, and $B_4^0 = -0.11$ G [4].

$Mn(C_3H_4N_2)_6X_2$ (X = Cl, Br, I) and **$Mn(C_3H_4N_2)_6(NCS)_2$** were prepared by addition of excess imidazole dissolved in warm ethanol (or acetone for the complex with X = I) to an ethanol solution of the coresponding manganese(II) salt. The precipitates (isothiocyanate initially an oil) were washed with ether and dried. The iodide and isothiocyanate were recrystallized from benzene [3]. The IR spectra were recorded as vaseline mulls, the Raman spectra (R) were taken from fine powders with an He-Ne laser, both in the 350 to 70 cm^{-1} range. Vibrational bands in cm^{-1}:

X	ν_1(R)	ν_2(R)	(R)	ν_3(IR)	ν_4(or ν_3, IR)	(IR), unassigned
Cl	169(s)	131(s)	106(sh)	205(s)	182(s)	230(sh), 152(s), 122(w)
Br	167(m)	122(sh)	105(sh)	205(s)	170(vs)	143(s), 112(w, br)
I	—	—	—	212(s)	178(vs)	152(m), 107(m)
NCS	179(s)	166(sh)	100(sh)	205(vs)	176(ms)	162(m)

The low-energy vibrational spectra suggest distorted octahedral structures for the $[Mn(C_3H_4N_2)_6]^{2+}$ ion with imidazole coordinated through the tertiary N(3) atom (just as for $Zn(C_3H_4N_2)_6Cl_2 \cdot 4H_2O$) [3].

$Mn(C_3H_4N_2)_4X_2$ (X = Cl, Br, I, NCS) and **$Mn(C_3H_4N_2)_4X_2 \cdot 2H_2O$** (X = Br, I). The anhydrous halides were obtained by reacting the calculated quantities of imidazole and MnX_2 ($MnCl_2$ in warm ethanol, $MnBr_2$ and MnI_2 in acetone) [3]. The bromo compound may also be prepared in ethanol with added ether. $Mn(C_3H_4N_2)_4(NCS)_2$ is obtained by refluxing an ethanol solution of $Mn(NCS)_2$ and imidazole in a 1:4 mole ratio for 30 min and keeping the solution overnight. The compounds were filtered off (under suction), washed with absolute ethanol and ether, and dried in vacuo [1]. Below are listed melting points (m.p.), molar electrical conductivities Λ of solutions in methanol, and effective magnetic moments μ_{eff} (from susceptibility measurements) for the bromo and the isothiocyanato complexes:

compound	m.p. in °C	Λ in $\Omega^{-1} \cdot cm^2 \cdot mol^{-1}$	μ_{eff} in μ_B
$Mn(C_3H_4N_2)_4Br_2$	196	19.4	5.64
$Mn(C_3H_4N_2)_4(NCS)_2$	194	20.4	5.94

Both compounds are nonelectrolytes in acetone and exhibit a high magnetic moment, indicating five unpaired d electrons [1]. The far-IR spectrum of $Mn(C_3H_4N_2)_4Cl_2$ recorded as vaseline mull in the range 400 to 70 cm^{-1} shows a strong and very broad absorption band at 175 cm^{-1} with two shoulders, at 208 and 235 cm^{-1}; the Raman spectrum (He-Ne laser) has only one shoulder, at 144 cm^{-1} [3]. The IR spectrum (Nujol or hexachlorobutadiene mulls) of $Mn(C_3H_4N_2)_4Br_2$ shows a broad ν(NH) band at 3320 cm^{-1} and a δ(NH) band at 1432 cm^{-1}. $Mn(C_3H_4N_2)_4(NCS)_2$ exhibits ν(NH) bands at 3370 and 3185 cm^{-1} and a δ(NH) band at 1432 cm^{-1}. (Frequencies of imidazole: ν(NH) at 3000 and 3165 cm^{-1}, δ(NH) at 1440 cm^{-1}.) The strong, sharp band of $Mn(C_3H_4N_2)_4(NCS)_2$ at 2080 cm^{-1} indicates a terminal NCS^- group, which rules out polymers [1].

$Mn(C_3H_4N_2)_4Br_2$ and $Mn(C_3H_4N_2)_4(NCS)_2$ are readily soluble in common organic solvents [1]. $Mn(C_3H_4N_2)_4Br_2$ and $Mn(C_3H_4N_2)_4I_2$ readily form the dihydrates when exposed to air. The IR spectra of the dihydrates are more complex than those of the anhydrous compounds. A distorted octahedral structure is assumed for both the anhydrous compounds and the dihydrates [3].

$Mn(C_3H_4N_2)_2X_2$ (X = Cl, Br, I, NCS). The compounds with X = Br and I were prepared by heating $Mn(C_3H_4N_2)_6X_2$ in vacuo at 100°C for several days. The compounds with X = Cl and NCS were obtained by mixing hot ethanol solutions of $MnCl_2$ or $Mn(NCS)_2$ and imidazole in a 1:2 mole ratio. The chloro compound precipitated only on addition of ether [3]. $Mn(C_3H_4N_2)_2(NCS)_2$ was also obtained by mixing cold methanol solutions of $Mn(NCS)_2$ and imidazole. The crystals, which precipitate after several days in a vacuum desiccator over P_2O_5, were filtered off, washed with a little methanol, and dried in a desiccator [5]. $Mn(C_3H_4N_2)_2I_2$ is pale yellow and very hygroscopic; $Mn(C_3H_4N_2)_2(NCS)_2$ forms green crystals [3], but it is light yellow according to [5]. The far-IR absorption spectra of $Mn(C_3H_4N_2)_2X_2$ (as vaseline mulls) in the range 400 to 70 cm^{-1} and the Raman spectrum (He-Ne laser) of powdered $Mn(C_3H_4N_2)_2(NCS)_2$ exhibit the following bands (in cm^{-1}):

X	spectrum	bands (relative intensities in parentheses)					
Cl	IR	250(sh)	214(s)	194(s)	159(m)	135(m)	105(mw)
Br	IR	250(m)	193(s)	176(s)	161(ms)	136(m)	100(w)
I	IR	200(sh)	182(s)	—	155(vs)	126(w)	100(m)
NCS	IR	244(s, br)	—	170(s, br)	—	128(s, br)	—
NCS	Raman	209(s)	147(s)	126(m)	105(sh)	—	—

The complexes $Mn(C_3H_4N_2)_2X_2$ have a polymeric octahedral structure. The IR band occurring at about 250 cm^{-1} (except for the unstable iodo compound) is assigned tentatively to an antisymmetric ($Mn\text{-}N_{ligand}$) stretching of the trans-octahedral $[Mn(C_3H_4N_2)_2X_2]$ unit. For the isothiocyanato compound this band is broad and may also contain a (Mn-NCS) stretching mode. A band at 464 cm^{-1} is attributed to the N-C-S bending vibration [3]; a strong IR band at 2110 cm^{-1} is assigned to ν(CN) stretching vibration. Magnetic susceptibility measurements yield the magnetic moment $\mu_{eff} = 5.95\ \mu_B$, which is consistent with a high-spin polymeric octahedral configuration for $Mn(C_3H_4N_2)_2(NCS)_2$, possibly with bridging NCS groups [5]. $Mn(C_3H_4N_2)_2I_2$ is very hygroscopic and rapidly turns brown in moist air. In dry air the compound decomposes more slowly (several weeks) [3].

$Mn(C_3H_4N_2)Cl_2$ was prepared from a mixture of $MnCl_2$ and imidazole (mole ratio 1:1) in warm ethanol. Its far-IR spectrum (recorded as for $Mn(C_3H_4N_2)_2X_2$) shows strong absorption bands at 250, 238, 178, 148, and 118 cm^{-1}. A (Mn-N_{ligand}) stretching vibration may contribute to the highest energy band (at 250 cm^{-1}). A polymeric octahedral structure is assumed for the compound [3].

$Mn(C_3H_4N_2)Br_2$ was prepared from stoichiometric quantities of $MnBr_2$ and imidazole in ethanol by adding ether. It is amorphous and melts at 300°C. Its IR spectrum exhibits two strong absorption bands, at 3445 and 1424 cm^{-1}, assigned to ν(NH) and δ(NH), respectively. Other principal absorption bands are at 1535(s), 1168(s), 1062(vs), 925(w), 746(s), and 716(s) cm^{-1}. The compound probably is polymeric like $Mn(C_3H_4N_2)_2Br_2$, having an octahedral configuration involving bridging Br atoms. The effective magnetic moment resulting from susceptibility measurements on the solid at room temperature is $\mu_{eff} = 6.09\ \mu_B$. $Mn(C_3H_4N_2)Br_2$ is only sparingly soluble in common organic solvents. Measurements of the electrical conductivity in acetone yield a molar conductivity of $\Lambda = 20.4\ \Omega^{-1} \cdot cm^2 \cdot mol^{-1}$, indicating a nonelctrolyte [1].

$3\,Mn(C_3H_3N_2)_2 \cdot 2\,C_3H_4N_2$. The compound was synthesized by heating a mixture of 0.5 g $Mn_2(CO)_{10}$ (1.3 mmol), 1 g imidazole (14.5 mmol), and 1 g tetramethylimidazole (8 mmol) in an evacuated reaction vessel at 160°C for 1 h. After cooling, 20 mL methanol was added by condensation. The colorless, very air-sensitive crystals were filtered off and washed with methanol. The crystals form almost quadratic platelets. According to Weissenberg and precession X-ray photographs the symmetry is monoclinic, and the space group is $P2_1/c$-C_{2h}^5 (No. 14). Lattice constants a = 10.775, b = 13.136, c = 10.677 Å, and β = 91.5°; Z = 2. The calculated density is 1.55 g/cm^2. The position of the atoms in the structure was found by means of Patterson (Mn), Fourier (C, N), and difference-Fourier (H) synthesis and refined to R = 0.046. The atomic parameters are listed in the original paper. The structure is shown in **Fig. 25**. The bonding angles N-Mn-N are 90 or 92° (2×3 angles) for Mn(II) and between 100 and 126°C (6 angles) for Mn(I). The (NH) and (C-H) bond lengths are within the range 0.78 to 1.12 Å [6].

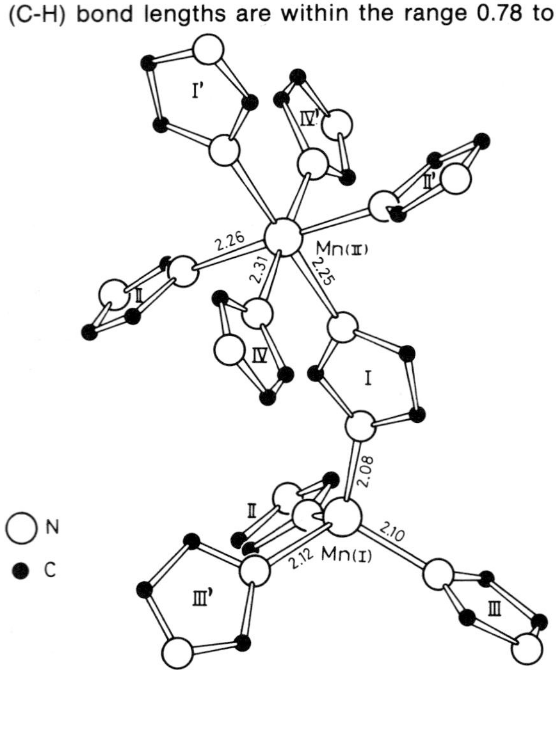

Fig. 25

Section of the crystal structure of $3\,Mn(C_3H_3N_2)_2 \cdot 2\,C_3H_4N$ with atomic distances in Å.

The atoms Mn(I) (at general positions) are tetrahedrally surrounded by bridging imidazole rings (I, II, III, III'). The atoms Mn(II) at the symmetry center are octahedrally coordinated by imidazole rings (I, I', II, II', IV, IV'). The Mn atoms I and II are linked together by two bidentate imidazole rings. The apices of the coordination octahedra of Mn(II) are occupied by the imidazole rings (IV). Thus there is a three-dimensional polyhedral framework consisting of octahedra which are linked to four tetrahedra, each tetrahedron being linked to two octahedra and two other tetrahedra. The tetrahedra are somewhat distorted, whereas the octahedra are almost regular. $3Mn(C_3H_3N_2)_2 \cdot 2C_3H_4N_2$ is isomorphous with the Fe^{II} complex [6].

References:

[1] A. K. Das, D. V. Ramana Rao (Indian J. Chem. **9** [1971] 480/1). – [2] J. B. Hodgson, G. C. Percy, D. A. Thornton (J. Mol. Struct. **66** [1980] 81/92, 82, 86). – [3] D. M. L. Goodgame, M. Goodgame, P. J. Hayward, G. W. Rayner-Canham (Inorg. Chem. **7** [1968] 2447/51). – [4] R. K. Sharma, K. V. Lingam, M. J. Rao, B. N. Bhattacharya (Proc. Nucl. Phys. Solid State Phys. Symp. C **21** [1978] 668; C.A. **92** [1980] No. 155568). – [5] A. D. Garnovskii, V. V. Skopenko, M. I. Knyazhanskii, O. A. Osipov, P. V. Gilyanovskii, T. D. Artyukhovich (Zh. Obshch. Khim. **42** [1972] 2068/73; J. Gen. Chem. [USSR] **42** [1972] 2064/8).

[6] R. Lehnert, F. Seel (Z. Anorg. Allgem. Chem. **464** [1980] 187/94).

16.2.11 With Derivatives of Imidazole

16.2.11.1 With 1-Methyl-1H-imidazole

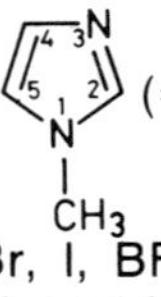

(= N-Methylimidazole = $C_4H_6N_2$)

$Mn(C_4H_6N_2)_6X_2$ (X = NO_3, Cl, ClO_4, Br, I, BF_4). The compounds were prepared by reacting stoichiometric amounts of MnX_2 and N-methylimidazole in ethanol or acetone with continuous stirring. The crystals appeared on concentrating the solution. They were filtered off, washed with a little ethanol and then some diethyl ether, and dried in vacuo [1]. An ethanol solution of $MnX_2 \cdot 6H_2O$ (X = ClO_4, BF_4) was dehydrated with sufficient $HC(OC_2H_5)_3$ and added to an ethanol solution of the ligand. The crystals, which appeared on standing or after addition of ether, were washed with sodium-dried ether and dried in vacuo (1 Torr) [2], also see [3]. The compounds $Mn(C_4H_6N_2)_6X_2$ are colorless or cream colored (X = Br and BF_4) [1]. $Mn(C_4H_6N_2)_6(ClO_4)_2$ is isomorphous with the perchlorates of Mg, Fe, Co, Ni, Zn, and Cd, while $Mn(C_4H_6N_2)_6(BF_4)_2$ is isomorphous with the tetrafluoroborates of Co, Ni, Zn, and Cd as shown by X-ray powder diagrams [3], also see [2]. Below are listed the melting points (m.p.), molar conductivities Λ of solutions in acetone at room temperature, and effective magnetic moments μ_{eff} from susceptibility measurements for $Mn(C_6H_4N_2)_6X_2$ [1] (values of [2, 3] in parentheses):

X	Cl	Br	I	NO_3	ClO_4	BF_4
m.p. in °C	125	142	215	168	>240 (281)	210 (241)
Λ in $\Omega^{-1} \cdot cm^2 \cdot mol^{-1}$	31.4	34.8	131.4	43.7	289	228.6
μ_{eff} in μ_B	5.8	6.0	6.0	5.9	6.0 (5.90)	6.2

The compounds of the type $Mn(C_4H_6N_2)_6X_2$ are high-spin Mn-d^5 complexes. The chlorides, bromides, and nitrates behave as poorly conducting nonelectrolytes in acetone, indicating that the anions are coordinated to Mn. Thus it is possible that two N-methylimidazole molecules dissociate in acetone and are replaced by the anions which enter the inner coordination sphere to form the poorly conducting species $[Mn(C_4H_6N_2)_4X_2]$. The iodide seems to be a 1:1 electrolyte; the perchlorate and tetrafluoroborate are probably 1:2 electrolytes.

In the IR spectrum (Nujol mulls, 4000 to 500 cm^{-1}) $Mn(C_4H_6N_2)_6(NO_3)_2$ shows a strong ν(NO) stretching band at 1390 cm^{-1} and two IR active deformation bands at 830 and 720 cm^{-1}, indicating an uncoordinated NO_3^- ion with D_{3h} symmetry. $Mn(C_4H_6N_2)_6(ClO_4)_2$ exhibits a strong, broad, unsplit band at 1050 cm^{-1} and a medium band at 940 cm^{-1} assigned to (OClO) bending, which indicate an uncoordinated ClO_4^- ion of T_d symmetry. $Mn(C_4H_6N_2)_6(BF_4)_2$ has a similar strong band at 1050 cm^{-1} indicating uncoordinated BF_4 ions with T_d symmetry. The complexes $Mn(C_4H_6N_2)_6X_2$ appear to have monomeric octahedral structures [1]. An IR band at 934 cm^{-1}, assigned to ring vibrations of the ligand N-methylimidazole, is shifted by 24 cm^{-1} towards higher frequencies from its free-ligand value [2, 3]. For other assigned IR and Raman bands in the 4000 to 200 cm^{-1} region, see [3]. In the far-IR and Raman (R) spectra of $Mn(C_4H_6N_2)_6X_2$ (400 to 150 cm^{-1}) there is a band at 366 (368 R) cm^{-1}, which is assigned to (CN) deformation. Two other bands, at 234 cm^{-1} (weak) and 211 cm^{-1} (IR, very strong), are tentatively assigned to ($Mn\text{-}N_{ligand}$) stretching; a band at 190 cm^{-1} (Raman, medium, broad) is unassigned. The ESR (X band) spectrum of powdered $Mn(C_4H_6N_2)_6(ClO_4)_2$ gives $g=2.01\pm0.01$. A sample diluted in solid $Zn(C_4H_6N_2)_6(ClO_4)_2$ gives $g=2.00\pm0.005$ and a hyperfine splitting constant $A=89\pm1$ Gauss, which is lower than those of similar 6-coordinate Mn^{2+} ions and thus indicates increased covalency of the $Mn\text{-}N_{ligand}$ bond [3].

$Mn(C_4H_6N_2)_4X_2$ (X = Br, I, NCS). The compounds were prepared in the same way as $Mn(C_4H_6N_2)_6X_2$ using MnX_2 and N-methylimidazole in a stoichiometric 1:4 ratio (also up to 1:6 for X = NCS). Below are listed the color, melting points (m.p.), molar conductivities Λ of solutions in acetone at room temperature, and effective magnetic moments μ_{eff} from susceptibility measurements [1]:

X	color	m.p. in °C	Λ in $\Omega^{-1}\cdot cm^2\cdot mol^{-1}$	μ_{eff} in μ_B
Br	cream	135	13	6.0
I	yellow	220	77.4	6.0
NCS	colorless	172	43	5.8

For the bromide a melting point of 169°C is also reported by [2]. The compounds are essentially nonelectrolytes in acetone though some partial dissociation (for X = I and NCS) may occur. The magnetic moments are those of high-spin $Mn(d^5)$ complexes and are in accord with a distorted octahedral structure. The IR absorption spectrum of the compounds in Nujol mulls from 4000 to 500 cm^{-1} shows ligand bands split of shifted due to lattice effects or departures from idealized symmetry. For $Mn(C_4H_6N_2)_4(NCS)_2$ a strong band at 2060 cm^{-1}, assigned to ν(CN) and a medium one at 840 cm^{-1}, assigned to ν(CS), are indicative of coordination through the N atom of the ambidentate NCS group [1]. The ESR spectrum of $Mn(C_4H_6N_2)_4Br_2$ is similar to that of the 5-methylpyrazole complex (p. 273) suggesting a distorted octahedral symmetry [2].

$Mn(C_4H_6N_2)_3X_2$ (X = NO_3, Cl, NCS). To prepare $Mn(C_4H_6N_2)_3(NO_3)_2$ an ethanol solution of $Mn(NO_3)_2\cdot6H_2O$ was dehydrated with $HC(OC_2H_5)_3$ and added to an ethanol solution of the ligand. The crystals, which appeared on standing or after addition of dry ether, were washed with dry ether and dried in vacuo [2]. $Mn(C_4H_6N_2)_3Cl_2$ was prepared by stirring

$MnCl_2$ and N-methylimidazole in a 1:3 mole ratio overnight at room temperature. $Mn(C_4H_6N_2)_3(NCS)_2$ was obtained by heating $Mn(C_4H_6N_2)_4(NCS)_2$ to constant weight at 130°C, which required several hours [1].

The complexes are colorless [1, 2]. The nitrato compound melts at 136°C. It probably has coordinated NO_3^- ions. As shown by X-ray powder diagrams, it is isomorphous with the analogous Cd complex [2]. For the chloro compound (m.p. 155°C) and the thiocyanato compound (m.p. 135°C) polymeric octahedral structures were suggested. Their magnetic moments, $\mu_{eff}=5.9$ μ_B (from susceptibilities), are those of high-spin Mn(d^5) complexes. The compounds are nonelectrolytes in acetone ($\Lambda=26.9$ and 36.6 $\Omega^{-1}\cdot cm^2\cdot mol^{-1}$). The IR absorption spectrum of $Mn(C_4H_6N_2)_3(NCS)_2$ exhibits the ν(NCS) band at 2110 cm^{-1} and the ν(CS) band at 770 cm^{-1} (the former at higher, the latter at lower frequencies than in $Mn(C_4H_6N_2)_4(NCS)_2$), thus indicating bridging NCS groups [1].

$Mn(C_4H_6N_2)_2X_2$ (X = N_3, Cl, Br) and **$Mn(C_4H_6N_2)_2(N_3)_2\cdot5H_2O$.** The halogeno complexes were obtained from $MnCl_2$ or $MnBr_2$ and N-methylimidazole in a 1:2 mole ratio in the same manner as for $Mn(C_4H_6N_2)_6X_2$ (p. 295). For preparing $Mn(C_4H_6N_2)_2(N_3)_2\cdot5H_2O$ an ethanol solution of $Mn(C_4H_6N_2)_2Cl_2$ (1 mol) was mixed with an aqueous solution of NaN_3 (2 mol). $Mn(C_4H_6N_2)_2(N_3)_2$ was obtained by heating the hydrated compound at 130°C for 4 h [1]. Below are listed color, melting points (m.p.), molar conductivities Λ in acetone, and magnetic moments μ_{eff} (from susceptibilities):

compound	color	m.p. in °C	Λ in $\Omega^{-1}\cdot cm^2\cdot mol^{-1}$	μ_{eff} in μ_B
$Mn(C_4H_6N_2)_2Cl_2$	colorless	168	17	5.9
$Mn(C_4H_6N_2)_2Br_2$	colorless	164	29	6.0
$Mn(C_4H_6N_2)_2(N_3)_2\cdot5H_2O$	brown	195*)	11.2	5.9

*) Complex dehydrated at 130°C.

The compounds are nonelectrolytes in acetone, their magnetic moments are those of high-spin Mn(d^5) complexes. A polymeric octahedral structure is suggested. The IR spectrum of $Mn(C_4H_6N_2)_2(N_3)_2\cdot5H_2O$ exhibits three absorption bands, at 2075, 1330, and 635 cm^{-1}, which are assigned to symmetric and antisymmetric stretching and a deformation mode of the azide (N_3) group, respectively, suggesting coordination of the N_3 groups to the Mn^{2+} ion. Absorption bands at about 3400 cm^{-1} confirm the presence of H_2O, which is lost completely at 130°C [1].

$Mn(C_4H_6N_2)Cl_2$ was obtained by heating $Mn(C_4H_6N_2)_2Cl_2$ to constant weight at 130°C, which requires several hours. The colorless compound melts at >240°C. It is insoluble in acetone. From measurements of the susceptibility a magnetic moment $\mu_{eff}=5.9$ μ_B results. The complex is supposed to have a polymeric octahedral structure [1].

$(CH_3)_4N[Mn(C_4H_6N_2)Cl_3]$ and **$(C_2H_5)_4N[Mn(C_4H_6N_2)Br_3]$.** The trimethylammonium compound was prepared by mixing an alcohol solution of hydrated $MnCl_2$ (1 mol) with an alcohol solution of N-methylimidazole (1 mol) and $(CH_3)_4NCl$ (1 mol) with vigorous stirring. $(C_2H_5)_4N[Mn(C_4H_6N_2)Br_3]$ was obtained by adding $MnBr_2\cdot4H_2O$ (1 mol) to a solution of $(C_2H_5)_4NBr$ (1 mol) and ligand (1 mol) in acetone with stirring. It crystallized if the concentrated solution was placed in a deep freeze. Color, melting points (m.p.), molar conductivities Λ in acetone, and magnetic moments μ_{eff} of the compounds (from susceptibilities) are listed on p. 298.

compound	color	m.p. in °C	Λ in $\Omega^{-1}\cdot cm^2\cdot mol^{-1}$	μ_{eff} in μ_B
$(CH_3)_4N[Mn(C_4H_6N_2)Cl_3]$	colorless	165	148	6.1
$(C_2H_5)_4N[Mn(C_4H_6N_2)Br_3]$	yellow	125	103	6.1

The alkylammonium salts are 1:1 electrolytes in acetone. Their magnetic moments are those of high-spin Mn(d^5) complexes. They have a tetrahedral structure [1].

References:

[1] G. R. Chaudhury [Roy Choudhury], K. C. Dash (Transition Metal. Chem. [Weinheim] **2** [1977] 253/6). – [2] J. A. Welleman, F.-B. Hulsbergen, J. Verbiest, J. Reedijk (J. Inorg. Nucl. Chem. **40** [1978] 143/7). – [3] J. Reedijk (Inorg. Chim. Acta **3** [1969] 517/22).

16.2.11.2 With 2-Methyl-1H-imidazole 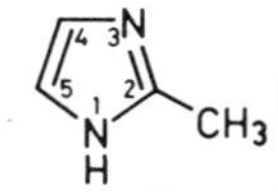 $(=C_4H_6N_2)$

$Mn(C_4H_6N_2)_6(NO_3)_2$ was prepared by reacting stoichiometric quantities of $Mn(NO_3)_2$ and 2-methylimidazole in ethanol and then adding ether. The brownish white crystalline compound melts at 162°C. Its IR spectrum exhibits two absorption bands, at 3385(m) and 3170(w) cm^{-1}, assigned to ν(NH) stretching modes. Other principal absorption bands are at 1566(vs), 1095(s), 1070(m), 937(w), 756(vs), and 723(w) cm^{-1}. A structure based on 6- or 8-coordinate Mn is suggested by the absence of the IR bands of free ligands. The effective magnetic moment from susceptibility measurements at room temperature is $\mu_{eff}=5.91\ \mu_B$, indicating five unpaired d-electrons. The compound is soluble in common organic solvents. A coordination of the NO_3 group to Mn is concluded from the low electrical conductivity in acetone ($\Lambda=28.3\ \Omega^{-1}\cdot cm^2\cdot mol^{-1}$) [1].

$Mn(C_4H_6N_2)_4X_2$ ($X=ClO_4$, BF_4). To a solution of 1 mmol $MnX_2\cdot 6H_2O$ in 5 mL of ethanol which had been dehydrated with 3 mL of $HC(OC_2H_5)_3$, a solution of 5 mmol of 2-methylimidazole in 5 mL of ethanol was added; O_2 and moisture were excluded. To get crystals some ethanol is evaporated, ether is added, and the solution is cooled to 5°C. The crystals were collected, washed with several portions of dry ether, and dried in vacuo at room temperature. Both compounds are buff colored and isomorphous with the corresponding Fe and Cd complexes as shown by X-ray powder diagrams. The perchlorate melts at 161 to 163°C; the tetrafluoroborate melts at 137 to 138°C. The IR spectra of the compounds (Nujol mulls, 4000 to 200 cm^{-1}) show rather broad anion absorption bands at 1090, 922, 622, and 463 cm^{-1} for ClO_4^- and 1055, 770, 518, and 355 cm^{-1} for BF_4^-. The second and fourth anion bands of each compound are due to forbidden absorptions and indicate a weak bond between Mn and these anions. The bands observed in the 3420 to 3300 cm^{-1} range can be assigned to the (NH) stretching vibrations of the ligand. They are about 30 cm^{-1} lower for the perchlorate than for the tetrafluoroborate compound. Other rather broad ligand bands are found between 750 and 650 cm^{-1} and at ~270 cm^{-1}, the latter masking Mn-ligand vibrations. The ESR spectra (X and Q bands) of $Mn(C_4H_6N_2)_4X_2$ exhibit many bands, ranging from $g_{eff}=20$ to $g_{eff}=0.5$, whose positions are nearly the same for the two compounds. Hyperfine Mn splittings in some bands occur in the Mn-doped Cd compounds. The ESR spectra can be assigned to six-coordinate Mn^{2+} ions in an environment distorted by an axial ligand field with additional rhombic splitting. The derived parameters are $D=0.52\pm0.01\ cm^{-1}$ and $\lambda=E/D=0.15\pm0.01$. (D is about as large as that found for the 5-methylpyrazole complex, for which λ is close to zero. The rather large values of λ for the imidazole complexes indicate that their symmetry is lower than tetragonal.) For 0.001M solutions in acetone the reported molar conductivities in $\Omega^{-1}\cdot cm^2\cdot mol^{-1}$ are $\Lambda = 270$ (ClO_4) and 230 (BF_4), indicating 1:2 electrolytes. The compounds are oxidized partially to Mn^{III} in air [2].

$Mn(C_4H_6N_2)_3Cl_2$. The compound forms from the components irrespective of mole ratio and crystallizes from ethanol solutions as well-formed pale yellow monoclinic prisms. According to X-ray diffraction the space group is $P2_1/c$-C_{2h}^5 (No. 14). Lattice constants: a = 8.189(1), b = 14.763(1), c = 14.419(1) Å, β = 108.69(1)°; Z = 4. For the atomic positions and the anisotropic thermal parameters see the tables in the paper. The bond lengths in Å are shown in **Fig. 26**. Bond angles at the Mn atom: Cl(1)-Mn-Cl(2) = 95.99(4)°, Cl(1)-Mn-N(11) = 168.03(10)°, Cl(1)-Mn-N(21) = 88.80(11)°, Cl(1)-Mn-N(31) = 86.88(11)°, Cl(2)-Mn-N(11) = 95.96(10)°, Cl(2)-Mn-N(21) = 118.83(9)°, Cl(2)-Mn-N(31) = 110.44(11)°, N(11)-Mn-N(21) = 86.05(15)°, N(11)-Mn-N(31) = 88.32(13)°, N(21)-Mn-N(31) = 130.72(14)°, etc. In $Mn(C_4H_6N_2)_3Cl_2$ the central Mn atom has a distorted trigonal-bipyramidal coordination (Fig. 26) with one Cl and one N_{ligand} atom in axial positions, while the other Cl and two N_{ligand} atoms occupy the equatorial positions. The complex thus has approximately C_s symmetry, where the "mirror plane" comprises the two Cl atoms, an organic ligand, and the central Mn atom. There is a considerable difference between the axial and equatorial bond lengths, which are 2.525 and 2.392 Å for Mn-Cl and 2.249 and 2.195 Å (mean) for Mn-N. The distortion of the trigonal bipyramid is significant: the axial Cl-Mn-N angle is 168° and the three equatorial angles are 110°, 119°, and 131°. The main cohesive forces between the complex molecules arise from hydrogen bonds of the type N-H···Cl. The N-Cl distances are 3.23 to 3.37 Å. One N atom of each ligand (see Fig. 26) takes part in these bonds, two of them involving the Cl(1) atoms and one involving the Cl(2) atom. The density measured by flotation is D = 1.51 g/cm³; the calculated density is 1.50 g/cm³ [3].

Fig. 26

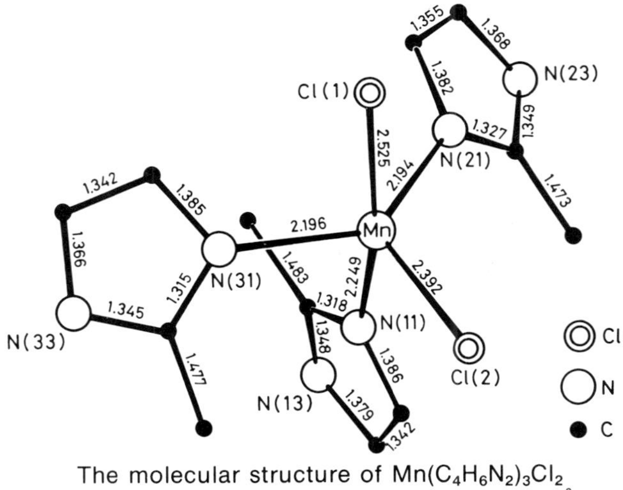

The molecular structure of $Mn(C_4H_6N_2)_3Cl_2$ with interatomic distances (bond length) in Å.

$Mn(C_4H_6N_2)_2X_2$ (X = Cl, Br) and **$Mn(C_4H_6N_2)Cl_2$.** The compounds were prepared by reacting stoichiometric quantities of $MnCl_2$ or $MnBr_2$ and 2-methylimidazole in ethanol like for the complex with imidazole, $Mn(C_3H_4N_2)_6(NO_3)_2$, p. 291. Melting points (in °C), molar conductivities Λ (in $\Omega^{-1}\cdot cm^2\cdot mol^{-1}$) of solutions in acetone, effective magnetic moments μ_{eff} (in μ_B) from susceptibility measurements, and assigned absorption bands (in cm^{-1}) of the IR spectra are listed below:

compound	m.p.	Λ	μ_{eff}	ν(NH)			δ(NH)
$Mn(C_4H_6N_2)_2Cl_2$	198	15.1	5.74	—	3345(s)	3240(m)	1424(vs)
$Mn(C_4H_6N_2)_2Br_2$	192	25.4	6.02	—	3410(m)	3245(m)	1428(m)
$Mn(C_4H_6N_2)Cl_2$	300	25.5	5.80	3610(s)	3365(m)	3220(m)	1421(br)

For unassigned bands, see the paper [1].

$Mn(C_4H_6N_2)Cl_2$ is amorphous and only sparingly soluble in common organic solvents. $Mn(C_4H_6N_2)_2Cl_2$ and $Mn(C_4H_6N_2)_2Br_2$ are crystalline. They are not freely soluble in the organic solvents. They all are nonelectrolytes in acetone and exhibit high magnetic moments, which indicate five unpaired d-electrons, which is compatible with 6-coordinate Mn in a polymeric octahedral structure [1].

References:

[1] A. K. Das, D. V. Ramana Rao (Indian J. Chem. **9** [1971] 480/1). – [2] J. Reedijk (Rec. Trav. Chim. **91** [1972] 507/16, 508/10, 513/4). – [3] F. L. Phillips, F. M. Shreeve, A. C. Skapski (Acta Cryst. B **32** [1976] 687/92).

16.2.11.3 With 4-Methyl-1H-imidazole

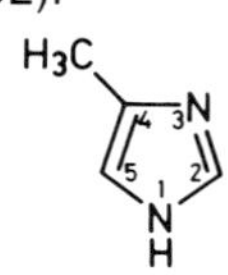

(= $C_4H_6N_2$)

$Mn(C_4H_6N_2)_nX_2$ (n = 3, 4, 6; X = NO_3, Cl, ClO_4, Br, I, NCS). The compounds $Mn(C_4H_6N_2)_6X_2$ (X = NO_3, ClO_4, Br, I, NCS) and $Mn(C_4H_6N_2)_4Cl_2$ were prepared by refluxing a stoichiometric mixture of the Mn^{II} salt and the ligand in ethanol 30 min after adding triethyl orthoformate for dehydration. The solution was then slowly concentrated to bring about crystallization, but sometimes it is necessary to refrigerate the mixture to about 0°C for 1 to 2 d. The compounds $Mn(C_4H_6N_2)_3Cl_2$, $Mn(C_4H_6N_2)_4Br_2$, and $Mn(C_4H_6N_2)_3(NCS)_2$ were obtained by heating $Mn(C_4H_6N_2)_4Cl_2$, $Mn(C_4H_6N_2)_6Br_2$, and $Mn(C_4H_6N_2)_6(NCS)_2$ in an oven at 130°C for 95 h.

$Mn(C_4H_6N_2)_4Br_2$ is hygroscopic. The complexes prepared are listed in the following table along with their properties. The molar conductivities Λ of 10^{-3} M solutions were measured in acetone at room temperature:

compound	color	m.p. in °C	μ_{eff} in μ_B	Λ in $\Omega^{-1}\cdot cm^2\cdot mol^{-1}$
$Mn(C_4H_6N_2)_6(NO_3)_2$	pale yellow	140	5.9	28.2
$Mn(C_4H_6N_2)_6(ClO_4)_2$	gray	173	6.1	242.2
$Mn(C_4H_6N_2)_6Br_2$	colorless	174	5.6	21.1
$Mn(C_4H_6N_2)_6I_2$	pale yellow	155	5.7	88.1
$Mn(C_4H_6N_2)_6(NCS)_2$	pale yellow	80	5.9	25.8
$Mn(C_4H_6N_2)_4Cl_2$	colorless	183	6.1	13.0
$Mn(C_4H_6N_2)_4Br_2$	brown	—	—	19.0*)
$Mn(C_4H_6N_2)_3Cl_2$	gray	175	6.1	17.1**)
$Mn(C_4H_6N_2)_3(NCS)_2$	brown	liquid	—	43.1

*) In methanol. – **) In nitromethane.

The complexes are high-spin, Mn having five unpaired electrons, and nonelectrolytes, except $Mn(C_4H_6N_2)_6(ClO_4)_2$, which is a 1:2 electrolyte. However, the IR spectra of $Mn(C_4H_6N_2)_6X_2$ (4000 to 400 cm^{-1}) reveal that the anions X = NO_3, ClO_4, and NCS are not coordinated to Mn because they show the absorption band of the ions. The electronic spectra of the complexes in methanol or acetone solutions show a weak absorption band in the 30000 to 33000 cm^{-1} region. This may be due to vibronic coupling.

$(CH_3)_4N[Mn(C_4H_6N_2)Cl_3]$ and $(CH_3)NH_3[Mn(C_4H_6N_2)Cl_3]$. The colorless compounds were prepared by adding a solution of hydrated $MnCl_2$ in a mixture of ethanol and triethyl orthoformate to a 1:1 mixture of the ligand and $(CH_3)_4NCl$ or $NH_3(CH_3)Cl$ in the same solvent. The reactants were vigorously stirred and then concentrated. The compounds melt above 240°C. The tetramethylammonium salt and the monomethylammonium salt have effective magnetic moments of 6.1 and 5.7 μ_B, respectively. Their molar conductivity is 118.7 $\Omega^{-1}\cdot cm^2\cdot mol^{-1}$ for the former in nitromethane and 178 $\Omega^{-1}\cdot cm^2\cdot mol^{-1}$ for the latter in methanol.

Reference:

G. R. Chaudhury, K. C. Dash (J. Inorg. Nucl. Chem. **43** [1981] 2189/90).

16.2.11.4 With 1,2-Dimethyl-1H-imidazole

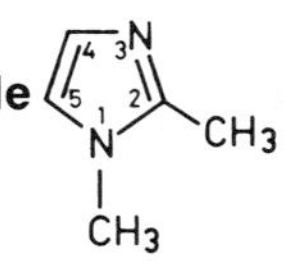

(= $C_5H_8N_2$)

$Mn(C_5H_8N_2)_4(NCS)_2$. The compound was prepared by adding the stoichiometric amount of the ligand to a 0.01M solution of $Mn(NCS)_2$ in ethanol which had been dehydrated by excess $HC(OC_2H_5)_3$. The colorless crystals are isomorphous with those of the corresponding compounds of Fe, Co, Ni, and Cd as shown by X-ray powder diagrams; they melt at 154 to 155°C. A tetragonally distorted octahedral structure of $Mn(C_5H_8N_2)_4(NCS)_2$ with trans-NCS groups coordinated to Mn by their N atoms is suggested by the electronic spectra of the related Fe, Co, and Ni compounds and a single (CN) stretching band in the IR spectrum at about 2080 cm^{-1} [1].

$Mn(C_5H_8N_2)_3(NO_3)_2$ was obtained by blending alcohol solutions of $Mn(NO_3)_2\cdot 6H_2O$ (dehydrated by $HC(OC_2H_5)_3$) and the ligand. The buff colored compound is isomorphous with the corresponding Cd complex as shown by X-ray powder diagrams; it melts at 147 to 148°C. The IR spectrum recorded for Nujol mulls in the 4000 to 200 cm^{-1} range shows the antisymmetric NO_3 stretching bands near 1460 and 1300 cm^{-1} and a very strong symmetric NO_3 stretching band near 1030 cm^{-1}. The antisymmetric NO_3 bending mode occurs at 818 cm^{-1}. These bands indicate two strongly coordinated NO_3^- anions. A structure with 5-coordinate Mn might be possible since the ESR spectrum indicates axial symmetry (D_{3h}). In the far-IR spectrum the $Mn\text{-}N_{ligand}$ stretching band probably is masked by a strong ligand band near 265 cm^{-1} having a shoulder at 290 cm^{-1}. The EPR spectrum taken in the X- and Q-band region gives g = 2.00(1), a zero-field splitting D = 0.065(5) cm^{-1}, and λ(= E/D) = 0.05(3). The small value of λ indicates a nearly axial symmetry. Thus a structure with three coplanar N_{ligand} atoms and two apical NO_3^- anions is suggested, but structures with bridging or bidentate NO_3^- anions cannot be ruled out [2].

$Mn(C_5H_8N_2)_2X_2$ (X = Cl, Br, I). The cream-colored compounds were prepared from stoichiometric amounts of $MnX_2\cdot nH_2O$ and ligand in the same manner as $Mn(C_5H_8N_2)_4(NCS)_2$. The chloro complex is isomorphous with the chloro complexes of Fe, Co, and Zn as shown by the X-ray powder diagrams; there is also isomorphism between $Mn(C_5H_8N_2)_2Br_2$ and $Mn(C_5H_8N_2)_2I_2$. Listed below are the melting points (m.p.), magnetic moments μ_{eff} (in μ_B) from susceptibility measurements at 100 and 300 K, and the zero-field splitting parameters, D, E, and λ(=E/D) from the ESR (X band) spectra of $Mn(C_5H_8N_2)_2X_2$:

X	m.p. in °C	μ_{eff} at 100 K	μ_{eff} at 300 K	D in cm^{-1}	E in cm^{-1}	λ
Cl	148 to 149	5.98	5.97	0.1	0.015	0.15
Br	142 to 143	5.97	5.95	0.4	0.08	0.20
I	166 to 167	—	—	0.6	0.12	0.20

All three compounds are believed to have distorted tetrahedral structures, with two organic ligands and two X^- ions coordinated to Mn; the distortion increasing from X = Cl to X = I as shown by parameter D. This is also confirmed by the far-IR spectra, which show rather high Mn-X stretching frequencies. A polymeric structure is excluded since the magnetic moments are invariant beween 100 and 300 K [1].

References:

[1] J. Reedijk (J. Inorg. Nucl. Chem. **35** [1973] 239/45, 241/2). – [2] J. Reedijk (Rec. Trav. Chim. **91** [1972] 1373/82, 1375/7).

16.2.11.5 With 1-Ethyl-1H-imidazole

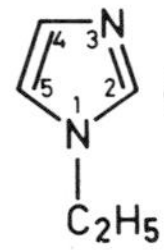

(=N-Ethylimidazole = $C_5H_8N_2$)

$Mn(C_5H_8N_2)_6X_2$ (X = NO_3, ClO_4, Br, I, BF_4). The compounds were prepared by reacting stoichiometric amounts of MnX_2 and N-ethylimidazole in dry ethanol in the same manner as the N-methylimidazole complexes (p. 295) [1]; for X = ClO_4 also see [2]. $Mn(C_5H_8N_2)_6(ClO_4)_2$ is isomorphous with $Cd(C_5H_8N_2)_6(ClO_4)_2$ as shown by X-ray powder diagrams [2]. Below are listed the properties of the complexes (molar conductivities Λ of solutions in acetone at room temperature and effective magnetic moments from the susceptibility) [1]:

X	NO_3	Br	I	ClO_4	BF_4
color	cream	cream	yellow	colorless	cream
m.p. in °C	105	130	158	168	105
Λ in $\Omega^{-1}\cdot cm^2\cdot mol^{-1}$	37.8	27	121.3	200	194.3
μ_{eff} in μ_B	6	5.8	5.8	6	6.1

The perchlorate was also reported to melt at 180°C [2]. All five compounds are high-spin $Mn(d^5)$ complexes. The bromide and nitrate behave as nonelectrolytes in acetone, indicating anions coordinated to Mn. The iodide seems to be a 1:1 electrolyte; the perchlorate and tetrafluoroborate are 1:2 electrolytes [1].

The IR spectra of $Mn(C_5H_8N_2)_6X_2$ (Nujol mulls, 4000 to 500 cm^{-1}) show absorption bands which can be interpreted like those of the corresponding N-methylimidazole complexes (p. 296). The complexes probably have a monomeric octahedral structure [1]. An IR band at 935 cm^{-1} assigned to ring vibrations of the ligand N-ethylimidazole in

$Mn(C_5H_8N_2)_6(ClO_4)_2$ is found to be shifted by 27 cm^{-1} towards higher frequencies with respect to the free ligand. In the far-IR spectrum there is only one strong band in the 265 to 160 cm^{-1} region. It can be assigned to $\nu(Mn\text{-}N_{ligand})$ stretching vibrations [2].

$Mn(C_5H_8N_2)_4X_2$ (X = Cl, Br, NCS). The colorless complexes were prepared by reacting stoichiometric amounts of MnX_2 and N-ethylimidazole in the same manner as for the N-methylimidazole complexes (p. 295) [1]; for X = Cl and Br see also [2]. The chloro and bromo compounds are isomorphous as shown by their X-ray powder diagrams. The complexes are also isomorphous with $Ni(C_5H_8N_2)_4Cl_2$ [2]. Other properties of $Mn(C_5H_8N_2)_4X_2$ are tabulated below (molar conductivities Λ of solutions in acetone at room temperature):

X	m.p. in °C [1,2]	m.p. in °C [2]	Λ in $\Omega^{-1}\cdot cm^2\cdot mol^{-1}$ [1]	μ_{eff} in μ_B [1]
Cl	123	125	19.2	5.8
Br	liquid	146	—	—
NCS	115	—	32.2	5.7

The compounds are essentially nonelectrolytes in acetone. Their magnetic moments are those of high-spin $Mn(d^5)$ complexes. The IR spectra of $Mn(C_5H_8N_2)_4X_2$ (Nujol mulls, 4000 to 500 cm^{-1}) show absorption bands which can be interpreted as for the N-methylimidazole complexes (p. 296). The compounds probably have a distorted octahedral structure; the NCS group of $Mn(C_5H_8N_2)_4(NCS)_2$ is coordinated to Mn via the N atom [1]. The ESR spectra of $Mn(C_5H_8N_2)_4X_2$ (X = Cl, Br) are similar to those of the corresponding 5-methylpyrazole complexes (p. 273), for which a distorted octahedral symmetry was suggested [2].

$Mn(C_5H_8N_2)_3X_2$ (X = NO_3, Cl, Br). The compounds were prepared by adding an ethanol solution of $MnX_2\cdot nH_2O$ (dehydrated with sufficient $HC(OC_2H_5)_3$) to an ethanol solution of N-ethylimidazole (stoichiometric amounts). The crystals which precipitate on standing or after addition of dry ether were washed with dry ether and dried in vacuo. The chloro and nitrato compounds are ismorphous with the Cd complexes (X-ray powder diagrams). Melting points in °C: 117 (NO_3), 105 (Cl), and 169 (Br). The ESR spectrum of $Mn(C_5H_8N_2)_3(NO_3)_2$ is similar to that of the 5-methylpyrazole complex $Mn(C_4H_6N_2)_4(NO_3)_2$ (p. 273), thus suggesting a distorted octahedral symmetry with NO_3 ions boded to 6-coordinate Mn [2].

$[MnF(C_5H_8N_2)_3]_4(BF_4)_4$. The compound was obtained as a colorless powder by reacting an ethanol solution of $Mn(BF_4)_2\cdot 6H_2O$ (with added $HC(OC_2H_5)_3$) and an ethanol solution of N-ethylimidazole. It is isomorphous with the corresponding compounds of Co and Cd, as shown by the X-ray powder diagrams. The IR spectrum indicates the presence of uncoordinated BF_4^- and bridging Mn-F-Mn units. The F ions are formed by slow decomposition of BF_4^- ions. The complex has a tetrameric cuban type cluster structure like $[CoF(C_5H_8N_2)_3]_4(BF_4)_4$, whose crystal structure (cubic, space group Pa3-T_h^6, No. 205) was determined [3].

The specific heat C of $[MnF(C_5H_8N_2)_3]_4(BF_4)_4$ measured between 0.2 and 100 K is shown as C/R in **Fig. 27a**, p. 304, up to 6 K together with the estimated lattice and magnetic contributions. The curve shows a rather broad maximum at 1 K as expected for the magnetic specific heat of a tetrameric antiferromagnetic compound. The height of this maximum falls short of the calculated one by a factor 1.23. Above 6 K (lattice) specific heat increases steadily to C = 200 $J\cdot mol^{-1}\cdot K^{-1}$ at 75 K. From the slope of C above 10 K a total specific heat of about 250 $J\cdot mol^{-1}\cdot K^{-1}$ ($\sim$30 R) may be deduced for room temperature.

The magnetic susceptibility χ of the complex in the 2 to 80 K region is shown in **Fig. 27b** together with that predicted for an S = $^5/_2$ Heissenberg tetramer with exchange constant J/k = −1.50 K. The magnetization curve yields J/k = −0.8 K. Because the experimental data give higher values of χ (and lower J/k) than predicted by theory, it is assumed that there are ferromagnetic interactions between the tetrameric clusters which decrease the field required to saturate the sample. Thus J/k was concluded to be 1.2±0.5 K [4].

Fig. 27

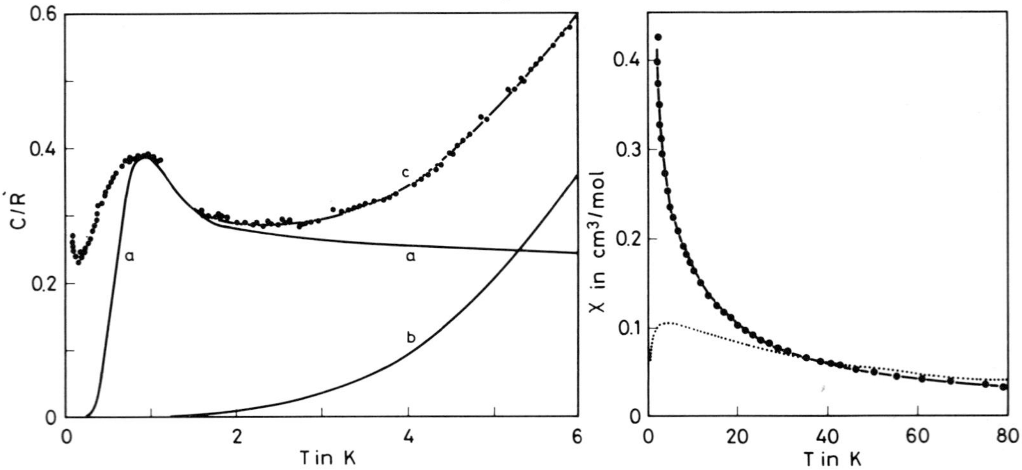

a) Specific heat data for $[MnF(C_5H_8N_2)_3]_4(BF_4)_4$. The solid curves a and b represent the estimated magnetic and lattice contributions, respectively; curve c is the sum of contributions a and b.

b) Magnetic susceptibility of $[MnF(C_5H_8N_2)_3]_4(BF_4)_4$ versus temperature. The dotted curve is the theoretical prediction for the antiferromagnetic S = $^5/_2$ Heisenberg tetramer with J/k = −1.50 K.

$(CH_3)_4N[Mn(C_5H_8N_2)Cl_3]$. The compound was obtained by adding an alcohol solution of hydrated $MnCl_2$ to a solution of $(CH_3)_4NCl$, and N-ethylimidazole in acetone (1:1:1 mole ratio) with stirring. The light yellow crystals melt above 240°C. Susceptibility measurements yield the effective magnetic moment $\mu_{eff} = 6.0\ \mu_B$. Electrical conductivity measurements for 0.001M solutions in nitromethane at room temperature give $\Lambda = 127.3$ $\Omega^{-1}\cdot cm^2\cdot mol^{-1}$, corresponding to a 1:1 electrolyte. The high-spin Mn(d^5) complex probably has a monomeric tetrahedral structure [1].

References:

[1] G. Chaudhury, K. C. Dash (Transition Metal Chem. [Weinheim] **2** [1977] 253/6, 254). – [2] J. A. Welleman, F. B. Hulsbergen, J. Verbiest, J. Reedijk (J. Inorg. Nucl. Chem. **40** [1978] 143/7, 144). – [3] J. C. Jansen, H. van Koningsveld, J. Reedijk (Nature **269** [1977] 318/9). – [4] J. J. Smit, G. M. Nap, L. J. De Jongh, J. A. C. Van Ooijen, J. Reedijk (Physica B+C [Amsterdam] **97** [1979] 365/76, 372/5).

16.2.11.6 With 1-Propyl-1H-imidazole 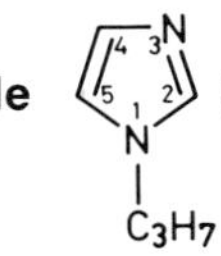 (= N-Propylimidazole = $C_6H_{10}N_2$)

$Mn(C_6H_{10}N_2)_6X_2$ (X = NO_3, Cl, ClO_4, Br, I, BF_4). The colorless compounds were prepared by reacting MnX_2 and N-propylimidazole (mole ratio 1:6) in stirred ethanol solution. The precipitate was filtered off, washed several times with alcohol, then with ether, and dried in vacuo [1]. The complexes may also be prepared following the method for the N-methylimidazole complexes (p. 295). The perchlorate and hexafluoroborate are isomorphous with the corresponding Cd complexes, and $Mn(C_6H_{10}N_2)_6(NO_3)_2$ is isomorphous with the Co and Ni nitrato complexes as shown by the X-ray diagrams [2]. Melting points (those of [2] in parentheses), molar conductivities Λ (in $\Omega^{-1}\cdot cm^2\cdot mol^{-1}$) of 0.001M solutions in acetone or nitromethane, and effective magnetic moments μ_{eff} from susceptibility measurements at room temperature for $Mn(C_6H_{10}N_2)_6X_2$ [1] are given below:

X	NO_3	Cl	Br	I	ClO_4	BF_4
m.p. in °C	90(109)	80	143	149	90(120)	78(114)
Λ in acetone	16.1	23.1	38.4	123	258.8	133
Λ in nitromethane	7.8	4.8	11.9	—	—	50.2
μ_{eff} in μ_B	5.8	6.0	5.8	5.7	5.8	5.8

All the compounds $Mn(C_6H_{10}N_2)_6X_2$ are high-spin Mn-d^5 complexes. The complexes for X = Cl^-, Br^-, and NO_3^- behave as nonelectrolytes; those for X = ClO_4 and BF_4 are 1:2 electrolytes in acetone and nitromethane. The IR absorption spectra of $Mn(C_6H_{10}N_2)_6X_2$ were recorded as Nujol mulls in the 4000 to 500 cm^{-1} region. For the tetrafluoroborate they show an unsplit $\nu_3(BF_4)$ band in the 1140 to 1050 cm^{-1} region and an unsplit $\nu_4(BF_4)$ band at 640 cm^{-1}. For the perchlorate there is a broad strong ν_3 band in the 1150 to 1050 cm^{-1} range and a sharp ν_4 band at 640 cm^{-1}. These bands indicate ionic BF_4^- and ClO_4^- (T_d symmetry). The complex $Mn(C_6H_{10}N_2)_6(NO_3)_2$ exhibits a strong $\nu_3(NO)$ stretching mode at about 1390 cm^{-1} and two IR-active deformation modes, at 830 cm^{-1} and 730 cm^{-1}, which indicate uncoordinated NO_3 ions having T_d symmetry [1]. An IR band at 933 cm^{-1} assigned to ring vibrations of N-propylimidazole in $Mn(C_6H_{10}N_2)_6(BF_4)_3$ is shifted 25 cm^{-1} to higher frequencies from its free ligand position. In the far-IR spectrum, 265 to 160 cm^{-1}, there is only one strong band which can be assigned to Mn-N_{ligand} stretching vibrations [2]. The electronic spectrum of $Mn(C_6H_{10}N_2)_6X_2$ dissolved in acetone or methanol has a weak absorption band in the 30000 to 33000 cm^{-1} region. Heating $Mn(C_6H_{10}N_2)_6Cl_2$ and $Mn(C_6H_{10}N_2)_6Br_2$ in air at 120 to 130°C to constant weight causes the loss of four moles of ligand. The complexes for X = I, NO_3, ClO_4, and BF_4^- decompose at 120°C to leave brown or black residues. Some of the complexes are fungicides [1].

$Mn(C_6H_{10}N_2)_4(NCS)_2$. The compound was prepared by stirring solutions of $Mn(NCS)_2$ and the ligand in a 1:4 (or higher) mole ratio for some time. The colorless complex melts at 70°C and has a magnetic moment μ_{eff} = 5.8 μ_B at room temperature. The molar conductivity of a 0.001M solution in acetone is Λ = 34.95 $\Omega^{-1}\cdot cm^2\cdot mol^{-1}$ at room temperature and indicates a nonelectrolyte. The IR absorption spectrum (recorded as Nujol mulls in the 4000 to 500 cm^{-1} region) shows two bands, at 2060 and 770 cm^{-1}, assigned to the (CN) and (CS) stretching modes, respectively, which indicate coordination of the NCS group to Mn through its N atom. On heating to 120°C the compound loses two moles of ligand [1].

$Mn(C_6H_{10}N_2)_3(NO_3)_2$. The compound was prepared from $Mn(NO_3)_2 \cdot 6H_2O$ and N-propylimidazole by the method for the N-ethylimidazole complex (p. 302). It melts at 91°C, and it is isomorphous with the corresponding Cd complexes as shown by an X-ray structure study. Probably the NO_3 ions are coordinated to Mn [2].

$[MnF(C_6H_{10}N_2)_3]_4[BF_4]_4$. The preparation and properties are similar to those of the N-ethylimidazole complex (p. 303) [3].

$Mn(C_6H_{10}N_2)_2X_2$ (X = Cl, Br, NCS). The compounds were obtained by heating $Mn(C_6H_{10}N_2)_6Cl_2$, $Mn(C_6H_{10}N_2)_6Br_2$, or $Mn(C_6H_{10}N_2)_4(NCS)_2$ at 120 to 130°C to constant weight. The brown chloro and the black bromo compounds are liquid at room temperature. $Mn(C_6H_{10}N_2)_2Cl_2$ is insoluble in acetone. The bromo and isothiocyanato complexes are nonelectrolytes in acetone ($\Lambda = 55$ and 45.7 $\Omega^{-1} \cdot cm^2 \cdot mol^{-1}$, respectively). The yellow $Mn(C_6H_{10}N_2)_2(NCS)_2$ is extremely hygroscopic [1].

$(CH_3)_4N[Mn(C_6H_{10}N_2)Cl_3]$. The colorless compound melts above 240°C and has a magnetic moment $\mu_{eff} = 6.1\ \mu_B$ at room temperature. The molar conductivity of a 0.001M solution in nitrobenzene is $\Lambda = 131.3\ \Omega^{-1} \cdot cm^2 \cdot mol^{-1}$ at room temperature and indicates a 1:1 electrolyte. The compound is a fungicide [1].

References:

[1] G. R. Chaudhury, K. C. Dash (J. Indian Chem. Soc. **56** [1979] 87/9). – [2] J. A. Welleman, F. B. Hulsbergen, J. Verbiest, J. Reedijk (J. Inorg. Nucl. Chem. **40** [1978] 143/7, 144). – [3] J. C. Jansen, H. van Koningsveld, J. Reedijk (Nature **269** [1977] 318/9).

16.2.11.7 With 1-Butyl-1H-imidazole

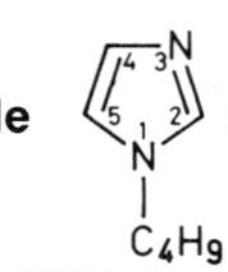

(= N-Butylimidazole = $C_7H_{12}N_2$)

$Mn(C_7H_{12}N_2)_6X_2$ (X = NO_3, ClO_4, I, BF_4). To prepare the compounds with X = NO_3, ClO_4, BF_4 0.01 mol $MnX_2 \cdot nH_2O$ was dissolved in 5 to 15 mL of ethanol and 0.1 mol $HC(OC_2H_5)_3$, and the solution is added to a 1:1 solution of 0.07 mol N-butylimidazole in ethanol. The crystals which precipitated upon standing or after addition of dry diethyl ether were collected, washed with dry diethyl ether (three portions), and dried in vacuo at room temperature [1]. The iodide was prepared similarly. It is isomorphous with $Fe(C_7H_{12}N_2)_6I_2$ as shown by X-ray powder diagrams [2]. There is also isomorphism of $Mn(C_7H_{12}N_2)_6(NO_3)_2$ with the corresponding complexes of Co, Ni, and Cd, of $Mn(C_7H_{12}N_2)_6(ClO_4)_2$ with the complexes of Fe and Cd, and of $Mn(C_7H_{12}N_2)_6(BF_4)_2$ with $Cd(C_7H_{12}N_2)_6(BF_4)_2$ [1, 3].

The colorless nitrate, perchlorate, and tetrafluoroborate melt at 105, 152, and 148°C, respectively [1, 3]. The pale yellow iodide melts at 166 to 167°C [2]. The IR spectra (4000 to 150 cm^{-1}) and the Raman spectra (2000 to 100 cm^{-1}) of $Mn(C_7H_{12}N_2)_6(ClO_4)_2$ and of the pure liquid ligand have been recorded, and the absorption bands were tentatively assigned. Nearly all ligand bands also appear in the Mn compound, and the bands are only slightly shifted. Also the ClO_4^- bands are observed. Thus the spectra of the coordination compounds appear to be the simple sum of the anion absorptions and the free-ligand bands. Only two bands appear to be strongly dependent on the kind of metal (Mn, Cd, Fe, Co, Ni, Cu): A very strong band at 202 cm^{-1}, which is not in the free-ligand spectra, was assigned to the antisymmetric $Mn\text{-}N_{ligand}$ stretching, and the ring bending band of the ligand at 908 cm^{-1}, which is shifted by 26 cm^{-1} towards higher frequencies on coordination to Mn [1].

In the ESR spectrum of the investigated $Mn(C_7H_{12}N_2)_6X_2$ complexes at both X- and Q-band frequencies, there is a broad band at $g=2.01\pm0.01$ having a line width of 550 Gauss. Samples diluted in solid $Cd(C_7H_{12}N_2)_6X_2$ (X = NO_3, ClO_4, BF_4) yield a six-line signal centered at $g=2.005\pm0.005$. The hyperfine coupling constant is $A=89\pm1$ Gauss. The vibrational and ESR spectra are in agreement with regular octahedral symmetry about the Mn^{2+} ions [1]. The ESR spectrum of $Mn(C_7H_{12}N_2)_6I_2$ shows a single line near $g_{eff}=2$ in agreement with a rather regular environment about the Mn^{2+} ion [2].

The compounds are quite stable in moist air, and most of them are insoluble in water but readily soluble in ethanol, acetone, CH_2Cl_2, and $CHCl_3$ [1, 2]. On reacting the compounds with "imidazolyzed" polystyrene in $CHCl_3$ (as the swelling agent) polymeric Mn complexes form. These look promising as catalysts for ester hydrolysis, oxidation of organic compounds, or reduction of O_2 or N-oxides [4].

$Mn(C_7H_{12}N_2)_4X_2$ (X = Cl, Br, NCS). The colorless compounds were prepared by mixing stoichiometric amounts of MnX_2 and ligand in ethanol solution following the method given for $Mn(C_7H_{12}N_2)_6X_2$ on p. 306 [2]. $Mn(C_7H_{12}N_2)_4(NCS)_2$ is isomorphous with the corresponding complexes of Co, Ni, and Cd as shown by X-ray powder diagrams [2]. Melting points: 62 to 64°C (Cl), 94 to 98°C (Br) [2, 3], and 78 to 79°C (NCS) [2].

The IR spectra were recorded as Nujol mulls in the 4000 to 150 cm^{-1} region; the Raman spectra of the solids were taken in the 2150 to 50 cm^{-1} region using He-Ne laser excitation (shown for the chloride in [2]). In the far-IR and Raman spectra the following bands are observed and assigned [2]:

X	ν(Mn-X) IR	ν(Mn-N) IR	ν(Mn-X) or ν(Mn-N) Raman	other bands, Raman
Cl	205(s)	193(m, sh)	200(w)	158(m)
Br	180(m)	200(m)	not investigated	—
NCS	207(vs)	207(vs)	250(w), 190(s)	150(w)

The IR and Raman ligand bands in the range from 4000 to 300 cm^{-1} are scarcely shifted upon coordination to Mn. The spectra suggest an octahedral symmetry about the Mn^{2+} ions since the $Mn\text{-}N_{ligand}$ and Mn-X frequencies are rather low. The anions are probably in trans positions. The unsplit (CN) stretching mode of the NCS^- anion at 2040(±5) cm^{-1} (IR) and 2060(±3) cm^{-1} (Raman) indicate coordination of NCS^- via the N atom.

The ESR spectra of the chloro and bromo compounds taken at X- and Q-band frequencies show absorptions far from $g_{eff}=2$, indicative of a distorted octahedral or tetrahedral environment, while for $Mn(C_7H_{12}N_2)_4(NCS)_2$ a single line near $g_{eff}=2$ is indicative of a regular environment around the Mn^{2+} ion. The distortion parameters D and $\lambda(=E/D)$ are $D=0.12$ cm^{-1}, $\lambda=0.1$ for X = Cl and $D=0.63$ cm^{-1}, $\lambda=0.02$ for X = Br (values similar to those of the methylpyrazole complex $Mn(C_4H_6N_2)_4Br_2$, p. 273). Thus the ESR suggests a trans octahedral structure of $Mn(C_7H_{12}N_2)_4Br_2$.

All compounds are stable in moist air and soluble in organic solvents such as ethanol, acetone, $CHCl_3$, and CH_2Cl_2 [2]. On reacting these complexes with "imidazolyzed" polystyrene in $CHCl_3$, the polymeric Mn complexes of that ligand form [4].

References:

[1] J. Reedijk (J. Inorg. Nucl. Chem. **33** [1971] 179/88, 181, 185). – [2] J. Reedijk (Rec. Trav. Chim. **90** [1971] 1249/64, 1251, 1256/60). – [3] J. A. Welleman, F. B. Hulsbergen, J. Verbiest, J. Reedijk (J. Inorg. Nucl. Chem. **40** [1978] 143/7, 144). – [4] F. B. Hulsbergen, J. Manassen, J. Reedijk, J. A. Welleman (J. Mol. Catal. **3** [1977/78] 47/50, 48).

16.2.11.8 With 1-Vinyl-1H-imidazole

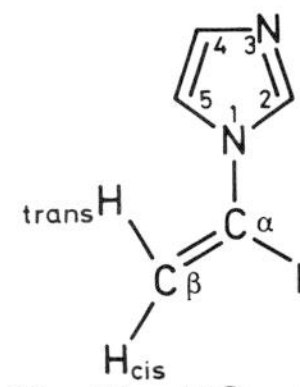

(= N-Vinylimidazole = $C_5H_6N_2$)

$Mn(C_5H_6N_2)_nX_2$ (n = 6, 4, 3, 2, 1.5; X = N_3, NO_3, Cl, ClO_4, Br, I, BF_4, NCS). The compounds $Mn(C_5H_6N_2)_6X_2$ (X = NO_3, ClO_4, I, BF_4) and $Mn(C_5H_6N_2)_4X_2$ (X = Cl, Br, BF_4, NCS) were prepared from the stoichiometric amounts of the Mn^{II} salts and the ligand in ethanol solution. The precipitates were washed with ethanol, then ether, and dried in vacuo [1]. $Mn(C_5H_6N_2)_4Cl_2$ was prepared by reacting solutions of $MnCl_2 \cdot 4H_2O$ and ligand in a mixture of alcohol and acetone at room temperature. Then the precipitate was washed with the solvent and dried at 60 to 70°C under a residual pressure of 8 to 10 Torr [2].

The compounds $Mn(C_5H_6N_2)_3(NCS)_2$, $Mn(C_5H_6N_2)_2Br_2$, and $Mn(C_5H_6N_2)_{1.5}Cl_2$ formed on heating $Mn(C_5H_6N_2)_4X_2$ at 120 to 130°C for 97 h (X = NCS), 60 h (X = Br), or 93 h (X = Cl). To prepare $Mn(C_5H_6N_2)_2(N_3)_2$ a slight excesss of NaN_3, dissolved in a minimum volume of water, was added to an ethanol solution of $Mn(C_5H_6N_2)_4Cl_2$, and the mixture stirred [1].

The following table lists the prepared $Mn(C_5H_6N_2)_nX_2$ compounds along with their color, melting points (m.p.), magnetic moments μ_{eff} (from susceptibility), and molar conductivities Λ of 0.001M solutions in acetone at room temperature [1]:

compound	color	m.p. in °C	μ_{eff} in μ_B	Λ in $\Omega^{-1}\cdot cm^2\cdot mol^{-1}$
$Mn(C_5H_6N_2)_6(NO_3)_2$	colorless	110	6.10	62.8, 65.7*)
$Mn(C_5H_6N_2)_6(ClO_4)_2$	pale yellow	165	5.83	222.0
$Mn(C_5H_6N_2)_6I_2$	colorless	160	6.12	116.0*)
$Mn(C_5H_6N_2)_6(BF_4)_2$	colorless	80	5.98	203.7, 157.7*)
$Mn(C_5H_6N_2)_4Cl_2$	colorless	165 (177 [4])	5.90	35.5
$Mn(C_5H_6N_2)_4Br_2$	colorless	198	5.87	19.6
$Mn(C_5H_6N_2)_4(BF_4)_2$	colorless	110	5.93	160.3, 106.9*)
$Mn(C_5H_6N_2)_4(NCS)_2$	colorless	122	5.93	31.6
$Mn(C_5H_6N_2)_3(NCS)_2$	pale yellow	103 (dec.)	5.98	insol.
$Mn(C_5H_6N_2)_2(N_3)_2$	gray	152 (dec.)	5.97	37.9
$Mn(C_5H_6N_2)_2Br_2$	cream	180	5.90	31.7
$Mn(C_5H_6N_2)_{1.5}Cl_2$	cream	>240	6.20	37.3

*) In nitromethane.

All complexes are high-spin octahedral or tetrahedral on the basis of their magnetic moments [1]. The ^{1}H NMR spectrum of $Mn(C_5H_6N_2)_4Cl_2$ was recorded from a 0.008M solution in nitromethane (also the internal standard) at 80°C. The shift of the vinyl protons is much less than that of the ring protons H(4) and H(5) but higher than in the complexes of Fe, Co, Ni, and Cu. Thus the unpaired spin densities within the π system of the imidazole ring appear to be notably delocalized in the complex [3]. A more recent investigation on a 0.05M solution of ligand and $Mn(C_5H_6N_2)_4Cl_2$ (mole ratio ~100:1) in dimethyl sulfoxide at 50, 90, 130, and 170°C yielded the resonance shifts δ(H) relative to $(CH_3)_2Si$ shown in **Fig. 28**, from which the following proton interaction constants, A (in 10^4 Øe), were calculated: H(2) = −26, H(4) = 67, H(5) = 76, $H_\alpha = -7$, $H_{\beta cis} = -5$, and $H_{\beta trans} = -13$. These A values can not be explained only by a π delocalization mechanism but should be considered the result of the transmission of coupling through both the σ and π systems of the ligand. Coordination by N(1) or N(3) atoms of 1-vinylimidazole to Mn^{2+} is discussed [4].

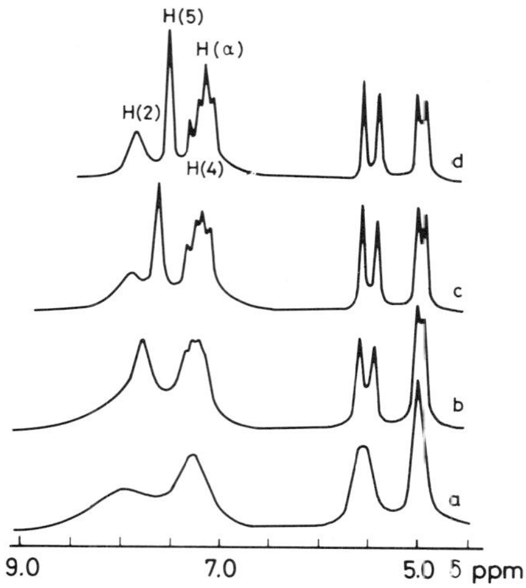

Fig. 28

^{1}H NMR spectra of 1-vinyl-1H-imidazole with added paramagnetic complex $Mn(C_5H_6N_2)_4Cl_2$ at 50°C (a), 90°C (b), 130°C (c), and 170°C (d). Mole ratio ligand: $Mn(C_5H_6N_2)_4Cl_2 \approx 100:1$.

The specific electrical conductivity of solid $Mn(C_5H_6N_2)_4Cl_2$, measured between 20 and 100°C ($\varkappa = 4.43 \times 10^{-14}$ to 1.38×10^{-11} $\Omega^{-1} \cdot cm^{-1}$), characterizes the complex as a dielectric [2]. All the $Mn(C_5H_6N_2)_4X_2$ compounds are nonelectrolytes in acetone and nitromethane except the iodo, perchlorato, and tetrafluoroborato complexes. The unexpected nonelectrolytic nature of $Mn(C_5H_6N_2)_6(NO_3)_2$ suggest dissociation of some vinylimidazole ligands and coordination of the NO_3^- ions on dissolution [1].

The electronic spectra of all these octahedral complexes show weak absorption bands at about 29850 and 33000 cm^{-1}, which are assigned to $^6A_{1g} \rightarrow {}^4E_g(D)$ and $^6A_{1g} \rightarrow {}^4T_{1g}(P)$ transitions, respectively.

The IR spectra recorded as Nujol mull or KBr discs in the 4000 to 200 cm^{-1} region show the following features: $Mn(C_5H_6N_2)_4Cl_2$ exhibits an absorption band at 1650 cm^{-1} assigned to the vinyl group, which shifts only slightly on coordination (~5 cm^{-1}). A pair of bands at 1511 and 1495 cm^{-1}, assigned to the heterocyclic ring, is shifted by 10 to 30 cm^{-1} towards higher frequencies. This indicates a coordination of the N(3) atoms of the imidazole ring to Mn [2]. The ν(Mn-Cl) band at 280 cm^{-1} and the ν(Mn-Br) band in $Mn(C_5H_6N_2)_4Br_2$ at 275 cm^{-1} indicate direct Mn-X bonding. The perchlorate and tetrafluoroborate complexes show strong unsplit ν_3 and ν_4 bands (ν_3 at ~1100 cm^{-1}, ν_4 at ~600 cm^{-1}) of the uncoordinated anions. $Mn(C_5H_6N_2)_4(BF_4)_2$ is probably tetrahedral, but $Mn(C_5H_6N_2)_6(ClO_4)_2$ and $Mn(C_5H_6N_2)_6(BF_4)_2$ apparently are octahedral complexes. Solid $Mn(C_5H_6N_2)_6(NO_3)_2$ shows a strong (NO) stretching band at 1390 cm^{-1} (ν_3) and the two deformation modes at 830 (ν_2) and 730 (ν_4) cm^{-1}, indicative of uncoordinated NO_3^- ions of D_{3h} symmetry. $Mn(C_5H_6N_2)_2(N_3)_2$ shows an intense band at 2120 cm^{-1} and a weak band around 1300 cm^{-1}, assigned to antisymmetric and symmetric stretching modes, respectively, of the N_3^- ion, and a bending mode at 642 cm^{-1}. The value agree well with a terminal coordination of the N_3^- ion. $Mn(C_5H_6N_2)_4(NCS)_2$ shows a ν(CN) band at 2048 cm^{-1} and ν(CS) band at 762 cm^{-1}, which indicate terminal coordination of the thiocyanate groups to Mn through the N atoms. In $Mn(C_5H_6N_2)_3(NCS)_2$ these bands appear at 2080 and

790 cm^{-1}, respectively, and are indicative of bridging thiocyanate groups bonded to Mn also through the S atom in a Mn-NCS-Mn bridge [1].

In the mass spectrum of $Mn(C_5H_6N_2)_4Cl_2$ only the peaks of the ligand molecule and its fragmentation products are observed (m/e = 94, 80, 67, and 41) [1]. The air-stable complex $Mn(C_5H_6N_2)_4Cl_2$ losses 2% by weight on heating to 185°C and 15 to 18% at 600°C. It is somewhat soluble in alcohol, acetone, and nitromethane but decomposes in acid and alkaline media [2].

$[Mn(C_5H_6N_2)_nCl_2]_\infty$ (n = 3 and 4). The colorless polymeric compound $[Mn(C_5H_6N_2)_3Cl_2]_\infty$ was prepared by reacting manganese(II) chloride and poly-1-vinylimidazole in absolute ethanol. The colorless polymeric compound $[Mn(C_5H_6N_2)_4Cl_2]_\infty$ was prepared by polymerizing a solution of 0.2 g $Mn(C_5H_6N_2)_4Cl_2$ in 5 mL ethanol in the presence of 0.008 g azoisobutyronitrile. The ampule with the reactant was sealed under N_2 and kept in a thermostat at 70 ± 0.5°C for 24 h. The precipitate was washed with ethanol and dried to constant weight in vacuo at 55°C. It can also be prepared by the simultaneous complex formation and polymerization of ethanol solutions of $MnCl_2$ and 1-vinylimidazole.

The two polymers melt with decomposition between 360 and 380°C. Thermogravimetric analysis shows that the polymeric ligand is somewhat destabilized on coordination to Mn. The polymeric complexes are insoluble in most organic solvents, partially soluble in water and dimethylformamide. They are decomposed by acids and alkalis [5]. They behave as dielectrics; the specific electrical conductivity is $\varkappa = 10^{-16}$ to 10^{-17} $\Omega^{-1} \cdot cm^{-1}$ at 20°C and increases by a factor of about 10^5 beween 130 and 180°C.

The IR absorption spectra were recorded for samples pelleted with KBr. The bands of the imidazole ring at 1505 and 920 cm^{-1} are shifted towards higher frequencies by about 20 cm^{-1}, while the doublet at 1115 and 1090 cm^{-1} is converted on coordination into a very strong band at 1110 cm^{-1}.

$(CH_3)_4N[Mn(C_5H_6N_2)Cl_3]$ and **$(C_2H_5)_4N[Mn(C_5H_6N_2)Br_3]$.** To prepare the colorless $(CH_3)_4N[Mn(C_5H_6N_2)Cl_3]$ an ethanol solution of $MnCl_2 \cdot 4H_2O$ was added to a stirred ethanol solution of equimolar amounts of $(CH_3)_4NCl$ and 1-vinylimidazole. Acetone is the solvent used to prepare the yellow $(C_2H_5)_4N[Mn(C_5H_6N_2)Br_3]$. Both compounds melt above 240°C. Their magnetic moments are 5.62 μ_B (methylammonium compound) and 5.77 μ_B (ethylammonium compound); their molar electrical conductivities in acetone at room temperature are $\Lambda = 171.0$ and 119.4 $\Omega^{-1} \cdot cm^2 \cdot mol^{-1}$, respectively. The IR spectrum of $(CH_3)_4N[Mn(C_5H_6N_2)Cl_3]$ shows an absorption band at 285 cm^{-1} assigned to ν(Mn-Cl) vibrations [1].

References:

[1] G. R. Chaudhury, K. C. Dash (Indian J. Chem. A **17** [1979] 364/7). – [2] G. G. Skvortsova, E. S. Domnina, Yu. N. Ivlev, N. N. Chipanina, V. I. Skorobogatova, Yu. A. Myachin (Zh. Obshch. Khim. **42** [1972] 596/9; J. Gen. Chem. [USSR] **42** [1972] 592/5). – [3] V. K. Voronov, Yu. N. Ivlev, E. S. Domnina, G. G. Skvortsova (Khim. Geterotsikl. Soedin. **1972** 994/5; Chem. Heterocycl. Compounds [USSR] **8** [1972] 904/5). – [4] V. K. Voronov, T. E. Moskovskaya, V. I. Glukhikh, A. I. Rezvukhin, E. S. Domnina, G. G. Skvortsova (Koord. Khim. **5** [1979] 218/27, 219, 221; Soviet J. Coord. Chem. **5** [1979] 166/74, 167/8). – [5] G. G. Skvortsova, E. S. Domnina, Yz. N. Ivlev, N. N. Chipanina (Zh. Obshch. Khim. **46** [1976] 1593/8; J. Gen. Chem. [USSR] **46** [1976] 1552/6).

16.2.11.9 With Other 1-Vinyl-1H-imidazoles

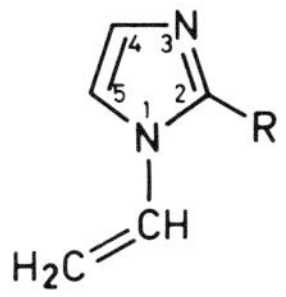

I) $R=CH_3$ ($=C_6H_8N_2$)
II) $R=CH_2OH$ ($=C_6H_8N_2O$)

Complexes in Solution. Polarography (dropping mercury cathode, SCE) of oxygen-free aqueous solutions of $MnCl_2$ and ligand II at ionic strength I = 0.05 (KCl) showed that the Mn^{2+} reduction wave is shifted towards more positive potentials upon addition of ligand (0.02 to 0.08 M). The observed half-wave potentials indicate the complexes $Mn(C_6H_8N_2O)Cl_2$ and $Mn(C_6H_8N_2O)_2Cl_2$ have stability constants $\log K_1 = 1.21 \pm 0.08$ and $\log K_2 = 2.33 \pm 0.80$ (K in L/mol). The polarographic reduction of the complexes at ligand concentrations from 0.001 to 0.2 M reveals the reduction to be a quasi-reversible two-electron reaction [1].

$Mn(C_6H_8N_2)_2Cl_2$ was prepared by reacting $MnCl_2 \cdot 4H_2O$ and ligand I in a 1:4 mole ratio in alcohol or a mixture of alcohol and ether. The colorless crystalline precipitate was isolated, washed repeatedly with the solvent, and dried in a vacuum. It is air stable, melting at 198 to 199°C.

The IR spectrum recorded for samples pelleted with KBr shows absorption bands at 1560, 1520, and 900 cm^{-1}, assigned to stretching or deformation modes of the imidazole ring, which on coordination are shifted toward higher frequencies by about 20 to 25 cm^{-1}. The introduction of a methyl group into 1-vinylimidazole increases the electron donating power of the ligand considerably as shown by the larger imidazole band shifts ($\Delta\nu \approx 10$ cm^{-1} for the 1-vinylimidazole). The elctronic spectrum of $Mn(C_6H_8N_2)_2Cl_2$ in aqueous solution (200 to 700 nm) has an absorption maximum at $\lambda = 228$ nm ($\varepsilon = 22500$). This is assigned to a $\pi \rightarrow \pi^*$ transition. The complex is soluble in water, alcohol, acetone, nitromethane, and dimethylformamide [2].

$Mn(C_6H_8N_2O)_2Cl_2$ and **$Mn(C_6H_8N_2O)Cl_2$** were prepared by reacting the calculated amounts of $MnCl_2$ and ligand II in a mixture of acetone and ether or acetone and ethanol. The resulting precipitate was filtered off, washed with the solvent used for the preparation, and dried at 60 to 70°C at a pressure of 10 to 15 Torr. $Mn(C_6H_8N_2O)_2Cl_2$ forms colorless air-stable crystals, which decompose at 128 to 130°C. The IR spectra suggest that coordination occurs through the N(3) atom with participation of the hydroxy methyl group. The compounds are monomeric. They are soluble in ethanol, acetone, and dimethyl sulfoxide [1].

References:

[1] E. S. Domnina, L. V. Baikalova, L. E. Protasova, N. M. Deriglasov, N. N. Chipanina, D. D. Taryashinova, V. I. Skorobogatova, G. G. Skvortsova (Koord. Khim. **5** [1979] 14/20; Soviet J. Coord. Chem. **5** [1979] 10/5). – [2] E. S. Domnina, G. G. Skvortsova, N. P. Glazkova, N. N. Chipanina, D. D. Taryashinova, L. E. Protasova (Zh. Obshch. Khim. **46** [1976] 168/71; J. Gen. Chem. [USSR] **46** [1976] 166/8).

16.2.11.10 With 4,4′-Methylenebis(1H-imidazole)

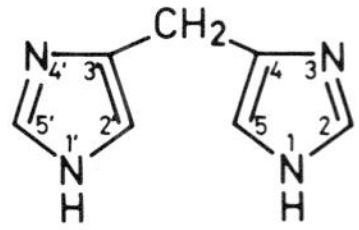

($=C_7H_8N_4$)

With exclusion of air by nitrogen, pH-potentiometric titrations in aqueous solutions of $Mn(NO_3)_2$ and 4,4′-methylenebis(1H-imidazole) (mole ratio 1:5) at 30°C and ionic strength I = 0.16 M (KNO_3) revealed the existence of two complexes $[Mn(C_7H_8N_4)]^{2+}$ and $[Mn(C_7H_8N_4)_2]^{2+}$, with the stability constants log $K_1 = 2.96 \pm 0.03$ and log $K_2 = 2.84 \pm 0.05$ (K_n in L/mol), log $\beta_2 = 5.80$ (β_2 in L^2/mol^2), C. N. C. Drey, J. S. Fruton (Biochemistry **4** [1965] 1258/63, 1261).

16.2.11.11 With 2-(1H-Imidazol-4-yl)ethylamine

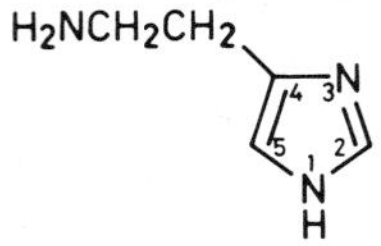

(= Histamine = $C_5H_9N_3$)

The pH of aqueous solutions of $Mn(NO_3)_2$ and histamine hydrochloride $C_5H_9N_3 \cdot 2HCl$ (mole ratio 1:5) at 15 and 40°C, and ionic strength I = 0.2 M (KNO_3) suggest the formation of a 1:1 complex with the formation constant log $K_1 = 2.98 \pm 0.1$ at 15°C and 2.95 ± 0.1 at 40°C (K_1 in L/mol). The complex-forming reaction enthalpy is then $\Delta H = -0.5(\pm 2)$ kcal/mol, B. Rao, H. B. Mathur (J. Inorg. Nucl. Chem. **33** [1971] 809/16, 811/2).

16.2.11.12 With 2-(1H-Imidazol-2-yl)pyridine

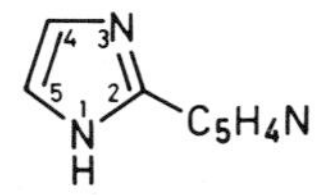

(= $C_8H_7N_3$)

$Mn(C_8H_7N_3)_2(ClO_4)_2 \cdot 2H_2O$ was obtained by mixing an aqueous solution of $Mn(ClO_4)_2$ and an ethanol solution of the ligand in a 1:2 mole ratio. On evaporation followed by cooling, the yellow solid complex was deposited, and it was recrystallized from aqueous ethanol. Susceptibility measurements at 19°C yielded the magnetic moment $\mu_{eff} = 5.91\ \mu_B$. A molar conductivity of $\Lambda = 56.0\ \Omega^{-1} \cdot cm^2 \cdot mol^{-1}$ was measured in nitrobenzene solutions at 25°C. The compound melts (or decomposes) above 300°C, B. Chiswell, F. Lions, B. S. Morris (Inorg. Chem. **3** [1964] 110/4).

16.2.11.13 With Miconazole

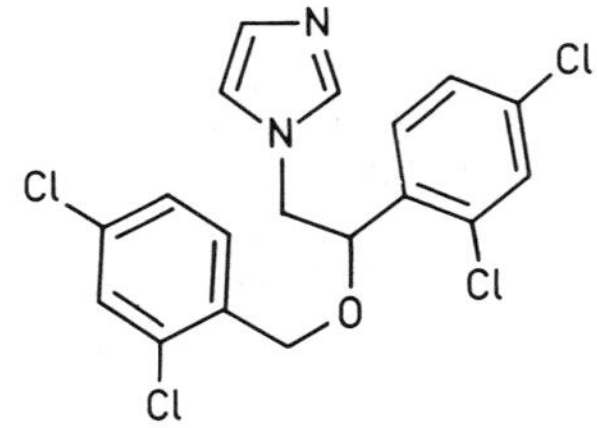

(= $C_{18}H_{14}Cl_4N_2O$)

The interaction of miconazole with Mn^{2+} was studied by means of polarographic measurements on 5×10^{-5} M $MnCl_2 \cdot 4H_2O$ to which amounts of the ligand were successively added. The maximum ligand-to-metal ratio was 400. Thus the following stepwise formation constants (K in L/mol) were determined:

K_1	K_2	K_3	K_4	K_5	K_6
4.93×10^2	2.60×10^2	1.37×10^2	72.3	38.1*)	20.1*)

*) Extrapolated.

Reference:

G. J. Williams, W. I. Jottier, C. J. de Rauter (Analusis **9** [1981] 327/32).

16.2.12 With 1H-Benzimidazole 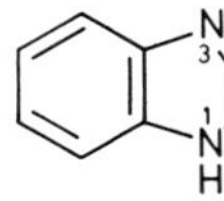 ($=C_7H_6N_2$)

Complexes in Solution. The pH-potentiometric measurements on aqueous solutions of a manganese(II) salt and benzimidazole at 25°C and ionic strength I=0.5M (KNO_3) revealed the formation of a 1:1 complex with stability constant log $K_1=0.88$ (K_1 in L/mol), B. Lenarcik, W. Maciejewski (Polish J. Chem. **55** [1981] 31/8, 34).

$Mn(C_7H_6N_2)_2(NCS)_2$. The compound was prepared by mixing methanol solutions of $Mn(SCN)_2$ and ligand in a 1:2 mole ratio. The light yellow crystals appeared after the solution has stood several days in a vacuum desiccator over P_4O_{10}. They were filtered off, washed with a little methanol, and dried in a desiccator. The IR spectrum shows an absorption band at 2090 cm^{-1}, assigned to ν(CN) of the NCS group, which together with the magnetic moment of $\mu_{eff}=5.79\ \mu_B$ suggests a polymeric octahedral structure, either due to NCS bridges or due to the ligands coordinated intermolecularly, A. D. Garnovskii, V. V. Skopenko, M. I. Knyazhanskii, O. A. Osipov, P. V. Gilyanovskii, T. D. Artyukhovich (Zh. Obshch. Khim. **42** [1972] 2068/73; J. Gen. Chem. [USSR] **42** [1972] 2064/8, 2065).

16.2.13 With Derivatives of 1H-Benzimidazole

16.2.13.1 With 1-Vinyl-1H-benzimidazole 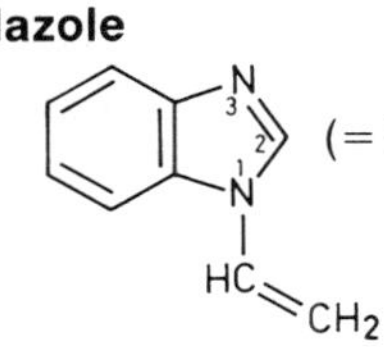 ($=C_9H_8N_2$) and Its Polymer

$Mn(C_9H_8N_2)_2Cl_2$. The compound was prepared by reacting $MnCl_2\cdot4H_2O$ and 1-vinylbenzimidazole (1:2 mole ratio) in a suitable solvent, such as water, alcohol, ether, or acetone. The precipitate was filtered off, washed repeatedly with the solvent and dried at 60 to 70°C under a residual pressure of 8 to 10 Torr. The white, air-stable complex melts at 236°C with decomposition. The IR absorption spectrum of the complex shows no notable shift of the band at 1650 cm^{-1}, assigned to the vinyl group or of the imidazole ring stretching bands in the 1600 to 1400 cm^{-1} range. This fact indicates a rather low basicity of 1-vinylbenzimidazole as compared with 1-vinylimidazole. Thus it is assumed that the unshared electron pair of the tertiary N atom in the 3 position of the imidazole ring participates in complexing with $MnCl_2$. The electronic spectrum (210 to 1000 nm) exhibits no charge-transfer bands. The electric conductivity indicates a dielectric. The solubility in organic solvents is low, but is best in acetone, alcohols, and nitromethane. The complex decomposes in alkaline and acid media [1].

$[Mn(C_9H_8N_2)_5Cl_2]_n$. The compound was obtained by reacting $MnCl_2$ and polymeric 1-vinylbenzimidazole in a 1:2 mole ratio in absolute alcohol. The white precipitate was washed repeatedly on the filter with hot alcohol and dried at 70°C. It melts at 345 to 357°C with decomposition. The IR spectrum of the Mn complex is identical with that of the original polymer. Conductivity measurements show that the complex is a typical dielectric whose specific electric conductivity rises from 10^{-16} to $10^{-17}\ \Omega^{-1}\cdot cm^{-1}$ at 20°C to about 10^{-13} to $10^{-11}\ \Omega^{-1}\cdot cm^{-1}$ between 130 and 180°C. The thermal stability of the polymer is somewhat reduced on complexing with $MnCl_2$ [2].

References:

[1] G. G. Skvortsova, E. S. Domnina, Yu. N. Ivlev, N. N. Chipanina, V. I. Skorobogatova, Yu. A. Myachin (Zh. Obshch. Khim. **42** [1972] 596/9; J. Gen. Chem. [USSR] **42** [1972] 592/5). – [2] G. G. Skvortsova, E. S. Domnina, Yu. N. Ivlev, N. N. Chipanina (Zh. Obshch. Khim. **46** [1976] 1593/8; J. Gen. Chem. [USSR] **46** [1976] 1552/6).

16.2.13.2 With 2-Methyl- and 2-Benzyl-1H-benzimidazole

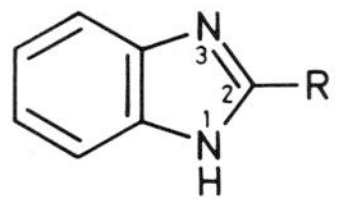

I) R = CH_3 (= $C_8H_8N_2$)
II) R = $CH_2C_6H_5$ (= $C_{14}H_{12}N_2$)

$Mn(C_8H_8N_2)_2X_2$ (X = Cl, Br, NO_3, NCS, $^1/_2SO_4$). The chloride was prepared by mixing saturated methanol solutions of $MnCl_2$ and 2-methylbenzimidazole (ligand I) in a 1:2 mole ratio and evaporating the mixture to a syrupy consistency. The crystallization was stimulated by repeated washing with ether. The pale pink crystals were dried at 70 to 80°C. The bromide was obtained by triturating 0.32 g of $MnBr_2$ and 0.3 g of ligand I in an agate mortar. The mixture was dissolved in a minimum quantity of methanol. The pink crystals which appeared on evaporation were collected, washed with benzene and ether, and dried as above. The nitrate and isothiocyanate were obtained by stirring the solid ligand I (0.3 g) in saturated solutions of $Mn(NO_3)_2$ (0.26 g) in methanol or $Mn(SCN)_2$ (0.25 g) in ethanol. The solutions were evaporated to a syrupy consistency, and crystallization was induced by rubbing with a glass rod. The pink crystals were washed and dried. The colorless sulfate was prepared from $MnSO_4$ (0.18 g) and ligand (0.3 g) in the same manner as the bromide [1].

The table lists the temperature of decomposition t_{dec} in °C, the effective magnetic moment μ_{eff} at 293 K, and the Curie-Weiss constant Θ_p in K:

X	Cl	Br	NO_3	NCS	$^1/_2SO_4$
t_{dec}	140	185	130	200	132
μ_{eff} in μ_B	5.60	5.60	—	5.00	—
Θ_p	+25	+23	—	−45	—

The magnetic susceptibilities obey the Curie-Weiss law between 288 and 423 K. The values of μ_{eff}, which remain practically constant in the temperature range studied, are those of high-spin Mn $3d^5$ complexes for X = Cl and Br. The lower magnetic moment and Curie-Weiss constant of the isothiocyanate may indicate a weak antiferromagnetism [1].

In the IR spectra a broad band at about 3250 cm^{-1} assigned to ν(NH) of the ligand remains nearly unchanged on complexation. The stretching vibration of the imidazole ring in the ligand, ν(CN), is shifted towards higher frequencies by only 10 to 15 cm^{-1}, indicating rather weak bonding. The band for the nitrate at about 1390 cm^{-1} is assigned to $\nu(NO_3^-)$; the band at 1135 cm^{-1} for the sulfate is assigned to $\nu(SO_4^{2-})$. Thus both nitrate and sulfate ions are in the outer sphere. The thiocyanate band is at 2060 cm^{-1}

All compounds are decomposed by water or on heating to 200°C (see table above) and are readily soluble in methanol [1].

$Mn(C_{14}H_{12}N_2)_2X_2$ (X = Cl, Br, I, NCS, $^1/_2SO_4$) and **$Mn(C_{14}H_{12}N_2)_2(NO_3)_2 \cdot 6H_2O$.** The chloride was prepared by mixing concentrated methanol solutions of $MnCl_2 \cdot 4H_2O$ and ligand II in a 1:2 mole ratio. The mixture was evaporated on a water bath to leave an oil which crystallizes on repeated washing with benzene and ether. The crystals were dried at 80 to 90°C. The bromide and the nitrate hexahydrate were obtained by grinding ligand II (0.3 g) and $MnBr_2$ (0.25 g) or $Mn(NO_3)_2 \cdot 6H_2O$ (0.24 g) in an agate mortar and dissolving the mixture in a minimum quantity of methanol. The crystallization was stimulated by rubbing with a glass rod. The crystals were washed with benzene and ether and dried as above. The isothiocyanate, sulfate, and iodide were obtained by dissolving dry 2-benzylbenzimidazole in methanol solutions of $Mn(NCS)_2$, $MnSO_4$, or MnI_2. The mixtures are evaporated slowly (the iodide at room temperature) to leave oils which crystallize on addition of ether. The crystals were washed with a minimum of methanol and then with a mixture of benzene and ether. The iodide crystallizes on washing the oily residue repeatedly with ether and acetone. All compounds are dried at 80 to 90°C [2].

The complexes are light pink, except the iodide, which is yellow. In the table below are listed the decomposition points t_{dec} in °C and the shift of the stretching frequencies of the CN band of the imidazole ring from their free-ligand position (at 1630 cm^{-1}):

X	Cl	Br	I	NO_3*)	NCS	$^1/_2SO_4$
t_{dec}	>210	165	76	148	130	180
$\Delta\nu(CN)$ in cm^{-1}	−25	−30	−5	−20	−30	−5

*) Refers to the hexahydrate.

The observed shifts indicate coordination of the ligand to Mn through the N atom in the 3 position of the imidazole ring. The NH group of 2-benzylbenzimidazole does not take part in complex formation. The broad band in the 3250 to 3350 cm^{-1} region, assigned to $\nu(NH)$ in the free ligand, is unchanged in all the complexes. The nitrate has a band at 1390 cm^{-1}, assigned to the NO_3^- ion, which must then be in the outer sphere. The thiocyanate exhibits a strong doublet band, with maxima at 2060 and 2100 cm^{-1}, which belong to the CN stretching vibrations of the NCS^- ion and a bridging NCS group joined to Mn by the N atom. A band at 740 cm^{-1} is assigned to $\nu(CS)$ of the NCS^- ion. As the NCS groups of the compound are nonequivalent a dimeric isothiocyanate complex having both bridging and ionic NCS groups and terminally bonded ligands was proposed. The sulfate shows a band at 1135 cm^{-1}, which is assigned to vibrations of the outer-sphere SO_4^{2-} ion [2].

The magnetic susceptibility of the chloride, bromide, and thiocyanate complexes obey the Curie-Weiss law between 30 and 120°C. Their magnetic moments, ranging between $\mu_{eff} = 5.4$ and 5.6 μ_B, are nearly those of high-spin Mn^{II} complexes. All compounds are air stable, readily soluble in alcohols but decomposing in water [2].

References:

[1] K. F. Slyusarenko, M. V. Artemenko, D. I. Lapina, P. A. Suprunenko, G. I. Kal'naya (Ukr. Khim. Zh. **42** [1976] 29/33, 31; Soviet Progr. Chem. **42** No. 1 [1976] 28/31). – [2] K. F. Slyusarenko, M. V. Artemenko, D. I. Lapina, G. A. Pokhodnaya (Ukr. Khim. Zh. **41** [1975] 566/70; Soviet Progr. Chem. **41** No. 6 [1975] 4/7).

16.2.13.3 With 2-Hydroxymethyl- and 2-Hydroxyphenyl Derivatives of 1H-Benzimidazole

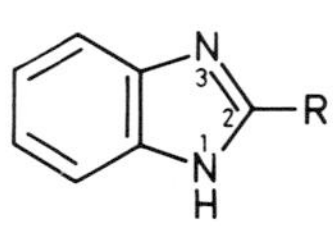

I) $R = CH_2OH$ ($= C_8H_8N_2O$)
II) $R = C_6H_4OH$ ($= C_{13}H_{10}N_2O$)

Complexes in Solution. pH-potentiometric measurements on solutions of $Mn(ClO_4)_2$ and ligand I in aqueous dioxane (50 vol%) at 25°C suggest the formation of 1:1 and 1:2 complexes with the stability constants $\log K_1 = 7.98$ and $\log K_2 = 6.32$ (K_n in L/mol). For comparison of log K_n with those of other metal complexes of ligand I, see [1]. Measurements with ligand II as the ligand under the same conditions indicate a 1:1 complex with log $K_1 = 6.58$ [2].

$Mn(C_8H_8N_2O)_3X_2$ (X = Cl, Br, I) and **$Mn(C_8H_8N_2O)_2X_2$** (X = NO_3, CH_3COO, NCS, $^1/_2SO_4$). The complexes were prepared as follows: A) MnX_2 (X = Cl, Br) and ligand I are ground together in a 1:3 mole ratio and dissolved in a minimum amount of methanol. The crystals are washed repeatedly with chloroform and ether and dried at 90 to 100°C. B) To a methanol solution of MnI_2 the ligand is added in a 1:3 mole ratio. The solution is evaporated slowly until crystals appear. These are washed with ether. C) A methanol solution containing the manganese(II) salt (nitrate or acetate) and ligand I in a 1:2 mole ratio is evaporated slowly until an oil forms. This is washed repeatedly with benzene and ether until crystals form. The crystals are dried. D) Close to saturated solutions of $Mn(SCN)_2$ and ligand I are mixed in a 1:2 mole ratio and the resulting crystals are washed with ether. E) To a solution of manganese(II) sulfate in a mixture of water and methanol, the ligand is added in a 1:2 mole ratio. Upon slow evaporation an oil forms. It is washed with benzene and ether to induce crystallization.

The complexes are all colorless, except the acetate, which is pale rose. All complexes are crystallinic as shown by X-ray diffraction. The atomic density radial distribution shows that the metal-to-halogenide bonds are ionic whereas one NCS group is coordinated to Mn and the other is ionic. In the following table are recorded the melting points (m.p.) and magnetic moments μ_{eff} in μ_B of the complexes at 289 and 354 K:

complex	preparation	m.p. in °C	μ_{eff} at 289 K	μ_{eff} at 354 K
$Mn(C_8H_8N_2O)_3Cl_2$	A	165	5.70	5.75
$Mn(C_8H_8N_2O)_3Br_2$	A	152	5.47 [1)]	5.57 [2)]
$Mn(C_8H_8N_2O)_3I_2$	B	155	5.68	5.70
$Mn(C_8H_8N_2O)_2(NO_3)_2$	C	175	5.27	5.28
$Mn(C_8H_8N_2O)_2(CH_3COO)_2$	C	103	5.63	5.65
$Mn(C_8H_8N_2O)_2(NCS)_2$	D	148	5.22	5.22
$Mn(C_8H_8N_2O)_2SO_4$	E	135	5.04	5.07

[1)] At 288 K. – [2)] At 353 K.

The magnetic moments of the complexes in the 15 to 150°C temperature range are largely independent on temperature but are considerably lower than expected for high-spin $Mn(d^5)$ complexes, except for $Mn(C_8H_8N_2O)_3X_2$. Such reduced moments, which apparently are due to weak antiferromagnetic interactions, indicate polymeric complexes with bridging ligands or anions. This is also confirmed by the IR spectrum of the thiocyanate, which shows two bands, at 2120 and 2080 cm^{-1}, assigned to (CN) stretching vibrations of bridging and ionic isothiocyanate groups. The nitrato complex shows a strong band due to the NO_3^- ion at 1390 cm^{-1}. The electric conductivity of the complexes in methanol (dilution 1:1000) corresponds to that of the manganese(II) salts, thus indicating extensive solvolysis [3].

References:

[1] T. J. Lane, J. M. Daly (J. Am. Chem. Soc. **81** [1959] 2953/5). – [2] W. D. Johnson, H. Freiser (Anal. Chim. Acta **11** [1954] 301/8, 303). – [3] K. F. Slyusarenko, M. V. Artemenko, D. I. Lapina, G. A. Pokhodnaya, T. M. Butsik, P. A. Suprunenko (Ukr. Khim. Zh. **46** [1980] 596/601; Soviet Progr. Chem. **46** No. 6 [1980] 36/40).

16.2.13.4 With 2-Hydrazino-1H-benzimidazole

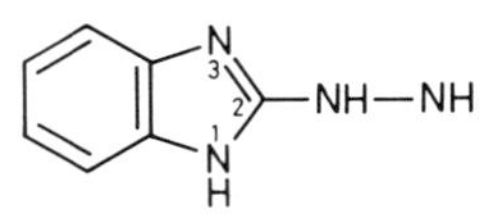

$(=C_7H_8N_4)$

$[Mn(C_7H_7N_4)_2(H_2O)_2]$. The complex was prepared by refluxing a methanol solution of manganese acetate and 2-hydrazinobenzimidazole. The physical properties (magnetic susceptibility, IR and electronic spectra, molar conductivity) show the complex to be a nonelectrolyte of octahedral structure, containing two coordinated water molecules. These are lost upon heating as shown by thermogravimetric analysis. The complex shows fungistatic activity, G. V. Reddy, D. S. Rao, M. C. Ganorkar (Natl. Acad. Sci. Letters [India] **4** [1981] 13/6 from C.A. **95** [1981] No. 143325).

16.2.13.5 With 2-(2-Pyridyl)-1H-benzimidazole

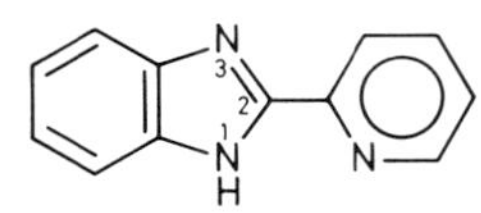

$(=C_{12}H_9N_3)$

Complexes in Solution. Solutions containing $Mn(ClO_4)_2$, $HClO_4$, and 2-(2-pyridyl)benzimidazole in aqueous dioxane (50 vol%) were studied by pH-potentiometry. Coordination of Mn^{2+} to the pyridine nitrogen of the imidazole ring tends to increase the acidity of the imino hydrogen. The manganese complex is half neutralized at pH 8.4. A formation constant of log $K_1 \sim 3.8$ at pH 8.4 results from a graph. The formation of the manganese complex is accompanied by a color change from pink at pH 2.8, to tan at pH 4.0, and to yellow tan at pH 8.5 [1].

$Mn(C_{12}H_9N_3)_2X_2$ (X = Br, ClO_4, NCS, $^1/_2SO_4$) and **$Mn(C_{12}H_9N_3)_2Cl_2 \cdot nH_2O$** (n = 1, 6). The halogeno compounds were obtained by refluxing ethanol solutions containing 1 mol of MnX_2 and slightly more than 2 mol of the ligand for 30 min. In this case the chloro compound is isolated as a monohydrate [3]. Mixing solutions of $MnCl_2$ in water and ligand in ethanol (mole ratio 1:2) yielded $Mn(C_{12}H_9N_3)_2Cl_2 \cdot 6H_2O$ [2]. The perchlorate was obtained on adding a solution of $NaClO_4$ to a boiling ethanol solution of equimolar amounts of $MnCl_2$ and ligand [3]. $Mn(C_{12}H_9N_3)_2(NCS)_2$ was precipitated when the deep wine-red solution of $Mn(CH_3COO)_3$ in a dilute HCl solution of the ligand was treated with a concentrated solution of KSCN or when a solution of the chloro or bromo complex was treated with NH_4SCN [3]. The compound is also obtained by mixing cold methanol solutions of $Mn(NCS)_2$ and ligand in a 1:1 or 1:2 mole ratio. Crystals appeared after the solution had been left several days in a vacuum desiccator [4]. $Mn(C_{12}H_9N_3)_2SO_4$ was prepared by reacting a hot ethanol solution of the ligand with an excess of a hot aqueous solution of $MnSO_4$ [3].

The complexes are yellow or yellowish white (X = $^1/_2SO_4$). The magnetic moment μ_{eff} (at T in K) and the molar conductivity Λ in DMF at 25 or 26°C are given for $Mn(C_{12}H_9N_3)_2X_2$ by [3]:

X	Cl[1)]	ClO_4	SCN	$^1/_2SO_4$
μ_{eff} in μ_B	6.14 (295.8)	—	5.99[2)] (304.4)	6.11 (296.8)
Λ in $\Omega^{-1} \cdot cm^2 \cdot mol^{-1}$	25.7	116.0	—	—

[1)] Monohydrate. – [2)] $\mu_{eff} = 5.94\ \mu_B$ [4].

The chloro and bromo compounds probably contain 6-coordinate Mn with the bidentate ligand 2-(2-pyridyl)benzimidazole, as suggested by the values of μ_{eff} and Λ of the chloride. That the perchlorate contains free ClO_4^- ions is suggested by a broad IR absorption band in the 1125 to 1050 cm^{-1} region and by the high value of its molar conductivity. The IR spectrum of $Mn(C_{12}H_9N_3)_2(NCS)_2$ shows an absorption band at 3050 cm^{-1}, which is assigned to the NH stretching vibration of the ligand. A strong band at 2075 cm^{-1} [3], observed as a doublet at 2090 and 2060 cm^{-1} by [4], is assigned to the CN stretching vibration of the thiocyanate group [3, 4]. The IR data and the value of μ_{eff} suggest a polymeric octahedral structure, which could arise by the further bonding of a nitrogen atom of the imidazole ring to manganese (intermolecular coordination of the ligand) or by the formation of a bridge structure with an Mn-NCS-Mn link [4].

The electronic absorption spectrum of a solution in dimethylformamide exhibits an absorption maximum at 32300 cm^{-1} with an extinction coefficient of 46500 $L\cdot mol^{-1}\cdot cm^{-1}$. The complex does not luminesce at 293 K. At 77 K the dimethylformamide solution shows a long-wave band having maxima at 23000 and 21200 cm^{-1} and shoulders at 20300 and 19300 cm^{-1}. It is concluded that excitation occurs in the π-electron system of the ligand, which is slightly perturbed by coordination to Mn. The transitions causing luminescence have a $\pi \rightarrow \pi^*$ character and are localized in the π-electron system of the ligand [4]. $Mn(C_{12}H_9N_3)_2SO_4$ could not be further investigated because it is insoluble in common solvents [3].

$Mn(C_{12}H_9N_3)X_2$ (X = Cl, Br) and **$Mn(C_{12}H_9N_3)(C_{12}H_8N_3)_2$.** The pale yellow halogeno complexes were prepared in the same way as the 1:2 complexes except that a large excess of MnX_2 was used. The yellow water-insoluble inner complex $Mn(C_{12}H_9N_3)(C_{12}H_8N_3)_2$ was obtained by adding diluted aqueous ammonia to an aqueous solution containing excess $MnCl_2$, ligand, and a large excess of NH_4Cl until the pH was ~6.

The magnetic moment μ_{eff} (at T in K) and the molar conductivity Λ in DMF at 25 or 26°C are given for $Mn(C_{12}H_9N_3)X_2$ by [3]:

X	Cl	Br	$C_{12}H_8N_3$
μ_{eff} in μ_B	—	5.91(296.5)	6.03(302.2)
Λ in $\Omega^{-1}\cdot cm^2\cdot mol^{-1}$	25.7	—	10.6

The μ_{eff} and Λ of the halgeno compounds suggest 4-coordinate manganese [3].

$Mn(C_{12}H_9N_3)_2(CH_3COO)_2(ClO_4)_2$. An aqueous suspension of ligand is dissolved by adding a minimum volume of 6 M HCl solution. Then a large excess of $Mn(CH_3COO)_3$ is added in small portions with constant stirring. The filtered solution is precipitated by adding a concentrated solution of $NaClO_4$ to yield a dark brown solid. The IR spectrum shows a strong absorption band near 3090 cm^{-1}, indicating that the NH group of imidazole is not affected by complexation. The bands at 1598 and 1370 cm^{-1} are due to the CO stretching vibrations of the monodentate acetate ions, and the band at 1130 to 1090 cm^{-1} indicates the presence of uncoordinated ClO_4 ions. The manganese appears to be in oxidation state +4, but the magnetic moment μ_{eff} = 3.38 μ_B at 303 K is rather low for an Mn^{IV} complex. The compound is instantaneously reduced by NCS^- ion to form yellow $Mn(C_{12}H_9N_3)_2(NCS)_2$ [3].

References:

[1] T. Harkins, H. Freiser (J. Am. Chem. Soc. **78** [1956] 1143/6). – [2] B. Chiswell, F. Lions, B. S. Morris (Inorg. Chem. **3** [1964] 110/4). – [3] S. P. Ghosh, A. Mishra (J. Indian Chem. Soc. **52** [1975] 791/3). – [4] A. D. Garnovskii, V. V. Skopenko, M. I. Knyazhanskii, O. A. Osipov, P. V. Gilyanovskii, T. D. Artyukhovich (Zh. Obshch. Khim. **42** [1972] 2068/73; J. Gen. Chem. [USSR] **42** [1972] 2064/8).

16.2.13.6 With 1H-Indolo[1,2-a]benzimidazole

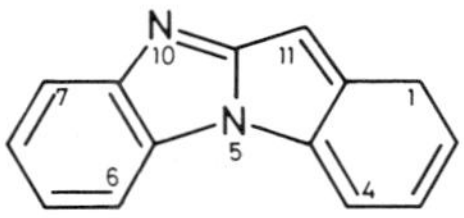

(= $C_{14}H_{10}N_2$)

$Mn(C_{14}H_{10}N_2)_2X_2$ (X = Cl, Br). The compounds were obtained by mixing the ethanol solutions of MnX_2 and ligand in a 2:5 mole ratio. The precipitate was filtered off, washed with ethanol, and dried in vacuo over $Mg(ClO_4)_2$. The far-IR spectra were recorded as polythene mulls in the 400 to 40 cm^{-1} range. Wave numbers of absorption bands in cm^{-1} and assignments:

X	ν(Mn-X)	ν(Mn-N)	δ(Mn-X)
Cl	322,298	212	100(sh)
Br*)	244,232	212	90

*) The ν(Mn-X) and ν(Mn-N) modes cannot be distinguished clearly for X = Br.

There is also a strong ligand band at 250 cm^{-1} and a medium one at 204 cm^{-1}. A shoulder appearing for $Mn(C_{14}H_{10}N_2)_2Cl_2$ at 278 cm^{-1} cannot be assigned but may indicate some decomposition as it becomes stronger the longer the complex is stored. The observed ν(Mn-X) and δ(Mn-X) modes are indicative of tetrahedral complexes, G. A. Melson, R. H. Nuttall (J. Mol. Struct. **1** [1967/68] 405/11, 407).

16.2.14 With 1H-Naphth[2,3-d]imidazole-2-methanol

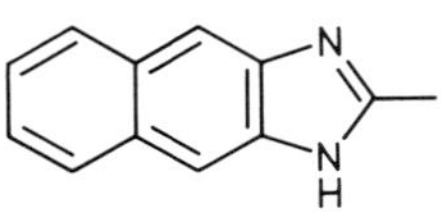

(= $C_{12}H_{10}N_2O$)

Potentiometric measurements (glass electrode) on solutions of $Mn(ClO_4)_2$ and ligand in aqueous dioxane (50 vol%) at 25°C suggest the formation of a 1:1 complex having the stability constant log K_1 = 7.05 (K_1 in L/mol). For comparison of log K_1 with other metal complexes of this ligand, see T. J. Lane, K. P. Quinlan (J. Am. Chem. Soc. **82** [1960] 2997/9).

16.2.15 With Derivatives of Imidazoline

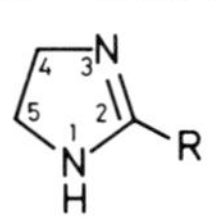

I) R = $C_{17}H_{35}$ (= $C_{20}H_{40}N_2$)
II) R = C_5H_4N (= $C_8H_9N_3$)

Complexes of composition $MnX_2 \cdot 3\,L$ and $MnX_2 \cdot 3\,L \cdot 3\,CH_3COOH$ with L = 2-heptadecyl-4,5-dihydro-1H-imidazole (ligand I) and X = NO_3, Cl, $^1/_2SO_4$, CH_3COO are reported in [1]. Potentiometric pH measurements at 25°C on solutions of $Mn(ClO_4)_2$ and 2-(4,5-dihydro-1H-imidazol-2-yl)pyridine (ligand II) revealed the existence of a 1:1 complex. An approximate formation constant log K_1 = 3.9 is given (K_1 in L/mol) [2].

References:

[1] Union Carbide Carbon Corp., J. N. Hogsett (U.S. 2881181 [1956/59]; C.A. **1959** 15098). – [2] T. R. Harkins, H. Freiser (J. Am. Chem. Soc. **78** [1956] 1143/6, 1144).

16.2.16 With 2-Imidazolidinone 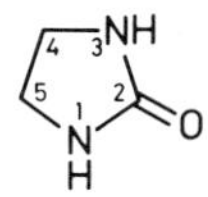 (= N, N′-Ethyleneurea = $C_3H_6N_2O$ = L)

MnL_6X_2 (X = ClO_4, BF_4), **$MnL_6Br_2 \cdot L$** and **MnL_4X_2** (X = NO_3, Cl, Br, NCS). All compounds were prepared by basically the same procedure. The stoichiometry depended upon the mole ratio of ligand and manganese(II) salt. Thus an ethanol solution of a manganese(II) salt (0.01 mol) dehydrated by $HC(OC_2H_5)_3$ was added to a hot ethanol solution of the ligand (0.06 or 0.04 mol). (Methanol often produced better crystalline powders.) The colorless crystals which appeared on cooling or addition of ether were washed with ethanol and sodium-dried ether and dried in a vacuum at room temperature.

Melting points in °C:

complex.	$MnL_6Br_2 \cdot L$	$MnL_6(ClO_4)_2$	$MnL_6(BF_4)_2$	$MnL_4(NO_3)_2$
m.p.	156 to 158	165 to 168	168 to 169	145 to 146
complex	MnL_4Cl_2	MnL_4Br_2	$MnL_4(NCS)_2$	
m.p.	148 to 150	164 to 165	188 to 189	

The X-ray powder diagams revealed the isomorphism of some manganese(II) complexes with complexes of other metals M:

compound	$ML_6Br_2 \cdot L$	$ML_6(ClO_4)_2$	$ML_6(BF_4)_2$	$ML_4(NCS)_2$
metal	Mg, Ni, Co	Cd	Zn, Cd, Ni, Co	Ni, Co

MnL_6 $(BF_4)_2$ is also isomorphous with $ML_6(ClO_4)_2$ for M = Mg, Zn, Ni, Co.

IR spectra in the 4000 to 400 cm^{-1} region (Nujol mulls) are given in the paper for complexes of general formulas ML_6X_2 (X = ClO_4, BF_4) and $ML_6Br_2 \cdot L$ and where M is not restricted to Mn. All complex compounds containing the ion MnL_6^{2+} showed very similar spectra with respect to the bands that could be ascribed to the ligand. However, the bonding sites of the ligand (CO or NH) are uncertain since the IR spectra showed very broad bands in the amide region (1500 to 1700 cm^{-1}). Moreover, the CO and NH frequencies are influenced by hydrogen bonding between the NH proton and the anions, which may shift the NH stretching and bending vibration frequencies more than does coordination of manganese. Because of the strong coupling between the CO stretching and NH bending mode, hydrogen bonding between the anions and the ligand NH group can also affect the position of the CO stretching. However, observations on related complexes indicated that the 2-imidazolidinone molecules are coordinated via the carbonyl oxygen. The presence of a seventh, uncoordinated ligand molecule in $MnL_6Br_2 \cdot L$ was seen in the IR spectrum (weak bands around 1700 and 1275 cm^{-1} and splitting of the bands at 1030 and 930 cm^{-1}). The spectra of MnL_6X_2 with X = ClO_4 and BF_4 showed some bands assigned to anion vibrations indicative of uncoordinated ClO_4^- or BF_4^-. These anion bands are rather broad, due to hydrogen bonding between the anions and the NH groups of the ligand.

In the far-IR spectrum (450 to 50 cm^{-1}) of $MnL_6(ClO_4)_2$, a very strong and broad absorption band at 240 to 205 cm^{-1} was tentatively assigned to the Mn-ligand stretching vibration. Two broad bands appeared at 144(s) and 100(m) cm^{-1}.

In the IR spectrum of $MnL_4(NCS)_2$ a single CN stretching vibration is observed near 2095 to 2080 cm^{-1}. This indicates coordination of the thiocyanate groups to Mn by the N atoms. The unsplit band suggests trans configuration.

ESR spectra at room temperature on powder samples of MnL_6X_2 (X = ClO_4, BF_4), $MnL_6Br_2 \cdot L$, and $MnL_4(NCS)_2$ all showed a single line centered around g = 2.0, which agrees with manganese ions in a rather regular octahedral environment. The compunds MnL_4X_2 ($X = NO_3$, Cl, Br) exhibited very complex ESR spectra at both X- and Q-band frequencies, thus indicating a distorted octahedral structure. The zero-field splitting parameters D and E (in cm^{-1}) were calculated:

X	NO_3	Cl	Br
D	0.065 ± 0.002	0.13 ± 0.01	0.12 ± 0.02
E	$\leqq 0.002$	0.013 ± 0.003	0.025 ± 0.008

The quite large rhombic splitting parameter E of the bromide may indicate cis arrangement of the Br^- anions in MnL_4Br_2.

Reference:

J. Reedijk, T. M. Mulder, J. A. Smit (Inorg. Chim. Acta **13** [1975] 219/27, 221/2, 224).

Formula Index of Ligands

The ligands treated in this volume are arranged in the index according to the system of A. Hill (J. Am. Chem. Soc. **22** [1900] 478/94). In this system, the first criterion for the location of a ligand is the number of C atoms; the second, the number of hydrogen atoms. Then come the numbers of atoms of the other elements, set in alphabetical order.

The first column contains the empirical formulas of the ligands. The second column shows the names of the ligands. Additional ligands occurring in mixed-ligand complexes and in adducts are placed as subheadings. Thus, the complex $[Mn(C_4H_{13}N_3)(C_{10}H_8N_2)]^{2+}$ derived from diethylenetriamine and 2,2′-bipyridyl, see pp. 215/7, may be found under the main entry $C_4H_{13}N_3$ and the subheading "and $C_{10}H_8N_2$". But it is also listed under the main entry $C_{10}H_8N_2$ and the subheading " and $C_4H_{13}N_3$". On the other hand, those additional ligands which are described as main ligands in volume **D 1** occur only as subheadings in the formula index of this volume. The last column lists the pertinent pages. The boldfaced **D 1** before the page numbers refers to those complex compounds described in volume "Mangan" **D 1** in which ligands of the present volume occur as additional ligands. Water and alcohols of solvates are not included as additional ligands.

Empirical formulas and names are based on the neutral ligand without regard of its state in the complex. The nomenclature corresponds to IUPAC rules. Trivial names are sometimes used to facilitate identification. Abbreviations for ligand names are listed only when they have been used in this volume.

Table of Conversion Factors

Following the notation in Landolt-Börnstein [7], values which have been fixed by convention are indicated by a bold-face last digit. The conversion factor between calorie and Joule that is given here is based on the thermochemical calorie, $cal_{th\,ch}$, and is defined as 4.184**0** J/cal. However, for the conversion of the „Internationale Tafelkalorie", cal_{IT}, into Joule, the factor 4.1868 J/cal is to be used [1, p. 147]. For the conversion factor for the British thermal unit, the Steam Table Btu, BTU_{ST}, is used [1, p. 95].

Force	N	dyn	kp
1 N (Newton)	1	10^5	0.1019716
1 dyn	10^{-5}	1	1.019716×10^{-6}
1 kp	9.8066**5**	$9.8066\mathbf{5} \times 10^5$	1

Pressure	Pa	bar	kp/m^2	at	atm	Torr	lb/in^2
1 Pa (Pascal) = 1 N/m^2	1	10^{-5}	1.019716×10^{-1}	1.019716×10^{-5}	0.986923×10^{-5}	0.750062×10^{-2}	145.0378×10^{-6}
1 bar = 10^6 dyn/cm^2	10^5	1	10.19716×10^3	1.019716	0.986923	750.062	14.50378
1 kp/m^2 = 1 mm H_2O	9.8066**5**	$0.98066\mathbf{5} \times 10^{-4}$	1	10^{-4}	0.967841×10^{-4}	0.735559×10^{-1}	1.422335×10^{-3}
1 at = 1 kp/cm^2	$0.98066\mathbf{5} \times 10^5$	0.98066**5**	10^4	1	0.967841	735.559	14.22335
1 atm = 76**0** Torr	$1.0132\mathbf{5} \times 10^5$	1.0132**5**	1.033227×10^4	1.033227	1	760	14.69595
1 Torr = 1 mm Hg	133.3224	1.333224×10^{-3}	13.59510	1.359510×10^{-3}	1.315789×10^{-3}	1	19.33678×10^{-3}
1 lb/in^2 = 1 psi	6.89476×10^3	68.9476×10^{-3}	703.069	70.3069×10^{-3}	68.0460×10^{-3}	51.7149	1

Work, Energy, Heat	J	kWh	kcal	Btu	MeV
1 J (Joule) = 1 Ws = 1 Nm = 10^7 erg	1	2.778×10^{-7}	2.39006×10^{-4}	9.4781×10^{-4}	6.242×10^{12}
1 kWh	$3.\mathbf{6} \times 10^{6}$	1	860.4	3412.14	2.247×10^{19}
1 kcal	4184.**0**	1.1622×10^{-3}	1	3.96566	2.6117×10^{16}
1 Btu (British thermal unit)	1055.06	2.93071×10^{-4}	0.25164	1	6.5858×10^{15}
1 MeV	1.602×10^{-13}	4.450×10^{-20}	3.8289×10^{-17}	1.51840×10^{-16}	1

Power	kW	h. p. (PS)	kp m/s	kcal/s
1 kW = 10^{10} erg/s	1	1.35962	101.972	0.239006
1 h. p. (PS)	0.73550	1	**75**	0.17579
1 kp m/s	$9.8066\mathbf{5} \times 10^{-3}$	0.01333	1	2.34384×10^{-3}
1 kcal/s	4.184**0**	5.6886	426.650	1

References:

[1] A. Sacklowski, Die neuen SI-Einheiten, Goldmann, München 1979. (Conversion tables in an appendix.)
[2] International Union of Pure and Applied Chemistry, Manual of Symbols and Terminology for Physicochemical Quantities and Units, Pergamon, London 1979; Pure Appl. Chem. **51** [1979] 1/41.
[3] The International System of Units (SI), National Bureau of Standards Spec. Publ. 330 [1972].
[4] H. Ebert, Physikalisches Taschenbuch, 5th Ed., Vieweg, Wiesbaden 1976.
[5] Kraftwerk Union Information, Technical and Economic Data on Power Engineering, Mülheim/Ruhr 1978.
[6] E. Padelt, H. Laporte, Einheiten und Größenarten der Naturwissenschaften, 3rd Ed., VEB Fachbuchverlag, Leipzig 1976.
[7] Landolt-Börnstein, 6th Ed., Vol. II, Pt. 1, 1971, pp. 1/14.